Microalgae for Environmental Biotechnology

This is the first book to present the idea of using Industry 4.0 and smart manufacturing in the microalgae industry for environmental biotechnology. It provides the latest developments on microalgae for use in environmental biotechnology, explains process analysis from an engineering point of view, and discusses the transition to smart manufacturing and how state of the art technologies can be incorporated. It covers applications, technologies, challenges, and future perspectives.

- Showcases how Industry 4.0 can be applied in algae industry
- Covers new ideas generated from Industry 4.0 for Industrial Internet of Things (IIoT)
- Demonstrates new technologies invented to cater to Industry 4.0 in microalgae
- Features worked examples related to biological systems

Aimed at chemical engineers, bioengineers, and environmental engineers, this is an essential resource for researchers, academics, and industry professionals in the microalgae biotechnology field.

Microalgae for Environmental Biotechnology

Smart Manufacturing and Industry 4.0 Applications

Edited by
Pau Loke Show
Wai Siong Chai
Tau Chuan Ling

CRC Press
Taylor & Francis Group
Boca Raton London

CRC Press is an imprint of the
Taylor & Francis Group, an **informa** business

First edition published 2023
by CRC Press
6000 Broken Sound Parkway NW, Suite 300, Boca Raton, FL 33487–2742

and by CRC Press
4 Park Square, Milton Park, Abingdon, Oxon, OX14 4RN

CRC Press is an imprint of Taylor & Francis Group, LLC

ISBN: 978-1-032-06411-6 (hbk)
ISBN: 978-1-032-06412-3 (pbk)
ISBN: 978-1-003-20219-6 (ebk)

DOI: 10.1201/9781003202196

Typeset in Times
by Apex CoVantage, LLC

Contents

Authors..vii
Contributors..ix
Preface..xiii

Chapter 1 Microalgae as a Source of Sustainability 1

Pik Han Chong, Jian Hong Tan and Joshua Troop

Chapter 2 The Potential of Microalgae for Environmental Biotechnology 67

Fazril Ideris, Mei Yin Ong, Jassinnee Milano, Mohd Faiz Muaz Ahmad Zamri, Saifuddin Nomanbhay, Abd Halim Shamsuddin, Teuku Meurah Indra Mahlia and Pau Loke Show

Chapter 3 Current Issues and Challenges of Applying Microalgae in Environmental Biotechnology.. 107

Xiao Gui Xing, Gao Ya Qian, Sook Sin Chan, Guo Rui Xin, Kit Wayne Chew and Pau Loke Show

Chapter 4 How Far Has the Development for Industrial Internet of Things (IoT) in Microalgae?.. 145

Vimal Angela Thiviyanathan, Hooi Ren Lim, Pei En Tham and Pau Loke Show

Chapter 5 Understanding Environmental Biotechnology 4.0 175

Wai Yan Cheah, Tengku Nilam Baizura Tengku Ibrahim, Nurul Syahirah Mat Aron, Wai Siong Chai and Pau Loke Show

Chapter 6 What Are Smart Microalgae?.. 205

Nur Azalina Suzianti Feisal, Noor Haziqah Kamaludin, Dingling Zhuang, Kit Wayne Chew and Pau Loke Show

Chapter 7 Towards the Digitalization of Environmental Biotechnology.......227

Viggy Tan Wee Gee, Kuan Shiong Khoo and Pau Loke Show

Chapter 8 Smart Factory of Microalgae in Environmental Biotechnology 263

Shazia Ali, Kuan Shiong Khoo, Hooi Ren Lim, Hui Suan Ng and Pau Loke Show

Chapter 9 Biomanufacturing in Microalgae Industry 293

Leong Wei, Wen Yi Chia, Kit Wayne Chew and Pau Loke Show

Chapter 10 Implementation of Microalgae 4.0 in Environmental Biotechnology 327

Akshara Ann Varghese, Doris Ying Ying Tang, Sze Shin Low and Pau Loke Show

Chapter 11 Industry Perspectives of Microalgae 4.0 363

Kreena Gada, Angela Paul Peter, Wai Siong Chai and Pau Loke Show

Chapter 12 Sustainability and Development of Microalgae 4.0 397

Henry Ng, Chung Hong Tan, Saifuddin Nomanbhay and Pau Loke Show

Chapter 13 Remaining Challenges and Uncertainties 421

Sze Yin Cheng, Zhi Ting Ang and Pau Loke Show

Index 445

Authors

Professor Ir. Ts. Dr. Pau Loke SHOW is Director of Sustainable Food Processing Research Centre and Co-Director of Future Food Malaysia, Beacon of Excellence in the University of Nottingham Malaysia, Selangor, Malaysia. He is Full Professor of Biochemical Engineering at the Department of Chemical and Environmental Engineering, Faculty of Science and Engineering, University of Nottingham Malaysia. He has successfully obtained his PhD in 2 years after obtaining his bachelor's degree from Universiti Putra Malaysia, Selangor, Malaysia. He is currently Professional Engineer (PEng) registered with the Board of Engineers Malaysia, Chartered Engineer (CEng) of the Engineering Council UK (MIChemE), and Professional Technologist (PTech) registered with the Malaysia Board of Technologists. Prof Ir. Ts. Dr. Show obtained the Post Graduate Certificate of Higher Education (PGCHE) in 2014 and is now a Fellow of the Higher Education Academy (FHEA) UK. Since he started his career in 2012, he has received numerous prestigious academic awards, including the Tan Sri Emeritus Professor Augustine S H Ong International Special Award on Innovations and Inventions in Palm Oil 2021; APEC Science Prize for Innovation, Research and Education ("ASPIRE") Malaysia Award 2020; Malaysia Young Scientist 2019 Award; ASEAN-India Research and Training Fellowship 2019; The DaSilva Award 2018; JSPS Fellowship 2018; Top 100 Asian Scientists 2017; Asia's Rising Scientists Award 2017; and Young Researcher in IChemE Award 2016. He has graduated more than 20 PhD and MSc students and is leading a team of 20 members consisting of postdoctoral, PhD, and MSc research students. Up to 2020, he has published more than 550 journal papers in less than 8 years of his career. His publications have been cited over 11,000 times over the past 5 years. His current h-index is 55, placing him among the top leaders of his chosen field (Microalgae Technology). He is also the Primary Project leader for more than 35 projects from international, national, and industry projects, with total amount more than MYR5.0 million. He has been recently awarded as Top Peer Reviewer (Top 1% of Reviewers in Global Top Peer Reviewer Awards in Engineering, Global Top Peer Reviewer Awards in Cross-Field, Global Top Peer Reviewer Awards in Chemistry, Global Top Peer Reviewer Awards in Biology and Biochemistry, Global Top Peer Reviewer Awards in Agricultural Sciences). Also, he has acted as Handling Editor for more than 1,000 submitted manuscripts in numerous journals.

Dr. Wai Siong CHAI is currently a postdoctoral research at the School of Mechanical Engineering and Automation, Harbin Institute of Technology, Shenzhen, China. He completed 2 years of postdoctoral training at Zhejiang University, involving in green propellant research. He developed the main framework and assisted the research team in obtaining an internal grant of RMB 200,000. He obtained his PhD from the University of Nottingham and was the sole recipient of scholarship in Malaysia to attend an international conference in 2015. He is Associate Member (AMIChemE) of the Institution of Chemical Engineers (IChemE). Dr. Chai's current field of research involves ammonia, environmental remediation, and energetic ionic liquid applications. He is also interested in studying the microalgae interaction with ammonia and its application toward biofuel production. Dr. Chai has published more than 20 journal papers and made several international conference presentations, with a current h-index of 13. His publications are of high quality, which are evidenced by Tier 1 publications such as *Renewable and Sustainable Energy Reviews*, *Chemical Engineering Journal*, *Journal of Cleaner Production*, and *Fuel*, where one of his papers is listed as Top Cited Paper. He currently serves as Editorial Board Member of *Bioengineered*. He has also been rated as an excellent reviewer on Publons.

Tau Chuan LING obtained his PhD from the University of Birmingham, Birmingham, UK, in 2002. He is Fellow Member of the Institution of Chemical Engineers (FIChemE). He is Full Professor of Biotechnology at the Institute of Biological Sciences, Faculty of Science, Universiti Malaya (UM), Kuala Lumpur, Malaysia. He has more than 20 years of research experience in the field of downstream processing, renewable energy, and biochemical engineering. Prof Dr. Ling has published more than 280 international peer-reviewed journals. He was awarded the Young Asian Biotechnologist Prize 2017 from The Society for Biotechnology, Japan.

Contributors

Shazia Ali
Department of Chemical and Environmental Engineering
Faculty of Science and Engineering
University of Nottingham Malaysia
Jalan Broga, Semenyih, Selangor Darul Ehsan, Malaysia

Zhi Ting Ang
Department of Chemical and Environmental Engineering
Faculty of Science and Engineering
University of Nottingham Malaysia
Jalan Broga, Semenyih, Selangor Darul Ehsan, Malaysia

Nurul Syahirah Mat Aron
Department of Chemical and Environmental Engineering
Faculty of Science and Engineering
University of Nottingham Malaysia
Jalan Broga, Semenyih, Selangor Darul Ehsan, Malaysia

Wai Siong Chai
School of Mechanical Engineering and Automation
Harbin Institute of Technology
Shenzhen, Guangdong, China

Sook Sin Chan
Institute of Biological Sciences
Faculty of Science
University of Malaya
Kuala Lumpur, Malaysia

Wai Yan Cheah
Centre of Research in Development, Social and Environment (SEEDS)
Faculty of Social Sciences and Humanities
Universiti Kebangsaan Malaysia
Bangi, Selangor, Malaysia

Sze Yin Cheng
Institute of Biological Sciences
Faculty of Science
University of Malaya
Kuala Lumpur, Malaysia

Kit Wayne Chew
School of Energy and Chemical Engineering
Xiamen University Malaysia
Jalan Sunsuria, Bandar Sunsuria, Sepang, Selangor Darul Ehsan, Malaysia

Wen Yi Chia
Department of Chemical and Environmental Engineering
Faculty of Science and Engineering
University of Nottingham Malaysia
Jalan Broga, Semenyih, Selangor Darul Ehsan, Malaysia

Pik Han Chong
School of Food Science and Biotechnology
Zhejiang Gongshang University
Hangzhou, Zhejiang, China

Nur Azalina Suzianti Feisal
Department of Environmental Health
Faculty of Health Sciences
MAHSA University
Bandar Saujana Putra, Jenjarom, Selangor, Malaysia

Kreena Gada
Department of Chemical and Environmental Engineering
Faculty of Science and Engineering
University of Nottingham Malaysia
Jalan Broga, Semenyih, Selangor Darul Ehsan, Malaysia

Viggy Tan Wee Gee
Department of Chemical Engineering
Faculty of Science and Engineering
University of Nottingham Malaysia
Semenyih, Selangor Darul Ehsan, Malaysia

Tengku Nilam Baizura Tengku Ibrahim
Department of Environmental Health
Faculty of Health Sciences
Universiti Teknologi MARA
Cawangan Pulau Pinang, Kampus Bertam, Kepala Batas Penang, Malaysia

Fazril Ideris
Institute of Sustainable Energy
Universiti Tenaga Nasional (UNITEN)
Jalan IKRAM-UNITEN
Kajang, Selangor, Malaysia
and
AAIBE Chair of Renewable Energy
Universiti Tenaga Nasional (UNITEN)
Kajang, Selangor, Malaysia

Noor Haziqah Kamaludin
Center of Environmental Health & Safety
Faculty of Health Sciences
Universiti Teknologi MARA
Puncak Alam, Selangor, Malaysia

Kuan Shiong Khoo
Faculty of Applied Sciences
UCSI University
Cheras, Kuala Lumpur, Malaysia

Hooi Ren Lim
Department of Chemical and Environmental Engineering
Faculty of Science and Engineering
University of Nottingham Malaysia
Jalan Broga, Semenyih, Selangor Darul Ehsan, Malaysia

Sze Shin Low
Research Centre of Life Science and Healthcare
China Beacons Institute
University of Nottingham Ningbo China
Ningbo, Zhejiang, China

Teuku Meurah Indra Mahlia
Centre for Green Technology
School of Civil and Environmental Engineering
University of Technology
Sydney, NSW, Australia

Jassinnee Milano
Institute of Sustainable Energy
Universiti Tenaga Nasional (UNITEN)
Jalan IKRAM-UNITEN
Kajang, Selangor, Malaysia
and
AAIBE Chair of Renewable Energy
Universiti Tenaga Nasional (UNITEN)
Kajang, Selangor, Malaysia

Henry Ng
Department of Chemical and Environmental Engineering
Faculty of Science and Engineering
University of Nottingham Malaysia
Jalan Broga, Semenyih, Selangor Darul Ehsan, Malaysia

Hui Suan Ng
Faculty of Applied Sciences
UCSI University
Cheras, Kuala Lumpur, Malaysia

Saifuddin Nomanbhay
Institute of Sustainable Energy
Universiti Tenaga Nasional (UNITEN)
Jalan IKRAM-UNITEN
Kajang, Selangor, Malaysia
and
AAIBE Chair of Renewable Energy
Universiti Tenaga Nasional (UNITEN)
Kajang, Selangor, Malaysia

Mei Yin Ong
Institute of Sustainable Energy
Universiti Tenaga Nasional (UNITEN)
Jalan IKRAM-UNITEN
Kajang, Selangor, Malaysia
and
AAIBE Chair of Renewable Energy
Universiti Tenaga Nasional (UNITEN)
Kajang, Selangor, Malaysia

Angela Paul Peter
Department of Chemical and Environmental Engineering
Faculty of Science and Engineering
University of Nottingham Malaysia
Jalan Broga, Semenyih, Selangor Darul Ehsan, Malaysia

Gao Ya Qian
Biology and Medicine
Faculty of Engineering
China Pharmaceutical University
Nanjing, China

Abd Halim Shamsuddin
AAIBE Chair of Renewable Energy
Universiti Tenaga Nasional (UNITEN)
Kajang, Selangor, Malaysia

Pau Loke Show
Department of Chemical and Environmental Engineering
Faculty of Science and Engineering
University of Nottingham Malaysia
Jalan Broga, Semenyih, Selangor Darul Ehsan, Malaysia

Chung Hong Tan
Institute of Sustainable Energy
Universiti Tenaga Nasional (UNITEN)
Jalan IKRAM-UNITEN
Kajang, Selangor, Malaysia
and
AAIBE Chair of Renewable Energy
Universiti Tenaga Nasional (UNITEN)
Kajang, Selangor, Malaysia

Jian Hong Tan
School of Biosciences
University of Nottingham Malaysia
Semenyih, Selangor, Malaysia

Doris Ying Ying Tang
Department of Chemical and Environmental Engineering
Faculty of Science and Engineering
University of Nottingham Malaysia
Jalan Broga, Semenyih, Selangor Darul Ehsan, Malaysia

Pei En Tham
Department of Chemical and Environmental Engineering
Faculty of Science and Engineering
University of Nottingham Malaysia
Jalan Broga, Semenyih, Selangor Darul Ehsan, Malaysia

Vimal Angela Thiviyanathan
Institute of Sustainable Energy
Universiti Tenaga Nasional
Kajang, Selangor, Malaysia

Joshua Troop
Dearne Valley College
Wath upon Dearne, United Kingdom

Akshara Ann Varghese
Department of Chemical and Environmental Engineering
Faculty of Science and Engineering
University of Nottingham Malaysia
Jalan Broga, Semenyih, Selangor Darul Ehsan, Malaysia

Leong Wei
Department of Chemical and Environmental Engineering
Faculty of Science and Engineering
University of Nottingham Malaysia
Jalan Broga, Semenyih, Selangor Darul Ehsan, Malaysia

Guo Rui Xin
Faculty of Engineering
China Pharmaceutical University
Nanjing, China

Xiao Gui Xing
Pharmaceutical Engineering
Faculty of Engineering
China Pharmaceutical University
Nanjing, China

Mohd Faiz Muaz Ahmad Zamri
Institute of Sustainable Energy
Universiti Tenaga Nasional (UNITEN)
Jalan IKRAM-UNITEN
Kajang, Selangor, Malaysia
and
AAIBE Chair of Renewable Energy
Universiti Tenaga Nasional (UNITEN)
Kajang, Selangor, Malaysia

Dingling Zhuang
Institute of Biological Sciences
Faculty of Science
Universiti Malaya
Kuala Lumpur, Malaysia

Preface

Industrial Revolution 4.0, which integrates digital revolution, plays a significant role in creating a sustainable computer-based economy through the creation of new products and systems from more streamlined process. The combination of Industrial Revolution and microalgal technology yields Microalgae 4.0, which will make a huge difference in their competitiveness with existing processes, significantly reducing their cost and use of resources. The book describes the technological breakthroughs and developments of Microalgae 4.0 in response to the Industrial Revolution 4.0 and sustainability issues, particularly on environmental biotechnology. Its aim is to provide the fundamentals so that beginners can understand the basic principles of the techniques, as well as to offer constructive opinions on the latest findings. This emerging technology is a simple yet efficient process and has high scalability, which makes it useful for applications in environmental biotechnology. This book is expected to encourage industries to adapt this technology for the bioremediation of environment to provide better cost efficiency and greener processing, as well as to attract the attention of researchers and professionals of microalgal biotechnology, particularly in environmental applications.

The book is organized into chapters, each expressing the principles of Microalgae 4.0 applications. The following chapters describe various aspects of digital technologies as well as challenges involved that have been applied to environmental biotechnology. Chapter 1 of this book provides a general introduction to the topics covered in this book and introduces microalgae as the source for sustainability. Chapters 2 and 3 reveal the potential of microalgae and current issues and challenges of applying microalgae in environmental biotechnology applications. Chapters 4, 5, 6, and 7 explain the development status of the Industrial Internet of Things in microalgae, environmental biotechnology 4.0, smart microalgae, and digitalization aspects of environmental biotechnology. Chapters 8 and 9 deal with the concept of Smart Factory of microalgae in environmental biotechnology and intelligent biomanufacturing in microalgae industry. Chapters 10 and 11 investigate the implementation of Microalgae 4.0 in environmental biotechnology and industry perspectives of Microalgae 4.0. Chapters 12 and 13 cover the sustainability and development of microalgae usage and the remaining challenges and uncertainties.

The content of this book comprehensively summarizes the important findings and achievements of the research done thus far concerning Microalgae 4.0. It is important to understand the basic working principles of Microalgae 4.0, so that the various digital technologies integrated into Microalgae 4.0 have greater benefits than the conventional process. The information provided in this book will open up future possibilities for incorporating Microalgae 4.0 with other technologies to improve the environmental quality and cost efficiency of the process.

1 Microalgae as a Source of Sustainability

Pik Han Chong[1], *Jian Hong Tan*[2] and *Joshua Troop*[3]

[1] School of Food Science and Biotechnology, Zhejiang Gongshang University, Hangzhou, Zhejiang, China

[2] School of Biosciences, University of Nottingham Malaysia, Semenyih, Selangor, Malaysia

[3] Dearne Valley College, Manvers Park, Wath upon Dearne, Swinton, Rotherham, United Kingdom

CONTENTS

1.1 Introduction........2
1.2 What Is Biotechnology?........3
1.3 Environmental Biotechnology........5
1.3.1 Waste Treatment, Pollution Control, Remediation, and Restoration........8
1.3.2 Pollution Detection and Monitoring—Biomonitoring, Biosensors, and Biomarkers........9
1.3.3 Conservation of Biodiversity and Environment........9
1.4 Sustainability........11
1.4.1 How Does Sustainability Become Popular Again?........11
1.4.2 The Concept of Sustainability........12
1.4.3 Why Is Sustainability Important?........14
1.4.4 Efforts on Achieving Sustainability........15
1.5 Introduction to Microalgae........17
1.6 Microalgae and Its Sustainable Applications........23
1.6.1 Food and Nutrition........23
1.6.2 Pharmaceuticals and Fine Chemicals........26
1.6.3 Biomass and Biofuel........27
1.6.4 Architecture and Urban........29
1.6.5 Materials........32
1.6.6 Waste Treatment........34
1.6.7 Environment and Conservation........36
1.6.7.1 Microalgae as an Indicator of Pollutants........37
1.6.7.2 Restoration of Degraded Environments through Microalgal Bioremediation........38
1.6.7.3 Carbon Sequestration with Microalgae........38

DOI: 10.1201/9781003202196-1

1.7 The Future of Microalgal Technology—Microalgae 4.0 39
1.8 Conclusions 40
Acknowledgment 41
References 41

1.1 INTRODUCTION

Different from the kelp and seaweeds you can usually find in the ocean while snorkeling, microalgae are microscopic algae. Sometimes, they are called cyanobacteria. Just like any bacteria, microalgae are invisible to the naked eye, only to be seen under the lens of a microscope. This book is about using microscopic algae to create all kinds of technology, their massive capability and their potential to be used as a renewable source of energy and energy production. Microalgae can be considered to be a waste-free unadulterated resource in both their use and production. They are also notable for being incredibly effective to easily get renewed due to their massive growth rate with the end goal of creating a fully renewable and environment-friendly oil. The ability for microalgae to create harvestable chemical compounds makes it a potentially astounding breakthrough in biotechnology, with some biotechnologists believing that the next milestone in renewable energy could be achieved based on microalgae. Microalgae are capable of propagating and reproducing in a large range of environmental properties such as temperature, pH value, light levels, nutrients, and salinity, granted that a water source is provided (Juneja, Ceballos, and Murthy 2013). Due to the ease of access, reproduction, and potential for microalgae, this book will cover the pros, cons, and theorized prospective for microalgae. In this chapter, a spotlight will be shed on the topics of biotechnology, environmental biotechnology, the basics of microalgae, their applications, and potentials.

Economic unions of different continents, as well as intergovernmental organizations, have recognized the need for the existence of relevant platforms of bioeconomy, such as the ECLAC, OECD, FAO of UN, EU Commission, ASEAN, and International Bioeconomy Forum (IBF) (Martinez-Goss 2007; Oborne 2010; Rodríguez, Rodrigues, and Sotomayor 2019; EUMOFA 2020; Teitelbaum, Boldt, and Patermann 2020; Bößner, Johnson, and Shawoo 2021; Cai et al. 2021). Bioeconomy describes a future economy that is closely associated with biological resources in every economic sector possible. In various fields such as production, industrial processes, and technology, we can see increasing biological innovations and solutions. The potential of microalgae and the role they will play in the future of a circular bioeconomy have also been recognized (Yarnold et al. 2019). Thus, microalgae have been included as part of the bio-economical innovation and solution. For example, the European Union promotes microalgae-based products as a new source of nutrition for human and animal consumption (Enzing et al. 2014; EUMOFA 2020). Alternatively, microalgae have the potential to work as biofuel, which would mean possible complete replacement for petrol. This could lead to a momentous change in the production of harmful greenhouse gases and a switch to a completely renewable energy source, potentially ending the ongoing and further upcoming fuel crisis and tensions regarding shortening amounts of fuels globally (Khan, Shin, and Kim 2018; Low et al. 2021). The United Kingdom also has plans regarding bioeconomy strategies. In 2016, the UK government released a roadmap with the aim of determining a bioeconomy strategy and further modified the plans with their "UK Synthetic Biology Strategy Plan" which was implemented in later 2016 (Teitelbaum, Boldt, and Patermann 2020).

A section will also be dedicated to the immense promising factors of microalgae directed toward the development of renewable biofuels to combat the worryingly harmful effects that the current fuel usage is causing to the environment, ecosystems, and atmosphere, including the fact that the currently most heavily used fuel sources are nonrenewable, leading to the eventual depletion of these fuels and a global fuel crisis that would cripple all forms of transport, power supply, and production. This chapter will also delve into the potential that microalgae hold by having an impressively high amount of nutrients such as carbohydrates, protein, vitamins, and minerals, which can be used as food supplements or even as food sources due to their high nutritional density. In addition, microalgae have antibiotic properties that could be used in antibiotics and potentially even be used in cancer treatment drugs. Other beneficial properties of microalgae to be explored in this chapter are their inherent ability to be utilized in wastewater treatment as well as in atmospheric carbon dioxide reduction. This chapter will include the potential role of microalgae in architecture and urban planning, and the possibilities of microalgae in their use directed toward materials and detoxifying land. Furthermore, the properties of microalgae will be summarized for their usage in many different fields with in-depth explanations of how microalgae have the possibility of inherently improving or replacing certain practices which have been already used.

1.2 WHAT IS BIOTECHNOLOGY?

The oldest and simplest form of biotechnology is the process known as selective breeding. It is widely applied in agriculture for both animals and plants (Ratledge and Kristiansen 2006; Seidel 1998). Humans selectively breed plants and animals to preserve or/and remove certain characteristics. Most of the vegetables and pets around us today are products of selective breeding. Among plants that were selectively bred for agriculture were soybean and maize. Soybean has been selectively bred to achieve better seed protein composition. However, a study showed that selective breeding did not significantly change the soybean protein composition and quality for the past 75 years (Mahmoud et al. 2006). In contrast to maize (*Zea mays*), it evolved from a plant with dry and extremely hard grains, and this maize ancestor is known as Teosinte (Figure 1.1).

It is a wild grass that originated from Mexico, which became an edible crop through selective breeding and now is largely available in supermarkets (Iltis 1983; Q. Chen et al. 2020)

Aquaculture, too, is no stranger to selective breeding. Blue mussels (*Mytilus edulis*) had also been altered artificially through selective breeding for better meat

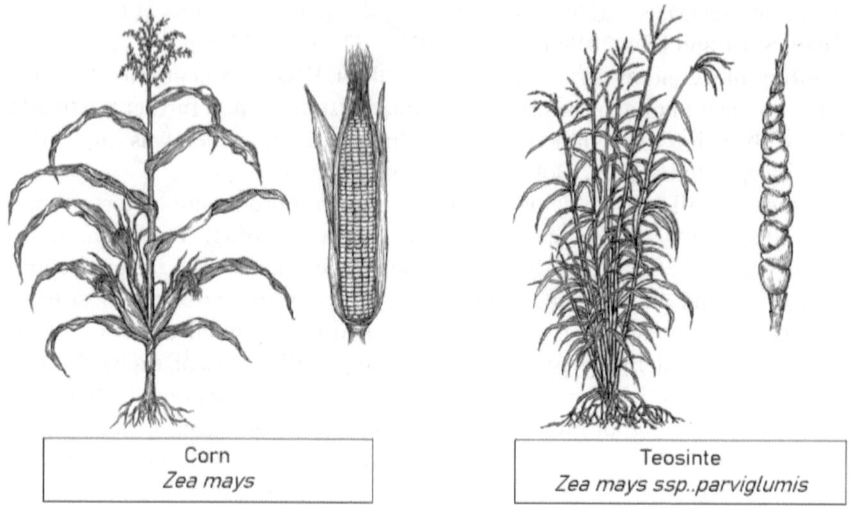

FIGURE 1.1 Illustration of plant and ear of corn and teosinte.

yield and shell shape (T. T. T. Nguyen, Hayes, and Ingram 2012). Finfish species, specifically the Atlantic salmon and rainbow trout, are among the fish that are farmed and selectively bred in Europe (Janssen et al. 2017). Human's greatest companions, dogs (*Canis lupus familiaris*), were highly selectively bred for various reasons such as livestock shepherding and hunting, which both aided humans in securing their livestock and prey (Lupo 2017; Cunningham-Smith and Emery 2020). Furthermore, farm animals such as pigs, chickens, and cows are selectively bred for common traits of interest that are better growth, milk production, meat quality, and sometimes for beneficial social behaviors (Andersson 2001; Turner 2011).

Another early bio-technology tool was the use of microorganisms to make new food or preserve them. Through different combinations of sugar sources and yeast types, different products of fermentation can be produced. Brewing and fermentation were practiced mainly, leading to the production of bread and alcohol (Meussdoerffer 2009). Dairy products and cheese which constitute a typical European diet and culture use lactic acid bacteria such as *Lactobacillus delbrueckii* (Benninga 1990). The extensive use of microorganisms led to the development of vaccines (Plotkin 2011; Lloyd and Cheyne 2017). The first few antibiotic was also discovered in the early 20th century treating infections and diseases (Walsh 2003). Biotechnology was therefore defined as: "The purposeful use of biological systems or the altering of living systems, organisms, or parts of organisms to develop products or systems that benefit humankind" (Bud 1994).

However, biotechnology, today, no longer limits its benefits only to humankind. Biotechnology can also benefit plants, animals, and even the environment. Through the early discovery of genes back in 1865, biotechnology evolved into its modern stage, where genetic technology was the center point of modern biotechnology (Gayon 2016). Due to its massive amounts of data, modern biotechnology is being commonly paired with the use of computers and information systems (Straiton et al. 2019). The first genome that was entirely sequenced was the Haemophilus influenzae back in 1995. Its genome size is around 1.83 million base pairs (Fleischmann et al. 1995). Sequencing it through manual DNA sequencing (also known as Sanger sequencing) would take 183 days (10 kilobases/day). Such methods would be time-consuming, laborious, not cost-effective, and even more so with the human genome (Metzker 2005). The size of the human genome is over three billion pairs, and it took 13 years to sequence it. Although the human genome has been completely sequenced, it was only possible through a massive national project undertaken by the United States and the technology from the age of information that we presently live in (Watson 1990).

As a result, modern biotechnology is now an interdisciplinary field of study. It often overlaps with other scientific fields which work closely with the Information Technology (IT) and Engineering field. This creates more specialized fields such as Bioinformatics and Bioengineering. With the industrial world slowly stepping into Industry 5.0, biotechnology could no longer be aloof from the industrial sector that will integrate bionics and synthetic biology (Sachsenmeier 2016). All the aforementioned instances are just the tip of the iceberg. Most of the biotechnologies stated before have been used mostly for food and medicine. In recent years, biotechnologies have been involved in the global environmental movement and climate change as well.

1.3 ENVIRONMENTAL BIOTECHNOLOGY

Biotechnology can be applied in almost any field of study, which could be divided into four major groups: Industrial, Medical, Food, and Environment. However, due to diverse applications of biotechnology, the four major groups are also further specified into different color codes. According to Dasilva (2004) and Kafarski (2012), there are ten colors of biotechnology as shown in Figure 1.2. All ten different colors represent specific fields and disciplines as elaborated as follows.

- Red: Medicine, Health, Biopharmacy
  - Also known as health biotech, it purses better health preservation and pharmaceuticals and the production and creation of new vaccines, antibiotics, drugs, regenerative therapies, and more (Sasson 2005).
- Yellow: Nutrition and Food Processing and Productions of Insects
  - Centers around the improvement of food and nutrition. Often overlaps with green biotechnology, fortification of nutrition in fresh produce (Mackey 2002), or food processor with the use of biomaterials and enzymes (Knorr and Sinskey 1985). Alternatively, yellow also represents insect biotechnology which aims to utilize insect-derived enzymes in food biotechnology and industrial processes (Mika, Zorn, and Rühl 2013).

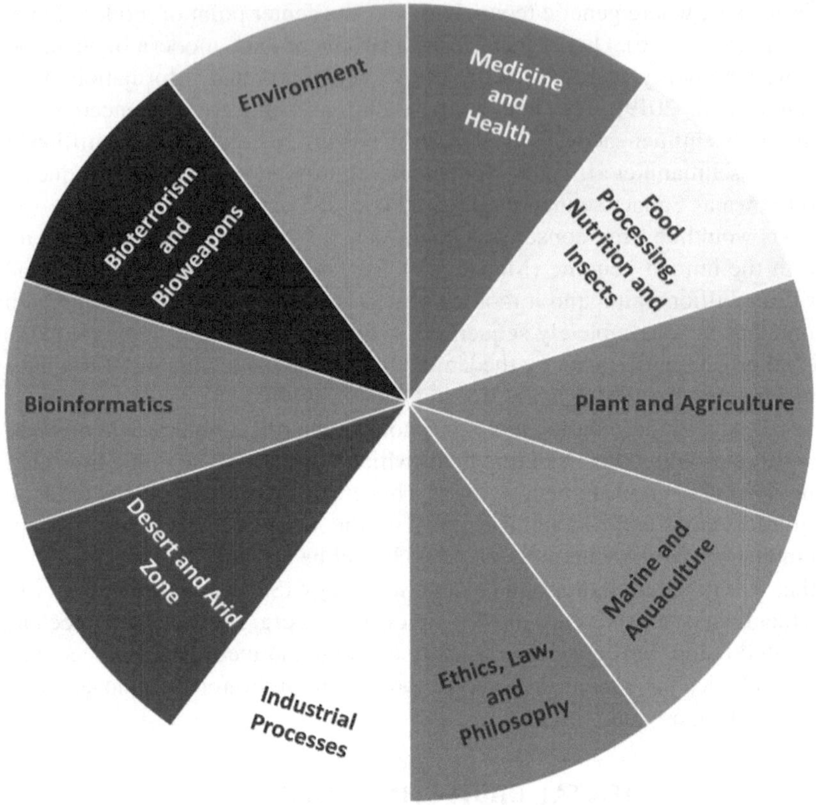

FIGURE 1.2 Simple color chart describing various fields and disciplines of biotechnology. There are ten colours: Red (Medicine), Yellow (Nutrition and Insects), Green (Plant), Blue (Marine), Purple (Ethics), White (Industrial), Brown (Desert), Gold (Bioinformatics), Black (Bioweapons), and Grey (Environment)

- Green: Plants, Food Crop, Agriculture
  - Genetically modified organisms (GMOs) are one of the main examples of green biotechnology. Other innovations include biopesticides, biofertilizers, and biopolymers like bioplastics (Barcelos et al. 2018)
- Blue: Marine, Aquaculture
  - Another biotechnology that is based on marine biodiversity and ecosystems, marine biotechnology utilizes marine bioactive compounds that can be applied in the food, cosmetic, and biofuel industries (Barcelos et al. 2018).
- Violet: Ethics, Law, and Philosophy
  - Violet encompasses the legal aspects of biotechnology such as Intellectual Property Rights, Patents, Bioethics on Genetic Experiments, or Biosecurity against Bioweapons from Black Biotechnology (Fitzsimons 2007; Twardowski 2011).

- White: Industrial Production and Processes
    - White represents biotechnology that is extensively applied in the industrial sector mainly in the processing and production of chemicals, materials, enzymes, and fuel. Its technology aids producing biocatalysts for biochemical production and aids in the industrial production of biomaterials using microorganisms (Barcelos et al. 2018).
- Brown: Arid Zone and Deserts Biotechnology
    - Brown Biotechnology is similar to Grey but is mainly applied on deserts and arid lands in Africa and the Middle East. Its focus is on creating new biotechnologies for areas such as bioenergy, bioprospecting, biomineralization, and greening deserts lands to help desert inhabitants utilize arid zones (Rodríguez-Núñez et al. 2020).
- Gold: Bioinformatics
    - Computational biotechnology, in both software and hardware, allows us to harness the functional information of omics studies such as Genomics, Proteomics, and Transcriptomics. Nanobiotechnology also falls under this category, which incorporates nanotechnology with biotechnology. It brings us nano-sized tools like Nanobiosensors and Nanotubes (Barabadi 2017; Kanwugu, Ivantsova, and Chidumaga 2018).
- Black/Dark: Bioterrorism and Bioweapons
    - The partially negative side of biotechnology involves its destructive use in areas such as bioterrorism using biological weapons and agents (Danciu 2011). However, it also means prevention and counter-terrorism of bioterrorism with the aid of police and military (Lachmayer 2006).
- Grey: Environmental Biotechnology
    - Which also equates to Grey Biotechnology, which is our main focus for this subchapter. The application of biotechnology to solve environmental problems and ecosystems results in the field called Environmental Biotechnology (Dasilva 2004; Kafarski 2012). The application of biotechnology is used to review and study the natural environment. Grey Biotechnology also focuses on the protection and restoration of the environment and, thus, on nature and the ecosystem itself. It largely utilizes living systems including plants, microorganisms, and animals as means to prevent, treat, or remediate pollution and waste (Gavrilescu 2010). A major advantage of environmental biotechnology is that it will help to maintain environmental safety and cleanliness for use by future generations. It helps nonhuman organisms and engineers adapt to changes in their environment and find useful ways to keep the environment clean and green (Mohapatra 2010). Environmental biotechnology helps preventing the use of natural resources in an abusive manner and prevents excessive production of harmful pollutants or wastes that affect the environment. Societal development must be done in such a way that it protects our

ecosystem along with our environment and also helps to advance it (van Hullebusch, Singh, and Mal 2021). In this field, the types of areas of application are waste treatment and management, pollution control and prevention, and the conservation of biodiversity and the environment (Zylstra and Kukor 2005). The technologies that aid in doing so are elaborated as given in the subsequent sections.

1.3.1 Waste Treatment, Pollution Control, Remediation, and Restoration

All living beings naturally produce wastes. An excess of this waste causes pollution with devastating environmental consequences. This potential issue had been recognized by generations before us, even as early as 500 BC. Historically, waste has been dealt with by using landfills, incinerators, and in more recent years via the process of recycling (Williams 2005). Unfortunately, these methods, except recycling, have expensive drawbacks. Landfills require a vast proportion of valuable land which subsequently causes infrastructural disruption and occupation of wildlife habitats. Landfills also contaminate groundwater and soil through leachate (Lisk 1991). Additionally, incinerators burn waste material as an energy source. However, the combustion of waste emits greenhouse gases and toxic gases containing dioxin, furans, heavy metals, and acid gases (R. Sharma et al. 2013). Mishandling of waste leads to pollution, and the main sources of pollution are from household activities, factories, agriculture, and transportation (Ferronato and Torretta 2019). Handling and treating waste from anthropogenic activities before releasing them are a form of pollution control and prevention. Unfortunately, many natural environments are contaminated and polluted, proving that efforts of pollution control and prevention are not effective and efficient enough (Strasser 1996). Bioremediation is considered a solution to pollution, as it is a process that utilizes soil, water, and materials that are polluted. The reason for using the polluted materials is that the process of bioremediation uses microorganisms to eat, destroy, dilute, or absorb the active pollutant within the material, which makes the once polluted material less toxic or no longer toxic (I. Sharma 2020).

This technology uses biological systems, which mostly use microorganisms but are not limited to plants and fungi. The process of using fungal mycelia in bioremediation is called Mycoremediation; Phytoremediation uses plants to treat waste or pollution; and, lastly, Microbial bioremediation utilizes microorganisms to perform bioremediation (Pilon-Smits 2005; H. Singh 2006; Karigar and Rao 2011). Bioremediation is usually applied to remove, degrade, or detoxify contaminated or polluted media. Bioremediation consists of the multitude of the biological facilitated processes (or bioprocesses) such as biotreatment, biodegradation, and biotransformation (Gavrilescu 2010). Destruction (biodegradation and biotransformation) that is at the core of environmental biotechnology is carried out by an organism or a combination of organisms, and it forms the major part of applied processes for environmental clean-up (Crawford 2014). Biotransformation processes use natural and recombinant microorganisms (yeasts, fungi, bacteria), enzymes, and whole cells. Biotransformation plays a key role in the area of foodstuff, pharmaceutical industry, vitamins, specialty chemicals, and animal feedstock (Parkinson et al. 2018; Doble, Kruthiventi, and Gajanan Gaikar 2004). Heavy metals are common contaminants and pollutants that are toxic to wildlife. Bioremediation mechanisms such as bioleaching, biosorption, biomineralization, enzyme-catalyzed biotransformation are bioremediation treatments that could be applied to heavy metals (Dixit et al. 2015; Kang, Kwon, and So 2016).

1.3.2 Pollution Detection and Monitoring—Biomonitoring, Biosensors, and Biomarkers

As society grows, so does pollution. But the relationship between population growth and pollution growth is not linear, and this is a desperate situation that requires immediate attention. The researchers are theorizing and inventing new ways to counter the devastating and terrifying growth in pollution and global warming. There is a method being used not to counter, but to measure the growth in pollution called biomonitoring. To simply explain, biomonitoring is the process of identifying changes or anomalies in the ecosystem, landscape, and biodiversity (Needham, Calafat, and Barr 2007). Biomonitoring can be also used to detect toxic chemicals that would be harmful or lethal to humans and wildlife. Biomonitoring can be performed in a plethora of ways such as observing changes or looking out for unusual ill-health in wildlife (Bonada et al. 2006), by measuring chemicals within the body tissues of an organism, or even through urine cycles of an organism that resides in the suspected area (Vandenberg et al. 2010). Soil or plant matter can also be gathered to detect harmful chemicals or changes (Manning and Feder 1980). A tool that can be used for biomonitoring is biosensors. Biosensors are devices that can detect a chemical substance and determine the properties of the substance, leading to the identification of a polluted or toxic area. They do this by creating a signal that would be similar to the concentration of the analytes in the reaction of the substance. These sensors are even capable of detecting ions, bacteria, and organic compounds (Malmqvist 1993; Vigneshvar et al. 2016). As such, these ions, bacteria, and organic compounds could be considered biomarkers. Biomarkers refer to the indications of objective and quantifiable characteristics of biological processes, substances, or structures (Strimbu and Tavel 2010). For example, if a person is suspected to be suffering from mercury poisoning from their environment, a biomonitoring test will be done on that person and their environment. The test will look out for mercury as the biomarker in their blood and urine sample as well as samples from their environment (Paustenbach and Galbraith 2006).

1.3.3 Conservation of Biodiversity and Environment

To avoid the extinction of endangered or threatened plant species, conservation of these may be carried out by establishing Gene Banks or Genetic Resource

Centers using biotechnology processes like plant tissue culture. Cryopreservation technology is the most popular and effective method for long-term storage of plant cells and tissues, which involves storage at a very low temperature. The method includes the raising of sterile tissue cultures and the addition of cryoprotectant, freezing, and storage (Priyanka et al. 2021). Biodiversity is crucial for keeping balanced, functional systems and cycles currently in place throughout our ecosystem. Damage of biodiversity could effectively lead to the damage of our medicines and even food production (Diaz 2011; S. L. Chen et al. 2016). Many methods are being utilized to aid in the preservation of biodiversity in our ecosystem, and one of the most known and blanket solutions are government laws and regulations for direct or indirect changes to natural habitats (Rands et al. 2010). However, the legislations put in place to preserve biodiversity are not the same in all countries, as some countries have heavy legalities based on ecosystem preservation, while others have next to none. An example of legislation that focuses on the preservation of ecosystems and biodiversity is the legislation of the United States: "Endangered Species Act" (Verschuuren 2004).

Another significant method of maintaining biodiversity is habitat restoration and maintaining the wilderness habitat (Mittermeier et al. 2003). Habitat restoration is essentially forcing a habitat to contain the same ecosystem that it would have if it was undisturbed by humans until the habitat returns to its effectively balanced ecosystem (J. R. Miller and Hobbs 2007; Török and Helm 2017). This method can be utilized in several ways, such as adding wildlife to an area, which was especially proven to be effective for the Yellowstone National Park in the United States. In the case of Yellowstone, wolves were reintroduced into the park to stabilize the large and out-of-control population growth of coyotes and elks due to the lack of a natural predator for them—that is, the absence of wolves (Barber-Meyer, Mech, and White 2008). Other methods of habitat restoration include the addition of nesting islands for birds and changing the constitution of the soil by bioremediation to aid in the growing requirements for plants (Jones 2004; Lemenih 2004). A factor that has heavily impacted biodiversity is the introduction of invasive species. Invasive species usually are sourced from a different region and are typically not suitable for the ecosystem of the region they have been moved to. The incompatibility can lead to different outcomes, with the worst being that the invasive species dominate an area (Allendorf and Lundquist 2003; Pyšek and Richardson 2010). An example of a damaging invasive species is the introduction of the grey squirrel into Britain, which has led to a tremendous population decrease of the red squirrel of Britain (Barbara et al. 2018). This is because the grey squirrel is physically larger than the red squirrel and has ease forcing a red squirrel out of its habitat; the grey squirrel also breeds much faster than the red squirrel, easily creating a more favorable environment for the grey squirrel. The grey squirrel also often carries a parapoxvirus which is usually considered lethal for the red squirrel, but is survivable for the grey squirrel (Barbara et al. 2018).

An alternative way of conserving the environment is through policies that manage the environment such as natural resource management (NRM). NRM is the practice of planning, developing, and managing natural resources in terms of resource quantity and quality. This management of natural resources often involves land, water, minerals, forests, fisheries, wildlife, and even genetic resources (National Research Council 1995; Karamidehkordi 2012). NRM aims to control the usage of natural resources at a sustainable scale, offering resilience to both the human economy and to our natural ecosystem (Holling and Meffe 1996). NRM can be practiced through various activities such as greenhouse gas reduction policies and watershed management (Hooper 2011; Höhne, den Elzen, and Escalante 2014). An analysis of 40 years of global natural resource flow and productivity has shown a strong relationship among sustainable natural resource management, economic prosperity, and human well-being (Schandl et al. 2018). However, such management requires strong, firm, and sustainable governing principles and policies. Lockwood et al. (2010) laid out eight principles that should be included in any NRM framework: Legitimacy, Transparency, Accountability, Inclusiveness, Fairness, Integration, Capability, and Adaptability (Lockwood et al. 2010). Case studies on community NRM in the United States, Nepal, and Kenya showed more successful NRM due to stronger legal and financial support (Kellert et al. 2000). These principles and policies keep governments and nongovernmental organizations in check, making sure they uphold and deliver their promises to protect the environment. Paris Agreement by the United Nations Framework Convention on Climate Change (UNFCCC) is a great example of a global policy framework that aims to help governments around the world to fight climate change by reducing emissions (Horowitz 2016).

1.4 SUSTAINABILITY

1.4.1 How Does Sustainability Become Popular Again?

The world is never short of issues, whether it concerns human civilization or the environment. In recent years, the fight for a sustainable future has been increasing: against climate change, social injustice, and global economic inequality. In the age of the Anthropocene, all of these issues have turned into a ticking time bomb. This particular issue not only concerns humanity, but also concerns all life forms on earth, which could be the end of human civilization and possibly the end of the earth (Besley and Peters 2020). This issue is known as climate change that the world often discusses. It became a massively growing concern as it could change the earth with no point of return. Before climate change, both wildlife conservation and pollution were major environmental causes between the 1800s and 1900s (Buttel and Flinn 1974).

The use of plastics has existed around the 1800s, but the harmful, synthetic, and nonbiodegradable plastic started to be used commercially around the 1920s. However, just like pollution and the protection of endangered animal species, it wasn't treated seriously as a global issue. Plastic pollution was then famously brought to the world's attention in the late 2010s (Geyer 2020). Through the video of a stuck plastic straw removed from a turtle's nose, the video sparked global outrage. The outrage snowballed into an anti-straw movement around the globe after almost 100 years of plastic use (Figgener 2018). Slowly, the anti-straw movement started a ban for straws and caused a ripple effect that prompted the corporate world to follow suit. Eventually, countries around the world have implemented laws and policies to replace or decrease the use of plastic straws, including other single-use plastics. Since then, sustainability has become a wildfire topic not just between environmentalists and ecologists, but for the world regardless of any aspects of age or nationality. The fight for sustainability and a greener world has always been going on since the 1900s when the First Industrial Revolution had occurred (Andrady 2005). Although the First Industrial Revolution was starting point of our modern world, its consequences or "side effects" could no longer be ignored.

The plastic straw movement was one of the biggest environmental movements since it was a global one after Earth Day in 1970 (Dunlap and Mertig 1991). Then came Greta Thunberg, a 15-year-old Swedish student environmental activist, who made her waves by addressing at the podium of the 2018 United Nations Climate Change Conference. The next year, to show her commitment to reducing gas emissions, she sailed across the oceans to North America where she attended the 2019 UN Climate Action Summit (Putkonen 2019; Jung et al. 2020). Greta alone inspired scientists and the young generation across the world to take action. And the world did take an action by starting a strike known as the School Strike for Climate happening in over 123 countries with almost 4 million youth protestors. This strike was an international movement of school students who skipped Fridays to protest against their government for the inaction against climate change (Kühne 2019). From all that has happened in these last few years (2015–2021), we can see that sustainability is heavily demanded. Climate change is being taken much more seriously throughout the world as compared to the 1900s. Humans regardless of age and race are coming together as a species and putting in the effort to clean up and repair the damage we have caused on earth within the last 200 years (Nissen, Wong, and Carlton 2021; von Storch, Ley, and Sun 2021).

1.4.2 The Concept of Sustainability

But, is sustainability only limited to environmental problems? What is sustainability? What does it apply to? By definition, sustainability is the ability to support ourselves and our economy and environment without hindering the future generations (Brundtland 1987). Sustainability is all about making sure that the energy being spent or resources being used are renewable with no diminishing returns. Complete sustainability determines that the energy or resource is completely inexhaustible with no or little drawbacks. Sustainability is a concept that can scale massively into all kinds of functions, settings, and industries. It leaks into several different factors other than a resource such as population, climate change, energy production, manufacturing of cars, and many more variables (Kuhlman and Farrington 2010). Just a simple example: A 16-ounce plastic bottle of concentrated bubble bath could last longer for one bottle every two months.

With this, we can save one-quarter of a pound of plastics and 10 USD to 20 USD (US dollars) every year (Rogers and Kostigen 2007). Sustainability does not apply only to natural resources, but also to social and economic resources. When it comes to the topic of sustainability, three pillars support sustainability (Figure 1.3). The three pillars of sustainability are Economic Pillar, Social Pillar, and Environmental Pillar. These three pillars are equally important; if any one of the pillars is weak, the system is no longer sustainable. Achieving sustainability in any of the pillars will affect other pillars (Purvis, Mao, and Robinson 2019).

The social pillar represents social sustainability with aims to build a global society that works and fulfills the needs of everyone. Access to healthcare, shelter, education, equity of opportunities is essential to this pillar. Topics related to well-being and rights including social equity, health equity, community development, social support, human rights, labor rights, and more are commonly encompassed in this pillar. The final goal is to achieve a stable and peaceful global community. The opposite of sustainability under this pillar includes the excessive population in poverty, recurring wars, and endemic political corruption (Aidt 2011; Murphy 2012; Brown et al. 2021).

Economic sustainability is the pillar that ensures that humans from all cultures achieve economic success indefinitely—in other words, it shows a good fiscal stewardship in society. It is a truth that each citizen who is indefinitely prepared for economic recession while possessing long-term investment has reached their economic sustainability. If a nation has an indefinite capability to reduce the national debt and regulate the financial industry adequately, then the country has reached the pinnacle of its economic success and sustainability. This also includes having an extremely low percentage of people living below the poverty line. If

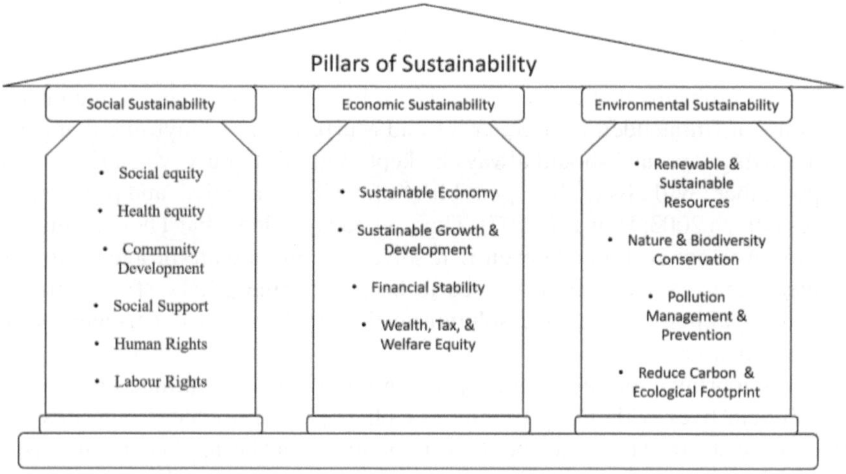

FIGURE 1.3 Representation of sustainability in form of three pillars. Adapted from Purvis et al. (2019).

excessive unemployment and recurring large recessions are common within an area, economical sustainability is affected (Jackson 2013; Atkinson and Morelli 2011; Geissdoerfer et al. 2017).

Environmental sustainability entails the integrity of nature and ecological cycles and systems uninterrupted by the changes we cause. An example of how we negatively impact environmental sustainability is deforestation or pollution. Problems that affect environmental sustainability and at the same time cause instability include climate damage/change, the collapse of the ecosystem, acidification of the ocean, pollution (most notably, air pollution), and ozone depletion. Scientists around the world have researched for solutions such as renewable energy, endangered species protection, reducing greenhouse gas emissions, and recycling (Goodland 1995; Harich 2010; Brown et al. 2021; Chai, Bao et al. 2021).

1.4.3 Why Is Sustainability Important?

In the 1920s, the refrigerator was killing people with leaked cooling gases in their own homes. Thomas Midgley Jr. invented the solution of using freon or Chlorofluorocarbons (CFC) in refrigerators, hairsprays, and, other consumer goods. It was thought to be a great refrigerant, propellant, and solvent because it was inert, colorless, odorless, and nontoxic. Fast-forward to today, it is planned through the Montreal Protocol, implemented in 1989, to eradicate CFC in our atmosphere and to reduce its production to preserve our Ozone Layer. The ozone layer is predicted to heal and repair itself sometime between 2040 and 2070 (Albrecht and Parker 2019; Sicard and Baker 2020). Thomas also invented the solution to the issue of engine knocking by adding tetraethyl lead into petrol. This solution was highly effective and economic; however, its drawbacks outweighed its benefits. The use of lead ultimately caused lead poisoning and air pollution. The production of leaded petrol was eventually phase-downed, and it was banned in the 1980s (Nriagu 1990; Newell and Rogers 2003). Both of Thomas' inventions only brought temporary solutions to the initial problems. Ultimately, they created more problems that negatively affected the environment and human health. The Law of Unintended Consequences and Murphy's Law "anything that can go wrong will go wrong" should always be kept in mind. For every solution that we propose, there will always be a possibility of positive, negative, and perverse outcomes (Bloch 2003; Howard 2007). Thus, we can conclude that Thomas' invented solution was not sustainable even if it solved the immediate problem. To solve problems that we face, we also need to keep the three pillars of sustainability in mind, making sure that the solution to the problem will not trigger another problem to arise.

The importance of sustainability lies in the ability to continue our comfortable human lives without destroying our planet, which includes our ecosystem and nature itself. This concept of sustainability can be applied in any aspect of our lives, be it about people, culture, society, industry, policies, or country. Sustainability is also important to businesses and the management of organizations. Because a sustainable business encompasses management and coordination of the three pillars, business sustainability ensures responsible, ethical, and ongoing success. It runs on the concept of perpetuity that makes sure that anything that is sustainable does not expire. Humans have not truly practiced sustainability in the past in the form of practices such as excessive consumerism and littering. However, the current and future generations have started to demand sustainability and a better future. The fight for sustainability did not stop at environmental issues alone. Humans have been fighting to solve myriad social and economic issues. This is due to the fact that humans not only need stability from our environment and nature but also need social and economic stability to sustain human civilization. All in all, we want to achieve sustainability because we want to survive indefinitely (Anderson 2006; Clift 2007).

1.4.4 Efforts on Achieving Sustainability

Gender equality, ending world hunger, and stopping climate change are some of the global issues that the world is trying to achieve. According to the governments of the world, we can fulfill this goal by the year 2030. The United Nations has rolled out their plans, which are called the sustainable development goals, with the mission of achieving an optimal and sustainable future for all life by 2030. Humans have been working on countless solutions that supposedly will solve our global issues (Halisçelik and Soytas 2019). Still, we have yet to solve the problems with the proposed solutions.

Renewable energy, for example, has been under the spotlight as the biggest potential sustainable source of energy. The idea of harvesting energy from nature itself without any gas emissions or wastage like solar energy and wind energy is very appealing to all people concerned with climate change. However, as opposed to the prior beliefs or public opinion, scientists and studies have shown that renewable energies have been damaging our environment (Harjanne and Korhonen 2019) mainly because the infrastructure needed for these energies requires a large amount of land. This would mean destroying natural habitats or relocating wildlife, just to build solar and wind farms. As a result, we would be sacrificing the ecosystem and wildlife that we want to protect in the first place. Sometimes, we are blinded by our goals of saving the planet while proposing unrealistic solutions that can do more harm than good. Even when there were ample arguments and evidence that radioactive energy from uranium could be a renewable, clean, and safe energy, nuclear energy remains heavily scrutinized (Shellenberger 2017). Since there is evidence that renewable energies can harm our planet, should renewable energies be considered sustainable? Should we consider solar and wind energy to be a part of the sustainability framework?

Growth of population is notably higher in developing countries, while consumption of resources and pollution are both higher in more developed countries. Back in the 1990s, there was an initiative of reducing global poverty by 50%. A significant number of the population in China and India was affected by poverty. Surprisingly, that goal of reducing global poverty levels was achieved due to sustainable economic growth, which also increased social sustainability. This shows that countries in poverty will flourish when there are increased GDP and economic growth (Besley and Peters 2020). However, the global population has exceeded 7 billion people in 2011, and, hence, sustainability will be harder to achieve. Only 100 years ago, there was an estimate of 1.6 billion people in the world, and in the 1960s the population was halved in proportion to the present. The population is predicted to continue rising in a nonlinear fashion. Every person requires vital resources to survive, and so the stress on the earth's supplies and resources will also grow. Although a bigger population generally means more resources to be used, the distribution of resources throughout the world remains unbalanced and uneven (Bevan et al. 2017; Crist, Mora, and Engelman 2017). A large indicator of unsustainability is wealth distribution. In our current society, the lack of wealth directly affects a person's ability to access energy, water, or food. Wealth often represents the ability to acquire resources which directly equates to survivability in the modern world (Bagchi and Svejnar 2015; Zucman 2019).

Over a third of the world's population still lives in poverty with limited access to energy, water, or food. Achieving sustainability is a key and crucial milestone toward the reduction of wealth inequality and poverty around the world and its different civilizations. It is estimated that over a third of the world's population is living in poverty or with access to limited or unsubstantial resources, clean water, supplies, and food security. A major problem regarding sustainability is the direction and control of wealth around the world, as it is estimated that the richest 10% of the population in Organisation for Economic Co-operation and Development (OECD) countries is around 9.5 times richer than the poorest 10% (Cingano 2014). Another comparable statistic is that the top 10% owns more than 70% of the total wealth in Europe, China, and the United States combined. The poorest bottom of 50% owns less than 2% of total wealth. The wealthy also have many opportunities and loopholes to avoid taxation (Zucman 2019). Also, it is shown that different countries consume energy in a nonlinear fashion to their population—for example, the United States is responsible for the consumption of 24% of the world's total consumed energy despite only making up 4.5% of the earth's population—these figures aid to display the disparity between the usage and owning of resources in between the wealth of countries and individuals (Asif and Muneer 2007).

Therefore, one could deduce that the key step toward obtaining sustainability is the insurance of fair distribution of resources to individuals and nations, the proper usage of resources, and the management of waste. As civilization develops, the methods and theories used toward obtaining sustainable practice will also develop, where new and more effective or efficient techniques could be discovered. Literature regarding economic development such as *Limits to Growth* had always argued that the modern capitalistic growth-based economy is unsustainable (D. H. Meadows and Meadows 2007; D. Meadows and Randers 2012). To properly ensure that our growth as a civilization does not damage our planet, a structural reform on our economy and governments is necessary. We can no longer afford to commercialize the finite resources to become an economic bargaining chip; we can no longer allow earth's resources to become abused in a manner that is unsustainable toward the future generation of all living beings (Purvis, Mao, and Robinson 2019). The business-as-usual approach where companies and governmental organizations would preach on "sustainable development" on the surface and then later return to conventional profitable but unsustainable business models must stop as soon as possible (Bocken and Short 2021).

All in all, solutions and plans for achieving sustainability have to be sustainable on their own. We have to make sure that our methods don't defeat their original purpose or create another problem. To achieve sustainability that fits all, different motivations for exploring sustainability also need to be considered. This motivation can come from different drivers depending on one's perspective of worldview, logic, emotion, and/or instinct. In general, they are based around concepts of humanity inherently striving for survival. Different viewpoints for sustainability will create different solutions, outcomes, and different futures. Achieving sustainability is neither an easy nor a simple feat that we can achieve in a short time. This book welcomes the emergence of the idea of applying microalgae in almost every aspect of our lives, which could bring us closer to sustainability.

1.5 INTRODUCTION TO MICROALGAE

In the efforts of searching for a source of sustainability, algae have shown their potential before the 2000s. Algae are eukaryotic organisms known to have a diverse range of morphological diversities because they are from a polyphyletic group (Metting 1996). Algae are considered to be lower plants since they are some of the oldest organisms and the earliest plant groups to evolve (Kaur and Sharma 2018). Algae do not have true roots, stems, or leaves which makes them thallophytes. Similar to mosses and hornworts, thallophytes are plants with bodies that are not differentiated into root, stem, and, leaves and are thus termed thallus. Algae could be separated into two groups by size: Microalgae and Macroalgae. Microalgae belong to the group of simplest forms of algae. They are usually unicellular, microscopic (up to 100 micrometers), and could possess motility (Waterbury et al. 1985; Borowitzka 2018). The other group of algae is known as macroalgae, which are multicellular and macroscopic in size (up to 50 meters). Macroalgae are marine plants like seaweeds and kelps (examples of macroalgae are shown in Figure 1.4) which have anatomy constituting the thallus. The thallus is the whole body of macroalgae that comes with lamina (leaf-like structure), stipe (stem-like structure), and holdfast (root-like structure) (Baweja et al. 2016). In contrast to macroalgae, microalgae have the anatomy of a simple plant cell that comes in various shapes. Microalgae can exist individually or in chains of cells or groups of cells. Algae can also be classified by their chlorophyll content or rather by their color. There are red, brown, and green algae which consist of both micro- and macroalgae.

The cell structure of microalgae is very simple, and the major components of the cell are the cell wall, cell membrane, nucleus, mitochondria, and chloroplast. The chemical constituents of microalgae are carbohydrates, lipids, and proteins

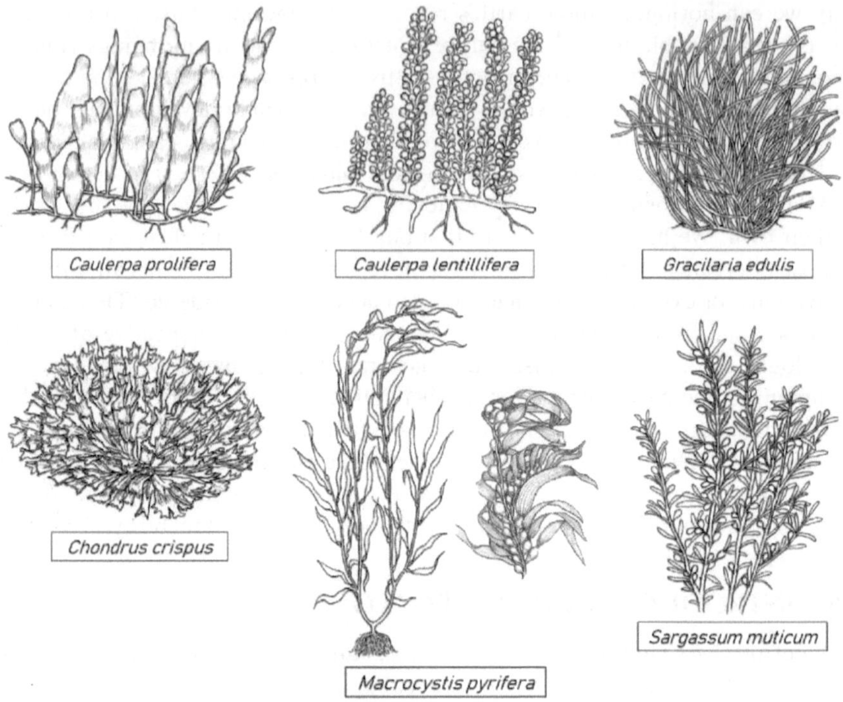

FIGURE 1.4 Examples of macroalgae.

which are reserved within the cell components (Arnold et al. 2015). Microalgae have simple growth requirements, needing only light, water, and carbon dioxide. They require phosphorous, sulfur, and nitrogen as their primary nutrition for development, along with other micronutrients such as iron, manganese, and zinc (Borowitzka 2018). Macroalgae and microalgae are heavily affected by environmental factors such as light levels, pH levels, carbon dioxide intake, temperature, and nutrient access. All these factors affect the cultivation and thriving of microalgae; light is essential for microalgae to allow them to photosynthesize for energy; carbon dioxide is required to allow for cell growth and maintenance within the microalgae (Lehmuskero, Skogen Chauton, and Boström 2018; Morales, Sánchez, and Revah 2018). Temperature and pH levels significantly influence the development of microalgae, as these conditions have to be suitable for the incredibly diverse array of species of microalgae (Wiencke, Amsler, and Clayton 2014). It is found that microalgae of different species and strains contain different levels of carbohydrates, proteins, and lipids (Renaud and Parry 1994; Gouveia et al. 2017). For example, *Chlamydomonas* spp. microalgae contain 24.7% lipid and 46.9% protein, while *Chlorella* spp. species consist of 16.1% lipid and 39% protein (Darwish et al. 2020). These differences allow them to flexibly adapt to many different environments with an abundance of different microalgae species. The adaptability of these microalgae species allows diverse environmental development and also heavily boosts their growth rate, and it is found that microalgae grow at a rate of five to ten times quicker than standard food crops would take to develop (Zullaikah et al. 2019).

Photosynthesis is a common process utilized by all organisms with chloroplast organelles, mainly composed of light dependent and light-independent reactions (refer to Figure 1.5). Photosynthesis, simply put, is a process required to convert solar energy produced by the sun into chemical-based energy that plants use for food and nutrition (Blankenship 2008). The light-dependent reaction uses energy from the sun to photodissociate water to synthesize hydrogen ions (H^+), energy in form of adenosine 5′-triphosphate (ATP), and oxygen. The light-independent reaction or Calvin cycle is the next crucial element for the process of photosynthesis, which fixes the carbon from carbon dioxide using the energy from ATP into sugars. To explain the Calvin cycle in a simple manner, it is the process of turning carbon dioxide in the atmosphere into sugars, which plants then utilize to grow (Raines 2003). The Calvin cycle is split into three different procedures. The first is the carbon fixation stage, which is where carbon dioxide is absorbed from the atmosphere to produce carboxylate ribulose 1,5-bisphosphate (RuBP) with the help of enzyme ribulose-1,5- bisphosphate carboxylase/oxygenase (RuBisCO) thus kick-starting the action of photosynthesis (Heureux et al. 2017). The second phase also known as the reduction phase prompts energy reaction with chemicals to create the essential sugar glyceraldehyde 3-phosphate (G-3-P). Finally, with the regenerative phase, utilizing the stored energy and sugar, the two can interrelate with each other, creating RuBisCO, ready to begin photosynthesis and repeat the cycle (Tamoi et al. 2005). The discovery of photosynthesis in algae was made in the nineteenth century, where microalgae cultures of *Chlorella vulgaris* went

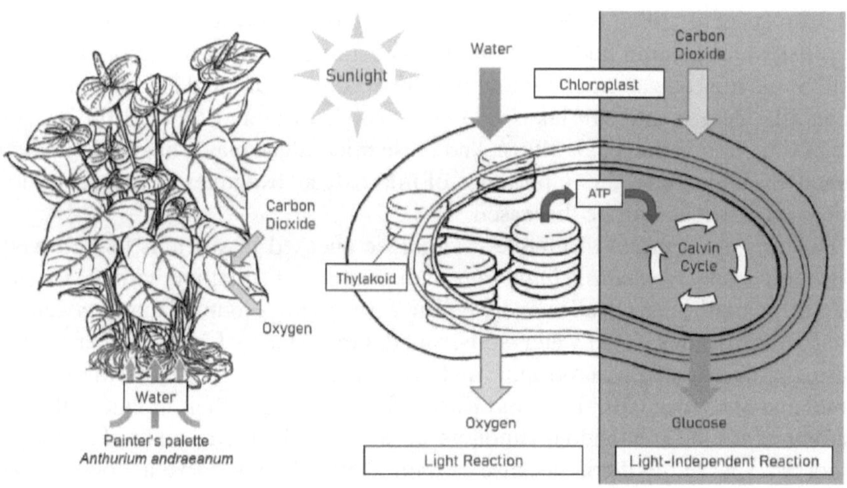

FIGURE 1.5 Simplistic representation of photosynthesis.

under trials for further research and grew by applying ^{14}C-labeled CO_2 in short bursts of photoassimilation (Bassham and Calvin 1960; Birmingham, Coleman, and Colman 1982).

Within the microscopic world, microalgae go by another name which is phytoplankton. Phytoplankton is part of the plankton community. Plankton are a special group of microscopic organisms that have no specific biological classification because of their diverse species and trophicity. Rather, plankton are classified and defined by their ecological niche (Brun et al. 2015; Leruste et al. 2018). Whereas in the case of microalgae, they are part of the phytoplankton trophic which is mainly autotrophic. Microalgae can produce their food and energy source primarily through photosynthesis (photoautotrophic) and sometimes through chemosynthesis (Richmond 2013; Fimbres-Acedo et al. 2020). Photosynthesis is the process of chemical energy being produced by the conversion of light energy as part of cellular respiration. This process is utilized by all organisms that possess chloroplast organelles and chlorophyll pigments in their cells (Raines 2003; Lehmuskero, Skogen Chauton, and Boström 2018). Most plants as well as algae perform photosynthesis. The chemical energy commonly derived from the sun is usually stored as carbohydrates. These carbohydrates are used to synthesize diverse novel compounds including carotenoids, antioxidants, fatty acids, enzymes, polymers, peptides, toxins, and sterols (Bilbao, Salvador, and Leonardi 2017). These diverse compounds have benefitted humans and the organisms of different levels in our ecosystem. An example of such a compound is the omega 3–6 fatty acid (DHA) produced by *Schizochytrium* sp. and *Crypthecodinium cohnii* and the vitamin tocopherol extracted from *Nannochloropsis oculate* (Bilbao, Salvador, and Leonardi 2017; Oliver et al. 2020).

The most common types of microalgae are diatoms, cyanobacteria, and dinoflagellates (Figure 1.6). They are typically found in freshwater and ocean water (Matsunaga et al. 2005; N. Wu et al. 2017). In these ecosystems, they can live in both the water column and sediment but sometimes inside a living organism as well. Some microalgae can be found in terrestrial soils, air, and extreme environments like deserts and salt lakes (Hu et al. 2002; Lyon and Mock 2014; Tesson et al. 2016; Abinandan et al. 2019). And these microalgae play a huge role in their habitat. To understand the importance of microalgae, examples of each common type of microalgae will be discussed.

Diatoms are a group of unique microalgae encased in silica as its cell well, called a frustule. *Navicula tripunctata* is an example of a diatom, with an appearance of a small boat (Bruder and Medlin 2008). This group of diatoms can be found in the oceans, fresh water, soils, and damp surfaces. Just like other microalgae, diatoms are photoautotrophic and generate about 20–50% of earth's oxygen (Smol and Stoermer 2010; Kale and Karthick 2015). Due to their unique cell wall, diatoms consume tremendous amounts of silicon, which can make diatoms an important source of silicon because of their shell. Diatoms serve as food for the aquatic food chain, and some of the diatom species are capable of fixing nitrogen (Battarbee et al. 2001; Benoiston et al. 2017).

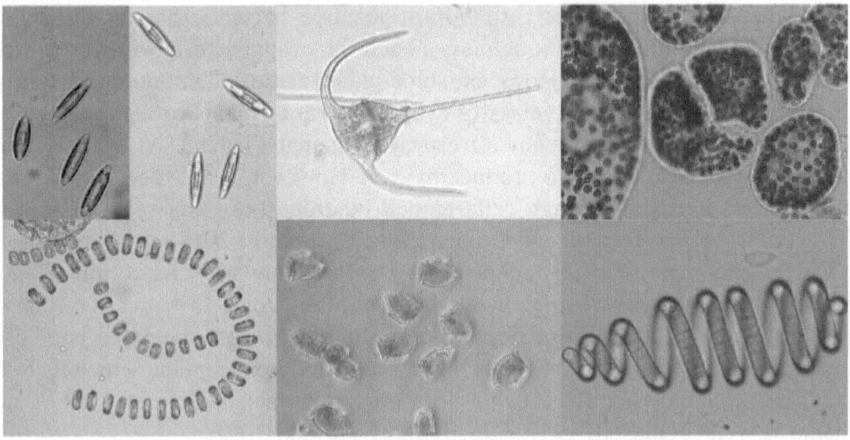

FIGURE 1.6 Photos of microalgae under the microscope. The left column, middle column, and right column represent diatoms, dinoflagellates, and cyanobacteria, respectively. The top row (left to right) shows *Navicula* sp., *Ceratium longipes*, *Microcystis wesenbergii*; the bottom row (left to right) shows *Thalassiosira nordenskioeldii*, *Dinophysis* sp., *Arthrospira fusiformis* (N. Lewis 2011; Moestrup 2016a, 2016b; Gouda 2011a, 2011b, 2011c). Reprinted from World Register of Marine Species (WoRMS) photogallery under the Creative Commons Attribution-Noncommercial-Share Alike 4.0 License.

Dinoflagellates are commonly found in both marine and freshwater habitats. They have amphiesma or cortex, which is a cell covering composed of a sequence of the membrane. It also possesses two flagella from its posterior and its ventral cell side (Taylor, Hoppenrath, and Saldarriaga 2008). *Symbiodinium* spp. is an example of a dinoflagellate, which largely lives in cnidarians and other marine species. Cnidarians are tropical marine organisms including corals, sea anemones, and jellyfish. These specific dinoflagellate species are considered to be endosymbiotic dinoflagellates because they live as mutualistic symbionts of the cnidarians. The *Symbiodinium* spp. enters its host through phagocytosis and gains shelter from its host. The symbiont photosynthesizes and creates by-products (fatty acids, pigments, and vitamins) that are beneficial for its host (LaJeunesse 2002; LaJeunesse et al. 2004; Davy, Allemand, and Weis 2012). However, the relationship between these microalgae and coral does not stop at mutualism. The presence and absence of *Symbiodinium* spp. are like life-and-death situation for the coral reef ecosystem. When the ocean water gets too warm and acidic, *Symbiodinium* spp. dies or leaves the coral. As a result, corals die and bleach when they do not receive essential nutrient from the microalgae (Pandolfi et al. 2011). Unfortunately, due to global warming and climate change, the bleaching of corals has reached up to 50% of Australia's Great Barrier Reef in 2016 (Hughes et al. 2018). On the other hand, some dinoflagellates can cause harm such as Ostreopsis to produce toxins that cause foodborne poisoning called ciguatera (Lehane and Lewis 2000).

At the end, cyanobacteria are Gram-negative bacteria that obtain energy through photosynthesis. Although it is a bacteria, cyanobacteria were recognized as being blue–green algae. However, some plant scientists and modern botanists might refrain or disagree to classify cyanobacteria as plants or algae (Reynolds 1984). They can be found in any terrestrial and aquatic habitat. As cyanobacteria are photosynthetic, they also produce oxygen. However, some species can repair atmospheric nitrogen through cells named heterocysts in anaerobic conditions (K. Kumar, Mella-Herrera, and Golden 2010; Castenholz 2015). *Arthrospira platensis* is one example of cyanobacteria, which is known by its common name Spirulina. For the *Arthrospira* genus bacteria, they come in helical coiling trichomes, with varied sizes and degrees of coiling. *Arthrospira* spp. has been used to make food supplements due to its high protein and lipid content, which includes other nutritional compounds such as vitamins, minerals, and photosynthetic pigments as well (Tomaselli 1997; Capelli and Cysewski 2010).

Currently, microalgae are being utilized minimally in human society in many different ways, as microalgae have diverse uses and untapped potential (Khan, Shin, and Kim 2018; Metsoviti et al. 2019). Novel compounds extracted from microalgae have been found to possess antioxidant, antibacterial, antiviral, anti-inflammatory, and anticancer properties (Breitling and Takano 2015). This allows commercial application in pharmaceutical and nutritional supplements. Novel compounds from microalgae are also used to repair skin damage in the cosmetic industry (Vieira, Pastrana, and Fuciños 2020). Microalgae-based feed also is used as a high nutritional feed for the animal and fish feed industry in a sustainable way (Yaakob et al. 2014). In the biofuel industry, algae have been under the spotlight as the next best choice for biofuel production (Ratha and Prasanna 2012). Research on microalgae is still ongoing to find the most suitable species and strain and also to maximize the yield of lipids and carbohydrates (Suganya et al. 2016). The fossil fuel that was formed was also derived from microalgae along with other organisms. Through photosynthesis, microalgae produce and store their energy. When they die, through sedimentation, extreme pressure, and heat exposure, these ancient dead microalgae form into petroleum and natural gas (Sato 1990). The same microscopic algae that formed fossil fuels are now primarily powering our cities. Unfortunately, fossil fuels have quite a lot of drawbacks. While they continue to fulfill 85% of our world's energy consumption, the gas emission from them continues to damage our world and ecosystem (Clift 2007; Landrigan et al. 2018). Examples of such damage include air pollution, ocean pollution, global warming, and most importantly climate change.

Microalgae have existed long ago, even long before humans evolved to walk. They have been producing oxygen through photosynthesis even before the existence of plants on earth. They are still doing so after green algae first evolved into earth's first land plant 470 million years ago (Metting 1996; Kaur and Sharma 2018). An estimated 50–80% of current earth's atmospheric oxygen is supplied by algae and phytoplankton from the ocean. The key qualities are their fast reproduction rate, ability to perform photosynthesis, diverse species, and their massive population around the world. It means that microalgae are massive consumers of greenhouse gases like carbon dioxide through photosynthesis. This makes microalgae one of the main reasons why the ocean is one of the biggest carbon sinks. Microalgae play a huge role in the world's carbon cycle other than the forest and soil. They also play an important role in our ecosystem as A) carbon sink, B) oxygen supplier, C) producer of the food web, D) crucial symbiont to other living organisms, and E) abundant and sustainable source of nutrients and chemical compounds.

In conclusion, through the information and examples given before, microalgae have shown various qualities that can be beneficial to human society and nature itself. However, microalgae also gave rise to fossil fuel energies which eventually brought harm to nature and human society. Therefore, we shall discuss further how microalgae will be able to contribute to sustainability and help us "clean" the environment.

1.6 MICROALGAE AND ITS SUSTAINABLE APPLICATIONS

Sustainability and its importance have been discussed in the earlier section, and microalgae and their potential in providing sustainable applications are discussed in this section. The potential of microalgae goes far beyond just providing energy as they are versatile. A summary of applications is shown in Table 1.1, which describes the purpose of using microalgae in the different economic and industrial sectors. Nonetheless, their potential still needs to fulfill the three pillars of sustainability. Therefore, in this section of this chapter, we shall explore and evaluate the potential of microalgae as a source of sustainability. A few questions that will be pondered include:

- How are microalgae effective to help to achieve sustainability in respective aspects/fields?
- Why is the application of microalgae important in the respective aspect?

1.6.1 Food and Nutrition

The human world population has been growing exponentially, reaching 8 billion soon and will reach 10 billion by the year 2050. Experts including the famous naturalist, Sir David Attenborough, have warned that humans have overpopulated the earth. With the current human population already putting our food supply and ecosystem at risk, the future poses a far greater challenge on securing food security and battling global malnutrition (George 2009; Perez-Escamilla et al. 2018). Humans have been searching through thick and thin for sustainable ways to produce and supply food, energy, and clean water to support 9.7 billion humans in the future. In response to climate change and declining resources, food trends such as veganism have emerged intending to reduce our "harmful" impact on the earth's ecosystem (Horrigan, Lawrence, and Walker 2002). Even so, raising today's production food supply by 50–70% in the next three decades would put a huge strain on the health of the soil and the environment (Borrelli et al. 2020),

TABLE 1.1
Application and the Corresponding Example of Microalgae Species

Areas of application	Example of species	Uses/benefits	Reference
Food and nutrition	*Arthrospira platensis, Chlorella vulgaris, Haematococcus pluvialis*	Techno-functional, coloring, nutrition	(Caporgno and Mathys 2018; Vieira, Pastrana, and Fuciños 2020)
Pharmaceutical	*Chlamydomonas reinhardtii, Arthrospira platensis, Haematococcus pluvialis*	Biorefinery, bioreactors, artificial chemical biosynthesis	(Pollak and Vouillamoz 2013; Yan et al. 2016; Vieira, Pastrana, and Fuciños 2020)
Biomass and biofuel	*Chlamydomonas* sp., *Nannochloropsis oculate, Scenedesmus obliquus*	Biomass production, biofuel feedstock	(Aratboni et al. 2019; Sajjadi et al. 2018; F. Hussain et al. 2021)
Architecture and urban	*Arthrospira platensis, Scenedesmus* sp.	Urban energy generation, ecological footprint reduction, urban improvement	(Decker, Hahn, and Harris 2016; Peruccio and Vrenna 2019; Carcassi 2021)
Materials	*Euglena gracilis, Chlamydomonas reinhardtii*	Material biosynthesis, biofabrication	(Lode et al. 2015; Dahoumane et al. 2016)
Waste treatment	*Nannochloropsis* sp., *Desmodesmus* sp., *Scenedesmus* sp.	Toxic waste treatment, nutrient recovery, biofuel feedstock	(Acién Fernández, Gómez-Serrano, and Fernández-Sevilla 2018; Emparan et al. 2020; F. Hussain et al. 2021)
Environment and conservation	*Ralstonia pickettii, Ulothrix* sp., *Volvox* sp.	Pollution indicator, pollution bioremediation, environment conservation	(Ryan, Pembroke, and Adley 2007; Gavrilescu 2010)

not to mention the global obstacles like climate change, soil erosion, and reducing water supply. As of now, malnutrition and nutrition disparities still plague countries over the globe, with low-income countries being affected the most (Crist, Mora, and Engelman 2017). Therefore, not only there is a need to feed 9.7 billion humans, but also to ensure they are well-nourished. If the energy source of the food chain was tracked to its origin, it shows that chemical energy in the form of calories comes from the sun. All plants are producers of the food web, who feed us and the animals as well. The same goes for microalgae.

Microalgae, being the primary producer in the ocean food chain, supply fish and whales with a good meal and nutrition. Feeding livestock with microalgae-based feed is also common, especially in aquaculture. Thus, one could imagine that microalgae can also be served on a plate to feed humans. Currently, there are food trends that are introducing microalgae in human food consumption. In a review, Caporgno et al. (2018) and Nova et al. (2020) reported examples of food products containing microalgae, including soup, yogurt, cheese, biscuits, bread, vegetarian food gels, pasta, and many more. The benefits of such application include food technological function, food coloration, and nutrition (Nova et al. 2020). Both reviews have recognized and agreed on their nutritional value and their potential as a food source. What makes microalgae valuable are their ability to produce high content of carbohydrates, lipids, and proteins, which are comparable to meat and milk (Chacón-Lee and González-Mariño 2010). Additionally, microalgae can produce secondary metabolites with bioactive potential for humans such as:

- Poly-unsaturated fatty acids (PUFAs): γ-linolenic (GLA), arachidonic (AA), docosahexaenoic (DHA), and eicosapentaenoic (EPA) acids)
- Vitamin A, B1, B2, B3, B5 B6, B9, B12, C, and E
- Carotenoids (β-carotene, astaxanthin, lutein, zeaxanthin, fucoxanthin)
- Chlorophylls
- Phycobiliproteins (phycoerythrin, phycocyanin)

Salmon is known for its nutritional fats and lipids, specifically DHA. However, salmon do not produce DHA themselves but gained it through consuming algae. Thus, consuming microalgae as part of a diet could mean getting DHA nutrients through its source (Kousoulaki et al. 2016). FAO recognized microalgae as a super food with its high nutritional value, which includes omega-3-fatty acids. Microalgae have also been reported to increase productivity in agriculture, aquaculture, livestock feed. There is an estimation of over 60,000 different species with different biochemical compositions and properties which have a high potential of applications (Cai et al. 2021). One of their merits is their low land requirement to produce protein. Algae are highly productive without the need for fertile soil or land. Just almost like plants, they require some essential minerals (Vadiveloo et al. 2016; Barbera et al. 2017). Microalgae are expected to be an inexhaustible source of biochemical such as proteins, lipids, carbohydrates and pigments such as carotenoids, vitamins, and hormones (Niccolai et al. 2019). Due to their biodiversity in estimates of 1 to 10 million species, there will be endless different potential, uses, and applications in almost every sector and any biotechnology field. With all these factors combined, microalgae are excellent choices for being a source of clean, sustainable, and inexhaustible raw material (N. K. Sharma and Rai 2011).

Just to paint a realistic picture: Arborea, a company invented in BioSolar Leaf, a device similar to a solar panel that can grow different microalgae with a supply of different nutrients (Arborea 2021; Evanson 2019). With their low use of space and high growth rate and efficiency, they use low energy. They will be able to produce high amounts of protein per area of land used, compared to plant- and animal-based food supply. They serve as a supplement for fortification and are just on their own. Versatile and abundant in diversity, microalgae would pique massive interest from food-processing companies. They can become organic food additives for flavoring or coloring, replacing their synthetic counterparts on top of adding health benefits (Vigani et al. 2015; Matos 2017). Imagine adding blue pigment extracted from cyanobacteria into plain white yogurt and turning it into a subjectively attractive blue yogurt. Microalgae and algae have been consumed traditionally in Africa, South America, and Asia in times as old as Inca Empire (c. 1400–1533 CE) and Jin Dynasty (AD 265–316). Even now, there are restaurants all over the globe that would provide *Nostoc sphaericum* (common name: cushuro) as being part of fine culinary arts. Interestingly, three-Michelin Star Spanish restaurants also use the phytoplankton *Tetraselmis chuii* as a key ingredient (Pérez-Lloréns 2020).

1.6.2 Pharmaceuticals and Fine Chemicals

Not limited to food, microalgae have a high value in the pharmaceutical industry as well. Pharmaceutical drugs are chemicals that are used to prevent, diagnose, treat, or cure a disorder or disease. Pharmaceutical drugs can be considered as fine chemicals, and in layman terms, they are known as "Medicines". Pharmaceutical drugs are either produced from chemical synthesis or through natural/biological sources including A) extraction of chemical substances or metabolic products from animals, microorganisms, or plants made by microorganisms or derived from plants and through B) biotechnological processes (Ma, Drake, and Christou 2003; Houdebine 2009; Breitling and Takano 2015). Drug discovery from natural sources has been found to provide diverse and unique molecular diversity and biological functionality. Especially in medicinal plants, drug discovery can produce simple to highly complex molecules, which contain metabolites with promising structural diversity (Lahlou 2013). These complex molecules and metabolites often come in different forms of organic compounds, which include polysaccharides, lipids, proteins, pigments, vitamins, and other compounds.

In the industry of fine chemicals, natural or modified enzymes are utilized in a process called Biocatalysis (same as biotransformation or bioconversion) to enhance or facilitate the production of molecules. Biocatalysis simplifies the operation of production and saves cost, without the need for living organisms, but it requires chemicals as starting materials. Biosynthesis, on the other hand, uses living organisms to convert organic material into fine chemicals from its cheap and natural feedstock. Recombinant DNA technology is widely applied to produce pharmaceuticals including proteins, antibodies, and hormones. Mammalian cell and plant cell cultures are the common two technologies used to produce fine chemicals. However, mammalian cell culture is much more demanding, requiring stringent operating parameters that cost higher with a factor of 10^5 (Pollak and Vouillamoz 2013). Mammalian cells cultures are also susceptible to the infection risk or contamination by pathogens from humans or animals. On the contrary, due to the lack of plasmodesmata, plant cell cultures have no means to be susceptible to both plant and animal pathogens, thus making plants a safer choice for pharmaceutical manufacturing. Even so, plant cell technology was not efficient enough to become industrially viable (Xu and Zhang 2014; R. B. Santos et al. 2016).

That has changed recently because Protalix Biotherapeutics created a proprietary ProCellEx™ plant cell-based protein expression system. This system is a novel bioreactor system that aims to improve efficiency. Since then, other similar companies such as Dow AgroScience and Phyton Biotech have produced similar systems (Xu and Zhang 2014). The bioreactor is polyethylene bags that are large and pliable to which a sterilized condition is applied with growth medium and air supplied from a central system. This method allows dozens of bioreactors to be congregated within a clean, confined facility, granting growth of thousands of liters of transformed plant cells producing the required recombinant protein on an industrial scale and at economical costs (Tekoah et al. 2015). If such a system could be made in the context of microalgae, pharmaceutical-grade molecular farms for microalgae novel bioactive and pharmaceutical compounds would be more economical and realistic with the use of plant cell culture systems. Bioactive compounds from microalgae that express anti-oxidant, anti-inflammatory, antimicrobial, anti-viral, antitumor, anticancer, and other properties and activities provide a great value to the industry. Microalgae have been hailed as an untapped and abundant source of these bioactive and pharmaceutical compounds. Such compounds that are not naturally present could also be produced with the help of recombinant DNA technology, producing recombinant proteins (Borowitzka 1995; Mimouni et al. 2012; Yan et al. 2016; Deniz, García-Vaquero, and Imamoglu 2017; Vieira, Pastrana, and Fuciños 2020; Zanella and Alam 2020).

1.6.3 Biomass and Biofuel

No living being can live without energy, and energy is needed by non-living objects like our shelter, societal facilities, and devices around us as well. Energy is integral to survivability and modern human society. Fossil fuels have been the main energy source for our society, even today. Our world has started to understand that fossil fuels are not sustainable, and the countries around the world have found innovative ways to replace fossil fuel with renewable energy. China, Japan, Germany, the United Kingdom, and the United States are among the countries that are playing a big role in the economy of renewable energy. These five countries are currently leading renewable energy development (Bhattacharya et al. 2016). Renewable energy refers to the energy that is collected through renewable resources. Renewable sources mostly come from natural sources including wind, geothermal heat, solar, tidal, and finally biomass energy (Asif and Muneer 2007; A. Hussain, Arif, and Aslam 2017). Biomass energy is also considered to be renewable energy owing to biomass being able to reproduce naturally in a relatively shorter timeframe compared to fossil fuels. The energy that is stored in the form of biomass usually comes in the form of biofuels such as bioethanol and biodiesel. Bioethanol is derived from starch and carbohydrate, whereas biodiesel is derived from fats and lipids. Both lipids and carbohydrate compounds store chemical energy that can be processed and transformed into biofuels. The main types of biofuels utilized globally are biodiesel and bioethanol (Somerville 2007; Hammond, Kallu, and McManus 2008). There are few other types of biofuels including biogas, bio-oils, bio-hydrogen which can be produced from plants, and cellulosic biomass (Bułkowska et al. 2016; Tiwari and Kiran 2018).

Microalgae are a great source of feedstock for biofuel production. There are microalgae with high oil content which are known as oleaginous microalgae. These microalgae can contain an oil content of 20–50% of their dry weight depending on the species, including *Botryococcus braunii* and *Aurantiochytrium* sp. (Yoshida et al. 2012). Triacylglycerol (TAG) and hydrocarbons are usually harvested from oleaginous microalgae as raw materials. To produce biodiesel, TAG must be converted into fatty acid methyl esters (FAMEs) through transesterification, and hydrocarbons such as botryococcene and squalene can be converted into octane and kerosene through hydrocracking (Taher et al. 2011). On the other hand, there are carbohydrate-rich microalgae that are more suitable for bioethanol production. *Chlorella vulgaris* and *Tetraselmis suecica* are examples of high-carbohydrate microalgae. The carbohydrate content of both microalgae can range up to 37–55% and 21–64% of their dry weight, respectively (Ho et al. 2013; Reyimu and Özçimen 2017). Along with other biofuels such as biomethanol, biohydrogen, and bio-syngas, biobutanol production is also possible with microalgae (Yeong et al. 2018).

However, economical and scalable microalgae-based biofuel production requires suitable and efficient cultivation, harvesting/dewatering, drying, and pretreatment methods for high biomass production which results in high biofuel yield. Microalgae can be cultivated almost anywhere, as long they have a suitable medium, sunlight, and carbon dioxide. Cultivation systems for microalgae often come with the main two types: Open Pond Systems (Circulation and Raceway) and Photobioreactors (Tubular, Flat Panel, and Biofilm) (Suparmaniam et al. 2019). Harvesting or dewatering of microalgae for biofuel production remains the process with the highest cost. There is a range of harvesting technology that could fall into two categories: A) Natural—Gravitational Sedimentation, B) Energy-intensive—Floatation, Centrifugation, Filtration, Reverse osmosis, and Coagulation/Flocculation (Kim et al. 2013; Suparmaniam et al. 2019). Among the technologies, coagulation/flocculation, centrifugation, and filtration are the preferred harvesting technologies based on six criteria including biomass quantity, biomass quality, cost, processing time, species, and waste toxicity. Upon combining the harvesting technologies, a combination of flocculation and sedimentation is an option with low cost and efficiency for harvesting microalgae (Singh and Patidar 2018).

Drying is the next process to remove the remaining water and dry out the microalgae biomass for conversion to biofuel. Drying technologies have various types: solar, convective, spray, drum/rolling, and freeze (C. L. Chen, Chang, and Lee 2015). Pretreatment is a process of enhancing lipid extraction efficiency by breaking the cell wall of microalgae. Microalgae cell wall breaking is an integral part of the downstream process to utilize microalgae biomass to create biofuels. There is a total of six types of pretreatment methods that are mechanical, physical, thermal, chemical, combined, and biological (Onumaegbu et al. 2018; Zabed et al. 2019). The last part of the downstream process is the conversion of biomass into biofuel. The type of conversion process depends on the type of biofuels produced. For biodiesel, lipid extraction is required, and there are two main types of technologies involved for this that are conventional solvent method and green solvent-based method. The conventional extraction methods include Soxhlet, Folch, Bligh and Dyer, whereas the more advanced green solvent-based method include supercritical fluids, ionic liquids, and switchable solvents (Khoo et al. 2020).

Currently, there are a few different generations of biofuels. First- and second-generation biofuels are those that were produced from higher plants. First-generation biofuels were produced from edible food crops, that is, mainly starch-based crops and oil crops (Luque et al. 2008). However, the use of food crops as biofuel feedstock was deemed unsustainable and unethical as it leads to competition on food crops for fuel instead of food, thus driving up the prices of food, land, and water. It has the potential to scale up like fossil fuels without competing with agriculture for water. Microalgae are suitable as a biofuel feedstock because they have a higher lipid content of up to 70% of their biomass, depending on its species and variants (D. Singh et al. 2020). On top of that, biofuels made from microalgae beat the most productive oil crops and oil palm in terms of production yield (Chisti 2008). Food shortage and global hunger have been a long-standing issue; thus, food crops shouldn't be utilized as biofuel feedstock. Nonfood feedstock has become the second-generation biofuel. Nonfood feedstock comprises energy crops, woody crops with lignocellulosic biomass, or just agricultural waste (Luque et al. 2008). Algae-based fuels belong to the third generation including microalgae. Microalgae are deemed to be the better feedstock compared to their predecessors due to their high growth rate, high biomass production, and low land and water requirements (Chisti 2007; Nascimento et al. 2014). Most importantly, both can never compete with biodiesel crops such as soybeans (46 gal/acre), sunflower (98 gal/acre), coconut (276 gal/acre), jatropha (194 gal/acre), and palm oil (610 gal/acre) by gallons per acre. As for microalgae produce from 6, 275 to 14, 635 gal/acre (Sajjadi et al. 2018), the United States itself uses an estimated 60 billion gallons of diesel per year, while soybean, canola, and palm produce only 50, 90, and 650 gallons of biodiesel per acre per year. Microalgae have the potential of producing 5,000–15,000 gallons of biodiesel per acre per year (Chisti 2007; Mulumba 2010; Nascimento et al. 2014; D. Singh et al. 2020).

1.6.4 Architecture and Urban

Since the realization of impending doom from climate change, renewable energy is becoming more popular and common day by day. The development of solar and wind energy has progressed massively. Many studies and analyses have concluded that both solar- and wind-based energies are good technologies with energy intensity comparable to fossil fuels and nuclear energy (Tremeac and Meunier 2009;

L. M. Miller and Keith 2018). However, though renewable energy has its advantages, it doesn't come without problems. Both wind and solar energies can be inconsistent as they depend on weather and day–night cycles. For example, in the nighttime and in shorter days during the winter season, solar energy production becomes negligible. Wind turbines have to be placed strategically in wind sites such as coastal areas, hills, and open plains with strong and reliable winds (Lakatos, Hevessy, and Kovács 2011; Al-Dousari, Al-Nassar, and Ahmed 2020). When they are run on the commercial scale of solar and wind farms, they can also cause negative impacts on the environment and wildlife. Wind farms have been found to kill bats and birds, impeding their migration and survival (Schirmacher et al. 2018). It is noted that wind turbine component needs to be recycled to reduce long-term environmental impacts. The transportation of turbine components has to be limited as much as possible to reduce their carbon footprint during the primary phase (Tremeac and Meunier 2009).

Large-scale solar farms have been shown to impact wildlife negatively and even destroying habitat and causing premature death (Lovich and Ennen 2011; Chock et al. 2021). Not to mention, as most solar panels are manufactured from China that uses production processes contributing twice as much to carbon footprint and greenhouse gas (GHG) emissions. Solar panels contain toxic chemicals, and they become electronic waste after finishing their life cycle (Yue, You, and Darling 2014). The infrastructure needed to harvest these renewable energies requires massive land use and especially solar power. Its infrastructure could predominantly create competition between agriculture and managed forests, but it still can be done sustainably when solar farms are integrated with other land use such that on a global scale, 27–54% of the land for solar energy will come indirectly from displaced unmanaged forests by the year 2050 (van de Ven et al. 2021). Thus, solar and wind energy technologies could defeat their purpose of fighting climate change, if they are not established with sustainability in mind.

On top of that, land is very precious, especially when the human population is growing, for whom more land is needed. This creates a competition of land between agriculture for food, solar farms for energy, and urban land for human shelter (Borrelli et al. 2020). Humans have been a major influence on the land on the earth. Between the years 1700 and 1980, the increase in land area used for crop plantation was 4.6 times, approximately from 2.65×10^6 km^2 to $15.01 \times 2.65\ 10^6$ km^2 (Meyer and Turner 1992). This increase was largely contributed by human population growth which needed more food. Today, 55% of the human population occupies urban land. In the year 2000, the global urban area was approximately 0.6 million km^2, and it's expected to grow up to 1.8–5.9 times by the year 2100 (Gao and O'Neill 2020). The organization for Economic Co-operation and Development (OECD) reported that GHG emissions from agricultural and land usage account for 23% of human-activity-related emissions. At the same time, it also caused the collapse and destruction of terrestrial ecosystems threatening 25% of both plant and animal species with extinction (OECD 2020). We have to make sure that land use for any human activity is efficient and doesn't harm wildlife and the environment.

Nowadays, the roofs of building around the urban jungle can be seen covered with solar panels. Wind turbines have started to appear more as micro-size and in small scale on city buildings. Such integration of solar and wind energy harvesting technology is part of an initiative for renewable energy micro-generation. These initiatives, including renewable energy in urban planning, IEA SHC Task 51, UK government's micro-generation strategy, and the Low Carbon Buildings Programme, have been undertaken by many countries (Walker 2011; Lobaccaro et al. 2019). As such, microalgae could be an alternative, innovative, and integrated part of urban planning. To achieve sustainability of our urban lands, efficiency and productiveness per square foot of the land used should be increased. Enormous areas of urban land could be further exploited to harvest energy, which otherwise just exist around us and are wasted, unused (Perea-Moreno, Hernandez-Escobedo, and Perea-Moreno 2018). To utilize the urban jungle and harvest renewable energy, architecture and microalgae can play important roles. The same infrastructure we live and work in could be covered with panels similar to solar panels. Instead of solar panels, panels filled with microalgae could cover our windows and buildings. Such panels are photobioreactors that can serve as urban vertical farms, which come with the advantage of year-round crop production, with no weather-related crop failures and reduced usage of water by 70–95%. In terms of feasibility, the concept of urban microalgae farming does work. Such microalgae-integrated architecture could be inefficient when the biomass produced is not certified by local authorities. It could also be costly when the urban farm does not receive enough sunlight (Schipfer and Matzenberger 2012; Peruccio and Vrenna 2019).

City Prosperity Index of UN-Habitat had determined that the prosperity of sustainable cities is completely intertwined with environmental sustainability. Cities are only able to maintain prosperity when environmental and social sustainability quotas are fully integrated with economic goals (Yigitcanlar, Dur, and Dizdaroglu 2015; Yigitcanlar et al. 2019). With the technology that integrates microalgae cultures into urban spaces, that plants that are needed in gardens and city parks can be condensed into architecture and urban structures. Examples include photobioreactor facade on building windows and photobioreactor park-gardens where this microalgae-integrated architecture can serve the purpose of the architecture and also produce biomass energy (Ilvitskaya and Chistyakova 2020). Using photoreactive systems in buildings and targeted implementations in architecture at micro- and macro-levels has many advantages. Energy generation (biofuels), carbon sequestration, reduction of GHG emissions, wastewater treatment, and recycling are the common advantages of such architecture (Oncel, Kose, and Oncel 2020; Cervera and Villalba 2021). This situation showcases a comprehensive approach to sustainability in development, which can blanket the whole city with photobioreactors developed by using microalgae (Sardá and Pioz 2015).

A great example of a microalgae-integrated building is BIQ (Bio-Intelligent Quotient) house in Hamburg, covered in microalgae bioreactors which capture the energy and generate renewable energy. Fifteen flats buildings are covered by 200 square meters of bioreactive/energy facade. The bioenergy facade is a window panel that contains culture medium and microalgae. Coupled with sophisticated automatic logic control, the system becomes integral and automatic in this BIQ house. It can supply heat and warm water without manual labor (Elrayies 2018). By equipping urban areas and concrete jungles with these bioenergy facades, such buildings become more productive by capturing solar and heat energy while generating biomass and capturing carbon dioxide. The external surface area of a building can be used to generate something useful. Given later are other examples of architecture and structures that are integrated with microalgae photobioreactors (Schipfer and Matzenberger 2012; Peruccio and Vrenna 2019; Ilvitskaya and Chistyakova 2020; Carcassi 2021):

1) Photo.Synth.Etica in Dublin, Ireland—A building facade designed by ecoLogicStudio.
2) BIQ House, Hamburg, Germany—A residential apartment designed by ARUP.
3) BIOtechHUT, Astana, Kazakhstan—A hut/pavilion designed by ecoLogicStudio.
4) Culture Urbaine, Geneva, Switzerland—A photobioreactor integrated into a viaduct and small highway designed by The Cloud Collective.
5) Algae Dome, Copenhagen, Denmark—A hut/pavilion designed by SPACE10.
6) Urban Algae Canopy, Milan, Italy—A canopy designed by ecoLogicStudio.

1.6.5 Materials

A growing global population means growing demands for more resources, materials, land, and water for higher living standards and industrialized lifestyles. A meta-analysis has shown that our current consumption and production formula, usage and design of infrastructure and technology, underlying paradigms and world views, and consumer behavior are among factors that result in unsustainable resource use. These resources will continue becoming scarcer and eventually be depleted, while environmental impacts associated with our consumption will worsen. Thus, our current resource use patterns on the global scale are unsustainable (Hirschnitz-Garbers et al. 2016). In similar regards to plastic, many industries utilize materials and resources that are not sustainable or are made with nonrenewable raw materials. The construction industry that is a massive consumer of concrete is one example that uses nonrenewable materials. As such, environmental global concerns from all around the world demanded a reduced usage and reliance on synthetic plastics and nonrenewable materials (Vroman and Tighzert 2009). Companies have found several ways through which materials with problematic characteristics can be managed sustainably by making strategic use of other characteristics of the materials. When approaching the material's end-of-life, problematic materials or materials with problematic characteristics can be repurposed or reused through closed-loop material management. Without strategic sustainability planning for material management, organizations risk phasing out materials perceived to be unsustainable (Lindahl et al. 2014).

Plastic is one of the greatest human-created polymers and materials. However, most of the plastic we use in the modern day is made synthetically with fossil fuels (Gilbert 2017). Fossil-fuel-based synthetic plastic takes too long to degrade and, coupled with massive global production and inability to dispose of them caused plastics to pollute our environment. However, we cannot exactly remove plastics entirely out of human modern life. Plastics made the development of electronic products such as computers and smartphones possible. Plastics play an equally important role in various industries including fast fashion, automotive, courier, and food today (Haefely 1947; Stauber 2007; Kirwan, Plant, and Strawbridge 2011; Pacheco-Torgal and Jalali 2012; Rabnawaz et al. 2017; Niinimäki et al. 2020). Biodegradable plastics can be made of biological raw materials or petroleum, which can be decomposed by living organisms. As biodegradable plastics are made from biological raw materials, it also means they are renewable. However, to prevent food scarcity, using food-based plant material is out of the question. Besides, growing plants for biodegradable bioplastics can cause pollution from fertilizers and divert the aim of land use for food production (Tabone et al. 2010). Thus, this is where microalgae could play a huge role as a new source of raw materials to produce plastics and other materials. *Chlamydomonas reinhardtii* can produce starch and triacylglycerol bioplastic beads, without extraction and purification that can withstand compressive stress to 1.7 megapascals (Kato 2019; Mathiot et al. 2019).

Besides, microalgae can produce a lot of other materials. Bio-cement production through microalgae using *Chlorella kessleri* and cement kiln dust as feedstock has found advantages over conventional cement production that are shorter preparation time, suitability for in-situ process, energy efficiency as compared to conventional cement production, and providing with ambient temperature and high pH value. Bio-cement production can contribute to energy savings and also minimize CO_2 emissions. Bio-cement is capable of repairing cracks in materials used in constructing buildings, developing different rock strengths, and regain strength within a month. It also supplements the durability of bricks by increasing compressive strength and reducing permeability (Irfan et al. 2019). Microalgae-based bioplastics are suitable and sustainable alternatives to petrochemical-based plastics. Blending microalgae biomass with different petroleum plastics of differing ratios can create different biocomposite plastics with different tensile strengths. To reduce cost and increase efficiency, microalgae could be used as bioreactors for polyhydroxyalkanoates (PHAs) production while utilizing wastewater as feedstock. With modern genetic engineering technology, the biosynthesis of microalgae biomass and its derivatives for bioplastic production will become economically viable (Franziska et al. 2011; Rahman and Miller 2017).

Bioprinting, biofabrication, or biomineralization are similar forms of biosynthesis that microalgae can perform. Bioprinting is the production of structures based on layer-by-layer biomaterial deposition. An experiment reported on successful extrusion-based bioprinting of hydrogel-based bioink which contained

Chlamydomonas reinhardtii live algae cells (Thakare et al. 2021). The applications of bioprinting using plant or algae cells include the production of metabolites for use in pharmaceutical, cosmetic, and food industries (Camere and Karana 2017). Bioprinting living materials has shown its potential when printing in usable 3D structures gets possible in materials such as bio-garments, adhesive labels, and blinds or curtains for windows. The fabricated living materials showed resilience to physical distortions and showed no leakage of the microalgal cells from the bioprinted living materials to the surrounding water, highlighting their biosecurity and ability to prevent environmental contamination (Balasubramanian et al. 2021). Fabrication of algae-laden hydrogel scaffolds for medical purposes is also possible with bioprinting alginate produced by microalgae in 3D geometrical shapes (Lode et al. 2015).

Biosynthesis of nanomaterials by microorganisms is gaining traction as a relatively new and environment-friendly approach in nanomanufacturing compared to traditional chemical and physical approaches. The mechanisms behind biomineralization have been extensively studied for the possible development of new nanomaterials (Grasso, Zane, and Dragone 2020). Diatoms as discussed in Section 1.5, usually come with intricate and ornated silica frustules that were biomineralized or biofabricated by the diatoms themselves (Vrieling et al. 2003). These frustules are often in nano-size that could be harvested to study or mimic how to biomanufacture these biosilica shells. These shells have found potential to be used in biosensors, medical diagnostics, and drug delivery (Ragni et al. 2018). Besides these frustules, microalgae have shown abilities to produce nano-sized precious metals including gold, silver, platinum, and palladium (Dahoumane et al. 2016; Grasso, Zane, and Dragone 2020). While microalgae can produce a myriad of nanomaterials, the use of nanoparticles can enhance the production of biomass and the accumulation of carbon-based compounds. Nanotechnology results in higher yield and efficiency of microalgae refineries. The extraction or harvesting of microalgae is often described as being tedious which makes them inefficient and makes harvesting at industrial-scale uneconomical. The employment of engineered nanoparticles in microalgae harvesting has helped to increase the efficiency of harvesting by up to 99% (M. K. Nguyen et al. 2019).

1.6.6 Waste Treatment

The conventional wastewater treatment system uses an activated bacterial sludge process in stirred ponds. Nonetheless, activated sludge has limited capabilities in completely removing nitrogen and phosphorus or heavy metals without chemicals (Wollmann et al. 2019). Theories regarding the potential of microalgae for the management of waste treatment have led to a large number of studies of the tiny algae, as the propagation of microalgae makes its cultivation very easy. In terms of water treatment, microalgae have quite a lot of applications in managing industrial waste, human sewage, livestock waste, food processing waste, and agricultural wastes (Rawat et al. 2016). Microalgae's ability to deal with waste is a wonderful way and an incredible opportunity to counter the devastating impact on the ecosystem and pollution that improper waste disposal can cause (Alam and Ahmade 2013).

Waste is typically found in excessive nutrients, which are sourced from waste such as livestock feeds and fertilizers from agricultural waste. Typically, such waste contains nitrogen (N), carbon (C), phosphorus (P) that are essential for plant growth (Rawat et al. 2016). With improper management, these chemicals can get released into our ecosystem and lead to excessive levels of pollution. When these chemicals are carelessly released or washed off by rain into aquatic environments, algae bloom or "eutrophication" occurs (Howarth et al. 2000). Eutrophication is an excessive plant and algal growth which can cause ecological changes and imbalance in marine and freshwater ecosystems. This results in impairing other aquatic plants and sometimes produces toxins harmful to aquatic animals (Howarth et al. 2000). Thus, nutrient recovery or removal would be a crucial process to prevent nutrient pollution. Phycoremediation is a bioremediation process that specifically utilizes algae, which in this case are the microalgae. It can prove to be cost-effective for removing nutrients from algae strains with special attributes such as extreme temperature tolerance or quick sedimentation behavior (Olguín 2003). It was discovered that municipal sludge treated with high-temperature water through hydrothermal treatments can allow selective nutrient recovery (Aida et al. 2016). Microalgae-based processes can reduce the energy consumption of wastewater treatment processes by half while recovering up to 90% of the nutrients contained in the wastewater. The nutrients eventually are used in the production of valuable biomass for biofuels or food (Acién Fernández, Gómez-Serrano, and Fernández-Sevilla 2018)

Microalgae, both aerobically and anaerobically, are being used for treating industrial effluents. It has been proved that microalgae have the potential to grow well in certain wastewater samples (Ramos-Suárez, Arroyo, and González-Fernández 2015; Yu, Kim, and Lee 2019; Rajitha et al. 2020; Chai, Tan et al. 2021). Therefore, these effluents can serve as an appropriate sustainable medium for biofuel feedstock. Several studies have shown that flue gas could be used as the source of carbon and wastewater as a nutrient source (Kothari et al. 2021). Microalgae such as *Chlorella* sp. and *Chlorococcum* sp. have shown capabilities of removing 90% of nitrogen from wastewater and carbon fixation of CO_2 from flue gas (G. Yadav, Dash, and Sen 2019). Bacterial anaerobic digestion (AD) of waste is one of the commonly applied processes for the conversion of biomass to biogases hydrogen (H_2) and methane (CH_4) (Ramos-Suárez, Arroyo, and González-Fernández 2015). However, AD is not much efficient, because its processes leave a significant amount of carbon, nitrogen, phosphorus, and CO_2. Thus, coupled with microalgae cultivation, fermentation effluents can be utilized to use up the remaining carbon, nitrogen, phosphorus, and CO_2, while producing biomass and reducing the carbon footprint (Y. di Chen et al. 2018). Such utilization of waste effluents for microalgae cultivation has limitless possibilities considering the species diversity of microalgae. Conventional microalgae (*Chlorella* ssp., *Arthrospira* ssp., *Scenedesmus* ssp., and *Nannochloropsis* ssp.) could be used to treat wastewater in normal conditions. As an extremophile, microalgae could be an option to treat wastewater in special and harsh conditions which include extreme temperatures, extreme pH, and high organic load. Such extremophile microalgae such as *Chlamydomonas acidophila* and *Galdieria sulphuraria* are suitable for wastewater treatment in extreme conditions (Howarth et al. 2000).

Microalgae are important when it comes to the treatment of heavy metals, xenobiotics, and other toxic compounds. Multiple studies have reported that wastewater from tannery, wet markets, palm oil mills, dairy factories, and the textile industry could be treated by microalgae. Effluents from these industries usually have a high organic load and high presence of heavy metals and toxic compounds (Kotteswari M. and Ranjith Kumar R 2012; Ajayan et al. 2015; Jais et al. 2017; Aragaw and Asmare 2018; Emparan et al. 2020). Microalgae-based treatment is crucial because conventional wastewater treatment system has limited capabilities of removing nutrients completely as well as removing heavy metals without additional chemicals (Wollmann et al. 2019). Municipal solid waste is usually dealt with landfilling, which generates landfill leachate filled with toxic compounds. A review has shown that the usage of microalgae to treat landfills is still in its infancy. Thus, microalgal bioremediation still requires more research to scale-up and maximize efficiency (Nawaz et al. 2020). Even so, microalgae still show potential in performing phycoremediation for tertiary treatment of landfill leachate and removing toxic metals, even with transgenic microalgae (Rajamani et al. 2007). A study found that the removal of heavy metals by percentage was observed as follows: manganese (97%), iron (97%), cobalt (92%), nickel (88%), chromium (75%), barium (72%), and zinc (32%) (Nair, Senthilnathan, and Nagendra 2019).

All in all, microalgae have shown their capabilities in treating waste effluents from different industries, even in extreme conditions. In terms of sustainability, studies have determined that microalgae wastewater treatment plant is more sustainable (Grönlund et al. 2004; A. M. dos Santos et al. 2020; F. Hussain et al. 2021). Therefore, it is more suitable for sustainable development when compared to the conventional wastewater treatment plant. To increase its efficiency, waste effluents such as municipal wastewater, sewage sludge, and agricultural waste are also used as biofuel feedstock (Solovchenko et al. 2013; Zuliani et al. 2016). A study has also shown that microalgae biomass cultivated from waste can be converted into biodiesel on a laboratory scale (Kumar, Kumar, and Nanda 2018). Since the 1960s, microalgae have been a candidate for wastewater treatment even till today. Due to the large land area requirement and the lack of technological understanding to up-scale, more research is required. However, there are clear advantages in using microalgae as the principal ingredient of treatment of waste, some of which are reduction of greenhouse gas, nutrient recovery, biomass generation for biofuel, and lower energy consumption compared to conventional methods (Acién et al. 2016; Y. Wang et al. 2016).

1.6.7 Environment and Conservation

With factories and cities covering the earth and human civilizations flourishing, pollution of environments worsens. As pollution covers headlines of the news today from time to time, it can be commonly found in the air, rivers, oceans, and soil. It comes from all parts of human civilization, from the plastics we use to the illegal industrial waste dumping and municipal sewage. Plastics pollute our ocean and feed the marine life with microplastics, which end up on the dinner plate (Ritchie and Roser 2018). Even in the case of agriculture, the overuse of fertilizers and pesticides can pollute the soil causing it to be too acidic or containing too many heavy metals which can poison food crops and affect their production (Yang et al. 2018). Studies suggest that long-term exposure to air pollution causes an increased risk of infection by airborne diseases such as COVID-19 and H5N1 avian influenza (Fattorini and Regoli 2020; X. Wu et al. 2020). Pollution has shown that it affects our health and the environment negatively and ultimately results in premature deaths of wildlife and humankind (Landrigan et al. 2018). The destruction of our environment by pollution raises the need for preventing and mitigating pollution. To combat pollution, prevention and mitigation efforts have been undertaken by countries around the world. Microalgae are part of the effort to solve the issue of pollution, where it can be a versatile living tool.

1.6.7.1 Microalgae as an Indicator of Pollutants

For waste treatment, the waste and pollutants are already known and expected. However, that is not the case in the natural environment. Therefore, there is a need to detect and find out the existing polluted environments. Biomonitors and biosensors are the tools to detect and indicate the existence of pollution (Cid et al. 2012). Microalgae have a wide range of utilizable properties across a large range of the ecology. Water pollution can be highlighted by microalgae owing to the thriving and response of microalgae in polluted waters. Microalgae are sensitive to their environmental conditions, where environmental changes can cause physiological responses (Chan et al. 2021). These responses can range from species composition, cell density, ash-free dry mass, chlorophyll, and enzyme activity (Omar 2010; Cid et al. 2012). A study used a battery of biomarkers including cell permeability, chlorophyll fluorescence, and enzyme (esterase and alkaline phosphatase) activity. It found chlorophyll fluorescence to be the most sensitive to toxicity, thus a suitable biomarker for in-situ biomonitoring of the environment (Gosset et al. 2019). Microalgae can serve as an indicator of the changes in habitats and wetlands. The ability to be a strong indicator of pollution in nutrients makes microalgae very effective for monitoring the quality of water.

Different species of algae have been found to react differently to pollution based on the exposure to pollution of each species (Chiellini et al. 2020). This is useful, as a certain growth or abundance in a certain type of algae could hint toward a river's degradation or pollution. Microalgae are usable in laboratory bioassays to study extracted water quality to test different species of algae within the water. The most notable microalgae for detecting toxins or pollution are cyanobacteria including *Anabaena* spp., *Microcystis* spp., *Oscillatoria* spp., *Nostoc* spp., *Mallamonas* spp., *Chroococcus* spp., *Dinobryon* spp., and *Staurastrum paradoxum* (Omar 2010; N. Yadav and Singh 2020). Several studies have shown and developed practical uses which demonstrate the bright future microalgae hold in their ability to detect pollution of seawater. An autonomous optical biosensor using the algae–protozoa symbiotic relationship of *Chlorella vulgaris* and *Tetrahymena pyriformis* was successfully developed for real-time monitoring of marine water and evaluation of biotoxicity (Turemis et al. 2018). Another example utilizes microalgae *Chlorella mirabilis* to create an optical bioassay to detect pesticides in a marine environment. It was installed onto marine buoys to detect and identify seawater pesticides within the coast, showing the potential of microalgae in detecting and screening marine water and biotoxicity (Moro et al. 2018).

1.6.7.2 Restoration of Degraded Environments through Microalgal Bioremediation

Microalgae can detect pollution and can also remediate it. Microalgae have shown the potential to remediate polluted environments through bioremediation which includes mechanisms such as bioaccumulation, biosorption, and biodegradation (Mustafa et al. 2021). These mechanisms have their differences but ultimately result in the removal of excess nutrients and pollutants. Biosorption in microalgae is a passive process that concentrates and binds contaminants to its cellular structure, which is commonly the cell wall (Ubando et al. 2021). The biosorption mechanism in microalgae has shown potential in binding with a multitude of pollutants such as heavy metals and radioactive metal waste with *Haematococcus pluvialis* and *Chlorella vulgaris* (Lee et al. 2019; Ubando et al. 2021). Bioaccumulation is an active process where an organism uptakes nutrients it needs to survive, along with the pollutants around it, thus removing the pollutants from the environment. The pollutants could be metabolized or simply accumulated and kept inside the organism with certain bioconcentration (Dwivedi et al. 2010; Mustafa et al. 2021). Biodegradation is another active process that degrades pollutants such as petroleum, herbicides, and organic compounds by the metabolic action of degradation, transformation, and mineralization (Subashchandrabose et al. 2013). Microalgae can produce enzymes capable of degrading harmful organic compounds and then transform into minerals or metabolites with the help of specific enzymes. Examples include laccases which can degrade anilines and dehalogenases that can degrade solvents and pesticides containing chlorine (Abdel-Shafy and Mansour 2018). In our environment, pollutants are diverse, where heavy metals, explosives, dyes, pesticides, and sometimes radioactive waste poison our world (Fukuda et al. 2014; ben Chekroun, Sánchez, and Baghour 2014). There are myriad of microalgae species and strains that can bioremediate a wide range of pollutants including hydrocarbons, fungicides, herbicides, chromium, and radioactive metals (cesium, iodine, and strontium) (Dosnon-Olette et al. 2010; ben Chekroun, Sánchez, and Baghour 2014; Fukuda et al. 2014; Plugaru et al. 2017; Satya et al. 2018; Tatarová et al. 2021; Moreira et al. 2021).

1.6.7.3 Carbon Sequestration with Microalgae

Carbon sequestration is currently seen as being one of the important methods to reduce GHG, namely capturing carbon dioxide. Naturally, we rely on the forest and plants to capture carbon and remove it from our atmosphere in hopes to mitigate climate change. Proper afforestation, forest management, and reduced deforestation can benefit us with a large carbon sink (S. L. Lewis et al. 2019; Favero, Daigneault, and Sohngen 2020). Currently, there are other artificial methods beyond the natural methods such as membrane separation and cryogenic distillation. However, artificial methods have certain drawbacks: the risk of storage leakage, expensive operation, and transportation (Anwar et al. 2018). Besides forest plantation and artificial carbon sequestration methods, the ocean filled with phytoplankton is another carbon sink. Ocean fertilization promotes the growth of phytoplankton, which in turn increases carbon dioxide uptake by the photosynthesis of phytoplankton. Though large-scale ocean fertilization is not recommended due to unintended consequences, ocean acidification and ecosystem changes happen (Williamson et al. 2012). Microalgae cultivation through biorefineries/photobioreactors is also effective for carbon sequestration. The ideal parameters for microalgae cultivation and carbon sequestration require warm tropical climates with an atmospheric temperature of 20°C to 25°C, making countries in the equator such as Argentina, Mexico, and Australia ideal for microalgae-based carbon sink (Aly and Balasubramanian 2017). Carbon that is sequestered by microalgae will be utilized by the microalgae itself to generate biomass. The biomass produced can be used to make biofuel, food, and materials, thus increasing the efficiency of microalgae cultivation (Singh and Dhar 2019).

1.7 THE FUTURE OF MICROALGAL TECHNOLOGY—MICROALGAE 4.0

In the world of Industry 4.0, manufacturing practices are revolutionized with the automation of traditional manufacturing and industrial practices while pairing with modern smart technology. Industry 4.0 is a synonym for smart manufacturing, which comes with technologies including advanced robotics, cloud computing, Internet of Things (IoT), and artificial intelligence supported by big data and analytics (Gargalo et al. 2020). The concept of smart manufacturing and Industry 4.0 can also be applied to the microalgae-based industry which brings us to the new concept of Microalgae 4.0. In Microalgae 4.0, the main technologies implemented in Industry 4.0 will be reciprocated in the microalgae industry. There are a few technologies that could increase sustainability through increased efficiency, reduced defects, and maintenance. Intelligence would be implemented in all levels of technology and automation with cyber-physical systems and IoT.

Sensor and process monitoring are important to biomanufacturing processes for ensuring increased productivity. Gargalo et al. (2020) reported on the recent advancement in sensors, process monitoring, and closed-loop process control supported with case studies. Advanced instruments such as oCelloScope and ParticleTech ApS allow real-time, automated online image data acquisition and analysis (Gargalo et al. 2020). These instruments emphasize the use of new monitoring and control approaches, such as Process Analytical Technology (PAT) tools and real-time monitoring and controlling of bioprocesses, which can improve and guarantee product quality to a certain extent. Smart sensor and image analysis are also crucial which can be used to assess the most crucial parameters such as microorganism growth. These various technologies could be highly advantageous for microalgae cultivation (García-Poza et al. 2020). Photobioreactors are essential infrastructure for microalgae cultivation. To ensure efficient and sustainable bioproduction, bioreactor modeling and computational fluid dynamics are employed to quantitatively predict the cell culture behavior and bioreactor performance. The mechanical modeling of microalgal cultures in bioreactors creates optimal model-based strategies which can facilitate economical bioproduction (Flevaris and Chatzidoukas 2021).

Besides improving the infrastructure, advancement on microalgae is also achievable. Genetic engineering of microalgae can help overcome or bypass the biological limitations of its metabolic capacity. Genetic traits such as higher accumulation of desired biomolecules, improved photosynthetic productivity, respective cellular production, production of value-added compounds, or just purely increasing algal biomass productivity (Fu et al. 2019) aim to improve the economic feasibility of the production process. There are recent advances in novel gene-editing tools such as zinc-finger nuclease (ZFN), TAL effector endonuclease (TALEN), and clustered regularly interspaced palindromic sequences (CRISPR/Cas9). These gene-editing tools can facilitate highly specific targeting of genes for editing, allowing us to alter microalgae genome toward designed properties for various applications (Wang et al. 2016; Jeon et al. 2017).

However, the current level of understanding the functional information of the microalgae genome is still insufficient. Microalgae genomes are complex just like any other genome, as they usually come with genome sequence, metabolic pathway maps, and the other omics and mutant resources (Nehme et al. 2018; Krassowski et al. 2020). Artificial intelligence can link up the knowledge gaps between microalgae genetic information and desired bioproducts. AI algorithms can provide analytics for understanding the large and complex genetic sequences in microalgae. Advances in computer vision have allowed for real-time microalgae identification which makes the process more efficient (Teng et al. 2020). AI algorithms could be applied in microalgae cultivation, system optimization, and other aspects of the supply chain (B. Wang 2020). The AI technology also can be applied to mutant strains of microalgae which sometimes hold a higher value than their common strain. Alternatively, high-throughput screening methodologies can isolate and allow easier selection of mutant strains. Mutant strains can be used for in-depth studies of the molecular basis of desired compound accumulation or for maximizing the production of compounds. One example is the *Chlamydomonas* high-lipid sorting (CHiLiS), which enables the isolation of mutants with high lipid content (G. Kumar et al. 2020).

1.8 CONCLUSIONS

This chapter has thoroughly explained the properties and utility of microalgae, as well as the prospects and potential throughout the usage of biotech-concerning microalgae. This chapter has explored the utilization of microalgae through the use of waste treatment and pollution control caused by the inherent toxin-reducing factors regarding waste treatment and the extremely carbon-dioxide-hungry nature of microalgae, allowing the possibility of carbon dioxide reduction throughout our atmosphere. Other properties of microalgae have been used to help toward sustainability efforts through the usage of microalgae as a biofuel, potentially able to completely replace petrol, which would eliminate the harmful greenhouse gas usage. Other sustainability factors lie in the ability to utilize microalgae toward nutrition as a food source/supplement, as well as the usage of microalgae in pharmaceuticals. This chapter has also explored the application of microalgae toward the well-being of the environment and in conservation methods, such as the potential to indicate pollutants and restore degraded environments by the removal of pollutants.

ACKNOWLEDGMENT

The authors of this chapter would like to thank Soon Oun Kang for the illustrations drawn in Figures 1.1, 1.4, and 1.5.

REFERENCES

Abdel-Shafy, Hussein I., and Mona S. M. Mansour. 2018. "Phytoremediation for the Elimination of Metals, Pesticides, PAHs, and Other Pollutants from Wastewater and Soil." In *Phytobiont and Ecosystem Restitution*, 101–36. doi:10.1007/978-981-13-1187-1_5.

Abinandan, Sudharsanam, Suresh R. Subashchandrabose, Kadiyala Venkateswarlu, and Mallavarapu Megharaj. 2019. "Soil Microalgae and Cyanobacteria: The Biotechnological Potential in the Maintenance of Soil Fertility and Health." *Critical Reviews in Biotechnology*. Frontiers Media S.A. doi:10.1080/07388551.2019.1654972.

Acién, F. G., C. Gómez-Serrano, M. M. Morales-Amaral, J. M. Fernández-Sevilla, and E. Molina-Grima. 2016. "Wastewater Treatment Using Microalgae: How Realistic a Contribution Might It Be to Significant Urban Wastewater Treatment?" *Applied Microbiology and Biotechnology*. doi:10.1007/s00253-016-7835-7.

Acién Fernández, Francisco Gabriel, Cintia Gómez-Serrano, and José María Fernández-Sevilla. 2018. "Recovery of Nutrients from Wastewaters Using Microalgae." *Frontiers in Sustainable Food Systems*. Frontiers Media S.A. doi:10.3389/fsufs.2018.00059.

Aida, Taku Michael, Toshiyuki Nonaka, Shinya Fukuda, Hiroki Kujiraoka, Yasuaki Kumagai, Ryoma Maruta, Masaki Ota, et al. 2016. "Nutrient Recovery from Municipal Sludge for Microalgae Cultivation with Two-Step Hydrothermal Liquefaction." *Algal Research* 18: 61–68. doi:10.1016/j.algal.2016.06.009.

Aidt, Toke S. 2011. "Corruption and Sustainable Development." In *International Handbook on the Economics of Corruption*, Vol. 2, 3–51. doi:10.4337/9780857936523.00007.

Ajayan, Kayil Veedu, Muthusamy Selvaraju, Pachikaran Unnikannan, and Palliyath Sruthi. 2015. "Phycoremediation of Tannery Wastewater Using Microalgae Scenedesmus Species." *International Journal of Phytoremediation* 17 (10). Taylor and Francis Inc.: 907–16. doi:10.1080/15226514.2014.989313.

Alam, Pervez, and Kafeel Ahmade. 2013. "Impact of Solid Waste on Health and the Environment." *International Journal of Sustainable Development and* 2 (1): 165–68. https://intelligentjo.com/images/Papers/general/waste/IMPACT-OF-SOLID-WASTE-ON-HEALTH-AND-THE-ENVIRONMENT.pdf.

Albrecht, Frederike, and Charles F. Parker. 2019. "Healing the Ozone Layer. The Montreal Protocol and the Lessons and Limits of a Global Governance Success Story." In *Great Policy Successes*, 304–22. https://www.sciencedirect.com/science/article/pii/S0360544219306000?casa_token=Fgl-WTVhtiUAAAAA:kwHznkiegInjO077nzeXXBSyelfI2v_Yt_F28Qusw7i_cIvWx_WSl6JD610X5oDGtLX3DZIfbQ.

Al-Dousari, Ali, Waleed Al-Nassar, and Modi Ahmed. 2020. "Photovoltaic and Wind Energy: Challenges and Solutions in Desert Regions." In *E3S Web of Conferences*, Vol. 166. doi:10.1051/e3sconf/202016604003.

Allendorf, Fred W., and Laura L. Lundquist. 2003. "Introduction: Population Biology, Evolution, and Control of Invasive Species." *Conservation Biology*. doi:10.1046/j.1523-1739.2003.02365.x.

Aly, Nazimdhine, and P. Balasubramanian. 2017. "Effect of Geographical Coordinates on Carbon Dioxide Sequestration Potential by Microalgae." *International Journal of Environmental Science and Development* 8 (2): 147–52. doi:10.18178/ijesd.2017.8.2.937.

Anderson, Dan R. 2006. "The Critical Importance of Sustainability Risk Management." *Risk Management* 53 (4). Sabinet Online: 66–72.

Andersson, Leif. 2001. "Genetic Dissection of Phenotypic Diversity in Farm Animals." *Nature Reviews Genetics*. doi:10.1038/35052563.

Andrady, A. L. 2005. An Environmental Primer. *Plastics and the Environment*: 1–75. John Wiley & Sons, Ltd. doi:10.1002/0471721557.ch1.

Anwar, M. N., A. Fayyaz, N. F. Sohail, M. F. Khokhar, M. Baqar, W. D. Khan, K. Rasool, M. Rehan, and A. S. Nizami. 2018. "CO2 Capture and Storage: A Way Forward for Sustainable Environment." *Journal of Environmental Management* 226: 131–44. doi:10.1016/j.jenvman.2018.08.009.

Aragaw, Tadele Assefa, and Abraham M. Asmare. 2018. "Phycoremediation of Textile Wastewater Using Indigenous Microalgae." *Water Practice and Technology* 13 (2): 274–84. doi:10.2166/wpt.2018.037.

Aratboni, Hossein Alishah, Nahid Rafiei, Raul Garcia-Granados, Abbas Alemzadeh, and José Rubén Morones-Ramírez. 2019. "Biomass and Lipid Induction Strategies in Microalgae for Biofuel Production and Other Applications." *Microbial Cell Factories*. BioMed Central Ltd. doi:10.1186/s12934-019-1228-4.

Arborea. 2021. "Arborea—Industrializing Photosynthesis." Accessed October 4. http://arborea.io/.

Arnold, Alexandre A., Bertrand Genard, Francesca Zito, Réjean Tremblay, Dror E. Warschawski, and Isabelle Marcotte. 2015. "Identification of Lipid and Saccharide Constituents of Whole Microalgal Cells by 13C Solid-State NMR." *Biochimica et Biophysica Acta—Biomembranes* 1848 (1): 369–77. doi:10.1016/j.bbamem.2014.07.017.

Asif, M., and T. Muneer. 2007. "Energy Supply, Its Demand and Security Issues for Developed and Emerging Economies." *Renewable and Sustainable Energy Reviews*. doi:10.1016/j.rser.2005.12.004.

Atkinson, Anthony B., and Salvatore Morelli. 2011. "Economic Crises and Inequality." *Papers.Ssrn.Com*. https://papers.ssrn.com/sol3/papers.cfm?abstract_id=2351471.

Bagchi, Sutirtha, and Jan Svejnar. 2015. "Does Wealth Inequality Matter for Growth? The Effect of Billionaire Wealth, Income Distribution, and Poverty." *Journal of Comparative Economics* 43 (3): 505–30. doi:10.1016/j.jce.2015.04.002.

Balasubramanian, Srikkanth, Kui Yu, Anne S. Meyer, Elvin Karana, and Marie Eve Aubin-Tam. 2021. "Bioprinting of Regenerative Photosynthetic Living Materials." *Advanced Functional Materials* 31 (31). John Wiley and Sons Inc. doi:10.1002/adfm.202011162.

Barabadi, Hamed. 2017. "Nanobiotechnology: A Promising Scope of Gold Biotechnology." *Cellular and Molecular Biology*. doi:10.14715/cmb/2017.63.12.2.

Barbara, Fadi, Valentina la Morgia, Valerio Parodi, Giuseppe Toscano, and Ezio Venturino. 2018. "Analysis of the Incidence of Poxvirus on the Dynamics between Red and Grey Squirrels." *Mathematics* 6 (7). doi:10.3390/math6070113.

Barbera, Elena, Ali Teymouri, Alberto Bertucco, Ben J. Stuart, and Sandeep Kumar. 2017. "Recycling Minerals in Microalgae Cultivation through a Combined Flash Hydrolysis-Precipitation Process." *ACS Sustainable Chemistry and Engineering* 5 (1). American Chemical Society: 929–35. doi:10.1021/acssuschemeng.6b02260.

Barber-Meyer, Shannon M., L. David Mech, and P. J. White. 2008. "Elk Calf Survival and Mortality Following Wolf Restoration to Yellowstone National Park." *Wildlife Monographs* 169 (1). John Wiley & Sons, Ltd: 1–30. doi:10.2193/2008-004.

Barcelos, Mayara C. S., Fernanda B. Lupki, Gabriela A. Campolina, David Lee Nelson, and Gustavo Molina. 2018. "The Colors of Biotechnology: General Overview and Developments of White, Green and Blue Areas." *FEMS Microbiology Letters* 365 (21). doi:10.1093/femsle/fny239.

Bassham, J. A., and M. Calvin. 1960. "The Path of Carbon in Photosynthesis." In *Die CO2-Assimilation/The Assimilation of Carbon Dioxide*, 884–922. Berlin, Heidelberg: Springer. doi:10.1007/978-3-642-94798-8_30.

Battarbee, Richard W., Vivienne J. Jones, Roger J. Flower, Nigel G. Cameron, Helen Bennion, Laurence Carvalho, and Stephen Juggins. 2001. "Diatoms." In *Tracking Environmental Change Using Lake Sediments*, 155–202. Dordrecht: Kluwer Academic Publishers. doi:10.1007/0-306-47668-1_8.

Baweja, P., S. Kumar, D. Sahoo, and I. Levine. 2016. "Biology of Seaweeds." In *Seaweed in Health and Disease Prevention*, 41–106. Academic Press. doi:10.1016/B978-0-12-802772-1.00003-8.

Benninga, H. 1990. "A History of Lactic Acid Making: A Chapter in the History of Biotechnology." In *Kluwer Academic Publishing*, 478. Netherlands: Springer. https://books.google.com/books/about/A_History_of_Lactic_Acid_Making.html?id=fdBMcYg_xGYC.

Benoiston, Anne Sophie, Federico M. Ibarbalz, Lucie Bittner, Lionel Guidi, Oliver Jahn, Stephanie Dutkiewicz, and Chris Bowler. 2017. "The Evolution of Diatoms and Their Biogeochemical Functions." In *Philosophical Transactions of the Royal Society B: Biological Sciences*. Royal Society Publishing. doi:10.1098/rstb.2016.0397.

Besley, Tina, and Michael A. Peters. 2020. "Life and Death in the Anthropocene: Educating for Survival Amid Climate and Ecosystem Changes and Potential Civilisation Collapse." In *Educational Philosophy and Theory*. Routledge. doi:10.1080/00131857.2019.1684804.

Bevan, Andrew, Sue Colledge, Dorian Fuller, Ralph Fyfe, Stephen Shennan, and Chris Stevens. 2017. "Holocene Fluctuations in Human Population Demonstrate Repeated Links to Food Production and Climate." *Proceedings of the National Academy of Sciences of the United States of America* 114 (49): E10524–31. doi:10.1073/pnas.1709190114.

Bhattacharya, Mita, Sudharshan Reddy Paramati, Ilhan Ozturk, and Sankar Bhattacharya. 2016. "The Effect of Renewable Energy Consumption on Economic Growth: Evidence from Top 38 Countries." *Applied Energy* 162 (January). Elsevier: 733–41. doi:10.1016/j.apenergy.2015.10.104.

Bilbao, Paola Scodelaro, Gabriela A. Salvador, and Patricia I. Leonardi. 2017. "Fatty Acids from Microalgae: Targeting the Accumulation of Triacylglycerides." In *Fatty Acids*. doi:10.5772/67482.

Birmingham, Brendan C., John R. Coleman, and Brian Colman. 1982. "Measurement of Photorespiration in Algae." *Plant Physiology* 69 (1): 259–62. doi:10.1104/pp.69.1.259.

Blankenship, Robert E. 2008. *Molecular Mechanisms of Photosynthesis. Molecular Mechanisms of Photosynthesis*. doi:10.1002/9780470758472.

Bloch, Arthur. 2003. "Murphy's Law." *Perigee*: 210. https://books.google.com/books/about/Murphy_s_Law.html?id=Huc56EBhvY0C.

Bößner, Stefan, Francis X. Johnson, and Zoha Shawoo. 2021. "Governing the Bioeconomy: What Role for International Institutions?" *Sustainability (Switzerland)* 13 (1): 1–24. doi:10.3390/su13010286.

Bocken, Nancy M. P., and Samuel W. Short. 2021. "Unsustainable Business Models—Recognising and Resolving Institutionalised Social and Environmental Harm." *Journal of Cleaner Production* 312. doi:10.1016/j.jclepro.2021.127828.

Bonada, Núria, Narcís Prat, Vincent H. Resh, and Bernhard Statzner. 2006. "Developments in Aquatic Insect Biomonitoring: A Comparative Analysis of Recent Approaches." *Annual Review of Entomology*. doi:10.1146/annurev.ento.51.110104.151124.

Borowitzka, Michael A. 1995. "Microalgae as Sources of Pharmaceuticals and Other Biologically Active Compounds." *Journal of Applied Phycology* 7 (1). Kluwer Academic Publishers: 3–15. doi:10.1007/BF00003544.

Borowitzka, Michael A. 2018. "Biology of Microalgae." In *Microalgae in Health and Disease Prevention*, 23–72. doi:10.1016/B978-0-12-811405-6.00003-7.

Borrelli, Pasquale, David A. Robinson, Panos Panagos, Emanuele Lugato, Jae E. Yang, Christine Alewell, David Wuepper, Luca Montanarella, and Cristiano Ballabio. 2020. "Land Use and Climate Change Impacts on Global Soil Erosion by Water (2015–2070)." *Proceedings of the National Academy of Sciences of the United States of America* 117 (36): 21994–91. doi:10.1073/pnas.2001403117.

Breitling, Rainer, and Eriko Takano. 2015. "Synthetic Biology Advances for Pharmaceutical Production." *Current Opinion in Biotechnology*. doi:10.1016/j.copbio.2015.02.004.

Brown, David, Doreen S. Boyd, Katherine Brickell, Christopher D. Ives, Nithya Natarajan, and Laurie Parsons. 2021. "Modern Slavery, Environmental Degradation and Climate Change: Fisheries, Field, Forests and Factories." *Environment and Planning E: Nature and Space* 4 (2). Sage Publications: 191–207. doi:10.1177/2514848619887156.

Bruder, Katrin, and Linda K. Medlin. 2008. "Morphological and Molecular Investigations of Naviculoid Diatoms. II. Selected Genera and Families." *Diatom Research* 23 (2): 283–329. doi:10.1080/0269249X.2008.9705759.

Brun, Philipp, Meike Vogt, Mark R. Payne, Nicolas Gruber, Colleen J. O'Brien, Erik T. Buitenhuis, Corinne le Quéré, Karine Leblanc, and Ya Wei Luo. 2015. "Ecological Niches of Open Ocean Phytoplankton Taxa." *Limnology and Oceanography* 60 (3). Wiley Blackwell: 1020–38. doi:10.1002/lno.10074.

Brundtland, Gro Harlem. 1987. "Report of the World Commission on Environment and Development: Our Common Future towards Sustainable Development." *World Commission and Development*. https://sustainabledevelopment.un.org/content/documents/5987our-common-future.pdf.

Bud, Robert. 1994. "The Uses of Life: A History of Biotechnology." *Technology and Culture* 35 (3). Cambridge University Press: 648. doi:10.2307/3106295.

Bułkowska, Katarzyna, Zygmunt Mariusz Gusiatin, Ewa Klimiuk, Artur Pawłowski, and Tomasz Pokój. 2016. *Biomass for Biofuels. Biomass for Biofuels*. CRC Press. doi:10.1201/9781315226422.

Buttel, Frederick, and William Flinn. 1974. "The Structure of Support for the Environmental Movement, 1968–1970." *Rural Sociology*. https://search.proquest.com/openview/bb6be42abd73045d4f0fc84c156acefd/1?pq-origsite=gscholar&cbl=1817355.

Cai, Junning, Alessandro Lovatelli, José Aguilar-Manjarrez, Lynn Cornish, Lionel Dabbadie, Anne Desrochers, Simon Diffey, et al. 2021. *Seaweeds and Microalgae: An Overview for Unlocking Their Potential in Global Aquaculture Development. FAO Fisheries and Aquaculture Circular*. Food and Agriculture Organization of the United Nations (FAO). doi:10.4060/CB5670EN.

Camere, Serena, and Elvin Karana. 2017. "Growing Materials for Product Design." In *Alive. Active. Adaptive: International Conference on Experiential Knowledge and Emerging Materials, EKSIG 2017*, 101–15. https://www.researchgate.net/profile/Serena-Camere/publication/319355171_Growing_materials_for_product_design/links/59a6c6fea6fdcc61fcfbbae7/Growing-materials-for-product-design.pdf.

Capelli, Bob, and Gerald R. Cysewski. 2010. "Potential Health Benefits of Spirulina Microalgae." *Nutrafoods* 9 (2). Springer Science and Business Media LLC: 19–26. doi:10.1007/bf03223332.

Caporgno, Martín P., and Alexander Mathys. 2018. "Trends in Microalgae Incorporation into Innovative Food Products with Potential Health Benefits." *Frontiers in Nutrition*. Frontiers Media S.A. doi:10.3389/fnut.2018.00058.

Carcassi, Olga Beatrice. 2021. "Nature Reloaded. Microalgae as Future Landscape Ecology." In *SpringerBriefs in Applied Sciences and Technology*, 105–14. Springer Science and Business Media Deutschland GmbH. doi:10.1007/978-3-030-54081-4_9.

Castenholz, Richard W. 2015. "General Characteristics of the Cyanobacteria." In *Bergey's Manual of Systematics of Archaea and Bacteria*, 1–23. American Cancer Society. doi:10.1002/9781118960608.cbm00019.

Cervera, Rosa, and Ma Rosa Villalba. 2021. "Will the Covid-19 Pandemic Transform Our Urban Habitat? A Microalgae Photobioreactors Labyrinth-Garden as an Answer for the Post-Covid Sustainable City." In *IOP Conference Series: Earth and Environmental Science*, Vol. 754. doi:10.1088/1755-1315/754/1/012027.

Chacón-Lee, T. L., and G. E. González-Mariño. 2010. "Microalgae for 'Healthy' Foods-Possibilities and Challenges." *Comprehensive Reviews in Food Science and Food Safety* 9 (6): 655–75. doi:10.1111/j.1541-4337.2010.00132.x.

Chai, Wai Siong, Yulei Bao, Pengfei Jin, Guang Tang, and Lei Zhou. 2021. "A Review on Ammonia, Ammonia-Hydrogen and Ammonia-Methane Fuels." *Renewable and Sustainable Energy Reviews* 147: 111254. doi:10.1016/J.RSER.2021.111254.

Chai, Wai Siong, Wee Gee Tan, Heli Siti Halimatul Munawaroh, Vijai Kumar Gupta, Shih Hsin Ho, and Pau Loke Show. 2021. "Multifaceted Roles of Microalgae in the Application of Wastewater Biotreatment: A Review." *Environmental Pollution* 269: 116236. doi:10.1016/J.ENVPOL.2020.116236.

Chan, Wing Yan, John G. Oakeshott, Patrick Buerger, Owain R. Edwards, and Madeleine J. H. van Oppen. 2021. "Adaptive Responses of Free-Living and Symbiotic Microalgae to Simulated Future Ocean Conditions." *Global Change Biology* 28 (5). Blackwell Publishing Ltd. doi:10.1111/gcb.15546.

Chekroun, Kaoutar ben, Esteban Sánchez, and Mourad Baghour. 2014. "The Role of Algae in Bioremediation of Organic Pollutants." *International Research Journal of Public and Environmental Health* 1 (2): 19–32. https://www.researchgate.net/profile/Mourad-Baghour-2/publication/285887445_The_role_of_algae_in_bioremediation_of_organic_pollutants/links/566c3ffa08ae62b05f0861b4/The-role-of-algae-in-bioremediation-of-organic-pollutants.pdf.

Chen, Ching Lung, Jo Shu Chang, and Duu Jong Lee. 2015. "Dewatering and Drying Methods for Microalgae." *Drying Technology* 33 (4). Taylor and Francis Inc.: 443–54. doi:10.1080/07373937.2014.997881.

Chen, Qiuyue, Luis Fernando Samayoa, Chin Jian Yang, Peter J. Bradbury, Bode A. Olukolu, Michael A. Neumeyer, Maria Cinta Romay, et al. 2020. "The Genetic

Architecture of the Maize Progenitor, Teosinte, and How It Was Altered during Maize Domestication." *PLoS Genetics* 16 (5). Public Library of Science. doi:10.1371/journal.pgen.1008791.

Chen, Shi Lin, Hua Yu, Hong Mei Luo, Qiong Wu, Chun Fang Li, and André Steinmetz. 2016. "Conservation and Sustainable Use of Medicinal Plants: Problems, Progress, and Prospects." *Chinese Medicine (United Kingdom).* BioMed Central Ltd. doi:10.1186/s13020-016-0108-7.

Chen, Yi di, Shih Hsin Ho, Dillirani Nagarajan, Nan qi Ren, and Jo Shu Chang. 2018. "Waste Biorefineries—Integrating Anaerobic Digestion and Microalgae Cultivation for Bioenergy Production." *Current Opinion in Biotechnology.* Elsevier Current Trends. doi:10.1016/j.copbio.2017.11.017.

Chiellini, Carolina, Lorenzo Guglielminetti, Sabrina Sarrocco, and Adriana Ciurli. 2020. "Isolation of Four Microalgal Strains from the Lake Massaciuccoli: Screening of Common Pollutants Tolerance Pattern and Perspectives for Their Use in Biotechnological Applications." *Frontiers in Plant Science* 11 (December). Frontiers Media S.A. doi:10.3389/FPLS.2020.607651/FULL.

Chisti, Yusuf. 2007. "Biodiesel from Microalgae." *Biotechnology Advances.* Elsevier. doi:10.1016/j.biotechadv.2007.02.001.

Chisti, Yusuf. 2008. "Biodiesel from Microalgae Beats Bioethanol." *Trends in Biotechnology* 26 (3). Elsevier Current Trends: 126–31. doi:10.1016/j.tibtech.2007.12.002.

Chock, Rachel Y., Barbara Clucas, Elizabeth K. Peterson, Bradley F. Blackwell, Daniel T. Blumstein, Kathleen Church, Esteban Fernández-Juricic, et al. 2021. "Evaluating Potential Effects of Solar Power Facilities on Wildlife from an Animal Behavior Perspective." *Conservation Science and Practice* 3 (2). Wiley. doi:10.1111/csp2.319.

Cid, Ángeles, Raquel Prado, Carmen Rioboo, Paula Suárez-Bregua, and Concepción Herrero. 2012. "Use of Microalgae as Biological Indicators of Pollution: Looking for New Relevant Cytotoxicity Endpoints." In *Microalgae: Biotechnology, Microbiology and Energy*, 311–24. https://ruc.udc.es/dspace/handle/2183/16634.

Cingano, Federico. 2014. "Trends in Income Inequality and Its Impact on Economic Growth." *OECD Social, Employment, and Migration Working Papers* 163. OECD Publishing: 15–59. http://dx.doi.org/10.1787/5jxrjncwxv6j-en.

Clift, Roland. 2007. "Climate Change and Energy Policy: The Importance of Sustainability Arguments." In *ECOS 05 - Proceedings of the 18th International Conference on Efficiency, Cost, Optimization, Simulation, and Environmental Impact of Energy Systems*, 11–18. https://www.sciencedirect.com/science/article/pii/S0360544206002192?casa_token=8HWc7mchxqkAAAAA:AGuIj1Xn2PLc4NWQnIFomtzK_kmBgvxVEl1Y356pmf3H9FA6ldy66N16BwPlH-oqzn9VmO8naA.

Crawford, Ronald L. 2014. "Overview: Biotransformation and Biodegradation." In *Manual of Environmental Microbiology*, 1051, 3rd ed. doi:10.1128/9781555815882.ch82.

Crist, Eileen, Camilo Mora, and Robert Engelman. 2017. "The Interaction of Human Population, Food Production, and Biodiversity Protection." *Science.* doi:10.1126/science.aal2011.

Cunningham-Smith, Petra, and Kitty Emery. 2020. "Dogs and People: Exploring the Human-Dog Connection." *Journal of Ethnobiology* 40 (4): 409–13. doi:10.2993/0278-0771-40.4.409.

Dahoumane, Si Amar, Claude Yéprémian, Chakib Djédiat, Alain Couté, Fernand Fiévet, Thibaud Coradin, and Roberta Brayner. 2016. "Improvement of Kinetics, Yield, and Colloidal Stability of Biogenic Gold Nanoparticles Using Living Cells of Euglena Gracilis Microalga." *Journal of Nanoparticle Research* 18 (3). Springer Verlag: 79. doi:10.1007/s11051-016-3378-1.

Danciu, Adrian. 2011. "The Implications of the Biotechnology for Bioterrorism." *Bulletin UASVM Agriculture* 68 (2): 1843–5386.

Darwish, Randa, Mohamed A. Gedi, Patchaniya Akepach, Hirut Assaye, Abdelrahman S. Zaky, and David A. Gray. 2020. "Chlamydomonas Reinhardtii Is a Potential Food Supplement with the Capacity to Outperform Chlorella and Spirulina." *Applied Sciences (Switzerland)* 10 (19): 1–17. doi:10.3390/app10196736.

Dasilva, Edgar J. 2004. "The Colours of Biotechnology: Science, Development and Humankind." *Electronic Journal of Biotechnology*. doi:10.4067/S0717-34582004000300001.

Davy, Simon K., Denis Allemand, and Virginia M. Weis. 2012. "Cell Biology of Cnidarian-Dinoflagellate Symbiosis." *Microbiology and Molecular Biology Reviews* 76 (2). American Society for Microbiology: 229–61. doi:10.1128/mmbr.05014-11.

Decker, Martina, George Hahn, and Libertad M. Harris. 2016. "Bio-Enabled Façade Systems Managing Complexity of Life through Emergent Technologies." *Complexity & Simplicity—Proceedings of the 34th ECAADe Conference* 1: 603–12. http://papers.cumincad.org/cgi-bin/works/Show?_id=ecaade2016_102.

Deniz, I., M. García-Vaquero, and E. Imamoglu. 2017. "Trends in Red Biotechnology: Microalgae for Pharmaceutical Applications." In *Microalgae-Based Biofuels and Bioproducts: From Feedstock Cultivation to End-Products*, 429–60. doi:10.1016/B978-0-08-101023-5.00018-2.

Diaz, Marlena A. 2011. *Plant Genetic Resources and Food Security. Plant Genetic Resources and Food Security*. doi:10.4324/9781849775762.

Dixit, Ruchita, Wasiullah, Deepti Malaviya, Kuppusamy Pandiyan, Udai B. Singh, Asha Sahu, Renu Shukla, et al. 2015. "Bioremediation of Heavy Metals from Soil and Aquatic Environment: An Overview of Principles and Criteria of Fundamental Processes." *Sustainability (Switzerland)*. doi:10.3390/su7022189.

Doble, Mukesh, Anil Kumar Kruthiventi, and Vilas Gajanan Gaikar. 2004. *Biotransformations and Bioprocesses. Biotransformations and Bioprocesses*. CRC Press. doi:10.1201/9780203026373.

Dosnon-Olette, Rachel, Patricia Trotel-Aziz, Michel Couderchet, and Philippe Eullaffroy. 2010. "Fungicides and Herbicide Removal in Scenedesmus Cell Suspensions." *Chemosphere* 79 (2): 117–23. doi:10.1016/j.chemosphere.2010.02.005.

Dunlap, Riley E., and Angela G. Mertig. 1991. "The Evolution of the U.S. Environmental Movement from 1970 to 1990: An Overview." *Society and Natural Resources* 4 (3): 209–18. doi:10.1080/08941929109380755.

Dwivedi, S., S. Srivastava, S. Mishra, A. Kumar, R. D. Tripathi, U. N. Rai, R. Dave, P. Tripathi, D. Charkrabarty, and P. K. Trivedi. 2010. "Characterization of Native Microalgal Strains for Their Chromium Bioaccumulation Potential: Phytoplankton Response in Polluted Habitats." *Journal of Hazardous Materials* 173 (1–3): 95–101. doi:10.1016/j.jhazmat.2009.08.053.

Elrayies, Ghada Mohammad. 2018. "Microalgae: Prospects for Greener Future Buildings." *Renewable and Sustainable Energy Reviews*. doi:10.1016/j.rser.2017.08.032.

Emparan, Quin, Yew Sing Jye, Michael K. Danquah, and Razif Harun. 2020. "Cultivation of Nannochloropsis Sp. Microalgae in Palm Oil Mill Effluent (POME) Media for Phycoremediation and Biomass Production: Effect of Microalgae Cells with and without Beads." *Journal of Water Process Engineering* 33. doi:10.1016/j.jwpe.2019.101043.

Enzing, Christien, Matthias Ploeg, Maria Barbosa, and Lolke Sijtsma. 2014. "Microalgae-Based Products for the Food and Feed Sector: An Outlook for Europe." *JRC Scientific and Policy Reports*, 19–37. doi:10.2791/3339.

EUMOFA. 2020. "Blue Bioeconomy: Situation Report and Perspectives." https://op.europa.eu/en/publication-detail/-/publication/487b1e66-47cc-11ea-b81b-01aa75ed71a1/language-en.

Evanson, Deborah. 2019. "World's First 'BioSolar Leaf' to Tackle Air Pollution in White City | Imperial News | Imperial College London." *Imperial College London*. https://www.imperial.ac.uk/news/191026/worlds-first-biosolar-leaf-tackle-pollution/.

Fattorini, Daniele, and Francesco Regoli. 2020. "Role of the Chronic Air Pollution Levels in the Covid-19 Outbreak Risk in Italy." *Environmental Pollution*. doi:10.1016/j.envpol.2020.114732.

Favero, Alice, Adam Daigneault, and Brent Sohngen. 2020. "Forests: Carbon Sequestration, Biomass Energy, or Both?" *Science Advances* 6 (13). American Association for the Advancement of Science. doi:10.1126/sciadv.aay6792.

Ferronato, Navarro, and Vincenzo Torretta. 2019. "Waste Mismanagement in Developing Countries: A Review of Global Issues." *International Journal of Environmental Research and Public Health*. doi:10.3390/ijerph16061060.

Figgener, Christine. 2018. "What I Learnt Pulling a Straw out of a Turtle's Nose." *Nature*. Nature Publishing Group. doi:10.1038/d41586-018-07287-z.

Fimbres-Acedo, Yenitze E., Paola Magallón-Servín, Rodolfo Garza-Torres, Maurício G. C. Emerenciano, Rosalía Servín-Villegas, Masato Endo, Kevin M. Fitzsimmons, and Francisco J. Magallón-Barajas. 2020. "Oreochromis Niloticus Aquaculture with Biofloc Technology, Photoautotrophic Conditions and Chlorella Microalgae." *Aquaculture Research* 51 (8). Blackwell Publishing Ltd: 3323–46. doi:10.1111/are.14668.

Fitzsimons, Peter John. 2007. "Biotechnology, Ethics and Education." *Studies in Philosophy and Education* 26 (1): 1–11. doi:10.1007/s11217-006-9011-5.

Fleischmann, Robert D., Mark D. Adams, Owen White, Rebecca A. Clayton, Ewen F. Kirkness, Anthony R. Kerlavage, Carol J. Bult, et al. 1995. "Whole-Genome Random Sequencing and Assembly of Haemophilus Influenzae Rd." *Science* 269 (5223): 496–512. doi:10.1126/science.7542800.

Flevaris, Konstantinos, and Christos Chatzidoukas. 2021. "Facilitating the Industrial Transition to Microbial and Microalgal Factories through Mechanistic Modelling within the Industry 4.0 Paradigm." *Current Opinion in Chemical Engineering*. doi:10.1016/j.coche.2021.100713.

Franziska, Hempel, Andrew Bozarth, Nicole Lindenkamp, Andreas Klingl, Stefan Zauner, Uwe Linne, Alexander Steinbüchel, and Uwe G. Maier. 2011. "Microalgae as Bioreactors for Bioplastic Production." *Microbial Cell Factories* 10. doi:10.1186/1475-2859-10-81.

Fu, Weiqi, David R. Nelson, Alexandra Mystikou, Sarah Daakour, and Kourosh Salehi-Ashtiani. 2019. "Advances in Microalgal Research and Engineering Development." *Current Opinion in Biotechnology*. doi:10.1016/j.copbio.2019.05.013.

Fukuda, Shin ya, Koji Iwamoto, Mika Atsumi, Akiko Yokoyama, Takeshi Nakayama, Ken ichiro Ishida, Isao Inouye, and Yoshihiro Shiraiwa. 2014. "Global Searches for Microalgae and Aquatic Plants That Can Eliminate Radioactive Cesium, Iodine and Strontium from the Radio-Polluted Aquatic Environment: A Bioremediation Strategy." *Journal of Plant Research* 127 (1). Springer: 79–89. doi:10.1007/s10265-013-0596-9.

Gao, Jing, and Brian C. O'Neill. 2020. "Mapping Global Urban Land for the 21st Century with Data-Driven Simulations and Shared Socioeconomic Pathways." *Nature Communications* 11 (1). doi:10.1038/s41467-020-15788-7.

García-Poza, Sara, Adriana Leandro, Carla Cotas, João Cotas, João C. Marques, Leonel Pereira, and Ana M. M. Gonçalves. 2020. "The Evolution Road of Seaweed Aquaculture: Cultivation Technologies and the Industry 4.0." *International Journal of Environmental Research and Public Health*. doi:10.3390/ijerph17186528.

Gargalo, Carina L., Isuru Udugama, Katrin Pontius, Pau C. Lopez, Rasmus F. Nielsen, Aliyeh Hasanzadeh, Seyed Soheil Mansouri, Christoph Bayer, Helena Junicke, and Krist V. Gernaey. 2020. "Towards Smart Biomanufacturing: A Perspective on Recent

Developments in Industrial Measurement and Monitoring Technologies for Bio-Based Production Processes." *Journal of Industrial Microbiology and Biotechnology* 47 (11): 947–64. doi:10.1007/s10295-020-02308-1.

Gavrilescu, Maria. 2010. "Environmental Biotechnology: Achievements, Opportunities and Challenges." *Dynamic Biochemistry, Process Biotechnology and Molecular Biology* 4 (1): 1–36.

Gayon, Jean. 2016. "From Mendel to Epigenetics: History of Genetics." *Comptes Rendus—Biologies* 339 (7–8): 225–30. doi:10.1016/j.crvi.2016.05.009.

Geissdoerfer, Martin, Paulo Savaget, Nancy M. P. Bocken, and Erik Jan Hultink. 2017. "The Circular Economy—A New Sustainability Paradigm?" *Journal of Cleaner Production*. doi:10.1016/j.jclepro.2016.12.048.

George, Alison. 2009. "David Attenborough on Our Crowded Planet." *New Scientist* 202 (2708): 28–29. doi:10.1016/s0262-4079(09)61324-5.

Geyer, Roland. 2020. "A Brief History of Plastics." In *Mare Elasticum—The Plastic Sea*, 31–47. Cham: Springer. doi:10.1007/978-3-030-38945-1_2.

Gilbert, Marianne. 2017. "Plastics Materials: Introduction and Historical Development." In *Brydson's Plastics Materials: Eighth Edition*, 2–18. Butterworth-Heinemann. doi:10.1016/B978-0-323-35824-8.00001-3.

Goodland, Robert. 1995. "The Concept of Environmental Sustainability." *Annual Review of Ecology and Systematics* 26 (1): 1–24. doi:10.4324/9781315241951-20.

Gosset, Antoine, Claude Durrieu, Pauline Barbe, Christine Bazin, and Rémy Bayard. 2019. "Microalgal Whole-Cell Biomarkers as Sensitive Tools for Fast Toxicity and Pollution Monitoring of Urban Wet Weather Discharges." *Chemosphere* 217: 522–33. doi:10.1016/j.chemosphere.2018.11.033.

Gouda, Rajashree. 2011a. "Dinophysis." *World Register of Marine Species Photogallery*. March 9. https://www.marinespecies.org/photogallery.php?album=1033&pic=39681.

Gouda, Rajashree. 2011b. "Navicula." *World Register of Marine Species Photogallery*. March 9. https://www.marinespecies.org/photogallery.php?album=4394&pic=39694.

Gouda, Rajashree. 2011c. "Thalassiosira Nordenskioeldii." *World Register of Marine Species Photogallery*. March 9. https://www.marinespecies.org/photogallery.php?album=4394&pic=39660#photogallery.

Gouveia, Joao Diogo, Jesus Ruiz, Lambertus A. M. van den Broek, Thamara Hesselink, Sander Peters, Dorinde M. M. Kleinegris, Alison G. Smith, Douwe van der Veen, Maria J. Barbosa, and Rene H. Wijffels. 2017. "Botryococcus Braunii Strains Compared for Biomass Productivity, Hydrocarbon and Carbohydrate Content." *Journal of Biotechnology* 248: 77–86. doi:10.1016/j.jbiotec.2017.03.008.

Grasso, Gerardo, Daniela Zane, and Roberto Dragone. 2020. "Microbial Nanotechnology: Challenges and Prospects for Green Biocatalytic Synthesis of Nanoscale Materials for Sensoristic and Biomedical Applications." *Nanomaterials* 10 (1): 11. doi:10.3390/nano10010011.

Grönlund, Erik, Anders Klang, Stefan Falk, and Jörgen Hanæus. 2004. "Sustainability of Wastewater Treatment with Microalgae in Cold Climate, Evaluated with Energy and Socio-Ecological Principles." *Ecological Engineering* 22 (3): 155–74. doi:10.1016/j.ecoleng.2004.03.002.

Haefely, G. 1947. "The Growing Importance of Plastics in the Electrical Industry." *Journal of the Institution of Electrical Engineers—Part II: Power Engineering* 94 (40). Institution of Engineering and Technology (IET): 301–8. doi:10.1049/ji-2.1947.0093.

Halisçelik, Ergül, and Mehmet Ali Soytas. 2019. "Sustainable Development from Millennium 2015 to Sustainable Development Goals 2030." *Sustainable Development* 27 (4). John Wiley and Sons Ltd: 545–72. doi:10.1002/sd.1921.

Hammond, G. P., S. Kallu, and M. C. McManus. 2008. "Development of Biofuels for the UK Automotive Market." *Applied Energy* 85 (6): 506–15. doi:10.1016/j.apenergy.2007.09.005.

Harich, Jack. 2010. "Change Resistance as the Crux of the Environmental Sustainability Problem." *System Dynamics Review* 26 (1): 35–72. doi:10.1002/sdr.431.

Harjanne, Atte, and Janne M. Korhonen. 2019. "Abandoning the Concept of Renewable Energy." *Energy Policy* 127: 330–40. doi:10.1016/j.enpol.2018.12.029.

Heureux, Ana M. C., Jodi N. Young, Spencer M. Whitney, Maeve R. Eason-Hubbard, Renee B. Y. Lee, Robert E. Sharwood, and Rosalind E. M. Rickaby. 2017. "The Role of Rubisco Kinetics and Pyrenoid Morphology in Shaping the CCM of Haptophyte Microalgae." *Journal of Experimental Botany* 68 (14): 3959–69. doi:10.1093/jxb/erx179.

Hirschnitz-Garbers, Martin, Adrian R. Tan, Albrecht Gradmann, and Tanja Srebotnjak. 2016. "Key Drivers for Unsustainable Resource Use—Categories, Effects and Policy Pointers." *Journal of Cleaner Production* 132: 13–31. doi:10.1016/j.jclepro.2015.02.038.

Ho, Shih Hsin, Shu Wen Huang, Chun Yen Chen, Tomohisa Hasunuma, Akihiko Kondo, and Jo Shu Chang. 2013. "Bioethanol Production Using Carbohydrate-Rich Microalgae Biomass as Feedstock." *Bioresource Technology* 135: 191–98. doi:10.1016/j.biortech.2012.10.015.

Höhne, Niklas, Michel den Elzen, and Donovan Escalante. 2014. "Regional GHG Reduction Targets Based on Effort Sharing: A Comparison of Studies." *Climate Policy* 14 (1). Taylor and Francis Ltd.: 122–47. doi:10.1080/14693062.2014.849452.

Holling, C. S., and Gary K. Meffe. 1996. "Command and Control and the Pathology of Natural Resource Management." *Conservation Biology* 10 (2). Blackwell Publishing Inc.: 328–37. doi:10.1046/j.1523-1739.1996.10020328.x.

Hooper, B. 2011. "Towards More Effective Integrated Watershed Management in Australia: Results of a National Survey, and Implications for Urban Catchment Management." *Journal of Contemporary Water Research and Education* 100 (1): 6. https://opensiuc.lib.siu.edu/cgi/viewcontent.cgi?article=1355&context=jcwre.

Horowitz, Cara A. 2016. "Paris Agreement." *International Legal Materials* 55 (4). Cambridge University Press: 740–55. doi:10.1017/s0020782900004253.

Horrigan, Leo, Robert S. Lawrence, and Polly Walker. 2002. "How Sustainable Agriculture Can Address the Environmental and Human Health Harms of Industrial Agriculture." *Environmental Health Perspectives*. Public Health Services, US Dept of Health and Human Services. doi:10.1289/ehp.02110445.

Houdebine, L. M. 2009. "Production of Pharmaceutical Proteins by Transgenic Animals." *Comparative Immunology, Microbiology and Infectious Diseases* 32 (2): 107–21. doi:10.20506/rst.37.1.2746.

Howard, Margaret. 2007. "The Law of Unintended Consequences." *Southern Illinois University Law Journal* 31 (3): 451–62. https://heinonline.org/hol-cgi-bin/get_pdf.cgi?handle=hein.journals/siulj31&section=25&casa_token=yVHFe9HGIlIAAAAA:NbIMlb41WkMypgY6JldtXGGq9cjZ4g0cKWODdLvtIU_TC9pcrwVkLLfvop8ch2PrN9N1Mdl7Zg.

Howarth, R. W., D. B. Anderson, J. E. Cloern, C. Elfring, C. Hopkinson, B. Lapointe, T. Malone et al. 2000. "Nutrient Pollution of Coastal Rivers, Bays, and Seas." *Issues in Ecology* 7 (7): 1–15. https://pubs.er.usgs.gov/publication/70174406.

Hu, Chunxiang, Yongding Liu, Lirong Song, and Delu Zhang. 2002. "Effect of Desert Soil Algae on the Stabilization of Fine Sands." *Journal of Applied Phycology* 14 (4): 281–92. doi:10.1023/A:1021128530086.

Hughes, T. P., J. T. Kerry, A. H. Baird, S. R. Connolly, A. Dietzel—Nature, and Undefined. 2018. "Global Warming Transforms Coral Reef Assemblages." *Nature* 556: 492–96. https://idp.nature.com/authorize/casa?redirect_uri=https://www.nature.com/articles/s41586-018-0041-2&casa_token=auGH4OwjLx4AAAAA:n0-_i-4POkmrdC-n1KY-y-vPo2G84tL8K0PDvjpR3NcQUj_qeEgW1a3rlIkUPnYltBtGwbFBynGc6uag5g.

Hullebusch, Eric D. van, Nand K. Singh, and Joyabrata Mal. 2021. "Biotechnological Intervention for Societal Development (BioSangam 2020)." *Environmental Science and Pollution Research*. Springer. doi:10.1007/s11356-021-14630-x.

Hussain, Akhtar, Syed Muhammad Arif, and Muhammad Aslam. 2017. "Emerging Renewable and Sustainable Energy Technologies: State of the Art." *Renewable and Sustainable Energy Reviews*. doi:10.1016/j.rser.2016.12.033.

Hussain, Fida, Syed Z. Shah, Habib Ahmad, Samar A. Abubshait, Haya A. Abubshait, A. Laref, A. Manikandan, Heri S. Kusuma, and Munawar Iqbal. 2021. "Microalgae an Ecofriendly and Sustainable Wastewater Treatment Option: Biomass Application in Biofuel and Bio-Fertilizer Production. A Review." *Renewable and Sustainable Energy Reviews* 137. doi:10.1016/j.rser.2020.110603.

Iltis, Hugh H. 1983. "From Teosinte to Maize: The Catastrophic Sexual Transmutation." *Science* 222 (4626): 886–94. doi:10.1126/science.222.4626.886.

Ilvitskaya, S. V., and A. G. Chistyakova. 2020. "Microalgae in Architecture as an Energy Source." In *IOP Conference Series: Materials Science and Engineering*, Vol. 944. doi:10.1088/1757-899X/944/1/012010.

Irfan, M. F., S. M. Z. Hossain, H. Khalid, F. Sadaf, S. Al-Thawadi, A. Alshater, M. M. Hossain, and S. A. Razzak. 2019. "Optimization of Bio-Cement Production from Cement Kiln Dust Using Microalgae." *Biotechnology Reports* 23. doi:10.1016/j.btre.2019.e00356.

Jackson, Tim. 2013. "Prosperity without Growth? The Transition to a Sustainable Economy." *Sustainable Development Commission*. doi:10.4324/9780203145593.

Jais, N. M., R. M. S. R. Mohamed, A. A. Al-Gheethi, and M. K. Amir Hashim. 2017. "The Dual Roles of Phycoremediation of Wet Market Wastewater for Nutrients and Heavy Metals Removal and Microalgae Biomass Production." *Clean Technologies and Environmental Policy*. doi:10.1007/s10098-016-1235-7.

Janssen, K., H. Chavanne, P. Berentsen, and H. Komen. 2017. "Impact of Selective Breeding on European Aquaculture." *Aquaculture* 472: 8–16. doi:10.1016/j.aquaculture.2016.03.012.

Jeon, Seungjib, Jong Min Lim, Hyung Gwan Lee, Sung Eun Shin, Nam Kyu Kang, Youn il Park, Hee Mock Oh, Won Joong Jeong, Byeong Ryool Jeong, and Yong Keun Chang. 2017. "Current Status and Perspectives of Genome Editing Technology for Microalgae." *Biotechnology for Biofuels*. BioMed Central Ltd. doi:10.1186/s13068-017-0957-z.

Jones, Carl G. 2004. "Conservation Management of Endangered Birds." In *Bird Ecology and Conservation*, 269–302. doi:10.1093/acprof:oso/9780198520863.003.0012.

Juneja, Ankita, Ruben Michael Ceballos, and Ganti S. Murthy. 2013. "Effects of Environmental Factors and Nutrient Availability on the Biochemical Composition of Algae for Biofuels Production: A Review." *Energies*. doi:10.3390/en6094607.

Jung, Jieun, Peter Petkanic, Dongyan Nan, and Jang Hyun Kim. 2020. "When a Girl Awakened the World: A User and Social Message Analysis of Greta Thunberg." *Sustainability (Switzerland)* 12 (7): 2707. doi:10.3390/su12072707.

Kafarski, Paweł. 2012. "Rainbow Code of Biotechnology." *Chemik* 66 (8): 814–16. http://pl.wikipedia.org/wiki/Biotechnologia.

Kale, Aditi, and Balasubramanian Karthick. 2015. "The Diatoms: Big Significance of Tiny Glass Houses." *Resonance* 20 (10). Springer: 919–30. doi:10.1007/s12045-015-0256-6.

Kang, Chang Ho, Yoon Jung Kwon, and Jae Seong So. 2016. "Bioremediation of Heavy Metals by Using Bacterial Mixtures." *Ecological Engineering* 89: 64–69. doi:10.1016/j.ecoleng.2016.01.023.

Kanwugu, Osman Nabayire, Maria N. Ivantsova, and Kingsley D. Chidumaga. 2018. "Gold Biotechnology: Development and Advancements." In *AIP Conference Proceedings*, Vol. 2015. American Institute of Physics Inc. doi:10.1063/1.5055107.

Karamidehkordi, Esmail. 2012. "Sustainable Natural Resource Management, a Global Challenge of This Century." In *Sustainable Natural Resources Management*. doi:10.5772/35035.

Karigar, Chandrakant S., and Shwetha S. Rao. 2011. "Role of Microbial Enzymes in the Bioremediation of Pollutants: A Review." *Enzyme Research*. doi:10.4061/2011/805187.

Kato, Naohiro. 2019. "Production of Crude Bioplastic-Beads with Microalgae: Proof-of-Concept." *Bioresource Technology Reports* 6: 81–84. https://www.sciencedirect.com/science/article/pii/S2589014X19300234?casa_token=PFWJYZOzldcAAAAA:PDjd1B849_LWy1VjNUamLLDvfw8Vy6IIIliDGOVWf2QlsIXFqa9tfvTiEtgEUSpVf02NDFtgxw.

Kaur, Kirandeep, and Pushp Sharma. 2018. "Hierarchy in Plants: Unicellular, Colonial and Multicellular." *The Journal of Plant Science Research* 34 (2): 207–20. doi:10.32381/jpsr.2018.34.02.9.

Kellert, S. R., J. N. Mehta, S. A. Ebbin, and L. L. Lichtenfeld. 2000. "Community Natural Resource Management: Promise, Rhetoric, and Reality." *Society and Natural Resources* 13 (8): 705–15. doi:10.1080/089419200750035575.

Khan, Muhammad Imran, Jin Hyuk Shin, and Jong Deog Kim. 2018. "The Promising Future of Microalgae: Current Status, Challenges, and Optimization of a Sustainable and Renewable Industry for Biofuels, Feed, and Other Products." *Microbial Cell Factories*. BioMed Central Ltd. doi:10.1186/s12934-018-0879-x.

Khoo, Kuan Shiong, Kit Wayne Chew, Guo Yong Yew, Wai Hong Leong, Yee Ho Chai, Pau Loke Show, and Wei Hsin Chen. 2020. "Recent Advances in Downstream Processing of Microalgae Lipid Recovery for Biofuel Production." *Bioresource Technology*. doi:10.1016/j.biortech.2020.122996.

Kim, Jungmin, Gursong Yoo, Hansol Lee, Juntaek Lim, Kyochan Kim, Chul Woong Kim, Min S. Park, and Ji Won Yang. 2013. "Methods of Downstream Processing for the Production of Biodiesel from Microalgae." *Biotechnology Advances*. doi:10.1016/j.biotechadv.2013.04.006.

Kirwan, Mark J., Sarah Plant, and John W. Strawbridge. 2011. "Plastics in Food Packaging." In *Food and Beverage Packaging Technology: Second Edition*, 157–212. doi:10.1002/9781444392180.ch7.

Knorr, Dietrich, and Anthony J. Sinskey. 1985. "Biotechnology in Food Production and Processing." *Science* 229 (4719): 1224–29. doi:10.1126/science.229.4719.1224.

Kothari, Richa, Shamshad Ahmad, Vinayak V. Pathak, Arya Pandey, Ashwani Kumar, Raju Shankarayan, Paul N. Black, and V. V. Tyagi. 2021. "Algal-Based Biofuel Generation through Flue Gas and Wastewater Utilization: A Sustainable Prospective Approach." *Biomass Conversion and Biorefinery*. Springer Science and Business Media Deutschland GmbH. doi:10.1007/s13399-019-00533-y.

Kotteswari, M., S. Murugesan, and R. Ranjith Kumar. 2012. "Phycoremediation of Dairy Effluent by Using the Microalgae Nostoc Sp." *International Journal of Environmental Research and Development* 2. ISSN 2249–3131.

Kousoulaki, K., T. Mørkøre, I. Nengas, R. K. Berge, and J. Sweetman. 2016. "Microalgae and Organic Minerals Enhance Lipid Retention Efficiency and Fillet Quality in Atlantic Salmon (Salmo Salar L.)." *Aquaculture* 451: 47–57. doi:10.1016/j.aquaculture.2015.08.027.

Krassowski, Michal, Vivek Das, Sangram K. Sahu, and Biswapriya B. Misra. 2020. "State of the Field in Multi-Omics Research: From Computational Needs to Data Mining and Sharing." *Frontiers in Genetics*. doi:10.3389/fgene.2020.610798.

Kuhlman, Tom, and John Farrington. 2010. "What Is Sustainability?" *Sustainability*. doi:10.3390/su2113436.

Kühne, Rainer Walter. 2019. "Climate Change: The Science Behind Greta Thunberg and Fridays for Future." *OSF Preprints*. doi:10.31219/osf.io/2n6kj.

Kumar, Gulshan, Ajam Shekh, Sunaina Jakhu, Yogesh Sharma, Ritu Kapoor, and Tilak Raj Sharma. 2020. "Bioengineering of Microalgae: Recent Advances, Perspectives, and Regulatory Challenges for Industrial Application." *Frontiers in Bioengineering and Biotechnology*. Frontiers Media S.A. doi:10.3389/fbioe.2020.00914.

Kumar, Krithika, Rodrigo A. Mella-Herrera, and James W. Golden. 2010. "Cyanobacterial Heterocysts." *Cold Spring Harbor Perspectives in Biology* 2 (4). doi:10.1101/cshperspect.a000315.

Kumar, Vinod, Akshay Kumar, and Manisha Nanda. 2018. "Pretreated Animal and Human Waste as a Substantial Nutrient Source for Cultivation of Microalgae for Biodiesel Production." *Environmental Science and Pollution Research* 25 (22). Springer Verlag: 22052–59. doi:10.1007/s11356-018-2339-x.

Lachmayer, Konrad. 2006. "Legal Aspects of 'Black Biotechnology'—Counter-Terrorism between Prevention and Use of Biotechnological Products." *Journal of International Biotechnology Law* 3 (6). Walter de Gruyter GmbH. doi:10.1515/jibl.2006.029.

Lahlou, Mouhssen. 2013. "The Success of Natural Products in Drug Discovery." *Pharmacology & Pharmacy* 4: 17–31. doi:10.4236/pp.2013.43A003.

LaJeunesse, T. C. 2002. "Diversity and Community Structure of Symbiotic Dinoflagellates from Caribbean Coral Reefs." *Marine Biology* 141 (2): 387–400. doi:10.1007/s00227-002-0829-2.

LaJeunesse, T. C., R. Bhagooli, M. Hidaka, L. DeVantier, T. Done, G. W. Schmidt, W. K. Fitt, and O. Hoegh-Guldberg. 2004. "Closely Related Symbiodinium Spp. Differ in Relative Dominance in Coral Reef Host Communities across Environmental, Latitudinal and Biogeographic Gradients." *Marine Ecology Progress Series* 284: 147–61. doi:10.3354/meps284147.

Lakatos, L., G. Hevessy, and J. Kovács. 2011. "Advantages and Disadvantages of Solar Energy and Wind-Power Utilization." *World Futures: Journal of General Evolution* 67 (6): 395–408. doi:10.1080/02604020903021776.

Landrigan, Philip J., Richard Fuller, Nereus J. R. Acosta, Olusoji Adeyi, Robert Arnold, Niladri (Nil) Basu, Abdoulaye Bibi Baldé, et al. 2018. "The Lancet Commission on Pollution and Health." *The Lancet*. doi:10.1016/S0140-6736(17)32345-0.

Lee, Keun Young, Sang Hyo Lee, Ju Eun Lee, and Seung Yop Lee. 2019. "Biosorption of Radioactive Cesium from Contaminated Water by Microalgae Haematococcus Pluvialis and Chlorella Vulgaris." *Journal of Environmental Management* 233: 83–88. doi:10.1016/j.jenvman.2018.12.022.

Lehane, Leigh, and Richard J. Lewis. 2000. "Ciguatera: Recent Advances but the Risk Remains." *International Journal of Food Microbiology*. doi:10.1016/S0168-1605(00)00382-2.

Lehmuskero, Anni, Matilde Skogen Chauton, and Tobias Boström. 2018. "Light and Photosynthetic Microalgae: A Review of Cellular- and Molecular-Scale Optical Processes." *Progress in Oceanography*. doi:10.1016/j.pocean.2018.09.002.

Lemenih, Mulugeta. 2004. "Effects of Land Use Changes on Soil Quality and Native Flora Degradation and Restoration in the Highlands of Ethiopia" (PhD Dissertation). Swedish University of Agricultural Sciences, Department of Forest Soils, Uppsala, Sweden. Sciences-New York. http://pub.epsilon.slu.se/560/.

Leruste, A., S. Villéger, N. Malet, R. de Wit, and B. Bec. 2018. "Complementarity of the Multidimensional Functional and the Taxonomic Approaches to Study Phytoplankton Communities in Three Mediterranean Coastal Lagoons of Different Trophic Status." *Hydrobiologia* 815 (1). Springer International Publishing: 207–27. doi:10.1007/s10750-018-3565-4.

Lewis, Nancy. 2011. "Ceratium Longipes." *World Register of Marine Species Photogallery*. March 29. https://www.marinespecies.org/photogallery.php?album=1033&pic=40252.

Lewis, Simon L., Charlotte E. Wheeler, Edward T. A. Mitchard, and Alexander Koch. 2019. "Restoring Natural Forests Is the Best Way to Remove Atmospheric Carbon." *Nature* 568 (7750): 25–28. doi:10.1038/d41586-019-01026-8.

Lindahl, Pia, Karl Henrik Robèrt, Henrik Ny, and Göran Broman. 2014. "Strategic Sustainability Considerations in Materials Management." *Journal of Cleaner Production* 64: 98–103. doi:10.1016/j.jclepro.2013.07.015.

Lisk, Donald J. 1991. "Environmental Effects of Landfills." *Science of the Total Environment, The* 100 (C): 415–68. doi:10.1016/0048-9697(91)90387-T.

Lloyd, John, and James Cheyne. 2017. "The Origins of the Vaccine Cold Chain and a Glimpse of the Future." *Vaccine*. doi:10.1016/j.vaccine.2016.11.097.

Lobaccaro, G., S. Croce, C. Lindkvist, M. C. Munari Probst, A. Scognamiglio, J. Dahlberg, M. Lundgren, and M. Wall. 2019. "A Cross-Country Perspective on Solar Energy in Urban Planning: Lessons Learned from International Case Studies." *Renewable and Sustainable Energy Reviews* 108: 209–37. doi:10.1016/j.rser.2019.03.041.

Lockwood, Michael, Julie Davidson, Allan Curtis, Elaine Stratford, and Rod Griffith. 2010. "Governance Principles for Natural Resource Management." *Society and Natural Resources* 23 (10): 986–1001. doi:10.1080/08941920802178214.

Lode, Anja, Felix Krujatz, Sophie Brüggemeier, Mandy Quade, Kathleen Schütz, Sven Knaack, Jost Weber, Thomas Bley, and Michael Gelinsky. 2015. "Green Bioprinting: Fabrication of Photosynthetic Algae-Laden Hydrogel Scaffolds for Biotechnological and Medical Applications." *Engineering in Life Sciences* 15 (2). Wiley-VCH Verlag: 177–83. doi:10.1002/elsc.201400205.

Lovich, Jeffrey E., and Joshua R. Ennen. 2011. "Wildlife Conservation and Solar Energy Development in the Desert Southwest, United States." *BioScience* 61 (12): 982–92. doi:10.1525/bio.2011.61.12.8.

Low, Sze Shin, Kien Xiang Bong, Muhammad Mubashir, Chin Kui Cheng, Man Kee Lam, Jun Wei Lim, Yeek Chia Ho, Keat Teong Lee, Heli Siti Halimatul Munawaroh, and Pau Loke Show. 2021. "Microalgae Cultivation in Palm Oil Mill Effluent (POME) Treatment and Biofuel Production." *Sustainability* 13 (6). Multidisciplinary Digital Publishing Institute: 3247. doi:10.3390/SU13063247.

Lupo, Karen D. 2017. "When and Where Do Dogs Improve Hunting Productivity? The Empirical Record and Some Implications for Early Upper Paleolithic Prey Acquisition." *Journal of Anthropological Archaeology* 47: 139–51. doi:10.1016/j.jaa.2017.05.003.

Luque, Rafael, Lorenzo Herrero-Davila, Juan M. Campelo, James H. Clark, Jose M. Hidalgo, Diego Luna, Jose M. Marinas, and Antonio A. Romero. 2008. "Biofuels: A Technological Perspective." *Energy and Environmental Science*. Royal Society of Chemistry. doi:10.1039/b807094f.

Lyon, Barbara R., and Thomas Mock. 2014. "Polar Microalgae: New Approaches towards Understanding Adaptations to an Extreme and Changing Environment." *Biology*. doi:10.3390/biology3010056.

Ma, Julian K. C., Pascal M. W. Drake, and Paul Christou. 2003. "The Production of Recombinant Pharmaceutical Proteins in Plants." *Nature Reviews Genetics*. doi:10.1038/nrg1177.

Mackey, Maureen. 2002. "The Application of Biotechnology to Nutrition: An Overview." *Journal of the American College of Nutrition* 21 (June): 157S–160S. doi:10.1080/07315724.2002.10719259.

Mahmoud, Ahmed A., Savithiry S. Natarajan, John O. Bennett, Thomas P. Mawhinney, William J. Wiebold, and Hari B. Krishnan. 2006. "Effect of Six Decades of Selective Breeding on Soybean Protein Composition and Quality: A Biochemical and Molecular Analysis." *Journal of Agricultural and Food Chemistry* 54 (11): 3916–22. doi:10.1021/jf060391m.

Malmqvist, M. 1993. "Biospecific Interaction Analysis Using Biosensor Technology." *Nature*. doi:10.1038/361186a0.

Manning, William J., and William A. Feder. 1980. "Biomonitoring Air Pollutants with Plants." *Applied Science Publishers*: 142.

Martinez-Goss, M. R. 2007. "6th Asia-Pacific Conference on Algal Biotechnology (APCAB): Makati City, Philippines, October 12–15, 2006." *Journal of Applied Phycology* 19: 603–5. doi:10.1007/s10811-007-9248-6.

Mathiot, Charlie, Pauline Ponge, Benjamin Gallard, Jean François Sassi, Florian Delrue, and Nicolas le Moigne. 2019. "Microalgae Starch-Based Bioplastics: Screening of Ten Strains and Plasticization of Unfractionated Microalgae by Extrusion." *Carbohydrate Polymers* 208: 142–51. doi:10.1016/j.carbpol.2018.12.057.

Matos, Ângelo Paggi. 2017. "The Impact of Microalgae in Food Science and Technology." *JAOCS, Journal of the American Oil Chemists' Society*. Springer Verlag. doi:10.1007/s11746-017-3050-7.

Matsunaga, Tadashi, Haruko Takeyama, Hideki Miyashita, and Hiroko Yokouchi. 2005. "Marine Microalgae." *Advances in Biochemical Engineering/Biotechnology*. doi:10.1007/b135784.

Meadows, Donella H., and Dennis Meadows. 2007. "The History and Conclusions of the Limits to Growth." *System Dynamics Review* 23 (2–3). John Wiley & Sons, Ltd: 191–97. doi:10.1002/sdr.371.

Meadows, D. H., and J. Randers. 2012. *The Limits to Growth: The 30-Year Update*, Vol. 43. Política y Sociedad. https://www.taylorfrancis.com/books/mono/10.4324/9781849775861/limits-growth-dennis-meadows-jorgan-randers.

Metsoviti, Maria N., Nikolaos Katsoulas, Ioannis T. Karapanagiotidis, and George Papapolymerou. 2019. "Current and Potential Applications of Microalgae: A Mini Review." *Oceanogr Fish Open Access Journal* 11 (3): 35–39. doi:10.19080/OFOAJ.2019.11.555811.

Metting, F. B. 1996. "Biodiversity and Application of Microalgae." *Journal of Industrial Microbiology and Biotechnology* 17 (5–6). Nature Publishing Group: 477–89. doi:10.1007/bf01574779.

Metzker, Michael L. 2005. "Emerging Technologies in DNA Sequencing." *Genome Research*. doi:10.1101/gr.3770505.

Meussdoerffer, Franz G. 2009. "A Comprehensive History of Beer Brewing." In *Handbook of Brewing: Processes, Technology, Markets*, 1–42. doi:10.1002/9783527623488.ch1.

Meyer, William B., and B. L. Turner. 1992. "Human Population Growth and Global Land-Use/Cover Change." *Annual Review of Ecology and Systematics* 23 (1). Annual Reviews Inc.: 39–61. doi:10.1146/annurev.es.23.110192.000351.

Mika, Nicole, Holger Zorn, and Martin Rühl. 2013. "Insect-Derived Enzymes: A Treasure for Industrial Biotechnology and Food Biotechnology." In *Yellow Biotechnology II*, Vol. 136, 1–17. Springer Science and Business Media Deutschland GmbH. doi:10.1007/10_2013_204.

Miller, James R., and Richard J. Hobbs. 2007. "Habitat Restoration—Do We Know What We're Doing?" *Restoration Ecology*. doi:10.1111/j.1526-100X.2007.00234.x.

Miller, Lee M., and David W. Keith. 2018. "Corrigendum: Observation-Based Solar and Wind Power Capacity Factors and Power Densities (Environmental Research Letters (2018) 13 (104008) doi: 10.1088/1748–9326/Aae102)." *Environmental Research Letters*. doi:10.1088/1748-9326/aaf9cf.

Mimouni, Virginie, Lionel Ulmann, Virginie Pasquet, Marie Mathieu, Laurent Picot, Gael Bougaran, Jean-Paul Cadoret, Annick Morant-Manceau, and Benoit Schoefs. 2012. "The Potential of Microalgae for the Production of Bioactive Molecules of Pharmaceutical Interest." *Current Pharmaceutical Biotechnology* 13 (15). Bentham Science Publishers: 2733–50. doi:10.2174/138920112804724828.

Mittermeier, R. A., C. G. Mittermeier, T. M. Brooks, J. D. Pilgrim, W. R. Konstant, G. A. B. da Fonseca, and C. Kormos. 2003. "Wilderness and Biodiversity Conservation." *Proceedings of the National Academy of Sciences of the United States of America* 100 (18): 10309–13. doi:10.1073/pnas.1732458100.

Moestrup, Øjvind. 2016a. "Arthrospira Fusiformis." *World Register of Marine Species—Photogallery*. February 29. https://www.marinespecies.org/photogallery.php?album=823&pic=111710.

Moestrup, Øjvind. 2016b. "Microcystis Wesenbergii." *World Register of Marine Species Photogallery*. June 28. https://www.marinespecies.org/photogallery.php?album=823&pic=111702.

Mohapatra, P. K. 2010. *Textbook of Environmental Physiology*. Delhi: I.K. International Publishing House Pvt. Ltd.

Morales, Marcia, León Sánchez, and Sergio Revah. 2018. "The Impact of Environmental Factors on Carbon Dioxide Fixation by Microalgae." *FEMS Microbiology Letters* 365 (3). doi:10.1093/femsle/fnx262.

Moreira, Ícaro Thiago Andrade, Célia Karina Maia Cardoso, Evelin Daiane Serafim Santos Franco, Isadora Machado Marques, Gisele Mara Hadlich, Antônio Fernando de Souza Queiroz, Ana Katerine de Carvalho Lima Lobato, and Olívia Maria Cordeiro de Oliveira. 2021. "Mangrove Ecosystem Restoration after Oil Spill: Bioremediation, Phytoremediation, Biofibers and Phycoremediation." In *Mangrove Ecosystem Restoration*. doi:10.5772/intechopen.95342.

Moro, Laura, Gianni Pezzotti, Mehmet Turemis, Josep Sanchís, Marinella Farré, Renata Denaro, Maria Grazia Giacobbe, Francesca Crisafi, and Maria Teresa Giardi. 2018. "Fast Pesticide Pre-Screening in Marine Environment Using a Green Microalgae-Based Optical Bioassay." *Marine Pollution Bulletin* 129 (1): 212–21. doi:10.1016/j.marpolbul.2018.02.036.

Mulumba, N. 2010. "Production of Biodiesel from Microalgae." https://search.proquest.com/openview/cbb6ad41b8ea772ee947f41185a74b90/1?pq-origsite=gscholar&cbl=18750.

Murphy, Kevin. 2012. "The Social Pillar of Sustainable Development: A Literature Review and Framework for Policy Analysis." *Sustainability: Science, Practice, and Policy* 8 (1). ProQuest: 15–29. doi:10.1080/15487733.2012.11908081.

Mustafa, Shazia, Haq Nawaz Bhatti, Munazza Maqbool, and Munawar Iqbal. 2021. "Microalgae Biosorption, Bioaccumulation and Biodegradation Efficiency for the Remediation of Wastewater and Carbon Dioxide Mitigation: Prospects, Challenges and Opportunities." *Journal of Water Process Engineering* 41. doi:10.1016/j.jwpe.2021.102009.

Nair, Abhilash T., Jaganathan Senthilnathan, and S. M. Shiva Nagendra. 2019. "Application of the Phycoremediation Process for Tertiary Treatment of Landfill Leachate and Carbon Dioxide Mitigation." *Journal of Water Process Engineering* 28: 322–30. doi:10.1016/j.jwpe.2019.02.017.

Nascimento, Iracema Andrade, Sheyla Santa Izabel Marques, Iago Teles Dominguez Cabanelas, Gilson Correia de Carvalho, Maurício A. Nascimento, Carolina Oliveira de Souza, Janice Isabel Druzian, Javid Hussain, and Wei Liao. 2014. "Microalgae Versus Land Crops as Feedstock for Biodiesel: Productivity, Quality, and Standard Compliance." *Bioenergy Research* 7 (3). Springer New York LLC: 1002–13. doi:10.1007/s12155-014-9440-x.

National Research Council. 1995. "Managing Global Genetic Resources—Agricultural Crop Issues and Policies." *National Academies Press* 21 (1): 104. doi:10.1016/0160-4120(95)90042-x.

Nawaz, Tabish, Ashiqur Rahman, Shanglei Pan, Kyleigh Dixon, Burgandy Petri, and Thinesh Selvaratnam. 2020. "A Review of Landfill Leachate Treatment by Microalgae: Current Status and Future Directions." *Processes* 8 (4). doi:10.3390/PR8040384.

Needham, Larry L., Antonia M. Calafat, and Dana B. Barr. 2007. "Uses and Issues of Biomonitoring." *International Journal of Hygiene and Environmental Health* 210 (3–4): 229–38. doi:10.1016/j.ijheh.2006.11.002.

Nehme, Ali, Zahraa Awada, Firas Kobeissy, Frédéric Mazurier, and Kazem Zibara. 2018. "Coupling Large-Scale Omics Data for Deciphering Systems Complexity." In *Systems Biology*, 153–72. Springer Science and Business Media Deutschland GmbH. doi:10.1007/978-3-319-92967-5_8.

Newell, Richard G., and Kristian Rogers. 2003. "The U.S. Experience with the Phasedown of Lead in Gasoline." *Resources for the Future*. https://web.mit.edu/ckolstad/www/Newell.pdf.

Nguyen, Minh Kim, Ju Young Moon, Vu Khac Hoang Bui, You Kwan Oh, and Young Chul Lee. 2019. "Recent Advanced Applications of Nanomaterials in Microalgae Biorefinery." *Algal Research* 41: 101522. doi:10.1016/j.algal.2019.101522.

Nguyen, Thuy T. T., Ben Hayes, and Brett A. Ingram. 2012. *Progress on Selective Breeding Program for Blue Mussel in Victoria*. Department of Primary Industries, Queenscliff, Vic. https://dro.deakin.edu.au/eserv/DU:30051553/nguyen-progressonselectivebreed-2012.pdf.

Niccolai, Alberto, Graziella Chini Zittelli, Liliana Rodolfi, Natascia Biondi, and Mario R. Tredici. 2019. "Microalgae of Interest as Food Source: Biochemical Composition and Digestibility." *Algal Research* 42. doi:10.1016/j.algal.2019.101617.

Niinimäki, Kirsi, Greg Peters, Helena Dahlbo, Patsy Perry, Timo Rissanen, and Alison Gwilt. 2020. "The Environmental Price of Fast Fashion." *Nature Reviews Earth and Environment*. doi:10.1038/s43017-020-0039-9.

Nissen, Sylvia, Jennifer H. K. Wong, and Sally Carlton. 2021. "Children and Young People's Climate Crisis Activism—a Perspective on Long-Term Effects." *Children's Geographies* 19 (3). Routledge: 317–23. doi:10.1080/14733285.2020.1812535.

Nova, Paulo, Ana Pimenta Martins, Carla Teixeira, Helena Abreu, Joana Gabriela Silva, Ana Machado Silva, Ana Cristina Freitas, and Ana Maria Gomes. 2020. "Foods with Microalgae and Seaweeds Fostering Consumers Health: A Review on Scientific and Market Innovations." *Journal of Applied Phycology*. doi:10.1007/s10811-020-02129-w.

Nriagu, Jerome O. 1990. "The Rise and Fall of Leaded Gasoline." *Science of the Total Environment* 92: 13–28. doi:10.1016/0048-9697(90)90318-O.

Oborne, Michael. 2010. "The Bioeconomy to 2030: Designing a Policy Agenda." *OECD Observer* 278: 35–37. https://search.proquest.com/openview/a51d482612ee60447 2c614b4b94f36c6/1?pq-origsite=gscholar&cbl=35885&casa_token=WSno4QyU

9bAAAAAA:AHpNUkVg0Tb4cMrVYy7BHIQuXRwFYWo0zwk2fAqw8LvYYMehTZfD24-zwXjV_8NJCNZhSNF_b8Y3bw.

OECD. 2020. *Towards Sustainable Land Use Aligning Biodiversity, Climate and Food Policies. OECD Publishing*. OECD. doi:10.1787/3809b6a1-en.

Olguín, Eugenia J. 2003. "Phycoremediation: Key Issues for Cost-Effective Nutrient Removal Processes." *Biotechnology Advances* 22: 81–91. doi:10.1016/S0734-9750(03)00130-7.

Oliver, Laura, Thomas Dietrich, Izaskun Marañón, Maria Carmen Villarán, and Ramón J. Barrio. 2020. "Producing Omega-3 Polyunsaturated Fatty Acids: A Review of Sustainable Sources and Future Trends for the EPA and DHA Market." *Resources*. Multidisciplinary Digital Publishing Institute. doi:10.3390/resources9120148.

Omar, Wan Maznah Wan. 2010. "Perspectives on the Use of Algae as Biological Indicators for Monitoring and Protecting Aquatic Environments, with Special Reference to Malaysian Freshwater Ecosystems." *Tropical Life Sciences Research* 21 (2). School of Medical Sciences, Universiti Sains Malaysia: 51–67. /pmc/articles/PMC3819078/.

Oncel, S. S., A. Kose, and D. S. Oncel. 2020. "Carbon Sequestration in Microalgae Photobioreactors Building Integrated." In *Start-Up Creation*, 161–200. doi:10.1016/b978-0-12-819946-6.00008-4.

Onumaegbu, C., J. Mooney, A. Alaswad, and A. G. Olabi. 2018. "Pre-Treatment Methods for Production of Biofuel from Microalgae Biomass." *Renewable and Sustainable Energy Reviews*. doi:10.1016/j.rser.2018.04.015.

Pacheco-Torgal, F., and Said Jalali. 2012. "Earth Construction: Lessons from the Past for Future Eco-Efficient Construction." *Construction and Building Materials*. doi:10.1016/j.conbuildmat.2011.10.054.

Pandolfi, J. M., S. R. Connolly, D. J. Marshall—Science, and Undefined. 2011. "Projecting Coral Reef Futures under Global Warming and Ocean Acidification." *Science.Sciencemag.Org*. doi:10.1126/science.1204794.

Parkinson, Andrew, Brian W. Ogilvie, David B. Buckley, Faraz Kazmi, Maciej Czerwinski, and Oliver Parkinson. 2018. "Chapter 6: Biotransformation of Xenobiotics." In *Casarett & Doull's Toxicology: The Basic Science of Poisons*, 9th ed. https://accessbiomedicalscience.mhmedical.com/content.aspx?bookid=2462&sectionid=202671272.

Paustenbach, Dennis J., and David Galbraith. 2006. "Biomonitoring and Biomarkers: Exposure Assessment Will Never Be the Same." *Environmental Health Perspectives*. doi:10.1289/ehp.8755.

Perea-Moreno, Miguel Angel, Quetzalcoatl Hernandez-Escobedo, and Alberto Jesus Perea-Moreno. 2018. "Renewable Energy in Urban Areas: Worldwide Research Trends." *Energies* 11 (3). doi:10.3390/en11030577.

Perez-Escamilla, Rafael, Odilia Bermudez, Gabriela Santos Buccini, Shiriki Kumanyika, Chessa K. Lutter, Pablo Monsivais, and Cesar Victora. 2018. "Nutrition Disparities and the Global Burden of Malnutrition." *BMJ (Online)* 361. doi:10.1136/bmj.k2252.

Pérez-Lloréns, José Lucas. 2020. "Microalgae: From Staple Foodstuff to Avant-Garde Cuisine." *International Journal of Gastronomy and Food Science* 21. doi:10.1016/j.ijgfs.2020.100221.

Peruccio, Pier Paolo, and Maurizio Vrenna. 2019. "Design and Microalgae. Sustainable Systems for Cities." *International Journal of Architecture, Art and Design*, no. 6: 218–27. doi:10.19229/2464-9309/6212019.

Pilon-Smits, Elizabeth. 2005. "Phytoremediation." *Annual Review of Plant Biology*. Annual Reviews. doi:10.1146/annurev.arplant.56.032604.144214.

Plotkin, Stanley A. 2011. *History of Vaccine Development. History of Vaccine Development*. Springer. doi:10.1007/978-1-4419-1339-5.

Plugaru, Sebastian Radu Cristian, Tudor Rusu, Katalin Molnar, and Laszlo Fodorpataki. 2017. "Chromium Removal from Polluted Water and Its Influence on Biochemical and Physiological Parameters in Algal Cells Used for Phytoremediation." *Studia Universitatis Babes-Bolyai Chemia* 62 (3): 225–38. doi:10.24193/subbchem.2017.3.19.

Pollak, Peter, and Raymond Vouillamoz. 2013. "Fine Chemicals." In *Ullmann's Encyclopedia of Industrial Chemistry*. Weinheim, Germany: Wiley-VCH Verlag GmbH & Co. KGaA, January. doi:10.1002/14356007.Q11_Q01.

Priyanka, Veerala, Rahul Kumar, Inderpreet Dhaliwal, and Prashant Kaushik. 2021. "Germplasm Conservation: Instrumental in Agricultural Biodiversity—a Review." *Sustainability (Switzerland)*. Multidisciplinary Digital Publishing Institute. doi:10.3390/su13126743.

Purvis, Ben, Yong Mao, and Darren Robinson. 2019. "Three Pillars of Sustainability: In Search of Conceptual Origins." *Sustainability Science* 14 (3). Springer: 681–95. doi:10.1007/s11625-018-0627-5.

Putkonen, Vesa. 2019. "Who Speaks for Earth? Climate Discourses and Voice in Greta Thunberg's Speech in UN Climate Action Summit on September 23, 2019." https://core.ac.uk/display/344909100.

Pyšek, Petr, and David M. Richardson. 2010. "Invasive Species, Environmental Change and Management, and Health." *Annual Review of Environment and Resources* 35 (November): 25–55. doi:10.1146/annurev-environ-033009-095548.

Rabnawaz, M., I. Wyman, R. Auras, and S. Cheng. 2017. "A Roadmap towards Green Packaging: The Current Status and Future Outlook for Polyesters in the Packaging Industry." *Green Chemistry*. doi:10.1039/c7gc02521a.

Ragni, Roberta, Stefania R. Cicco, Danilo Vona, and Gianluca M. Farinola. 2018. "Multiple Routes to Smart Nanostructured Materials from Diatom Microalgae: A Chemical Perspective." *Advanced Materials*. Wiley-VCH Verlag. doi:10.1002/adma.201704289.

Rahman, A., and C. D. Miller. 2017. "Microalgae as a Source of Bioplastics." In *Algal Green Chemistry: Recent Progress in Biotechnology*, 121–38. doi:10.1016/B978-0-444-63784-0.00006-0.

Raines, Christine A. 2003. "The Calvin Cycle Revisited." *Photosynthesis Research* 75 (1): 1–10. doi:10.1023/A:1022421515027.

Rajamani, Sathish, Surasak Siripornadulsil, Vanessa Falcao, Moacir Torres, Pio Colepicolo, and Richard Sayre. 2007. "Phycoremediation of Heavy Metals Using Transgenic Microalgae." In *Advances in Experimental Medicine and Biology*. New York, NY: Springer. doi:10.1007/978-0-387-75532-8_9.

Rajitha, K., M. Sarvajith, V. P. Venugopalan, and Y. V. Nancharaiah. 2020. "Development and Performance of Halophilic Microalgae-Colonized Aerobic Granular Sludge for Treating Seawater-Based Wastewater." *Bioresource Technology Reports* 11 (September). Elsevier: 100432. doi:10.1016/j.biteb.2020.100432.

Ramos-Suárez, J. L., N. Carreras Arroyo, and C. González-Fernández. 2015. "The Role of Anaerobic Digestion in Algal Biorefineries: Clean Energy Production, Organic Waste Treatment, and Nutrient Loop Closure." In *Algae and Environmental Sustainability*, 53–76. New Delhi: Springer. doi:10.1007/978-81-322-2641-3_5.

Rands, Michael R. W., William M. Adams, Leon Bennun, Stuart H. M. Butchart, Andrew Clements, David Coomes, Abigail Entwistle, et al. 2010. "Biodiversity Conservation: Challenges beyond 2010." *Science*. doi:10.1126/science.1189138.

Ratha, S. K., and R. Prasanna. 2012. "Bioprospecting Microalgae as Potential Sources of 'Green Energy'-Challenges and Perspectives (Review)." *Applied Biochemistry and Microbiology*. Springer. doi:10.1134/S000368381202010X.

Ratledge, Colin., and B. (Bjørn) Kristiansen, eds. 2006. *Basic Biotechnology*. Cambridge: Cambridge University Press.

Rawat, Ismail, Sanjay K. Gupta, Amritanshu Shriwastav, Poonam Singh, Sheena Kumari, and Faizal Bux. 2016. "Microalgae Applications in Wastewater Treatment." In *Algae Biotechnology*, 249–68. Springer International Publishing. doi:10.1007/978-3-319-12334-9_13.

Renaud, S. M., and D. L. Parry. 1994. "Microalgae for Use in Tropical Aquaculture II: Effect of Salinity on Growth, Gross Chemical Composition and Fatty Acid Composition of Three Species of Marine Microalgae." *Journal of Applied Phycology* 6 (3). Kluwer Academic Publishers: 347–56. doi:10.1007/BF02181949.

Reyimu, Zubaidai, and Didem Özçimen. 2017. "Batch Cultivation of Marine Microalgae Nannochloropsis Oculata and Tetraselmis Suecica in Treated Municipal Wastewater toward Bioethanol Production." *Journal of Cleaner Production* 150 (May). Elsevier: 40–46. doi:10.1016/j.jclepro.2017.02.189.

Reynolds, Colin S. 1984. *The Ecology of Freshwater Phytoplankton*. Cambridge University Press. https://books.google.com/books/about/The_Ecology_of_Freshwater_Phytoplankton.html?id=0nyrasgaTwMC.

Richmond, Amos. 2013. "Biological Principles of Mass Cultivation of Photoautotrophic Microalgae." In *Handbook of Microalgal Culture: Applied Phycology and Biotechnology*, 171–204, 2nd ed. John Wiley & Sons, Ltd. doi:10.1002/9781118567166.ch11.

Ritchie, Hannah, and Max Roser. 2018. "Plastic Pollution." *Our World in Data*. September. https://ourworldindata.org/plastic-pollution.

Rodríguez, A. G., M. S. Rodrigues, and O. Sotomayor. 2019. "Towards a Sustainable Bioeconomy in Latin America and the Caribbean: Elements for a Regional Vision." In *Natural Resources and Development Series N°193*, 1–52. https://repositorio.cepal.org/handle/11362/44994.

Rodríguez-Núñez, K., F. Rodríguez-Ramos, D. Leiva-Portilla, and C. Ibáñez. 2020. "Brown Biotechnology: A Powerful Toolbox for Resolving Current and Future Challenges in the Development of Arid Lands." *SN Applied Sciences*. Springer. doi:10.1007/s42452-020-2980-0.

Rogers, Elizabeth, and Thomas Kostigen. 2007. *The Green Book: The Everyday Guide to Saving the Planet One Simple Step at a Time*. http://www.loc.gov/catdir/toc/ecip0714/2007013222.html.

Ryan, M. P., J. T. Pembroke, and C. C. Adley. 2007. "Ralstonia Pickettii in Environmental Biotechnology: Potential and Applications." *Journal of Applied Microbiology*. John Wiley & Sons, Ltd. doi:10.1111/j.1365-2672.2007.03361.x.

Sachsenmeier, Peter. 2016. "Industry 5.0—The Relevance and Implications of Bionics and Synthetic Biology." *Engineering* 2 (2). Elsevier: 225–29. doi:10.1016/J.ENG.2016.02.015.

Sajjadi, Baharak, Wei Yin Chen, Abdul Aziz Abdul Raman, and Shaliza Ibrahim. 2018. "Microalgae Lipid and Biomass for Biofuel Production: A Comprehensive Review on Lipid Enhancement Strategies and Their Effects on Fatty Acid Composition." *Renewable and Sustainable Energy Reviews*. doi:10.1016/j.rser.2018.07.050.

Santos, Aline Meireles dos, Mariany Costa Deprá, Alberto Meireles dos Santos, Alexandre José Cichoski, Leila Queiroz Zepka, and Eduardo Jacob-Lopes. 2020. "Sustainability Metrics on Microalgae-Based Wastewater Treatment System." *Desalination and Water Treatment* 185: 51–61. doi:10.5004/dwt.2020.25397.

Santos, Rita B., Rita Abranches, Rainer Fischer, Markus Sack, and Tanja Holland. 2016. "Putting the Spotlight Back on Plant Suspension Cultures." *Frontiers in Plant Science*. Frontiers Media S.A. doi:10.3389/fpls.2016.00297.

Sardá, Rosa Cervera, and Javier Gómez Pioz. 2015. "Architectural Bio-Photo Reactors: Harvesting Microalgae on the Surface of Architecture." In *Biotechnologies and Biomimetics for Civil Engineering*, 163–79. Springer International Publishing. doi:10.1007/978-3-319-09287-4_7.

Sasson, Albert. 2005. *Medical Biotechnology: Achievements, Prospects and Perceptions*. United Nations University Press. https://books.google.com/books?id=WjC_e27xgUEC&pgis=1.

Sato, Motoaki. 1990. "Thermochemistry of the Formation of Fossil Fuels." *Special Publication—The Geochemical Society* 2. Fluid-Miner. Interact: 271–83. https://www.geochemsoc.org/files/6214/1261/1770/SP-2_271-284_Sato.pdf.

Satya, Awalina, Fachmijany Sulawesty, Ardiyan Harimawan, and Tjandra Setiadi. 2018. "Correlation of Aquatic Parameters to the Cadmium Bioaccumulation Capability onto Microalgae Biomass in an Urban Lake." *Journal of Water Sustainability* 2: 59–72. doi:10.11912/jws.2018.8.2.59-72.

Schandl, Heinz, Marina Fischer-Kowalski, James West, Stefan Giljum, Monika Dittrich, Nina Eisenmenger, Arne Geschke, et al. 2018. "Global Material Flows and Resource Productivity Forty Years of Evidence." *Journal of Industrial Ecology* 22 (4). John Wiley & Sons, Ltd: 827–38. doi:10.1111/jiec.12626.

Schipfer, F., and J. Matzenberger. 2012. "Assessment of Microalgae Production in Photo-Bioreactors for Vertical Farming in Urban Areas." *European Biomass Conference and Exhibition Proceedings* (June). ETA-Florence Renewable Energies: 18–22. doi:10.5071/20THEUBCE2012-2CV.1.15.

Schirmacher, Michael R., Alex Prichard, Todd Mabee, and Cris D. Hein. 2018. "Evaluating a Novel Approach to Optimize Operational Minimization to Reduce Bat Fatalities at the Pinnacle Wind Farm, Mineral County, West Virginia, 2015." An Annual Report Submitted to NRG Energy and the Bats and Wind Energy Cooperative.

Seidel, George E. 1998. "Biotechnology in Animal Agriculture." In *Animal Biotechnology and Ethics*, 50–68. Boston, MA: Springer. doi:10.1007/978-1-4615-5783-8_4.

Sharma, Indu. 2020. "Bioremediation Techniques for Polluted Environment: Concept, Advantages, Limitations, and Prospects." In *Trace Metals in the Environment—New Approaches and Recent Advances*. IntechOpen. doi:10.5772/intechopen.90453.

Sharma, Naveen K., and Ashwani K. Rai. 2011. "Biodiversity and Biogeography of Microalgae: Progress and Pitfalls." *Environmental Reviews* 19 (1). NRC Research Press: 1–15. doi:10.1139/A10-020.

Sharma, Raman, Meenakshi Sharma, Ratika Sharma, and Vivek Sharma. 2013. "The Impact of Incinerators on Human Health and Environment." *Reviews on Environmental Health*. De Gruyter. doi:10.1515/reveh-2012-0035.

Shellenberger, Michael. 2017. *The Nuclear Option Renewables Can't Save the Planet-but Uranium Can Energy and Civilization: A History*, Vol. 552. MIT Press. https://www.jstor.org/stable/44821880.

Sicard, Alexandre J., and R. Tom Baker. 2020. "Fluorocarbon Refrigerants and Their Syntheses: Past to Present." *Chemical Reviews*. American Chemical Society. doi:10.1021/acs.chemrev.9b00719.

Singh, Digambar, Dilip Sharma, S. L. Soni, Sumit Sharma, Pushpendra Kumar Sharma, and Amit Jhalani. 2020. "A Review on Feedstocks, Production Processes, and Yield for Different Generations of Biodiesel." *Fuel*. Elsevier. doi:10.1016/j.fuel.2019.116553.

Singh, Gulab, and S. K. Patidar. 2018. "Microalgae Harvesting Techniques: A Review." *Journal of Environmental Management*. Academic Press. doi:10.1016/j.jenvman.2018.04.010.

Singh, Harbhajan. 2006. "Fungal Biodegradation and Biodeterioration." In *Mycoremediation: Fungal Bioremediation*, 1–3. John Wiley & Sons. https://www.wiley.com/en-bd/Mycoremediation%3A+Fungal+Bioremediation-p-9780471755012.

Singh, Jyoti, and Dolly Wattal Dhar. 2019. "Overview of Carbon Capture Technology: Microalgal Biorefinery Concept and State-of-the-Art." *Frontiers in Marine Science*. Frontiers. doi:10.3389/fmars.2019.00029.

Smol, John P., and Eugene F. Stoermer. 2010. "Applications and Uses of Diatoms: Prologue." In *The Diatoms: Applications for the Environmental and Earth Sciences*, 3–7, 2nd ed. Cambridge University Press. doi:10.1017/CBO9780511763175.002.

Solovchenko, A. E., A. A. Lukyanov, S. G. Vasilieva, Ya V. Savanina, O. V. Solovchenko, and E. S. Lobakova. 2013. "Possibilities of Bioconversion of Agricultural Waste with the Use of Microalgae." *Moscow University Biological Sciences Bulletin* 68 (4): 206–15. doi:10.3103/S0096392514010118.

Somerville, Chris. 2007. "Biofuels." *Current Biology*. Elsevier. doi:10.1016/j.cub.2007.01.010.

Stauber, Rudolf. 2007. "Plastics in Automotive Engineering." *ATZ Worldwide* 109 (3). Springer: 2–4. doi:10.1007/BF03224916.

Storch, Lilian von, Lukas Ley, and Jing Sun. 2021. "New Climate Change Activism: Before and after the Covid-19 Pandemic." *Social Anthropology*. John Wiley & Sons, Ltd. doi:10.1111/1469-8676.13005.

Straiton, Jenny, Tristan Free, Abigail Sawyer, and Joseph Martin. 2019. "From Sanger Sequencing to Genome Databases and Beyond." *BioTechniques* 66 (2). Future Science Ltd London, UK: 60–63. doi:10.2144/btn-2019-0011.

Strasser, Jurt A. 1996. "Preventing Pollution." *Fordham Environmental Law Review* 8 (1). https://ir.lawnet.fordham.edu/elr/vol8/iss1/13.

Strimbu, Kyle, and Jorge A. Tavel. 2010. "What Are Biomarkers?" *Current Opinion in HIV and AIDS*. Curr Opin HIV AIDS. doi:10.1097/COH.0b013e32833ed177.

Subashchandrabose, Suresh R., Balasubramanian Ramakrishnan, Mallavarapu Megharaj, Kadiyala Venkateswarlu, and Ravi Naidu. 2013. "Mixotrophic Cyanobacteria and Microalgae as Distinctive Biological Agents for Organic Pollutant Degradation." *Environment International*. Pergamon. doi:10.1016/j.envint.2012.10.007.

Suganya, T., M. Varman, H. H. Masjuki, and S. Renganathan. 2016. "Macroalgae and Microalgae as a Potential Source for Commercial Applications along with Biofuels Production: A Biorefinery Approach." *Renewable and Sustainable Energy Reviews*. Pergamon. doi:10.1016/j.rser.2015.11.026.

Suparmaniam, Uganeeswary, Man Kee Lam, Yoshimitsu Uemura, Jun Wei Lim, Keat Teong Lee, and Siew Hoong Shuit. 2019. "Insights into the Microalgae Cultivation Technology and Harvesting Process for Biofuel Production: A Review." *Renewable and Sustainable Energy Reviews*. Pergamon. doi:10.1016/j.rser.2019.109361.

Tabone, Michaelangelo D., James J. Cregg, Eric J. Beckman, and Amy E. Landis. 2010. "Sustainability Metrics: Life Cycle Assessment and Green Design in Polymers." *Environmental Science and Technology* 44 (21). American Chemical Society: 8264–69. doi:10.1021/es101640n.

Taher, Hanifa, Sulaiman Al-Zuhair, Ali H. Al-Marzouqi, Yousef Haik, and Mohammed M. Farid. 2011. "A Review of Enzymatic Transesterification of Microalgal Oil-Based Biodiesel Using Supercritical Technology." *Enzyme Research*. doi:10.4061/2011/468292.

Tamoi, Masahiro, Miki Nagaoka, Yukinori Yabuta, and Shigeru Shigeoka. 2005. "Carbon Metabolism in the Calvin Cycle." *Plant Biotechnology*. Japanese Society for Plant Biotechnology. doi:10.5511/plantbiotechnology.22.355.

Tatarová, Dominika, Dušan Galanda, Jozef Kuruc, and Barbora Gaálová. 2021. "Phytoremediation of 137Cs, 60Co, 241Am, and 239Pu from Aquatic Solutions Using

Chlamydomonas Reinhardtii, Scenedesmus Obliquus, and Chlorella Vulgaris." *International Journal of Phytoremediation*: 1–6. doi:10.1080/15226514.2021.1900061.

Taylor, F. J. R., Mona Hoppenrath, and Juan F. Saldarriaga. 2008. "Dinoflagellate Diversity and Distribution." *Biodiversity and Conservation*. Springer. doi:10.1007/s10531-007-9258-3.

Teitelbaum, L., C. Boldt, and C. Patermann. 2020. "Global Bioeconomy Policy Report (IV): A Decade of Bioeconomy Policy Development around the World." https://gbs2020.net/wp-content/uploads/2020/11/GBS-2020_Global-Bioeconomy-Policy-Report_IV_web.pdf.

Tekoah, Yoram, Avidor Shulman, Tali Kizhner, Ilya Ruderfer, Liat Fux, Yakir Nataf, Daniel Bartfeld, et al. 2015. "Large-Scale Production of Pharmaceutical Proteins in Plant Cell Culture-the Protalix Experience." *Plant Biotechnology Journal*. John Wiley & Sons, Ltd. doi:10.1111/pbi.12428.

Teng, Sin Yong, Guo Yong Yew, Kateřina Sukačová, Pau Loke Show, Vítězslav Máša, and Jo Shu Chang. 2020. "Microalgae with Artificial Intelligence: A Digitalized Perspective on Genetics, Systems and Products." *Biotechnology Advances*. Elsevier Inc. doi:10.1016/j.biotechadv.2020.107631.

Tesson, Sylvie V. M., Carsten Ambelas Skjøth, Tina Šantl-Temkiv, and Jakob Löndahl. 2016. "Airborne Microalgae: Insights, Opportunities, and Challenges." *Applied and Environmental Microbiology*. American Society for Microbiology (ASM). doi:10.1128/AEM.03333-15.

Thakare, Ketan, Laura Jerpseth, Hongmin Qin, and Zhijian Pei. 2021. "Bioprinting Using Algae: Effects of Extrusion Pressure and Needle Diameter on Cell Quantity in Printed Samples." *Journal of Manufacturing Science and Engineering* 143 (1). American Society of Mechanical Engineers Digital Collection. doi:10.1115/1.4048853.

Tiwari, Archana, and Thomas Kiran. 2018. "Biofuels from Microalgae." *Advances in Biofuels and Bioenergy* (July). IntechOpen. doi:10.5772/INTECHOPEN.73012.

Tomaselli, Luisa. 1997. "Morphology, Ultrastructure and Taxonomy of Arthrospira (Spirulina Maxima and Arthrospira (Spirulina) Platensis." In *Spirulina Platensis (Arthrospira): Physiology, Cell-Biology and Biotechnology*, 1–16. doi:10.1201/9781482272970-9.

Török, Peter, and Aveliina Helm. 2017. "Ecological Theory Provides Strong Support for Habitat Restoration." *Biological Conservation*. doi:10.1016/j.biocon.2016.12.024.

Tremeac, Brice, and Francis Meunier. 2009. "Life Cycle Analysis of 4.5 MW and 250 W Wind Turbines." *Renewable and Sustainable Energy Reviews*. Pergamon. doi:10.1016/j.rser.2009.01.001.

Turemis, Mehmet, Silvia Silletti, Gianni Pezzotti, Josep Sanchís, Marinella Farré, and Maria Teresa Giardi. 2018. "Optical Biosensor Based on the Microalga-Paramecium Symbiosis for Improved Marine Monitoring." *Sensors and Actuators, B: Chemical* 270 (October). Elsevier: 424–32. doi:10.1016/j.snb.2018.04.111.

Turner, Simon P. 2011. "Breeding against Harmful Social Behaviours in Pigs and Chickens: State of the Art and the Way Forward." *Applied Animal Behaviour Science*. Elsevier. doi:10.1016/j.applanim.2011.06.001.

Twardowski, Tomasz. 2011. "Biotechnology in Medicine in the 21st Century: Natural Medicine and Intellectual Property Rights." *Asian Biotechnology and Development Review* 13 (3): 27–36. www.ris.org.in.

Ubando, Aristotle T., Aaron Don M. Africa, Marla C. Maniquiz-Redillas, Alvin B. Culaba, Wei Hsin Chen, and Jo Shu Chang. 2021. "Microalgal Biosorption of Heavy Metals: A Comprehensive Bibliometric Review." *Journal of Hazardous Materials* 402 (January). doi:10.1016/j.jhazmat.2020.123431.

Vadiveloo, Ashiwin, Navid R. Moheimani, Ramzy Alghamedi, Jeffrey J. Cosgrove, Kamal Alameh, and David Parlevliet. 2016. "Sustainable Cultivation of Microalgae by an

Insulated Glazed Glass Plate Photobioreactor." *Biotechnology Journal* 11 (3): 363–74. doi:10.1002/biot.201500358.

Vandenberg, Laura N., Ibrahim Chahoud, Jerrold J. Heindel, Vasantha Padmanabhan, Francisco J. R. Paumgartten, and Gilbert Schoenfelder. 2010. "Urinary, Circulating, and Tissue Biomonitoring Studies Indicate Widespread Exposure to Bisphenol A." *Environmental Health Perspectives*. National Institute of Environmental Health Sciences. doi:10.1289/ehp.0901716.

Ven, Dirk Jan van de, Iñigo Capellan-Peréz, Iñaki Arto, Ignacio Cazcarro, Carlos de Castro, Pralit Patel, and Mikel Gonzalez-Eguino. 2021. "The Potential Land Requirements and Related Land Use Change Emissions of Solar Energy." *Scientific Reports* 11 (1). Nature Publishing Group: 1–12. doi:10.1038/s41598-021-82042-5.

Verschuuren, Jonathan. 2004. "Effectiveness of Nature Protection Legislation in the European Union and the United States: The Habitats Directive and the Endangered Species Act." In *Cultural Landscapes and Land Use*, 39–67. Dordrecht: Springer. doi:10.1007/1-4020-2105-4_4.

Vieira, Marta v., Lorenzo M. Pastrana, and Pablo Fuciños. 2020. "Microalgae Encapsulation Systems for Food, Pharmaceutical and Cosmetics Applications." *Marine Drugs*. Multidisciplinary Digital Publishing Institute. doi:10.3390/md18120644.

Vigani, Mauro, Claudia Parisi, Emilio Rodríguez-Cerezo, Maria J. Barbosa, Lolke Sijtsma, Matthias Ploeg, and Christien Enzing. 2015. "Food and Feed Products from Micro-Algae: Market Opportunities and Challenges for the EU." *Trends in Food Science and Technology*. Elsevier. doi:10.1016/j.tifs.2014.12.004.

Vigneshvar, S., C. C. Sudhakumari, Balasubramanian Senthilkumaran, and Hridayesh Prakash. 2016. "Recent Advances in Biosensor Technology for Potential Applications—an Overview." *Frontiers in Bioengineering and Biotechnology*. Frontiers. doi:10.3389/fbioe.2016.00011.

Vrieling, Engel G., Sandra Hazelaar, Winfried W. C. Gieskes, Qianyao Sun, Theo P. M. Beelen, and Rutger A. Van Santen. 2003. "Silicon Biomineralisation: Towards Mimicking Biogenic Silica Formation in Diatoms." In, 301–34. Berlin, Heidelberg: Springer. doi:10.1007/978-3-642-55486-5_12.

Vroman, Isabelle, and Lan Tighzert. 2009. "Biodegradable Polymers." *Materials*. Molecular Diversity Preservation International. doi:10.3390/ma2020307.

Walker, Sara Louise. 2011. "Building Mounted Wind Turbines and Their Suitability for the Urban Scale-A Review of Methods of Estimating Urban Wind Resource." *Energy and Buildings* 43 (8): 1852–62. doi:10.1016/j.enbuild.2011.03.032.

Walsh, Christopher Thomas. 2003. *Antibiotics: Actions, Origins, Resistance*. American Society for Microbiology (ASM). https://www.cabdirect.org/cabdirect/abstract/2004 3133125.

Wang, Bo. 2020. *Simulation and Artificial Intelligent Methodologies for End-to-End Bio-Pharmaceutical Manufacturing and Supply Chain Risk Management*. (Doctoral dissertation, Northeastern University).

Wang, Qintao, Yandu Lu, Yi Xin, Li Wei, Shi Huang, and Jian Xu. 2016. "Genome Editing of Model Oleaginous Microalgae Nannochloropsis Spp. by CRISPR/Cas9." *Plant Journal* 88 (6). John Wiley & Sons, Ltd: 1071–81. doi:10.1111/tpj.13307.

Wang, Yue, Shih Hsin Ho, Chieh Lun Cheng, Wan Qian Guo, Dillirani Nagarajan, Nan Qi Ren, Duu Jong Lee, and Jo Shu Chang. 2016. "Perspectives on the Feasibility of Using Microalgae for Industrial Wastewater Treatment." *Bioresource Technology*. doi:10.1016/j.biortech.2016.09.106.

Waterbury, John B., Joanne M. Willey, Diana G. Franks, Frederica W. Valois, and Stanley W. Watson. 1985. "A Cyanobacterium Capable of Swimming Motility." *Science* 230 (4721): 74–76. doi:10.1126/science.230.4721.74.

Watson, James D. 1990. "The Human Genome Project: Past, Present, and Future." *Science* 248 (4951): 44–49. doi:10.1126/science.2181665.

Wiencke, Christian, Charles D. Amsler, and Margeret N. Clayton. 2014. "Macroalgae." In *Biogeographic Atlas of the Southern Ocean*, edited by Claude de Broyer and Philippe Koubbi, 66–73. The Scientific Committee on Antarctic Research. www.biodiversity.aq.

Williams, Paul T. 2005. *Waste Treatment and Disposal. John Wiley & Sons*. Wiley. https://www.wiley.com/en-co/Waste+Treatment+and+Disposal,+2nd+Edition-p-9780470849125.

Williamson, Phillip, Douglas W. R. Wallace, Cliff S. Law, Philip W. Boyd, Yves Collos, Peter Croot, Ken Denman, Ulf Riebesell, Shigenobu Takeda, and Chris Vivian. 2012. "Ocean Fertilization for Geoengineering: A Review of Effectiveness, Environmental Impacts and Emerging Governance." *Process Safety and Environmental Protection* 90 (6). Elsevier: 475–88. doi:10.1016/j.psep.2012.10.007.

Wollmann, Felix, Stefan Dietze, Jörg Uwe Ackermann, Thomas Bley, Thomas Walther, Juliane Steingroewer, and Felix Krujatz. 2019. "Microalgae Wastewater Treatment: Biological and Technological Approaches." *Engineering in Life Sciences*. Wiley-Blackwell. doi:10.1002/elsc.201900071.

Wu, Naicheng, Xuhui Dong, Yang Liu, Chao Wang, Annette Baattrup-Pedersen, and Tenna Riis. 2017. "Using River Microalgae as Indicators for Freshwater Biomonitoring: Review of Published Research and Future Directions." *Ecological Indicators*. Elsevier. doi:10.1016/j.ecolind.2017.05.066.

Wu, Xiao, Rachel C. Nethery, M. Benjamin Sabath, Danielle Braun, and Francesca Dominici. 2020. "Exposure to Air Pollution and COVID-19 Mortality in the United States: A Nationwide Cross-Sectional Study." *Science Advances*. Cold Spring Harbor Laboratory Preprints. doi:10.1101/2020.04.05.20054502.

Xu, Jianfeng, and Ningning Zhang. 2014. "On the Way to Commercializing Plant Cell Culture Platform for Biopharmaceuticals: Present Status and Prospect." *Pharmaceutical Bioprocessing* 2 (6). NIH Public Access: 499–518. doi:10.4155/pbp.14.32.

Yaakob, Zahira, Ehsan Ali, Afifi Zainal, Masita Mohamad, and Mohd Sobri Takriff. 2014. "An Overview: Biomolecules from Microalgae for Animal Feed and Aquaculture." *Journal of Biological Research (Greece)*. BioMed Central. doi:10.1186/2241-5793-21-6.

Yadav, Geetanjali, Sukanta K. Dash, and Ramkrishna Sen. 2019. "A Biorefinery for Valorization of Industrial Waste-Water and Flue Gas by Microalgae for Waste Mitigation, Carbon-Dioxide Sequestration and Algal Biomass Production." *Science of the Total Environment* 688 (October): 129–35. doi:10.1016/j.scitotenv.2019.06.024.

Yadav, Nisha, and D. P. Singh. 2020. "Microalgae and Microorganisms: Important Regulators of Carbon Dynamics in Wetland Ecosystem." In *Restoration of Wetland Ecosystem: A Trajectory Towards a Sustainable Environment*, 179–93. Singapore: Springer. doi:10.1007/978-981-13-7665-8_12.

Yan, Na, Chengming Fan, Yuhong Chen, and Zanmin Hu. 2016. "The Potential for Microalgae as Bioreactors to Produce Pharmaceuticals." *International Journal of Molecular Sciences* 17 (6): 962. doi:10.3390/ijms17060962.

Yang, Qianqi, Zhiyuan Li, Xiaoning Lu, Qiannan Duan, Lei Huang, and Jun Bi. 2018. "A Review of Soil Heavy Metal Pollution from Industrial and Agricultural Regions in China: Pollution and Risk Assessment." *Science of the Total Environment*. doi:10.1016/j.scitotenv.2018.06.068.

Yarnold, Jennifer, Hakan Karan, Melanie Oey, and Ben Hankamer. 2019. "Microalgal Aquafeeds as Part of a Circular Bioeconomy." *Trends in Plant Science*. doi:10.1016/j.tplants.2019.06.005.

Yeong, Tong Kai, Kailin Jiao, Xianhai Zeng, Lu Lin, Sharadwata Pan, and Michael K. Danquah. 2018. "Microalgae for Biobutanol Production—Technology Evaluation and Value Proposition." *Algal Research*. Elsevier. doi:10.1016/j.algal.2018.02.029.

Yigitcanlar, Tan, Fatih Dur, and Didem Dizdaroglu. 2015. "Towards Prosperous Sustainable Cities: A Multiscalar Urban Sustainability Assessment Approach." *Habitat International* 45 (P1). Pergamon: 36–46. doi:10.1016/j.habitatint.2014.06.033.

Yigitcanlar, Tan, Md Kamruzzaman, Marcus Foth, Jamile Sabatini-Marques, Eduardo da Costa, and Giuseppe Ioppolo. 2019. "Can Cities Become Smart without Being Sustainable? A Systematic Review of the Literature." *Sustainable Cities and Society*. Elsevier. doi:10.1016/j.scs.2018.11.033.

Yoshida, Masaki, Yuuhiko Tanabe, Natsuki Yonezawa, and Makoto M. Watanabe. 2012. "Energy Innovation Potential of Oleaginous Microalgae." *Biofuels* 3 (6). Future Science LtdLondon, UK: 761–81. doi:10.4155/bfs.12.63.

Yu, Hyeonjung, Jaai Kim, and Changsoo Lee. 2019. "Potential of Mixed-Culture Microalgae Enriched from Aerobic and Anaerobic Sludges for Nutrient Removal and Biomass Production from Anaerobic Effluents." *Bioresource Technology* 280 (May): 325–36. doi:10.1016/j.biortech.2019.02.054.

Yue, Dajun, Fengqi You, and Seth B. Darling. 2014. "Domestic and Overseas Manufacturing Scenarios of Silicon-Based Photovoltaics: Life Cycle Energy and Environmental Comparative Analysis." *Solar Energy* 105 (July). Pergamon: 669–78. doi:10.1016/j.solener.2014.04.008.

Zabed, Hossain M., Suely Akter, Junhua Yun, Guoyan Zhang, Faisal N. Awad, Xianghui Qi, and J. N. Sahu. 2019. "Recent Advances in Biological Pretreatment of Microalgae and Lignocellulosic Biomass for Biofuel Production." *Renewable and Sustainable Energy Reviews*. Elsevier Ltd. doi:10.1016/j.rser.2019.01.048.

Zanella, Lorenzo, and Md. Asraful Alam. 2020. "Extracts and Bioactives from Microalgae (Sensu Stricto): Opportunities and Challenges for a New Generation of Cosmetics." In *Microalgae Biotechnology for Food, Health and High Value Products*, 295–349. Singapore: Springer. doi:10.1007/978-981-15-0169-2_9.

Zucman, Gabriel. 2019. "Global Wealth Inequality." *Annual Review of Economics*. doi:10.1146/annurev-economics-080218-025852.

Zuliani, Luca, Nicola Frison, Aleksandra Jelic, Francesco Fatone, David Bolzonella, and Matteo Ballottari. 2016. "Microalgae Cultivation on Anaerobic Digestate of Municipalwastewater, Sewage Sludge and Agro-Waste." *International Journal of Molecular Sciences* 17 (10). Multidisciplinary Digital Publishing Institute: 1692. doi:10.3390/ijms17101692.

Zullaikah, Siti, Adi Tjipto Utomo, Medina Yasmin, Lu Ki Ong, and Yi Hsu Ju. 2019. "Ecofuel Conversion Technology of Inedible Lipid Feedstocks to Renewable Fuel." *Advances in Eco-Fuels for a Sustainable Environment* (January). Woodhead Publishing: 237–76. doi:10.1016/B978-0-08-102728-8.00009-7.

Zylstra, Gerben J., and Jerome J. Kukor. 2005. "What Is Environmental Biotechnology?" *Current Opinion in Biotechnology*. Elsevier Current Trends. doi:10.1016/j.copbio.2005.05.001.

2 The Potential of Microalgae for Environmental Biotechnology

Fazril Ideris[1,2], Mei Yin Ong[1,2], Jassinnee Milano[1,2], Mohd Faiz Muaz Ahmad Zamri[1,2], Saifuddin Nomanbhay[1,2,3], Abd Halim Shamsuddin[2], Teuku Meurah Indra Mahlia[4] and Pau Loke Show[5]

[1] Institute of Sustainable Energy, Universiti Tenaga Nasional (UNITEN), Jalan IKRAM-UNITEN, Kajang, Selangor, Malaysia

[2] AAIBE Chair of Renewable Energy, Universiti Tenaga Nasional (UNITEN), Kajang, Selangor, Malaysia

[3] AAIBE Chair of Renewable Energy Universiti Tenaga Nasional (UNITEN) Kajang, Selangor, Malaysia

[4] Centre for Green Technology, School of Civil and Environmental Engineering, University of Technology, Sydney, NSW, Australia

[5] Department of Chemical and Environmental Engineering, Faculty of Science and Engineering, University of Nottingham Malaysia, Jalan Broga, Semenyih, Selangor Darul Ehsan, Malaysia

CONTENTS

2.1 Introduction 68
2.2 Microalgae in Carbon Dioxide Sequestration 69
2.3 Wastewater Treatment Using Microalgae 71
2.4 Biofuels from Microalgae 76
2.5 Microalgae (and Cyanobacteria) in Agriculture 82
2.6 Microalgal Bioprocessing Up-Scaling Challenges 86
2.7 Conclusions 88
Acknowledgments 88
References 89

DOI: 10.1201/9781003202196-2

2.1 INTRODUCTION

Historically, humanity took more than 2 million years to reach 1 billion. In retrospective, only 200 years more was all it took for global population to mushroom to 7 billion. Now, we are looking at the 8 billion mark in a matter of few years. The growth in human population leads to rapid urbanization and industrialization. The world has seen an unprecedented emission of anthropogenic greenhouse gases such as methane and carbon dioxide, where 62% of the total emission comes from the burning of fossil fuels from various industries, while transportation sector contributes about 14% of the share (Eickemeier et al. 2014). These anthropogenic gases are the primary drivers of global warming. The catastrophic effects of global warming include erratic weather, heat waves, floods, droughts, declining freshwater supplies, and the rise in sea level (Jeffry et al. 2021; A. A. Zamri et al. 2021). The rise in sea level would displace more than a billion people residing in coastal cities such as Hong Kong, Miami, Perth, New York, and Bangkok (Neumann et al. 2015; World Economic Forum 2019). As of this moment, New Orleans is sinking by approximately 2 inches annually, while the city of Jakarta has sunk by 2.5 meter over a period of 10 years.

In addition to causing air pollution, the increase in human population also leads to water and land pollutions. Waste originating from both domestic and industrial sectors has become a huge concern. Since these wastes are unavoidable, proper waste management techniques need to be addressed to reduce the damaging impacts of waste disposal. Regulations set by various governing bodies are one of the measures to tackle these. Effluents containing pollutants need to be treated before being released to the environment. One of the options to treat effluents is by using various chemicals. However, the usage of chemicals in treating the effluents is indeed proven to be less economical and inefficient. With the advancements in knowledge, awareness, and technology, researchers and scientists are now turning their attentions to biotechnology in developing greener and sustainable methods in managing waste.

Biotechnology can be briefly described as the development and utilization of biological systems (animals, plants, bacteria, fungi, and algae) as detoxifications for various types of pollutions (water, land, and air) and for other environmental-friendly processes (Bhatia 2018). To achieve this, biotechnology integrates multiple fields such as biochemistry, chemical engineering, genetic engineering, and microbiology (Panesar and Kennedy 2006). The usage of these biological systems often resulted in numerous additional benefits, such as providing sources of renewable energy, food, and nutrients, where the waste of a process would work as a feedstock for other processes.

Currently, microalgae are gaining popularity as a potential source of biomass in meeting the insatiable demands of fuels, feed, food, and chemical productions (Markou, Vandamme, and Muylaert 2014). Microalgae provide an abundance of advantages when compared with other crops, such as ability to thrive in saline water or wastewater, high growth rate, and the ability to provide high amount of carbohydrate or lipid content (Branco-Vieira et al. 2020). Their ability to grow in

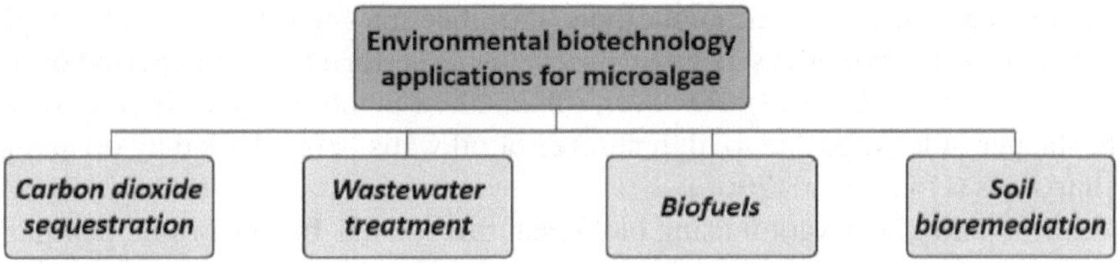

FIGURE 2.1 Environmental biotechnology applications for microalgae.

saline water is deemed to be very desirable, hence avoiding the usage of limited freshwater and fertile land resources. Therefore, these two precious resources can be utilized in cultivating crops for ensuring the global food security. With an estimation of more than 100,000 species, microalgae are highly potential candidates to be used in many biotechnology fields. The wide variety of species offers various morphologies, taxonomic lineages, and chemical compositions (Borowitzka 2018). For decades, scientists have been researching on the potential of microalgae as sustainable biotechnological means for numerous industrial and environmental applications (Moshood, Nawanir, and Mahmud 2021).

Just like other photosynthetic plants, microalgae survive by producing their own nutrients via photosynthesis process. Via this process, carbon dioxide is absorbed, while carbohydrates, water, and oxygen are being produced. Even though these microorganisms are viewed as being economically valuable, they have a bigger role in maintaining the overall global ecosystem (Chisti 2018a). The potential of microalgae can be divided into two general stages, which becomes apparent during cultivation stage and the during subsequent applications of the harvested biomass. The environmental biotechnology fields involving microalgae are illustrated in Figure 2.1.

2.2 MICROALGAE IN CARBON DIOXIDE SEQUESTRATION

The accumulation of greenhouse gases in the atmosphere is considered to be the main culprit responsible for global warming. This is considered as the main threat to humanity in the coming years. Carbon dioxide is the primary portion of greenhouse gases, accounting for about 77% of the total greenhouse emission from both industrial and domestic activities (Ruiz-Ruiz, Estrada, and Morales 2020). Sequestration is the best option in reducing the damaging impact of carbon dioxide, as it is almost impossible to stop global urbanization. Currently, there are chemical, physical, and biological methods available for sequestrating carbon dioxide. Chemical method involves the usage of absorbing liquids such as carbonate and ammonia solutions. However, this method is known to be expensive, with low absorption capacity, excessive energy usage, and low stability (Ochedi et al. 2021). Physical technology includes the injection of liquid carbon dioxide (normally originated from industries) deep into ocean beds. This method is considered effective due to the high viscosity and density of the injected carbon dioxide, creating a negative buoyancy condition. The injection site is then sealed, trapping carbon dioxide under the seabed and securing its storage for a long period of time (Y. Teng and Zhang 2018). However, this method involves a very high cost with the danger of leakage due to high number of oil wells being drilled by oil and gas companies (House et al. 2006).

Carbon dioxide fixation using biological method has become more popular in the recent years. This is done by utilizing the photosynthesis feature in plants, where atmospheric carbon dioxide is absorbed during the process (Jeffry et al. 2021). Photosynthesis happens naturally and is considered to be safe for the environment, while producing various valuable organic products. For using in biotechnical technologies, microalgae are among the best choices for plants. With high growth yield, no arable land usage, and possibilities of using the biomass for producing other valuable products, the usage of microalgae has its advantages over terrestrial plants. Absorption of carbon dioxide from the atmosphere by microalgae and cyanobacteria is considered to be very significant. About 50 billion tons of carbon dioxide is absorbed annually, and this mostly takes place in the ocean (Chisti 2018b). This accounts for half the amount of atmospheric carbon dioxide being fixated by plants. The process is so effective that there is a mechanism being proposed to utilize microalgae in an advanced way. The concept is to use microalgae as a 'biological carbon pump', where organic carbon is being produced after the absorption of carbon dioxide. The produced organic carbon is then sank in deep ocean as an effort to sequester carbon dioxide from the atmosphere (Guidi et al. 2016).

Unicellular microalgae are more effective in converting carbon dioxide as compared to other terrestrial plants (Aslam et al. 2018). They have the ability to absorb significant amount of carbon dioxide during photosynthesis, where 1.8 kg of carbon dioxide is fixed for each kg of microalgae biomass (De Bhowmick et al. 2014). Converted carbon is stored in various forms, such as triacylglycerides, free fatty acids, sterols, and wax esters (Q. Hu et al. 2008). The application of microalgae as a carbon dioxide sequester has been studied extensively (Eloka-Eboka and Inambao 2017; Ayatollahi, Esmaeilzadeh, and Mowla 2021; Dineshbabu et al. 2020).

In addition, microalgae have been used for treating flue gas originating from coal-fired power plants (Chia et al. 2021). Flue gas normally contains high concentration of carbon dioxide, which serves as an excellent source for microalgae growth. An indigenous species of *Isochrysis* sp. microalgae has been studied in the sequestration of carbon dioxide contained in actual coal-fired flue gas. In the study, a fixation rate of 350 $mgCO_2/(L{\cdot}day)$ was obtained under optimized conditions (Yahya, Harun, and Abdullah 2020). Meanwhile, Aslam et al. (2018) investigated the growth of microalgae under several carbon dioxide concentrations in coal-fired flue gas (1%, 3.0%, and 5.5%). It was revealed that lipid and fatty acid methyl ester contents of the microalgae are higher (14.03 μg/mL/day and 280.3 μg/mL, respectively) at 1% carbon dioxide concentration, compared to the other two concentrations (Aslam et al. 2018). In another study, it was found that *Scenedesmus obliquus* and *Chlorella pyrenoidosa* microalgae grow best in the culture with 10% carbon dioxide concentration, which is within the range of concentrations in flue gases for natural gas, coal-fired, oil-fired, and combined cycles turbines (Tang et al. 2011; X. Wang and Song 2020).

Apart from that, carbon dioxide originating from the farming industry is also another growing concern among environmentalists. Carbon dioxide contained in the wastewater can be fixated using microalgae biotechnology. Various microalgae species had been used in fixing carbon dioxide in wastewater, including *Spirulina*, *Chlorella kessleri*, *Chlorella fusca*, *Chlorella pyrenoidosa*, *Scenedesmus obliquus*, *Phormidium valderianum*, *Microcystis aeruginosa*, *Synechocystis salina*, *Pseudokirchneriella subcapitata*, and *Chlorella vulgaris* (Rosa et al. 2011; Deamici, Santos, and Costa 2019; A L Gonçalves, Simões, and Pires 2014; Dineshbabu et al. 2020; Tang et al. 2011). Those studies were focused on the performance of microalgae in fixating carbon dioxide from wastewater. However, it was found that *Chlorella* sp. is excellent not only for sequestrating carbon dioxide, but also for performing outstandingly in treating wastewater and providing good-quality biomass for producing biofuel. This species is known for its high lipid content, high adaptability to various environment conditions, and high tolerance to heavy metals (Znad et al. 2018). In a study, a carbon sequestration rate of 454 mg/(L·day) was obtained when treating swine slurry using *Chlorella vulgaris* microalgae (Zheng et al. 2020). In another study, sequestration rate and carbon dioxide (gas-phase) removal of 318 mg/(L·day) and 76.92%, respectively, were reported by Ayatollahi et al. by using *Chlorella vulgaris* in treating municipal wastewater (Ayatollahi, Esmaeilzadeh, and Mowla 2021). Quite recently, it was also discovered that the application of magnetic field for 1 hour per day would increase carbon dioxide fixation of *Chlorella fusca* by as much as 50% (Deamici, Santos, and Costa 2019). This shows the vast potential of microalgae to be further explored by using the currently available technologies. Several studies involving usage of microalgae in carbon dioxide sequestration are shown in Table 2.1. The role of microalgae in treating various forms of wastewater is covered more extensively in the next section.

2.3 WASTEWATER TREATMENT USING MICROALGAE

Wastewater can be defined as water after it has been used for various applications. These applications had made the water lose part of its original characteristics, and it is approximated that more than 4,155 km^3 of wastewater is generated annually (Martínez-Roldán and Cañizares-Villanueva 2020; Schwarzenbach et al. 2006). The released untreated wastewater with excessive nutrients from agriculture, domestic, and industrial activities would result in eutrophication in aquatic ecosystem. In addition, the wide array of pollutants contained in wastewater would pose significant damage to human and other living things. It has been reported that contaminated drinking water is responsible for daily deaths of more than 2,000 children aged five and below worldwide (Al-Gheethi et al. 2018). United Nations has put an emphasis on this matter via its Sustainable Development Goal 6—Clean Water and Sanitation (SDG-6) to reduce the release of untreated water

TABLE 2.1
Several Studies on Carbon Dioxide Sequestration Using Microalgae

Microalgal strain	CO_2 concentration	CO_2 fixation	References
Chlorella fusca	10%	42.8%	(da Silva Vaz, Costa, and de Morais 2016)
Chlorella minutissima	11%	80.74 mg·L^{-1}d^{-1}	(De Bhowmick, Sarmah, and Sen 2019)
	10%	247.98 mg·L^{-1}d^{-1}	(Freitas et al. 2017)
Chlorella sp.	20%	510 mg·L^{-1}d^{-1}	(Zhao and Su 2014)
Chlorella vulgaris	6%	72%	(P. K. Kumar et al. 2018)
	5%	140.91 mg·L^{-1}d^{-1}	(Chaudhary, Dikshit, and Tong 2018)
	5%	454 mg·L^{-1}d^{-1}	(Zheng et al. 2020)
	6.5%	222 mg·L^{-1}d^{-1}	(Anjos et al. 2013)
	5%	76.92% (318 mg·m^{-2}d^{-1})	(Ayatollahi, Esmaeilzadeh, and Mowla 2021)
	10%	430 mg·L^{-1}d^{-1}	(Jain et al. 2019)
	5%	187.65 mg·L^{-1}d^{-1}	(Yadav, Dash, and Sen 2019)
Chlorococcum infusionum	5%	94.68 mg·L^{-1}d^{-1}	(Yadav, Dash, and Sen 2019)
Nannochloropsis oculata	12%	31.9 g·m^{-2}d^{-1}	(Cheng et al. 2018)
Scenedesmus obliquus	10%	87.7% (148 mg·L^{-1}d^{-1})	(Ma et al. 2019)
	5%	129.82 mg·L^{-1}d^{-1}	(Chaudhary, Dikshit, and Tong 2018)
Scenedesmus sp.	10.6%	66%	(de Godos et al. 2014)
Spirulina sp.	10%	8.9%	(da Silva Vaz, Costa, and de Morais 2016)
Consortia of *Chlorella* sp., *Pediastrum* sp., *Phormidium* sp., and *Scenedesmus* sp.	12%	48%	(Van Den Hende et al. 2011)

by 50% by year 2030 (Allen, Metternicht, and Wiedmann 2019; Sutherland et al. 2021).

In treating wastewater, various contaminants (organic matter, phosphorus, and nitrogen) and pollutants (heavy metals, hormones, and colorants) need to be eliminated before it can be released to natural bodies of water such as lake, sea, and river. Due to rapid urbanization, the composition of wastewater has changed tremendously over the decades. More and more pollutants, which used to exist only in small amounts, are present at higher concentrations today. Conventional wastewater treatment methods such as coagulation--sedimentation, phosphorus and nitrogen removal, aerobic activated sludge-based treatment, and nitrification–denitrification are known to consume a lot of energy and time while discharging excessive amount of sludge and wasting resources which can be otherwise recycled (R. Kumar and Pal 2015). Hence, a more sustainable treatment process is needed to leave a smaller carbon print while fully utilizing the available resources. In addition, pollutant removal via conventional methods could lead to the formation of other forms of secondary pollutions. Hence, bioremediation via utilization of microalgae is considered as a breakthrough in removing these pollutants (Ana L Gonçalves, Pires, and Simões 2017). The usage of microalgae in treating wastewater is a promising strategy which offers numerous benefits while solving environmental issues. The adaptive features of microalgae can be used to treat wastewater with specific type of pollutants, without involving high setup cost or energy consumption. Generally, bioremediations using microalgae are done metabolically via mechanisms such as accumulation, transformation, mineralization, and degradation (Luyun Wang et al. 2019).

Wastewater treatment can be divided into three main stages, namely primary (solid removal), secondary (bacterial decomposition), and tertiary (extra filtration) ("Wastewater and Sewage Treatment" 2010). Microalgae are mainly used to remove excess nutrients in secondary effluents, hence reducing the risk of eutrophication in natural water bodies (K. Li et al. 2019; Chai, Tan et al. 2021). However, tertiary effluents might still contain high level of phosphorus and nitrogen, where microalgae could once again be used. Implementing microalgae in treating agricultural, domestic, and industrial wastewaters has been proven to be very effective in reducing nutrient content in the effluent (Leong, Huang, and Chang 2021). In fact, nitrogen, phosphorus, and other substances contained in wastewater serve as nutrition aiding the growth of microalgae. In other words, the polluted effluent actually serves as perfect breeding place for the microalgae. As a result, eutrophication can be avoided, while reducing the content of pathogens in the wastewater. The implementation of microalgae in wastewater treatment also reduces both biochemical oxygen demand (BOD) and chemical oxygen demand (COD) of the effluent (Rani et al. 2020). This is achieved when oxygen is produced by microalgae via photosynthesis process. The mechanism of wastewater treatment by microalgae is actually based on a symbiotic relationship between microalgae and bacteria. Carbon dioxide is absorbed by microalgae during photosynthesis, while generating oxygen from the process. The generated oxygen is then used by bacteria to degrade carbon, phosphorus, and nitrogen contained in the effluent. Apart from that, phosphorus and nitrogen also serve as nutrients aiding the growth of microalgae (K. Li et al. 2019). This process continues in a cycle, until nutrients in the wastewater are reduced to a very low level, meeting the stringent discharge requirement (Whitton et al. 2015).

Wastewater can be generally categorized into municipal, agricultural, and industrial wastewater. Municipal wastewater usually contains lower nutrient, toxin, and pollutant levels compared to other classes of wastewater. The manner of the usage of microalgae in treating municipal wastewater can be divided into three general categories, namely raw sewage (main pond), secondary effluent, and centrate (after the sludge has been dewatered). In their study, Wang et al.

(2010) discovered that *Chlorella* sp. microalgae grew best when centrate is used as the cultivation medium. This is due to higher nutrient content in the centrate, compared to raw sewage and secondary effluent (Liang Wang et al. 2009). In a separate study, it is found that COD removal and microalgae growth are relatively low for municipal wastewater due to low content of carbon in it (Otondo et al. 2018). However, this can be remedied with the addition of carbon dioxide in the concentration of 15% in order to encourage microalgae growth and the removal of nutrient from the medium (Ji et al. 2013). Furthermore, carbon dioxide aeration would also increase lipid content of the cultivated microalgae. In a study to evaluate the effectiveness of four different microalgae strains (*Nannochloropsis gaditana*, *Chlorella sorokiniana*, *Chlorella* sp. (*Pozzillo*) and *Dunaliella tertiolecta*), it was discovered that microalgae from *Chlorella* genus are the best for treating municipal wastewater (Lima et al. 2020).

On the other hand, wastewater originating from agriculture activities normally contains high level of nutrients, which could cause disastrous pollution of groundwater if not treated in a timely manner. The abundance of phosphorus, magnesium, nitrogen, potassium, and other active substances in the slurry provides an excellent source of nutrients for microalgae growth (Moheimani et al. 2018). However, the usage of microalgae directly in treating undiluted agriculture wastewater is often inhibited by the high amount of chroma, ammonia concentration, and suspended organic matter. This would lead to the risk of contamination and low efficiency in removing the pollutant (Franchino et al. 2016; K. Li et al. 2019). Nevertheless, *Chlorella vulgaris* was successfully used in treating undiluted swine slurry (Zheng et al. 2020). The results show tremendous improvement in the quality of the treated wastewater, where nitrogen removal rate of 74%, ammonia nitrogen removal of 78%, and phosphorus removal rate of 87% were reported. In a breakthrough study, real undiluted swine slurry was successfully treated by using *Chlamydomonas* sp. strain with high growth rate and showing high tolerance to swine wastewater, achieving removal efficiency of almost 100% phosphate, 96% nitrogen, and 81% oxygen (Qu et al. 2020). Moreover, wastewater from aquaculture was successfully treated using *Chlorella minutissima* using Recirculation Aquaculture System (RAS), resulting in total decrease of 88% and 99% for nitrogen and phosphorus, respectively (Hawrot-Paw et al. 2020).

Microalgae has also been used to remediate wastewater in form of palm oil mill effluent (POME) (Hariz et al. 2018; Ahmad, Buang, and Bhat 2016). This effluent contains a high level of suspended organic matter that turns out to be one of the major hindrances with this application. This leads to the lack of light transmission through the effluent, which inhibits the growth of microalgae (Jasni et al. 2020). In addition, even though the effluent is rich in nitrogen and phosphorus, most of the nutrients appear in particulate phase, instead of in liquid state. This makes it difficult for the remediation process by microalgae. Hence, the selection of microalgae that are able to thrive under these conditions is particularly important. In an investigation involving six microalgae strains, it was discovered that *Chlorella sorokiniana* is the most suitable to be cultivated using POME, where biomass of more than 100 mg/(L·day) can be accumulated (Khalid et al. 2019). In a separate comparative study of three microalgae species, it was also discovered that microalgae from *Chlorella* sp. are the best in bioremediating carbon dioxide and removing nutrients from POME (Ding et al. 2020; Low et al. 2021).

Apart from that, microalgae are also used in treating wastewater from industries, such as brewery (Amenorfenyo et al. 2019), textile (Fazal et al. 2021; Oyebamiji et al. 2019), leather tannery (Pena et al. 2020), pharmaceutical (J. K. Nayak and Ghosh 2019), and food (S. Gupta and Pawar 2018; X. Hu, Meneses, and Aly Hassan 2020). However, there is no universal practice for this due to the wide variety in characteristics of wastewater from different industries. Treatment system is often designed depending on the nature of the industry. Furthermore, microalgae are often used to remediate wastewater that is polluted with heavy metals, such as cadmium, mercury, arsenic, and lead (S. Tripathi and Poluri 2021). Addressed as a global threat to the ecosystem, the removal of heavy metals is essential due to their high level of toxicity to both human and other living things. Even though the heavy metals are nonbiodegradable, their removal from wastewater is done via the interactions between microalgae's cell wall and the metal ions. This adsorption process is so effective that it can happen in a matter of minutes (Martínez-Roldán and Cañizares-Villanueva 2020; Chai, Cheun et al. 2021). Microalgae also serve as an economical means for heavy metal bioremediation compared to the conventional wastewater treatment. In particular, brown algae have been identified as a promising agent for this purpose due to the presence of alginate and fucoidan in them (Rizwan et al. 2018). In a separate study, it was found that *Euglena gracilis* microalgae are able to remove iron, chromium, copper, and cadmium from water polluted with heavy metals, while thriving under the harsh toxic environment (Jasso-Chávez et al. 2021). It was also reported that *Dunaliella salina* microalgae are a strong candidate in bioremediating saline water polluted with inorganic arsenics such as arsenite, As(III), and arsenate, As(V). This is due to the unique characteristics of *Dunaliella salina* with a strong tolerance of salt and this particular metalloid. These inorganic arsenics would be converted to less toxic compounds such as arsenosugars, monomethylarsonate (MMA), and arsenolipids (S. Tripathi and Poluri 2021). It was also observed that the absorption effectiveness is further increased with the deprivation of phosphate in the culture (Y. Wang et al. 2017). Meanwhile, *Dunaliella bardawil* was found to be an effective species in treating wastewater polluted with high levels of aluminum (Akbarzadeh and Shariati 2014). Microalgae are also effective in removing unwanted drug residues and hormones contained in wastewater. A surfactant known as alkylphenol ethoxylates is widely used in numerous household, farming, and industrial applications, typically as a high-performance detergent. Nonylphenol obtained from the degradation of this surfactant is classified as a type of environmental hormone which is carcinogenic to humans. In a study conducted by Luyun Wang et al. (2019), it was discovered that four marine microalgae (*Phaeocystis globosa*, *Nannochloropsis oculata*, *Dunaliella salina*, and *Platymonas subcordiformis*) were able to biodegrade between 43.43% and 90.94% of the hormone after 120 hours of exposure. In a separate study, it was reported that estradiol (estrogen steroid hormone), diclofenac (painkiller), and triclosan (antibacterial) were effectively bioremediated by 91.73%, 74.68%, and 78.47%, respectively, from wastewater using a microalgae consortium (Bano, Malik, and Ahammad 2021).

One of the most efficient ways in treating wastewater from various origins (municipal, agricultural, and industry) is the implementation of high rate algal pond (HRAP). This system consists of shallow raceway ponds (often less than 50-cm depth) equipped with paddle wheels, designed to provide a better sunlight exposure, mixing, and aeration in the pond (Martínez-Roldán and Cañizares-Villanueva 2020). Consequently, shallow pond depth and addition of mixing facilitate the breakdown of organic waste and nutrient intake by the microalgae (J. Li et al. 2021). With low power consumption (less than 1 W/m^2), the implementation of HRAP can be viewed as a combination of secondary and partial tertiary treatments, where wastewater is treated at a higher efficiency compared to the traditional facultative ponds (Acién et al. 2016; Sutherland et al. 2021).

The capability of microalgae in treating a wide array of agricultural wastewater shows the versatility and vast potential of these microorganisms. However, the selection of microalgae species is important, since the ability to thrive in specific kind of wastewater depends on the microalgae species (Hwang et al. 2016). Some are highly adaptable in a particular wastewater sample, while others may struggle to survive in the harsh conditions. Due to the unique advantages of different microalgae in absorbing pollutants, the usage of microalgae consortia is probably the option in treating wastewater from various origins with a combination of various pollutants. Microalgae have reached their milestones in treating wastewater due to multiple additional advantages such as simple cultivation techniques, robustness, and rapid growth. Furthermore, the usage of microalgae in treating wastewater is considered as 'killing two birds with one stone', where wastewater can be treated, and microalgae biomass can be later harvested to produce high-value products. Recent investigations on the utilization of microalgae in treating wastewater are presented in Table 2.2.

2.4 BIOFUELS FROM MICROALGAE

Energy stored in microalgae is known as chemical energy, which is converted from the solar energy captured during photosynthesis. When cultivated under specific conditions, certain chemical composition of microalgae can be manipulated according to the objective of the cultivation (Low et al. 2021). Cultivated microalgae need to be harvested in bulk, before the biomass can be used for the next process. Several harvesting methods are found to be suitable for this purpose, such as flotation, centrifugation, filtration, and flocculation (G. Singh and Patidar 2018). Even though there is no single universal method to harvest microalgae, flocculation is known to be simple, fast, and cost effective. Flocculation method comprises physical, chemical, and biological approaches. Biological based flocculation is found to be among the most desirable due to its low cost and environment-friendly nature, while yielding more than 90% of the biomass

TABLE 2.2
Recent Studies on Wastewater Treatment Using Microalgae

Wastewater type	Microalgal strain	Removal efficiency	Microalgal productivity	Reference
Agricultural				
Palm oil mill effluent	*Chlorella sorokiniana*	TN: 87.1%, TP: 78.2%, NH_4^+: 88.3%, PO_4^{3-}: 78.5%	107.5 mg/L·d	(Khalid et al. 2019)
Real swine slurry	*Chlamydomonas* sp.	COD: 81%, TN: 96%, TP: 100%	1,445 mg/L·d	(Qu et al. 2020)
Undiluted swine slurry	*Chlorella vulgaris*	TN: 74%, TP: 87%, ammonia nitrogen: 78%	152.5 mg/L·d	(Zheng et al. 2020)
Industrial				
Pharmaceutical wastewater	Consortia of *Chlorella* sp., *Merismopedia* sp., *Closteriopsis* sp., *Scenedesmus* sp.	.Estradiol: 91.73%, diclofenac: 74.68%, triclosan: 78.47%	Not available	(Bano, Malik, and Ahammad 2021)
Rare earth element tailings	*Chlorococcum robustum*	Total inorganic nitrogen: 4.45 mg/L·h^{-1}, ammonium-nitrogen: 3.5 mg/ L·h^{-1}	533 mg/L·d	(Geng et al. 2022)
Textile and food processing wastewater	*Chlorella vulgaris* *Chlorococcum infusionum*	COD: 91.9%, TOC: 84.3 mg/L %, PO_4: 98.8%, NH_4: 95.9%, NO_3: 100%. COD: 85.2%, TOC: 72.9 mg/L %, PO_4: 85.8%, NH_4: 75.5%, NO_3: 100%	208.93 mg/L·d 105.4 mg/L·d	(Yadav, Dash, and Sen 2019)
Municipal				
Domestic wastewater	*Chlorella vulgaris*	COD: 93%, NH_4^+: 98%, PO_4^{3-}: 89%	87.4 mg/L·d	(Amini et al. 2020)
Landfill leachate	*Chlorella vulgaris*	Ca: 110 mg/L, Mg: 47 mg/L, K: 447 mg/L, Na: 1197 mg/L	87.06 mg/L·d	(Zhang et al. 2021)
Tertiary treatment plant	*Neochloris oleoabundans*	TN: 100%, TP: 32%	92 mg/L·d	(Razzak 2017)

TOC: total organic carbon, COD: chemical oxygen demand, TN: total nitrogen, TP: total phosphorus.

(Muhammad et al. 2021). Harvested microalgae biomass can be used as a feedstock in obtaining various biofuels as illustrated in Figure 2.2.

Liquid biofuels, such as bioethanol and biodiesel, have garnered a lot of interest in the recent years. These biofuels are considered to be as among the solutions in reducing the dependency on fossil fuels, while abating carbon dioxide emissions

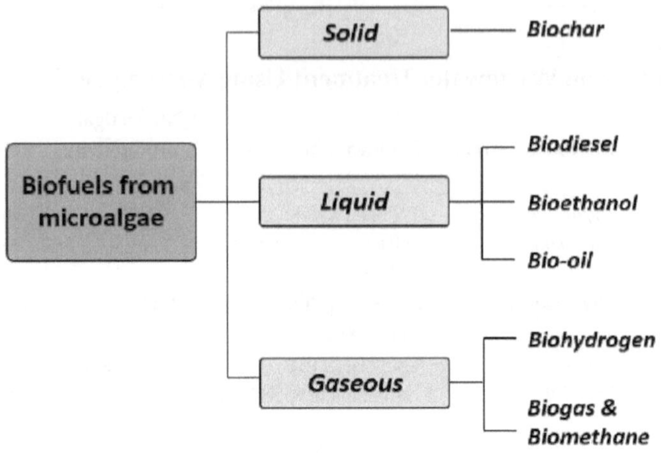

FIGURE 2.2 Biofuels derived from microalgal biomass.

to the atmosphere (Ideris et al. 2021). In addition, biodiesel and bioethanol blends can be used in current diesel and petrol engines, respectively, without any or little modifications of the engines. More and more automotive makers, such as BMW, John Deere, Volvo, Mercedes, and Volkswagen, have approved the usage of these biofuel blends in their engines (Atabani, Silitonga, and Ong 2013; Ashraful et al. 2014; M. Suresh, Jawahar, and Richard 2018). In the production of liquid biofuels, microalgae are classified as third- and fourth-generation feedstock (Kostas et al. 2021). First-generation feedstock consists of edible crops, while second-generation feedstock is based on nonedible plants, waste cooking oil, and agricultural/animal waste. Unlike first-generation feedstock, microalgae do not pose any threats to the global food supply. Furthermore, among microalgae's advantages over second-generation feedstock their high growth rate and ability to thrive in water (both freshwater and saline) are included. Experts also believed that microalgae could represent as much as 5% of the total global fuel consumption by year 2030 (Ribeiro et al. 2015).

Microalgae cell contains lipid, which is the main ingredient for biodiesel. The robust cell wall of microalgae needs to be broken down in order to extract the accumulated lipid. The extraction of lipid from microalgae can be performed using mechanical, biological, or chemical methods. The widely used mechanical methods involve the usage of ultrasound, microwave, bead milling, and hydrothermal liquefaction. On the other hand, biological methods involve the usage of enzymes, while chemical methods employ solvents such as hexane, petroleum ether, chloroform, and dimethyl carbonate (Mubarak, Shaija, and Suchithra 2015; Ong et al. 2019). The extracted lipid can then be transformed to biodiesel via processes such as transesterification and hydrothermal liquefaction (Fazril et al. 2020). Lipid content is among the most critical factors in determining the suitability of a feedstock sample in producing biofuels. Comparatively, lipid obtained

TABLE 2.3
Potential of Various Feedstocks for Biodiesel Production

Feedstock	Oil content (% of dry weight)	Oil yield (L oil /ha/year)	Land area requirement (m^2/year/kg biodiesel)	Biodiesel productivity (kg biodiesel/ ha/year)
Castor (Ricinus communis)	48	1,307	9	1,156
Corn/maize (Zea mays L.)	44	172	66	152
Jatropha (Jatropha curcas L.)	28	741	15	656
Microalgae (high oil content)	70	136,900	0.1	121,104
Microalgae (medium oil content)	50	97,800	0.1	86,515
Microalgae (low oil content)	30	58,700	0.2	51,927
Palm oil (Elaeis guineensis)	36	6,366	2	4,747
Peanut oil (Arachis hypogaea L.)	50	1,059	–	1,425–1,782
Rapeseed (Brassica napus L.)	41	974	12	862
Soybean (Glycine max L.)	18	636	18	562
Sunflower (Helianthus annuus L.)	40	1,070	11	946

from 1 hectare microalgae farm can be anywhere between 10 and 100 times more in quantity compared to other terrestrial oil crops. As shown in Table 2.3, microalgae have a huge potential to be used as feedstock for biodiesel production (Mata, Martins, and Caetano 2010; Ribeiro et al. 2015).

In addition, the biomass chemical composition of microalgae can be manipulated by placing them under stress conditions. For example, the amount of lipid increased by two folds when *Desmodesmus* sp. was cultivated under conditions of nitrogen deficiency (Nagappan and Kumar 2021). Other works also found that nitrogen suppression during the cultivation plays a crucial role in increasing the lipid content. As much as 70% increase of lipid of the biomass dried weight has been reported by some researchers (Dasgupta et al. 2015; Wan, Bai, and Zhao 2013). There are a few microalgae species that have been recommended to be used as biodiesel feedstock due to their high lipid content. Islam et al. (2013) conducted a study on fatty acid compositions of 12 microalgae species. From the study, it was discovered that among the best microalgae for biodiesel feedstock are *Nannochloropsis oculata*, *Extubocellulus* sp., *Biddulphia* sp., and *Chlorella vulgaris* (Islam et al. 2013). In a separate investigation involving *Spirulina maxima*, *Nannochloropsis* sp., *Neochloris oleoabundans*, *Chlorella vulgaris*, *Dunaliella tertiolecta*, and *Scenedesmus obliquus*, it was concluded that both *Nannochloropsis* sp. and *Neochloris oleoabundans* outperformed others in producing higher biodiesel yield with superior quality. In the study, both microalgae species managed to produce in excess of 50% lipid under nitrogen starvation (Gouveia and Cristina 2009).

Carbohydrate in the form of starch and cellulose is accumulated in the microalgae cell, and this compound is very useful for the production of bioethanol which can be obtained via fermentation process (Chisti 2020). The structure of starch obtained from microalgae consists of polymerized glucose that is chained by alpha-1,4 glycosidic bonds and branched at the alpha-1,6 positions, which is similar to other terrestrial plants commonly used as feedstock in bioethanol production (Ran et al. 2019). Harvested microalgae biomass needs to undergo hydrolysis in order to break the carbohydrate compound into monosaccharide, before the fermentation process. During the fermentation process, monosaccharide is converted into alcohol, commonly with the usage of yeast (Costa et al. 2017). Compared to other lignocellulosic materials, it is easier to hydrolyze carbohydrate contained in microalgae due to the absence of lignin. Hence, this is an advantage for microalgae biorefineries, as there is no need for additional pretreatment process which is normally needed for lignocellulosic feedstock (Suarez Ruiz et al. 2020). Microalgae from *Chlorella*, *Spirulina*, *Cholorococcum*, and *Chlamydomonas* genus have been identified to have excellent bioethanol yields (de Morais et al. 2019; Braga et al. 2018; Vieira Salla et al. 2016). In a study conducted by Braga et al. (2018), it was reported that nitrogen and phosphorus deficiency resulted in higher level of accumulated carbohydrate (59.1%, w/w) in *Spirulina* biomass. Another significant finding from the study is that protein accumulation is inversely proportional to the carbohydrate content in the biomass (Braga et al. 2018). In another study, it was discovered that carbohydrate content in *Chlorella zofingiensis* microalgae increases to 66.9% in just 5 days of cultivation under nitrogen deprivation (S. Zhu et al. 2014). A work carried out by Olia et al. (2019) showed that carbohydrate accumulation was maximized in a low-phosphate (0.031 g/L) medium. It was also found that the addition of organic carbon source from whey protein residues resulted in a higher carbohydrate content in *Spirulina platensis* microalgae (Vieira Salla et al. 2016).

Bio-oil can be obtained both from fresh microalgae biomass and its residue after compounds such as lipid and carbohydrate have been extracted or after the saccharification of starch (Francavilla et al. 2015). Dark in color with high viscosity and boiling point, bio-oil is considered as a promising alternative to petroleum-based liquid fuels. Heating value of microalgal bio-oil obtained is normally higher than of those obtained from lignocellulosic feedstock (Yang et al. 2019). It has also been reported that the yield of bio-oil from microalgae reached 75 wt.%, with high heating value as compared to fossil fuels (approximately 42 MJ/kg) (Azizi, Keshavarz Moraveji, and Abedini Najafabadi 2018). Bio-oil is often referred to as bio-crude, and it needs to be further upgraded before it can be properly used as a liquid fuel. Among the common processes to upgrade bio-oil are hydrogenation, cracking, deoxygenation, adsorption, distilling, and denitrogenation (López Barreiro et al. 2016). Unlike petroleum-based fuel, microalgae bio-oil is less hazardous to the environment as the amount of carbon dioxide generated during its combustion is the same as the absorbed amount during photosynthesis (Y. Guo et al. 2015). Examples of microalgae species which have been studied for bio-oil productions are *Galdieria sulphuraria* (Cui et al. 2020), *Pavlova* sp. (Aysu, Fermoso, and Sanna 2018), *Spirulina platensis* (Jafarian and Tavasoli 2018), *Scenedesmus almeriensis* (López Barreiro et al. 2016), *Nannochloropsis*, *Tetraselmis*, *Isochrysis galbana* (Azizi et al. 2020), and *Chlorella vulgaris* (Belotti et al. 2014).

Bio-oil can also be obtained via either hydrothermal liquefaction (HTL) or pyrolysis of microalgae biomass (Kan et al. 2014). Wet microalgae biomass can be utilized through HTL, while only dried biomass can be used via pyrolysis pathway. When microalgae with high moisture are used to produce bio-oil, there will be significant savings in cost as dewatering and drying steps can be circumvented. Hence, this is an advantage of hydrothermal liquefaction (HTL) over pyrolysis in obtaining bio-oil from microalgae biomass (Cano-Pleite et al. 2020). HTL of microalgae biomass constitutes physical and chemical conversion using water at temperatures higher than 250°C and pressures between 5 and 40 MPa, with or without catalysts (Y. Guo et al. 2015; de Morais et al. 2019). Properties of water change under these conditions, leading to the degradation of microalgae's macromolecules and polymerization of the small fragments which form products including bio-oil (Peterson et al. 2008). On the other hand, pyrolysis is an endothermic process that involves heating microalgae biomass under atmospheric pressure at 350–700°C in the absence of oxygen (Brindhadevi et al. 2021). Both processes convert biomass into three types of products, which are bio-oil, biogas/syngas, and biochar. Biochar is widely applied in numerous environmental and industrial applications. It is a high energy density solid biofuel suitable to be used as co-feedstock in conventional coal boilers without affecting the thermal efficiency (X. J. Lee et al. 2020; Giostri, Binotti, and Macchi 2016). Some species with high energy content such as *Enteromorpha* and *Chlorella* have been found to be most suitable for this purpose (Zhao et al. 2016). In addition, biochar also functions as a biofertilizer and carbon sequestration agent and is an excellent source for energy recovery (Bass et al. 2016). Microalgae biomass is also capable in producing more biochar compared to other feedstock. In a comparative study involving three microalgae species (*Chlorella vulgaris*, *Spirulina* sp., and *Nannochloropsis* sp.) and a lignocellulosic feedstock (nutshell), it is found that the amount of biochar obtained from all three microalgae species exceeded that obtained from nutshell (Binda et al. 2020). Biochar will be covered more extensively in the following subchapter 2.5.

Apart from that, biogas in form of methane (CH_4), water vapor (H_2O), hydrogen (H_2), nitrogen (N_2), and carbon dioxide (CO_2) blend can be obtained from microalgae biomass through anaerobic digestion (M. F. M. A. Zamri et al. 2021; Yukesh Kannah et al. 2021). Anaerobic digestion occurs in a confined space where oxygen is unavailable. This combustible gas is renewable and is a suitable substitute for fossil fuel. Not only it reduces the emission of harmful greenhouse gases, it also is cost effective as is evident from the reduction of electricity bills. In addition, carbon dioxide and other traces of gas can be extracted from biogas, leaving biomethane. For every 1 kg of microalgae biomass, there is a potential of

200 L biomethane to be obtained (Ahrens and Weiland 2007). This form of gas is a clean substitute for natural gas, which can be utilized as vehicle fuel and can be supplied into the gas grid or can provide electricity to remote areas and industries (Z. Teng et al. 2014). Biohydrogen is another beneficial gas which can be obtained through microbial photosynthesis and bioconversion of microalgae. Biohydrogen is sustainable, which serves as an alternative to fossil fuel. When burned, hydrogen leaves only water. This led to the belief that hydrogen is the chosen fuel for the future (Hosseinpour et al. 2017). Biohydrogen can also be directly supplied to a fuel cell for providing electrical energy (Elshobary et al. 2021). Biohydrogen production from microalgae can be further improved via genetic modification (Majidian et al. 2018). In a study, *Chlamydomonas reinhardtii* was genetically modified using atmospheric and room temperature plasma (ARTP) in the effort to increase the biohydrogen production. As a result, the modified species was able to produce up to 5.2 times more biohydrogen compared to its unmutated counterpart (Ban et al. 2019).

By manipulating the cultivation environments to have higher lipid, starch, or protein concentration in microalgae, the final desired product can be maximized accordingly (Aslam et al. 2018). However, cultivation conditions need to be set properly by considering the overall benefits of a project in order to optimize the economic benefits (de Farias Silva and Bertucco 2016). Microalgae can be conveniently integrated in a sustainable development process of circular economic models in treating wastewater and mitigating carbon dioxide, while the harvested biomass can be used as source of various biofuels (Serrà et al. 2020).

2.5 MICROALGAE (AND CYANOBACTERIA) IN AGRICULTURE

Utilization of synthetic fertilizers is a common way to increase agricultural yield. These fertilizers supply elements such as nitrogen and phosphorus to the crops to facilitate their growth and yield. The demand for these fertilizers continue to raise globally, even with the increase in price (Dineshkumar et al. 2018; Chisti 2020). However, excessive usage of synthetic fertilizers causes numerous adverse impacts to the environment, such as soil erosion, accumulation of heavy metal, pollution of groundwater sources, and eutrophication (Rahman and Zhang 2018; Kulkarni and Goswami 2019). In this context, biofertilizers are becoming more and more important in agriculture (Kholssi et al. 2021).

Phosphorus is a nonrenewable source which is vital to modern agriculture (Elser 2012). Phosphorus is mined from natural phosphate rocks, available only in few countries in the world. However, the insatiable demand for this element has led to an excessive mining of phosphate rocks. It is expected that the world will run out of this source within this century (Cordell and White 2014; Withers et al. 2020). Furthermore, the efficiency of phosphorus usage is only 20%, and the remaining 80% end up in wastewater originating from farm irrigation (Solovchenko et al. 2016). The process of making fertilizers from nitrogen and phosphorus is also energy-extensive. Nitrogen gas is converted into ammonia and nitrate through Haber process (Chai, Chew et al. 2021). During this reaction, a significant amount of energy (10–15 kWh) is needed for per kg of reacted nitrogen. Meanwhile, about 5–10 kWh of energy is required to produce 1 kg of phosphorus fertilizer from phosphate rocks (Yap et al. 2021). Hence, the usage of microalgae is an excellent way not only to accumulate these elements from wastewater, but also to provide these elements to the crops in a more sustainable way. This is further supported by many biofertilizers explored, where it is found that photosynthetic microorganisms (cyanobacteria and microalgae) have the most potential to remediate soil conditions and improve crops yields (S. Guo et al. 2020). In a study conducted by Chu et al. (2021), it was reported that *Chlorella vulgaris* and *Microcystis* sp. managed to remove 78.7% and 88.4% of phosphorus, respectively, from poultry farm wastewater. Harvested algal biomass was then used as a biofertilizer in the cultivation of wheat, resulting in 21.6% more grain yield compared to synthetic chemical fertilizer (Chu et al. 2021). This shows the excellent ability of microalgae in recovering phosphorus from wastewater and its function as biofertilizer. In another study, it was revealed that the usage of biofertilizer from algal biomass managed to improve rice yield as much as 22.99%, compared to the control study. Moreover, it was also discovered that the usage of this biofertilizer would significantly increase the nutritious free amino acid contents of rice by 46.26% (Yu et al. 2019).

The application of microalgae and cyanobacteria in agriculture can be done in various ways. One of the practical applications is by using biochar obtained from microalgae. Biochar that is formed via pyrolysis has a potential to be used for various environmental applications. Due to the high level of heteroatoms (such as P and N) contained in algal biochar, it is suitable to be used for soil treatment. In addition, the concentration of these intrinsic heteroatoms is in tenfold values compared to lignocellulosic biomass (Binda et al. 2020). This bioremediation of soil is done by increasing the nutrient (such as nitrogen and phosphorus), surface area, organic matter content, and electrical conductivity within the soil (Bornø et al. 2018). Biochar also functions by improving the overall nutrient absorption and water retention of the soil, thanks to its microporous structure and intrinsic mineral content (Chu et al. 2020). Consequently, this contributes to the increase in crop growth and yield. Apart from that, biochar also plays an important role in bioremediating arid and semiarid land. In a study, biochar was used together with cyanobacteria in creating soil biocrust to stabilize arid soil surface, hence providing fertile conditions for growing crops (A. Kumar and Singh 2020).

As microalgae are excellent at capturing carbon dioxide from the atmosphere, the direct application of microalgae biomass in agriculture would help crops to have a better growth and productivity (Salih and Salih 2011). The use of microalgae biomass as a soil additive increases the level of organic carbon content in soil. This is important as the depletion of organic carbon leads to degradation of the soil's fertility, which is common for agricultural lands (Stavi and Lal 2015). Apart from that, the addition of microalgae biomass to soil would improve water retention and structure of the soil. Acting as soil conditioners, the spread of land desertification may be reversed with direct application of microalgae biomass (Rossi et al. 2017). Dry *Acutodesmus dimorphus* biomass was used as a biofertilizer for tomato plants, and it had shown encouraging effects on seed germination, plant growth, and fruit yield (Garcia-Gonzalez and Sommerfeld 2016). In another study, the effects of application of liquid fertilizer prepared from *Chorococcum* sp. biomass to four different crops (*Capsicum annuum*, *Solanum lycopersicum*, *Vigna radiata*, and *Cucumis sativus*) were studied. It was discovered that the application of low-concentration liquid fertilizer (20% concentration) is enough to provide significant results on the growth, root and shoot lengths, number of leaves, and number of lateral roots of all four plants. This proves that low-cost liquid biofertilizer could be a potential substitute to the expensive synthetic fertilizer (Deepika and MubarakAli 2020). Moreover, dried biomasses from six microalgae/cyanobacteria species were used as slow-release biofertilizers in the cultivation of rice seedlings, resulting in an improved leaf width and shoot growth (Mukherjee et al. 2016). Cyanobacteria-based biofertilizers are known as diazotrophes (capable of fixing nitrogen), while enhancing the aeration within soil and supplying vitamin B12 to the crops. Nitrogen is fixed in their cell which is known as heterocyst, before providing the important element to crops (Berman-Frank, Lundgren, and Falkowski 2003). Among the most common diazotrophes are *Anabaena variabilis*, *Calothrix* sp., *Tolypothrix* sp., *Aulosira fertilisima*, and *Nostoc linckia*, which are normally found in rice fields. It is reported that as much as 60 kg/ha/season of nitrogen can be fixed by *Anabaena* (Chittora et al. 2020). Furthermore, these biofertilizers were also used in cultivations of various crops such as oat, chili, corn, lettuce, tomato, barley, oats, cotton, and sugarcane (Thajuddin and Subramanian 2004). Recent studies involving the usage of these microorganisms in agriculture are presented in Table 2.4.

It is known that both abiotic and biotic stresses are the common adversities that limit crop yield globally. Abiotic stresses are associated with the environment conditions, such as extreme heat, cold, soil salinity, and heavy metal toxicity (Zaidi et al. 2014). On the other hand, biotic stresses are caused by other living organisms such as parasitic nematodes, insects, weeds, and diseases (Saijo and Loo 2020). Normally, biotic stresses are dealt with chemical pesticides, which result in severe environmental degradation. In this case, microalgae can also be used as biocontrol agents in limiting the risks imposed by both abiotic and biotic stresses. The abilities of microalgae and cyanobacteria in increasing germination, removing heavy metals, and increasing carbon, nitrogen and phosphorus contents contribute in assisting crops to adapt in harsh environmental conditions. Meanwhile, cyanobacteria and microalgae are excellent at producing various enzymes and activating defense mechanisms in plants against attacks from biotic stresses. Allelopathic chemicals produced by these microorganisms are used as herbicides, fungicides, nematicides, and insecticides in fighting the attacks. Cyanobacteria can directly act against pests as biopesticides. Their mechanism to function as biopesticides is by creating a competition for nutrients and space with the pathogens (V. V. Kumar 2018). In a study, *Anabaena* sp. was successfully used as a fungicide in deterring the attack of *Podosphaera xanthii* fungus on zucchini plant. Direct antifungal activity from the cyanobacteria extract was evident by the reduction of the sporulation of the fungus (Roberti et al. 2015). Meanwhile, it was

TABLE 2.4
Example of Studies on the Usage of Microalgae and Cyanobacteria in Agriculture

Species	Crop	Method of application	Results	References
Anabaena torulosa, Anabaena doliolum, Anabaena laxa	Chrysanthemum	Inoculated to stem cuttings	Increased chlorophyll, root and short tissues	(Bharti et al. 2019)
Anabaena variabilis, Nostoc calcicola	Rice, corn, sorghum, black-eyed pea, Kodo millet	Immersion treatment of seeds	Increased plant height, root weight, leaves quantity	(A. Suresh et al. 2019)
***Chlorella* sp., *Scenedesmus* sp.**	Tomato	Dried deoiled biomass as biofertilizer in soil	Increased shoot length, root length, weight, yield	(Silambarasan et al. 2021)
Chlorella vulgaris	Ladies' fingers	Inoculated to seeds and dirt	Increased yield, germination, earlier maturity	(Agwa, Ogugbue, and Williams 2017)
Chlorella vulgaris, Spirulina platensis	Corn	Dried biomass as biofertilizer in soil	Increased early growth, yield, germination	(Dineshkumar et al. 2019)
***Chorococcum* sp.**	Tomato, mung bean, cucumber, bell pepper	Immersion treatment of seeds	-increased shoot length, root weight, leaves quantity by as much as 20%	(Deepika and MubarakAli 2020)
Navicula* sp., *Chlorella vulgaris	Corn	Biomass as organomineral biofertilizer	Increased nitrogen, boron, sodium content in plant	(Pereira et al. 2021)
***Nostoc* sp.**	Wheat	Dried biomass as biofertilizer in soil	Increased number of branches, more compacted root in plant	(Do Nascimento et al. 2019)
***Scenedesmus* sp.**	Rice	Dried deoiled biomass as biofertilizer in soil	Increased plant height, stem numbers, weight, yield	(M. Nayak, Swain, and Sen 2019)
Scenedesmus subspicatus	Onion	Immersion treatment of seedlings	Increased early growth, yield, germination, bulb size, sugar and protein contents	(Gemin et al. 2019)

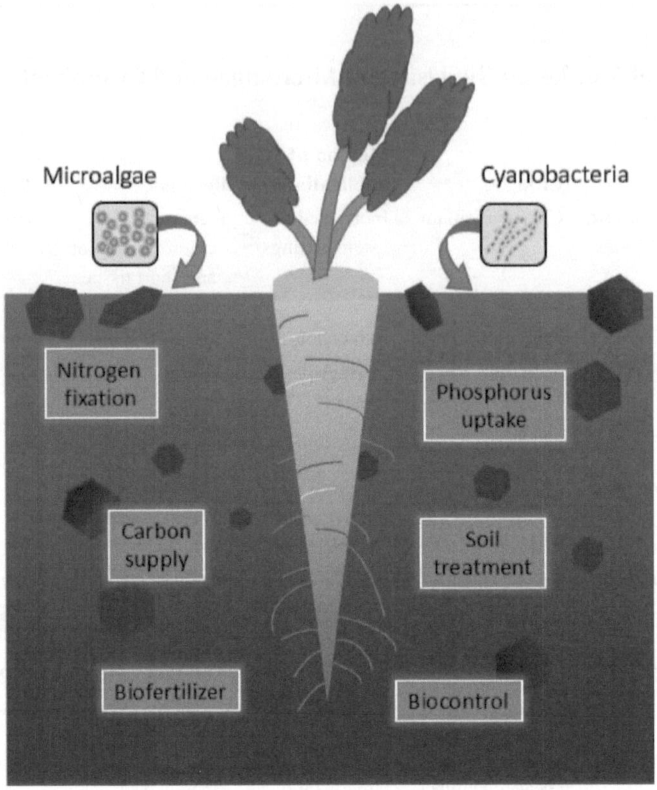

FIGURE 2.3 Cyanobacteria and microalgae role in soil bioremediation.

reported that *Calothrix elenkinii* cyanobacteria managed to show 61–66% disease control efficacy against *Fusarium solani* attacks on tomato plants (Mahawar, Prasanna, and Gogoi 2019). Figure 2.3 illustrates the various roles of microalgae and cyanobacteria in soil remediation. A comprehensive review on the effectiveness of these microorganism as biocontrol agents is available in the literature (Poveda 2021; Alvarez et al. 2021; Garlapati et al. 2019).

2.6 MICROALGAL BIOPROCESSING UP-SCALING CHALLENGES

Even though microalgae offer numerous advantages, their exploitation in bioprocessing is still to be opened for further research. Upstream processes, such as cultivation and harvesting the biomass, are among initial challenges that need to be addressed. There are a wide variety of factors, such as conditions of the surroundings, cultivation modes, possibility of microbial contamination, pond/photobioreactor designs, and growth monitoring, which need to be evaluated prior to setting up a particular cultivation system (Daneshvar et al. 2021). All of these factors are unique, depending on the utilizations and objectives of the set-up. Apart from that, limited knowledge on genetic engineering which includes molecular mechanism of microalgae is another biotechnical hurdle which hinders the scaling up processes (Xue et al. 2021). Additional information on microalgae species is sometimes needed in order to address the need of a cultivation system. For example, microalgae which thrive in open pond systems cannot be assumed to perform the same when cultivated in photobioreactors (Ali et al. 2021). This is supported by the fact that only a handful microalgae species, namely from the genus of *Chlorella*, *Spirulina*, and *Dunaliella*, have been cultivated successfully in open pond systems for commercial purposes (ElFar et al. 2020).

Another major hindrance in scaling up is the expensive processing cost for harvesting and drying of microalgae (Muylaert et al. 2017). As an example, it is reported that 70% of microalgal biodiesel production cost revolves around harvesting and drying stages (Skorupskaite, Makareviciene, and Gumbyte 2016). It is estimated that the costs to produce a liter of microalgal biodiesel are $2.00 and $2.59 via raceway pond and photobioreactor cultivations, respectively, highlighting the urgent need to improve the overall biofuel production method (Huang et al. 2017; Im, Kim, and Lee 2015; L. Zhu et al. 2017). Production of microalgal gaseous biofuels also faces almost similar challenges, in addition to low biomethane and biohydrogen yield (less than 40% and 12% v/v, respectively) and excessive tar residue from high nitrogen content in microalgae (Gholkar, Shastri, and Tanksale 2021; Majidian et al. 2018). All of these challenges have created a bottleneck in producing microalgal biofuels at commercial scale. Due to this, a positive energy balance cannot be obtained in a mass cultivation of microalgae (Ananthi et al. 2021). Other significant challenges hindering the development of microalgae bioprocessing plants are large initial investments and high maintenance cost (Rizwan et al. 2018). These are considered as the main economic barriers for the mass scale usage of microalgae in industry (Fazal et al. 2021).

Lack of success of previous microalgal bioprocessing companies is also a contributing factor which could hinder future microalgal biotechnology companies in securing funding from potential investors. There is no significant success of microalgal biotech companies in producing biofuels so far, despite hundreds of millions of dollars being invested by both government and private sectors since early 2000s. For example, a biotechnology company called Algenol Biotech LLC managed to secure a USD 25 million grant from the US Department of Energy to produce 400,000 liters of bioethanol annually. The company also received a funding of USD 93.5 million from Reliance, a petrochemical conglomerate. However, the results have been a disappointment even after being in operation for more than 15 years. The company had to switch their business into producing food additives and nutraceuticals (Novoveska, Brown, and Liner 2018). The same scenario happened to other microalgal biofuel companies, such as Solazyme and Sapphire Energy. These companies had to change their initial objectives due to huge hurdles faced during the operational period (Ganesan et al. 2020). At the moment, mass production of microalgae-based liquid biofuels is not possible, since it is not economically viable by using 'liquid biofuel production only' objective (O. K. Lee et al. 2015). Another factor that contributes to this failure is the wide discrepancies between theoretical modeling and actual operational plant. Various simulations for microalgae bioprocessing models are built on too many optimistic assumptions and simplifications. This led to oversimplified and unrealistic models, which do not represent the real-life operating conditions of a biorefinery (Silva et al. 2021). The situation is especially true when comparing the theoretical microalgal growth rates with actual cultivations (Kenny and Flynn 2016). Hence, these theoretical models need to be further improved to present the actual operation of a biorefinery (Yap et al. 2021). Although there are continuous efforts in developing integrated industrial processing using a biorefinery concept, the development of an energy-efficient plant with low-cost cultivation and harvesting methods still remains a huge challenge at the moment. Therefore, it is imperative for scientists to develop better technologies and techniques in improving aspects related to both upstream and downstream processes.

2.7 CONCLUSIONS

Microalgae (and cyanobacteria) have been studied and utilized in numerous applications. The versatility of these photosynthetic microorganisms has made them suitable for numerous aspects in environmental biotechnology, such as wastewater treatment, biofuels production, carbon dioxide sequestration, and soil bioremediation. Nevertheless, the wide diversity of microalgae is still waiting to be further explored. There are many unknowns that need to be addressed to fully understand the function of microalgae in different cultivation conditions. Advancement in microalgal genetic engineering is imperative to developing more robust species in tackling both environmental and industrial aspects. From various studies, it is also highlighted that the optimum utilization of microalgae biomass, energy, and chemicals can only be achieved via a consolidated biorefinery approach. Ideally, this starts from cultivating microalgae for detoxification of various pollutions and using the biomass to produce valuable products. However, existing technologies and techniques need to be improved for both upstream and downstream operations to ensure the sustainability of the biorefinery model. Moving forward, more thorough investigations involving many aspects in circular economic analysis are direly needed prior to the establishment of a biorefinery plant.

ACKNOWLEDGMENTS

This research was funded by AAIBE Chair of Renewable Energy (Project code: 202003KETTHA and 202006KETTHA). Furthermore, the authors would like to acknowledge UNITEN for the research facilities. Fazril Ideris would like to express his gratitude to UNITEN for the BOLD Scholarship. Mei Yin Ong would also like to thank UNITEN for the UNITEN Postgraduate Excellence Scholarship 2019.

REFERENCES

Acién, F. Gabriel, C. Gómez-Serrano, M. M. Morales-Amaral, J. M. Fernández-Sevilla, and E. Molina-Grima. 2016. "Wastewater Treatment Using Microalgae: How Realistic a Contribution Might It Be to Significant Urban Wastewater Treatment?" *Applied Microbiology and Biotechnology* 100 (21): 9013–22. doi:10.1007/s00253-016-7835-7.

Agwa, Obioma Kenechukwu, Chimezie Jason Ogugbue, and Enechojo Eunice Williams. 2017. "Field Evidence of Chlorella Vulgaris Potentials as a Biofertilizer for Hibiscus Esculentus." *International Journal of Agricultural Research* 12 (4): 181–89. doi:10.3923/ijar.2017.181.189.

Ahmad, Ashfaq, Azizul Buang, and A. H. Bhat. 2016. "Renewable and Sustainable Bioenergy Production from Microalgal Co-Cultivation with Palm Oil Mill Effluent (POME): A Review." *Renewable and Sustainable Energy Reviews* 65: 214–34. https://doi.org/10.1016/j.rser.2016.06.084.

Ahrens, Thorsten, and Peter Weiland. 2007. "Biomethane for Future Mobility." *Landbauforschung Volkenrode* 57 (March).

Akbarzadeh, Neda, and Mansour Shariati. 2014. "Aluminum Remediation from Medium by Dunaliella." *Ecological Engineering* 67: 76–79. https://doi.org/10.1016/j.ecoleng.2014.03.014.

Al-Gheethi, A. A., A. N. Efaq, J. D. Bala, I. Norli, M. O. Abdel-Monem, and M. O. Ab. Kadir. 2018. "Removal of Pathogenic Bacteria from Sewage-Treated Effluent and Biosolids for Agricultural Purposes." *Applied Water Science* 8 (2): 74. doi:10.1007/s13201-018-0698-6.

Ali, Shazia, Angela Paul Peter, Kit Wayne Chew, Heli Siti Halimatul Munawaroh, and Pau Loke Show. 2021. "Resource Recovery from Industrial Effluents through the Cultivation of Microalgae: A Review." *Bioresource Technology* 337: 125461. https://doi.org/10.1016/j.biortech.2021.125461.

Allen, Cameron, Graciela Metternicht, and Thomas Wiedmann. 2019. "Prioritising SDG Targets: Assessing Baselines, Gaps and Interlinkages." *Sustainability Science* 14 (2): 421–38. doi:10.1007/s11625-018-0596-8.

Alvarez, Adriana L., Sharon L. Weyers, Hannah M. Goemann, Brent M. Peyton, and Robert D. Gardner. 2021. "Microalgae, Soil and Plants: A Critical Review of Microalgae as Renewable Resources for Agriculture." *Algal Research* 54: 102200. https://doi.org/10.1016/j.algal.2021.102200.

Amenorfenyo, David Kwame, Xianghu Huang, Yulei Zhang, Qitao Zeng, Ning Zhang, Jiajia Ren, and Qiang Huang. 2019. "Microalgae Brewery Wastewater Treatment: Potentials, Benefits and the Challenges." *International Journal of Environmental Research and Public Health* 16 (11). MDPI: 1910. doi:10.3390/ijerph16111910.

Amini, Elham, Azadeh Babaei, Mohammad Reza Mehrnia, Jalal Shayegan, and Mohammad-Saeed Safdari. 2020. "Municipal Wastewater Treatment by Semi-Continuous and Membrane Algal-Bacterial Photo-Bioreactors." *Journal of Water Process Engineering* 36: 101274. https://doi.org/10.1016/j.jwpe.2020.101274.

Ananthi, V., P. Balaji, Raveendran Sindhu, Sang-Hyoun Kim, Arivalagan Pugazhendhi, and A. Arun. 2021. "A Critical Review on Different Harvesting Techniques for Algal Based Biodiesel Production." *Science of the Total Environment* 780: 146467. https://doi.org/10.1016/j.scitotenv.2021.146467.

Anjos, Mariana, Bruno D. Fernandes, António A. Vicente, José A. Teixeira, and Giuliano Dragone. 2013. "Optimization of CO2 Bio-Mitigation by Chlorella Vulgaris." *Bioresource Technology* 139: 149–54. https://doi.org/10.1016/j.biortech.2013.04.032.

Ashraful, A. M., H. H. Masjuki, M. A. Kalam, I. M. Rizwanul Fattah, S. Imtenan, S. A. Shahir, and H. M. Mobarak. 2014. "Production and Comparison of Fuel Properties, Engine Performance, and Emission Characteristics of Biodiesel from Various Non-Edible Vegetable Oils: A Review." *Energy Conversion and Management* 80: 202–28. https://doi.org/10.1016/j.enconman.2014.01.037.

Aslam, Ambreen, Skye R. Thomas-Hall, Maleeha Manzoor, Faiza Jabeen, Munawar Iqbal, Qamar uz Zaman, Peer M. Schenk, and M. Asif Tahir. 2018. "Mixed Microalgae Consortia Growth under Higher Concentration of CO2 from Unfiltered Coal Fired Flue Gas: Fatty Acid Profiling and Biodiesel Production." *Journal of Photochemistry and Photobiology B: Biology* 179: 126–33. https://doi.org/10.1016/j.jphotobiol.2018.01.003.

Atabani, A. E., A. S. Silitonga, and H. C. Ong. 2013. "Non-Edible Vegetable Oils: A Critical Evaluation of Oil Extraction, Fatty Acid Compositions, Biodiesel Production, Characteristics, Engine Performance and Emissions." *Renewable and Sustainable Energy Reviews* 18: 211–45.

Ayatollahi, Seyedeh Zeinab, Fereidun Esmaeilzadeh, and Dariush Mowla. 2021. "Integrated CO2 Capture, Nutrients Removal and Biodiesel Production Using Chlorella Vulgaris." *Journal of Environmental Chemical Engineering* 9 (2): 104763. https://doi.org/10.1016/j.jece.2020.104763.

Aysu, Tevfik, Javier Fermoso, and Aimaro Sanna. 2018. "Ceria on Alumina Support for Catalytic Pyrolysis of Pavlova Sp. Microalgae to High-Quality Bio-Oils." *Journal of Energy Chemistry* 27 (3): 874–82. https://doi.org/10.1016/j.jechem.2017.06.014.

Azizi, Kolsoom, Mostafa Keshavarz Moraveji, Aitor Arregi, Maider Amutio, Gartzen Lopez, and Martin Olazar. 2020. "On the Pyrolysis of Different Microalgae Species in a Conical Spouted Bed Reactor: Bio-Fuel Yields and Characterization." *Bioresource Technology* 311: 123561. https://doi.org/10.1016/j.biortech.2020.123561.

Azizi, Kolsoom, Mostafa Keshavarz Moraveji, and Hamed Abedini Najafabadi. 2018. "A Review on Bio-Fuel Production from Microalgal Biomass by Using Pyrolysis Method." *Renewable and Sustainable Energy Reviews* 82: 3046–59. https://doi.org/10.1016/j.rser.2017.10.033.

Ban, Shidong, Weitie Lin, Zhiwei Luo, and Jianfei Luo. 2019. "Improving Hydrogen Production of Chlamydomonas Reinhardtii by Reducing Chlorophyll Content via Atmospheric and Room Temperature Plasma." *Bioresource Technology* 275: 425–29. https://doi.org/10.1016/j.biortech.2018.12.062.

Bano, Farhat, Anushree Malik, and Shaikh Z. Ahammad. 2021. "Removal of Estradiol, Diclofenac, and Triclosan by Naturally Occurring Microalgal Consortium Obtained from Wastewater." *Sustainability (Switzerland)* 13 (14). doi:10.3390/su13147690.

Bass, Adrian M., Michael I. Bird, Gavin Kay, and Brian Muirhead. 2016. "Soil Properties, Greenhouse Gas Emissions and Crop Yield under Compost, Biochar and Co-Composted Biochar in Two Tropical Agronomic Systems." *Science of The Total Environment* 550: 459–70. https://doi.org/10.1016/j.scitotenv.2016.01.143.

Belotti, Gianluca, Benedetta de Caprariis, Paolo De Filippis, Marco Scarsella, and Nicola Verdone. 2014. "Effect of Chlorella Vulgaris Growing Conditions on Bio-Oil Production via Fast Pyrolysis." *Biomass and Bioenergy* 61: 187–95. https://doi.org/10.1016/j.biombioe.2013.12.011.

Berman-Frank, Ilana, Pernilla Lundgren, and Paul Falkowski. 2003. "Nitrogen Fixation and Photosynthetic Oxygen Evolution in Cyanobacteria." *Research in Microbiology* 154 (3): 157–64. https://doi.org/10.1016/S0923-2508(03)00029-9.

Bharti, Asha, Radha Prasanna, Gunjeet Kumar, Arun Kumar, and Lata Nain. 2019. "Co-Cultivation of Cyanobacteria for Raising Nursery of Chrysanthemum Using a Hydroponic System." *Journal of Applied Phycology* 31 (6): 3625–35. doi:10.1007/s10811-019-01830-9.

Bhatia, Saurabh. 2018. "History, Scope and Development of Biotechnology." *Introduction to Pharmaceutical Biotechnology* 1. IOP Publishing. doi:10.1088/978-0-7503-1299-8ch1.

Bhowmick, Goldy De, Ajit K. Sarmah, and Ramkrishna Sen. 2019. "Performance Evaluation of an Outdoor Algal Biorefinery for Sustainable Production of Biomass, Lipid and Lutein Valorizing Flue-Gas Carbon Dioxide and Wastewater Cocktail." *Bioresource Technology* 283: 198–206. https://doi.org/10.1016/j.biortech.2019.03.075.

Bhowmick, Goldy De, Ganeshan Subramanian, Sandhya Mishra, and Ramkrishna Sen. 2014. "Raceway Pond Cultivation of a Marine Microalga of Indian Origin for Biomass and Lipid Production: A Case Study." *Algal Research* 6: 201–9. https://doi.org/10.1016/j.algal.2014.07.005.

Binda, Gilberto, Davide Spanu, Roberta Bettinetti, Luca Magagnin, Andrea Pozzi, and Carlo Dossi. 2020. "Comprehensive Comparison of Microalgae-Derived Biochar from Different Feedstocks: A Prospective Study for Future Environmental Applications." *Algal Research* 52: 102103. https://doi.org/10.1016/j.algal.2020.102103.

Bornø, Marie Louise, Joseph Osafo Eduah, Dorette Sophie Müller-Stöver, and Fulai Liu. 2018. "Effect of Different Biochars on Phosphorus (P) Dynamics in the Rhizosphere of Zea Mays L. (Maize)." *Plant and Soil* 431 (1): 257–72. doi:10.1007/s11104-018-3762-y.

Borowitzka, Michael A. 2018. "Chapter 3 - Biology of Microalgae." In, edited by Ira A. Levine and Joël B T—Microalgae in Health and Disease Prevention Fleurence, 23–72. Academic Press. https://doi.org/10.1016/B978-0-12-811405-6.00003-7.

Braga, Vagner da Silva, Duna Joanol da Silveira Mastrantonio, Jorge Alberto Vieira Costa, and Michele Greque de Morais. 2018. "Cultivation Strategy to Stimulate High Carbohydrate Content in Spirulina Biomass." *Bioresource Technology* 269: 221–26. https://doi.org/10.1016/j.biortech.2018.08.105.

Branco-Vieira, M., T. M. Mata, A. A. Martins, M. A. V. Freitas, and N. S. Caetano. 2020. "Economic Analysis of Microalgae Biodiesel Production in a Small-Scale Facility." *Energy Reports* 6: 325–32. https://doi.org/10.1016/j.egyr.2020.11.156.

Brindhadevi, Kathirvel, Susaimanickam Anto, Eldon R. Rene, Manigandan Sekar, Thangavel Mathimani, Nguyen Thuy Lan Chi, and Arivalagan Pugazhendhi. 2021. "Effect of Reaction Temperature on the Conversion of Algal Biomass to Bio-Oil and Biochar through Pyrolysis and Hydrothermal Liquefaction." *Fuel* 285: 119106. https://doi.org/10.1016/j.fuel.2020.119106.

Cano-Pleite, Eduardo, Mariano Rubio-Rubio, Néstor García-Hernando, and Antonio Soria-Verdugo. 2020. "Microalgae Pyrolysis under Isothermal and Non-Isothermal Conditions." *Algal Research* 51: 102031. https://doi.org/10.1016/j.algal.2020.102031.

Chai, Wai Siong, Jie Ying Cheun, P. Senthil Kumar, Muhammad Mubashir, Zahid Majeed, Fawzi Banat, Shih-Hsin Ho, and Pau Loke Show. 2021. "A Review on Conventional and Novel Materials towards Heavy Metal Adsorption in Wastewater Treatment Application." *Journal of Cleaner Production* 296: 126589. doi:10.1016/j.jclepro.2021.126589.

Chai, Wai Siong, Chee Hong Chew, Heli Siti Halimatul Munawaroh, Veeramuthu Ashokkumar, Chin Kui Cheng, Young-Kwon Park, and Pau-Loke Show. 2021. "Microalgae and Ammonia: A Review on Inter-Relationship." *Fuel* 303: 121303. doi:10.1016/j.fuel.2021.121303.

Chai, Wai Siong, Wee Gee Tan, Heli Siti Halimatul Munawaroh, Vijai Kumar Gupta, Shih-Hsin Ho, and Pau Loke Show. 2021. "Multifaceted Roles of Microalgae in the Application of Wastewater Biotreatment: A Review." *Environmental Pollution* 269: 116236. doi:10.1016/j.envpol.2020.116236.

Chaudhary, Ramjee, Anil Kumar Dikshit, and Yen Wah Tong. 2018. "Carbon-Dioxide Biofixation and Phycoremediation of Municipal Wastewater Using Chlorella Vulgaris and Scenedesmus Obliquus." *Environmental Science and Pollution Research* 25 (21): 20399–406. doi:10.1007/s11356-017-9575-3.

Cheng, Jun, Zongbo Yang, Junhu Zhou, and Kefa Cen. 2018. "Improving the CO2 Fixation Rate by Increasing Flow Rate of the Flue Gas from Microalgae in a Raceway Pond." *Korean Journal of Chemical Engineering* 35 (2): 498–502. doi:10.1007/s11814-017-0300-1.

Chia, Shir Reen, Kit Wayne Chew, Hui Yi Leong, Shih-Hsin Ho, Heli Siti Halimatul Munawaroh, and Pau Loke Show. 2021. "CO2 Mitigation and Phycoremediation of Industrial Flue Gas and Wastewater via Microalgae-Bacteria Consortium: Possibilities and Challenges." *Chemical Engineering Journal* 425: 131436. https://doi.org/10.1016/j.cej.2021.131436.

Chisti, Yusuf. 2018a. "Chapter 2 - Society and Microalgae: Understanding the Past and Present." In, edited by Ira A. Levine and Joël B T—Microalgae in Health and Disease Prevention Fleurence, 11–21. Academic Press. https://doi.org/10.1016/B978-0-12-811405-6.00002-5.

Chisti, Yusuf. 2018b. "This Issue Featuring a Chapter from Microalgal Biotechnology : Potential and Production." (January 2013). doi:10.1515/green-2013-0018.

Chisti, Yusuf. 2020. *Microalgae Biotechnology: A Brief Introduction. Handbook of Microalgae-Based Processes and Products*. Elsevier Inc. doi:10.1016/b978-0-12-818536-0.00001-4.

Chittora, Deepali, Mukesh Meena, Tansukh Barupal, Prashant Swapnil, and Kanika Sharma. 2020. "Cyanobacteria as a Source of Biofertilizers for Sustainable Agriculture." *Biochemistry and Biophysics Reports* 22: 100737. https://doi.org/10.1016/j.bbrep.2020.100737.

Chu, Qingnan, Tao Lyu, Lihong Xue, Linzhang Yang, Yanfang Feng, Zhimin Sha, Bin Yue, Robert J. G. Mortimer, Mick Cooper, and Gang Pan. 2021. "Hydrothermal Carbonization of Microalgae for Phosphorus Recycling from Wastewater to Crop-Soil Systems as Slow-Release Fertilizers." *Journal of Cleaner Production* 283: 124627. https://doi.org/10.1016/j.jclepro.2020.124627.

Chu, Qingnan, Lihong Xue, Bhupinder Pal Singh, Shan Yu, Karin Müller, Hailong Wang, Yanfang Feng, Gang Pan, Xuebo Zheng, and Linzhang Yang. 2020. "Sewage Sludge-Derived Hydrochar That Inhibits Ammonia Volatilization, Improves Soil Nitrogen Retention and Rice Nitrogen Utilization." *Chemosphere* 245: 125558. https://doi.org/10.1016/j.chemosphere.2019.125558.

Cordell, Dana, and Stuart White. 2014. "Life's Bottleneck: Sustaining the World's Phosphorus for a Food Secure Future." *Annual Review of Environment and Resources* 39 (October): 161–88. doi:10.1146/annurev-environ-010213-113300.

Costa, Jorge Alberto Vieira, Luiza Moraes, Juliana Botelho Moreira, Gabriel Martins da Rosa, Adriano Seizi Arruda Henrard, and Michele Greque de Morais. 2017. "Microalgae-Based Biorefineries as a Promising Approach to Biofuel Production BT—Prospects and Challenges in Algal Biotechnology." In, edited by Bhumi Nath Tripathi and Dhananjay Kumar, 113–40. Singapore: Springer Singapore. doi:10.1007/978-981-10-1950-0_4.

Cui, Zheng, Jonah M. Greene, Feng Cheng, Jason C. Quinn, Umakanta Jena, and Catherine E. Brewer. 2020. "Co-Hydrothermal Liquefaction of Wastewater-Grown Algae and Crude Glycerol: A Novel Strategy of Bio-Crude Oil-Aqueous Separation and Techno-Economic Analysis for Bio-Crude Oil Recovery and Upgrading." *Algal Research* 51: 102077. https://doi.org/10.1016/j.algal.2020.102077.

Daneshvar, Ehsan, Yong Sik Ok, Samad Tavakoli, Binoy Sarkar, Sabry M. Shaheen, Hui Hong, Yongkang Luo, Jörg Rinklebe, Hocheol Song, and Amit Bhatnagar. 2021. "Insights into Upstream Processing of Microalgae: A Review." *Bioresource Technology* 329: 124870. https://doi.org/10.1016/j.biortech.2021.124870.

Dasgupta, Chitralekha Nag, M. R. Suseela, S. K. Mandotra, Pankaj Kumar, Manish K. Pandey, Kiran Toppo, and J. A. Lone. 2015. "Dual Uses of Microalgal Biomass : An Integrative Approach for Biohydrogen and Biodiesel Production." *Applied Energy* 146. Elsevier Ltd: 202–8. doi:10.1016/j.apenergy.2015.01.070.

Deamici, Kricelle Mosquera, Lucielen Oliveira Santos, and Jorge Alberto Vieira Costa. 2019. "Use of Static Magnetic Fields to Increase CO2 Biofixation by the Microalga Chlorella Fusca." *Bioresource Technology* 276: 103–9. https://doi.org/10.1016/j.biortech.2018.12.080.

Deepika, P., and D. MubarakAli. 2020. "Production and Assessment of Microalgal Liquid Fertilizer for the Enhanced Growth of Four Crop Plants." *Biocatalysis and Agricultural Biotechnology* 28: 101701. https://doi.org/10.1016/j.bcab.2020.101701.

Dineshbabu, Gnanasekaran, Vaithyalingam Shanmugasundaram Uma, Thangavel Mathimani, Dharmar Prabaharan, and Lakshmanan Uma. 2020. "Elevated CO2 Impact on Growth and Lipid of Marine Cyanobacterium Phormidium Valderianum BDU 20041—towards Microalgal Carbon Sequestration." *Biocatalysis and Agricultural Biotechnology* 25: 101606. https://doi.org/10.1016/j.bcab.2020.101606.

Dineshkumar, R., R. Kumaravel, J. Gopalsamy, Mohammad Nurul Azim Sikder, and P. Sampathkumar. 2018. "Microalgae as Bio-Fertilizers for Rice Growth and Seed Yield Productivity." *Waste and Biomass Valorization* 9 (5): 793–800. doi:10.1007/s12649-017-9873-5.

Dineshkumar, R., J. Subramanian, J. Gopalsamy, P. Jayasingam, A. Arumugam, S. Kannadasan, and P. Sampathkumar. 2019. "The Impact of Using Microalgae as Biofertilizer in Maize (Zea Mays L.)." *Waste and Biomass Valorization* 10 (5): 1101–10. doi:10.1007/s12649-017-0123-7.

Ding, Gong Tao, Nazlina Haiza Mohd Yasin, Mohd Sobri Takriff, Kamrul Fakir Kamarudin, Jailani Salihon, Zahira Yaakob, and Noor Irma Nazashida Mohd Hakimi. 2020. "Phycoremediation of Palm Oil Mill Effluent (POME) and CO2 Fixation by Locally Isolated Microalgae: Chlorella Sorokiniana UKM2, Coelastrella Sp. UKM4 and Chlorella Pyrenoidosa UKM7." *Journal of Water Process Engineering* 35: 101202. https://doi.org/10.1016/j.jwpe.2020.101202.

Eickemeier, Patrick, Steffen Schlömer, Ellie Farahani, Susanne Kadner, Steffen Brunner, Ina Baum, and Benjamin Kriemann. IPCC, 2014. *Climate Change 2014: Mitigation of Climate Change. Contribution of Working Group III to the Fifth Assessment Report of the Intergovernmental Panel on Climate Change* [Edenhofer, O., R. Pichs-Madruga, Y. Sokona, E. Farahani, S. Kadner, K. Seyboth, A. Adler, I. Baum, S. Brunner, P. Eickemeier, B. Kriemann, J. Savolainen, S. Schlömer, C. von Stechow, T. Zwickel and J.C. Minx (eds.)]. Cambridge University Press, Cambridge, United Kingdom and New York, NY, USA.

ElFar, Omar Ashraf, Chih Kai Chang, Hui Yi Leong, Angela Paul Peter, Kit Wayne Chew, and Pau Loke Show. 2020. "Prospects of Industry 5.0 in Algae: Customization of Production and New Advance Technology for Clean Bioenergy Generation." *Energy Conversion and Management* 10 (June). Elsevier Ltd: 100048. doi:10.1016/j.ecmx.2020.100048.

Eloka-Eboka, Andrew C., and Freddie L. Inambao. 2017. "Effects of CO2 Sequestration on Lipid and Biomass Productivity in Microalgal Biomass Production." *Applied Energy* 195: 1100–111. https://doi.org/10.1016/j.apenergy.2017.03.071.

Elser, James J. 2012. "Phosphorus: A Limiting Nutrient for Humanity?" *Current Opinion in Biotechnology* 23 (6): 833–38. https://doi.org/10.1016/j.copbio.2012.03.001.

Elshobary, Mostafa E., Hossain M. Zabed, Junhua Yun, Guoyan Zhang, and Xianghui Qi. 2021. "Recent Insights into Microalgae-Assisted Microbial Fuel Cells for Generating

Sustainable Bioelectricity." *International Journal of Hydrogen Energy* 46 (4): 3135–59. https://doi.org/10.1016/j.ijhydene.2020.06.251.

Farias Silva, Carlos Eduardo de, and Alberto Bertucco. 2016. "Bioethanol from Microalgae and Cyanobacteria: A Review and Technological Outlook." *Process Biochemistry* 51 (11): 1833–42. https://doi.org/10.1016/j.procbio.2016.02.016.

Fazal, Tahir, Muhammad Saif Ur Rehman, Fahed Javed, Mueed Akhtar, Azeem Mushtaq, Ainy Hafeez, Aamir Alaud Din, Javed Iqbal, Naim Rashid, and Fahad Rehman. 2021. "Integrating Bioremediation of Textile Wastewater with Biodiesel Production Using Microalgae (Chlorella Vulgaris)." *Chemosphere* 281: 130758. https://doi.org/10.1016/j.chemosphere.2021.130758.

Fazril, I., A. H. Shamsuddin, S. Nomanbhay, F. Kusomo, M. Hanif, M. F. M. Ahmad Zamri, A. Akhiar, and M. F. Ismail. 2020. "Microwave-Assisted in Situ Transesterification of Wet Microalgae for the Production of Biodiesel: Progress Review." *IOP Conference Series: Earth and Environmental Science* 476. IOP Publishing: 12078. doi:10.1088/1755-1315/476/1/012078.

Francavilla, M., P. Kamaterou, S. Intini, M. Monteleone, and A. Zabaniotou. 2015. "Cascading Microalgae Biorefinery: Fast Pyrolysis of Dunaliella Tertiolecta Lipid Extracted-Residue." *Algal Research* 11: 184–93. https://doi.org/10.1016/j.algal.2015.06.017.

Franchino, Marta, Valeria Tigini, Giovanna Cristina Varese, Rocco Mussat Sartor, and Francesca Bona. 2016. "Microalgae Treatment Removes Nutrients and Reduces Ecotoxicity of Diluted Piggery Digestate." *Science of The Total Environment* 569–570: 40–45. https://doi.org/10.1016/j.scitotenv.2016.06.100.

Freitas, B. C. B., A. P. A. Cassuriaga, M. G. Morais, and J. A. V. Costa. 2017. "Pentoses and Light Intensity Increase the Growth and Carbohydrate Production and Alter the Protein Profile of Chlorella Minutissima." *Bioresource Technology* 238: 248–53. https://doi.org/10.1016/j.biortech.2017.04.031.

Ganesan, Ramya, S. Manigandan, Melvin S. Samuel, Rajasree Shanmuganathan, Kathirvel Brindhadevi, Nguyen Thuy Lan Chi, Pham Anh Duc, and Arivalagan Pugazhendhi. 2020. "A Review on Prospective Production of Biofuel from Microalgae." *Biotechnology Reports* 27: e00509. https://doi.org/10.1016/j.btre.2020.e00509.

Garcia-Gonzalez, Jesus, and Milton Sommerfeld. 2016. "Biofertilizer and Biostimulant Properties of the Microalga Acutodesmus Dimorphus." *Journal of Applied Phycology* 28 (2): 1051–61. doi:10.1007/s10811-015-0625-2.

Garlapati, Deviram, Muthukumar Chandrasekaran, ArulAnanth Devanesan, Thangavel Mathimani, and Arivalagan Pugazhendhi. 2019. "Role of Cyanobacteria in Agricultural and Industrial Sectors: An Outlook on Economically Important Byproducts." *Applied Microbiology and Biotechnology* 103 (12): 4709–21. doi:10.1007/s00253-019-09811-1.

Gemin, Luiz Gabriel, Átila Francisco Mógor, Juliana De Oliveira Amatussi, and Gilda Mógor. 2019. "Microalgae Associated to Humic Acid as a Novel Biostimulant Improving Onion Growth and Yield." *Scientia Horticulturae* 256: 108560. https://doi.org/10.1016/j.scienta.2019.108560.

Geng, Yanni, Dan Cui, Liming Yang, Zhensheng Xiong, Spyros G. Pavlostathis, Penghui Shao, Yakun Zhang, Xubiao Luo, and Shenglian Luo. 2022. "Resourceful Treatment of Harsh High-Nitrogen Rare Earth Element Tailings (REEs) Wastewater by Carbonate Activated Chlorococcum Sp. Microalgae." *Journal of Hazardous Materials* 423: 127000. https://doi.org/10.1016/j.jhazmat.2021.127000.

Gholkar, Pratik, Yogendra Shastri, and Akshat Tanksale. 2021. "Renewable Hydrogen and Methane Production from Microalgae: A Techno-Economic and Life Cycle Assessment Study." *Journal of Cleaner Production* 279: 123726. https://doi.org/10.1016/j.jclepro.2020.123726.

Giostri, A., M. Binotti, and E. Macchi. 2016. "Microalgae Cofiring in Coal Power Plants: Innovative System Layout and Energy Analysis." *Renewable Energy* 95. Elsevier Ltd: 449–64. doi:10.1016/j.renene.2016.04.033.

Godos, I. de, J. L. Mendoza, F. G. Acién, E. Molina, C. J. Banks, S. Heaven, and F. Rogalla. 2014. "Evaluation of Carbon Dioxide Mass Transfer in Raceway Reactors for Microalgae Culture Using Flue Gases." *Bioresource Technology* 153: 307–14. https://doi.org/10.1016/j.biortech.2013.11.087.

Gonçalves, A. L., José C. M. Pires, and Manuel Simões. 2017. "A Review on the Use of Microalgal Consortia for Wastewater Treatment." *Algal Research* 24: 403–15. https://doi.org/10.1016/j.algal.2016.11.008.

Gonçalves, A. L., M. Simões, and J. C. M. Pires. 2014. "The Effect of Light Supply on Microalgal Growth, CO2 Uptake and Nutrient Removal from Wastewater." *Energy Conversion and Management* 85: 530–36. https://doi.org/10.1016/j.enconman.2014.05.085.

Gouveia, Luisa, and Ana Cristina. 2009. "Microalgae as a Raw Material for Biofuels Production." 269–74. doi:10.1007/s10295-008-0495-6.

Guidi, Lionel, Samuel Chaffron, Lucie Bittner, Damien Eveillard, Abdelhalim Larhlimi, Simon Roux, Youssef Darzi, et al. 2016. "Plankton Networks Driving Carbon Export in the Oligotrophic Ocean." *Nature* 532 (7600): 465–70. doi:10.1038/nature16942.

Guo, Suolian, Ping Wang, Xinlei Wang, Meng Zou, Chunxue Liu, and Jihong Hao. 2020. "Microalgae as Biofertilizer in Modern Agriculture BT—Microalgae Biotechnology for Food, Health and High Value Products." In, edited by Md. Asraful Alam, Jing-Liang Xu, and Zhongming Wang, 397–411. Singapore: Springer. doi:10.1007/978-981-15-0169-2_12.

Guo, Yang, Thomas Yeh, Wenhan Song, Donghai Xu, and Shuzhong Wang. 2015. "A Review of Bio-Oil Production from Hydrothermal Liquefaction of Algae." *Renewable and Sustainable Energy Reviews* 48: 776–90. https://doi.org/10.1016/j.rser.2015.04.049.

Gupta, Suvidha, and Sanjay B. Pawar. 2018. "An Integrated Approach for Microalgae Cultivation Using Raw and Anaerobic Digested Wastewaters from Food Processing Industry." *Bioresource Technology* 269: 571–76. https://doi.org/10.1016/j.biortech.2018.08.113.

Hariz, Harizah Bajunaid, Mohd Sobri Takriff, Muneer M. Ba-Abbad, Nazlina Haiza Mohd Yasin, and Noor Irma Nazashida Mohd Hakim. 2018. "CO2 Fixation Capability of Chlorella Sp. and Its Use in Treating Agricultural Wastewater." *Journal of Applied Phycology* 30 (6): 3017–27. doi:10.1007/s10811-018-1488-0.

Hawrot-Paw, Małgorzata, Adam Koniuszy, Małgorzata Gałczynska, Grzegorz Zajac, and Joanna Szyszlak-Bargłowicz. 2020. "Production of Microalgal Biomass Using Aquaculture Wastewater as Growth Medium." *Water (Switzerland)* 12 (1). doi:10.3390/w12010106.

Hende, Sofie Van Den, Han Vervaeren, Sem Desmet, and Nico Boon. 2011. "Bioflocculation of Microalgae and Bacteria Combined with Flue Gas to Improve Sewage Treatment." *New Biotechnology* 29 (1). Elsevier B.V.: 23–31. doi:10.1016/j.nbt.2011.04.009.

Hosseinpour, Soleiman, Mortaza Aghbashlo, Meisam Tabatabaei, Habibollah Younesi, Mehdi Mehrpooya, and Seeram Ramakrishna. 2017. "Multi-Objective Exergy-Based Optimization of a Continuous Photobioreactor Applied to Produce Hydrogen Using a Novel Combination of Soft Computing Techniques." *International Journal of Hydrogen Energy* 42 (12): 8518–29. https://doi.org/10.1016/j.ijhydene.2016.11.090.

House, Kurt Zenz, Daniel P. Schrag, Charles F. Harvey, and Klaus S. Lackner. 2006. "Permanent Carbon Dioxide Storage in Deep-Sea Sediments." *Proceedings of the National Academy of Sciences* 103 (33): 12291 LP–95. doi:10.1073/pnas.0605318103.

Hu, Qiang, Milton Sommerfeld, Eric Jarvis, Maria Ghirardi, Matthew Posewitz, Michael Seibert, and Al Darzins. 2008. "Microalgal Triacylglycerols as Feedstocks for Biofuel Production: Perspectives and Advances." *The Plant Journal* 54 (4). John Wiley & Sons, Ltd: 621–39. https://doi.org/10.1111/j.1365-313X.2008.03492.x.

Hu, Xinjuan, Yulie E. Meneses, and Ashraf Aly Hassan. 2020. "Integration of Sodium Hypochlorite Pretreatment with Co-Immobilized Microalgae/Bacteria Treatment of Meat Processing Wastewater." *Bioresource Technology* 304: 122953. https://doi.org/10.1016/j.biortech.2020.122953.

Huang, Qingshan, Fuhua Jiang, Lianzhou Wang, and Chao Yang. 2017. "Design of Photobioreactors for Mass Cultivation of Photosynthetic Organisms." *Engineering* 3 (3). Elsevier LTD on behalf of Chinese Academy of Engineering and Higher Education Press Limited Company: 318–29. doi:10.1016/J.ENG.2017.03.020.

Hwang, Jae-Hoon, Jared Church, Seung-Jin Lee, Jungsu Park, and Woo Hyoung Lee. 2016. "Use of Microalgae for Advanced Wastewater Treatment and Sustainable Bioenergy Generation." *Environmental Engineering Science* 33 (11). Mary Ann Liebert, Inc., Publishers: 882–97. doi:10.1089/ees.2016.0132.

Ideris, Fazril, Abd Halim Shamsuddin, Saifuddin Nomanbhay, Fitranto Kusumo, Arridina Susan Silitonga, Mei Yin Ong, Hwai Chyuan Ong, and Teuku Meurah Indra Mahlia. 2021. "Optimization of Ultrasound-Assisted Oil Extraction from Canarium Odontophyllum Kernel as a Novel Biodiesel Feedstock." *Journal of Cleaner Production* 288: 125563. https://doi.org/10.1016/j.jclepro.2020.125563.

Im, Hanjin, Bora Kim, and Jae W. Lee. 2015. "Bioresource Technology Concurrent Production of Biodiesel and Chemicals through Wet in Situ Transesterification of Microalgae." *Bioresource Technology* 193. Elsevier Ltd: 386–92. doi:10.1016/j.biortech.2015.06.122.

Islam, Muhammad A., Marie Magnusson, Richard J. Brown, Godwin A. Ayoko, Md. N. Nabi, and Kirsten Heimann. 2013. "Microalgal Species Selection for Biodiesel Production Based on Fuel Properties Derived from Fatty Acid Profiles." *Energies*. doi:10.3390/en6115676.

Jafarian, Sajedeh, and Ahmad Tavasoli. 2018. "A Comparative Study on the Quality of Bioproducts Derived from Catalytic Pyrolysis of Green Microalgae Spirulina (Arthrospira) Plantensis over Transition Metals Supported on HMS-ZSM5 Composite." *International Journal of Hydrogen Energy* 43 (43): 19902–17. https://doi.org/10.1016/j.ijhydene.2018.08.171.

Jain, Deepti, Supriya S. Ghonse, Tanmay Trivedi, Genevieve L. Fernandes, Larissa D. Menezes, Samir R. Damare, S. S. Mamatha, Sanjay Kumar, and Vishal Gupta. 2019. "CO2 Fixation and Production of Biodiesel by Chlorella Vulgaris NIOCCV under Mixotrophic Cultivation." *Bioresource Technology* 273: 672–76. https://doi.org/10.1016/j.biortech.2018.09.148.

Jasni, Jannatulhawa, Shalini Narayanan Arisht, Nazlina Haiza Mohd Yasin, Peer Mohamed Abdul, Sheng-Kai Lin, Chun-Min Liu, Shu-Yii Wu, Jamaliah Md Jahim, and Mohd Sobri Takriff. 2020. "Comparative Toxicity Effect of Organic and Inorganic Substances in Palm Oil Mill Effluent (POME) Using Native Microalgae Species." *Journal of Water Process Engineering* 34: 101165. https://doi.org/10.1016/j.jwpe.2020.101165.

Jasso-Chávez, Ricardo, M. Lorena Campos-García, Alicia Vega-Segura, Gregorio Pichardo-Ramos, Mayel Silva-Flores, Michel Geovanni Santiago-Martínez, R. Daniela Feregrino-Mondragón, et al. 2021. "Microaerophilia Enhances Heavy Metal Biosorption and Internal Binding by Polyphosphates in Photosynthetic Euglena Gracilis." *Algal Research* 58: 102384. https://doi.org/10.1016/j.algal.2021.102384.

Jeffry, Luqman, Mei Yin Ong, Saifuddin Nomanbhay, M. Mofijur, Muhammad Mubashir, and Pau Loke Show. 2021. "Greenhouse Gases Utilization: A Review." *Fuel* 301: 121017. https://doi.org/10.1016/j.fuel.2021.121017.

Ji, Min-Kyu, Reda A. I. Abou-Shanab, Seong-Heon Kim, El-Sayed Salama, Sang-Hun Lee, Akhil N. Kabra, Youn-Suk Lee, Sungwoo Hong, and Byong-Hun Jeon. 2013. "Cultivation of Microalgae Species in Tertiary Municipal Wastewater Supplemented with CO2 for Nutrient Removal and Biomass Production." *Ecological Engineering* 58: 142–48. https://doi.org/10.1016/j.ecoleng.2013.06.020.

Kan, Tao, Scott Grierson, Rocky de Nys, and Vladimir Strezov. 2014. "Comparative Assessment of the Thermochemical Conversion of Freshwater and Marine Micro- and Macroalgae." *Energy & Fuels* 28 (1). American Chemical Society: 104–14. doi:10.1021/ef401568s.

Kenny, Philip, and Kevin J. Flynn. 2016. "Coupling a Simple Irradiance Description to a Mechanistic Growth Model to Predict Algal Production in Industrial-Scale Solar-Powered Photobioreactors." *Journal of Applied Phycology* 28 (6): 3203–12. doi:10.1007/s10811-016-0892-6.

Khalid, Azianabiha A. Halip, Zahira Yaakob, Siti Rozaimah Sheikh Abdullah, and Mohd Sobri Takriff. 2019. "Assessing the Feasibility of Microalgae Cultivation in Agricultural Wastewater: The Nutrient Characteristics." *Environmental Technology & Innovation* 15: 100402. https://doi.org/10.1016/j.eti.2019.100402.

Kholssi, Rajaa, Priscila Vogelei Ramos, Evan A. N. Marks, Olimpio Montero, and Carlos Rad. 2021. "2Biotechnological Uses of Microalgae: A Review on the State of the Art and Challenges for the Circular Economy." *Biocatalysis and Agricultural Biotechnology* 102114. https://doi.org/10.1016/j.bcab.2021.102114.

Kostas, Emily T., Jessica M. M. Adams, Héctor A. Ruiz, Gabriela Durán-Jiménez, and Gary J. Lye. 2021. "Macroalgal Biorefinery Concepts for the Circular Bioeconomy: A Review on Biotechnological Developments and Future Perspectives." *Renewable and Sustainable Energy Reviews* 151: 111553. https://doi.org/10.1016/j.rser.2021.111553.

Kulkarni, Sunil, and Ajaygiri Goswami. 2019. "Effect of Excess Fertilizers and Nutrients: A Review on Impact on Plants and Human Population." *SSRN Electronic Journal* 2094–99. doi:10.2139/ssrn.3358171.

Kumar, Arun, and Jay Shankar Singh. 2020. "Biochar Coupled Rehabilitation of Cyanobacterial Soil Crusts: A Sustainable Approach in Stabilization of Arid and Semiarid Soils BT—Biochar Applications in Agriculture and Environment Management." In, edited by Jay Shankar Singh and Chhatarpal Singh, 167–91. Cham: Springer International Publishing. doi:10.1007/978-3-030-40997-5_8.

Kumar, Panga Kiran, S. Vijaya Krishna, Kavita Verma, K. Pooja, D. Bhagawan, and V. Himabindu. 2018. "Phycoremediation of Sewage Wastewater and Industrial Flue Gases for Biomass Generation from Microalgae." *South African Journal of Chemical Engineering* 25: 133–46. https://doi.org/10.1016/j.sajce.2018.04.006.

Kumar, Ramesh, and Parimal Pal. 2015. "Assessing the Feasibility of N and P Recovery by Struvite Precipitation from Nutrient-Rich Wastewater: A Review." *Environmental Science and Pollution Research* 22 (22): 17453–64. doi:10.1007/s11356-015-5450-2.

Kumar, Vankayalapati Vijaya. 2018. "Biofertilizers and Biopesticides in Sustainable Agriculture BT—Role of Rhizospheric Microbes in Soil: Volume 1: Stress Management and Agricultural Sustainability." In, edited by Vijay Singh Meena, 377–98. Singapore: Springer. doi:10.1007/978-981-10-8402-7_14.

Lee, Ok Kyung, Dong Ho Seong, Choul Gyun Lee, and Eun Yeol Lee. 2015. "Sustainable Production of Liquid Biofuels from Renewable Microalgae Biomass."

Journal of Industrial and Engineering Chemistry 29: 24–31. https://doi.org/10.1016/j.jiec.2015.04.016.

Lee, Xin Jiat, Hwai Chyuan Ong, Yong Yang Gan, Wei-Hsin Chen, and Teuku Meurah Indra Mahlia. 2020. "State of Art Review on Conventional and Advanced Pyrolysis of Macroalgae and Microalgae for Biochar, Bio-Oil and Bio-Syngas Production." *Energy Conversion and Management* 210: 112707. https://doi.org/10.1016/j.enconman.2020.112707.

Leong, Yoong Kit, Chi-Yu Huang, and Jo-Shu Chang. 2021. "Pollution Prevention and Waste Phycoremediation by Algal-Based Wastewater Treatment Technologies: The Applications of High-Rate Algal Ponds (HRAPs) and Algal Turf Scrubber (ATS)." *Journal of Environmental Management* 296: 113193. https://doi.org/10.1016/j.jenvman.2021.113193.

Li, Jun, Lila Otero-Gonzalez, Joris Michiels, Piet N. L. Lens, Gijs Du Laing, and Ivet Ferrer. 2021. "Production of Selenium-Enriched Microalgae as Potential Feed Supplement in High-Rate Algae Ponds Treating Domestic Wastewater." *Bioresource Technology* 333: 125239. https://doi.org/10.1016/j.biortech.2021.125239.

Li, Kun, Qiang Liu, Fan Fang, Ruihuan Luo, Qian Lu, Wenguang Zhou, Shuhao Huo, et al. 2019. "Microalgae-Based Wastewater Treatment for Nutrients Recovery: A Review." *Bioresource Technology* 291. Elsevier BV: 121934. doi:10.1016/j.biortech.2019.121934.

Lima, Serena, Valeria Villanova, Franco Grisafi, Giuseppe Caputo, Alberto Brucato, and Francesca Scargiali. 2020. "Autochthonous Microalgae Grown in Municipal Wastewaters as a Tool for Effectively Removing Nitrogen and Phosphorous." *Journal of Water Process Engineering* 38: 101647. https://doi.org/10.1016/j.jwpe.2020.101647.

López Barreiro, Diego, Blanca Ríos Gómez, Frederik Ronsse, Ursel Hornung, Andrea Kruse, and Wolter Prins. 2016. "Heterogeneous Catalytic Upgrading of Biocrude Oil Produced by Hydrothermal Liquefaction of Microalgae: State of the Art and Own Experiments." *Fuel Processing Technology* 148: 117–27. https://doi.org/10.1016/j.fuproc.2016.02.034.

Low, Sze Shin, Kien Xiang Bong, Muhammad Mubashir, Chin Kui Cheng, Man Kee Lam, Jun Wei Lim, Yeek Chia Ho, Keat Teong Lee, Heli Siti Halimatul Munawaroh, and Pau Loke Show. 2021. "Microalgae Cultivation in Palm Oil Mill Effluent (POME) Treatment and Biofuel Production." *Sustainability* 13 (6): 3247. doi:10.3390/su13063247.

Ma, Shanshan, Da Li, Yanling Yu, Dianlin Li, Ravi S. Yadav, and Yujie Feng. 2019. "Application of a Microalga, Scenedesmus Obliquus PF3, for the Biological Removal of Nitric Oxide (NO) and Carbon Dioxide." *Environmental Pollution* 252: 344–51. https://doi.org/10.1016/j.envpol.2019.05.084.

Mahawar, Himanshu, Radha Prasanna, and Robin Gogoi. 2019. "Elucidating the Disease Alleviating Potential of Cyanobacteria, Copper Nanoparticles and Their Interactions in Fusarium Solani Challenged Tomato Plants." *Plant Physiology Reports* 24 (4): 533–40. doi:10.1007/s40502-019-00490-8.

Majidian, Parastoo, Meisam Tabatabaei, Mehrshad Zeinolabedini, Mohammad Pooya Naghshbandi, and Yusuf Chisti. 2018. "Metabolic Engineering of Microorganisms for Biofuel Production." *Renewable and Sustainable Energy Reviews* 82: 3863–85. https://doi.org/10.1016/j.rser.2017.10.085.

Markou, Giorgos, Dries Vandamme, and Koenraad Muylaert. 2014. "Microalgal and Cyanobacterial Cultivation: The Supply of Nutrients." *Water Research* 65: 186–202. https://doi.org/10.1016/j.watres.2014.07.025.

Martínez-Roldán, Alfredo de Jesús, and Rosa Olivia Cañizares-Villanueva. 2020. "Chapter 7 - Wastewater Treatment Based in Microalgae." In, edited by Eduardo Jacob-Lopes,

Mariana Manzoni Maroneze, Maria Isabel Queiroz, and Leila Queiroz B T—Handbook of Microalgae-Based Processes and Products Zepka, 165–84. Academic Press. https://doi.org/10.1016/B978-0-12-818536-0.00007-5.

Mata, Teresa M., António A. Martins, and Nidia S. Caetano. 2010. "Microalgae for Biodiesel Production and Other Applications: A Review." *Renewable and Sustainable Energy Reviews* 14 (1): 217–32. doi:10.1016/j.rser.2009.07.020.

Moheimani, Navid Reza, Ashiwin Vadiveloo, Jeremy Miles Ayre, and John R. Pluske. 2018. "Nutritional Profile and in Vitro Digestibility of Microalgae Grown in Anaerobically Digested Piggery Effluent." *Algal Research* 35: 362–69. https://doi.org/10.1016/j.algal.2018.09.007.

Morais, Michele Greque de, Bárbara Catarina Bastos de Freitas, Luiza Moraes, Aline Massia Pereira, and Jorge Alberto Vieira Costa. 2019. "Chapter 18 - Liquid Biofuels From Microalgae: Recent Trends." In *Woodhead Publishing Series in Energy*, edited by Majid B T—Advanced Bioprocessing for Alternative Fuels Hosseini Biobased Chemicals, and Bioproducts, 351–72. Woodhead Publishing. https://doi.org/10.1016/B978-0-12-817941-3.00018-8.

Moshood, Taofeeq D., Gusman Nawanir, and Fatimah Mahmud. 2021. "Microalgae Biofuels Production: A Systematic Review on Socioeconomic Prospects of Microalgae Biofuels and Policy Implications." *Environmental Challenges* 5: 100207. https://doi.org/10.1016/j.envc.2021.100207.

Mubarak, M., A. Shaija, and T. V. Suchithra. 2015. "Review Article A Review on the Extraction of Lipid from Microalgae for Biodiesel Production." *ALGAL* 7. Elsevier B.V.: 117–23. doi:10.1016/j.algal.2014.10.008.

Muhammad, Gul, Asraful Alam, M. Mofijur, M. I. Jahirul, Yongkun Lv, Wenlong Xiong, Hwai Chyuan, and Jingliang Xu. 2021. "Modern Developmental Aspects in the Field of Economical Harvesting and Biodiesel Production from Microalgae Biomass." *Renewable and Sustainable Energy Reviews* 135 (July 2020). Elsevier Ltd: 110209. doi:10.1016/j.rser.2020.110209.

Mukherjee, Chandan, Rajojit Chowdhury, Tapan Sutradhar, Momtaj Begam, Sejuti Magdalene Ghosh, Sandip Kumar Basak, and Krishna Ray. 2016. "Parboiled Rice Effluent: A Wastewater Niche for Microalgae and Cyanobacteria with Growth Coupled to Comprehensive Remediation and Phosphorus Biofertilization." *Algal Research* 19: 225–36. https://doi.org/10.1016/j.algal.2016.09.009.

Muylaert, K., L. Bastiaens, D. Vandamme, and L. Gouveia. 2017. "5 - Harvesting of Microalgae: Overview of Process Options and Their Strengths and Drawbacks." In *Woodhead Publishing Series in Energy*, edited by Cristina Gonzalez-Fernandez and Raúl B T—Microalgae-Based Biofuels and Bioproducts Muñoz, 113–32. Woodhead Publishing. https://doi.org/10.1016/B978-0-08-101023-5.00005-4.

Nagappan, Senthil, and Gopalakrishnan Kumar. 2021. "Investigation of Four Microalgae in Nitrogen Deficient Synthetic Wastewater for Biorefinery Based Biofuel Production." *Environmental Technology & Innovation* 23: 101572. https://doi.org/10.1016/j.eti.2021.101572.

Nascimento, Mauro Do, Marina E. Battaglia, Lara Sanchez Rizza, Rafael Ambrosio, Andres Arruebarrena Di Palma, and Leonardo Curatti. 2019. "Prospects of Using Biomass of N2-Fixing Cyanobacteria as an Organic Fertilizer and Soil Conditioner." *Algal Research* 43: 101652. https://doi.org/10.1016/j.algal.2019.101652.

Nayak, Jagdeep K., and Uttam K. Ghosh. 2019. "Post Treatment of Microalgae Treated Pharmaceutical Wastewater in Photosynthetic Microbial Fuel Cell (PMFC) and Biodiesel Production." *Biomass and Bioenergy* 131: 105415. https://doi.org/10.1016/j.biombioe.2019.105415.

Nayak, Manoranjan, Dillip Kumar Swain, and Ramkrishna Sen. 2019. "Strategic Valorization of De-Oiled Microalgal Biomass Waste as Biofertilizer for Sustainable and Improved Agriculture of Rice (Oryza Sativa L.) Crop." *Science of The Total Environment* 682: 475–84. https://doi.org/10.1016/j.scitotenv.2019.05.123.

Neumann, Barbara, Athanasios T. Vafeidis, Juliane Zimmermann, and Robert J. Nicholls. 2015. "Future Coastal Population Growth and Exposure to Sea-Level Rise and Coastal Flooding—A Global Assessment." *PLoS One* 10 (3). Public Library of Science. doi:10.1371/journal.pone.0118571.

Novoveska, Lucie, Morgan Brown, and Barry Liner. 2018. "Rise and Fall of the Algae Biofuel Industry: What the Water Resource Recovery Facilities Can Learn." *World Water* 41 (February): 14–15, 27.

Ochedi, Friday O., Jianglong Yu, Hai Yu, Yangxian Liu, and Arshad Hussain. 2021. "Carbon Dioxide Capture Using Liquid Absorption Methods: A Review." *Environmental Chemistry Letters* 19 (1): 77–109. doi:10.1007/s10311-020-01093-8.

Olia, Mahroo Seyed Jafari, Mehrdad Azin, Abbas Akhavan Sepahy, and Nasrin Moazami. 2019. "Feasibility of Improving Carbohydrate Content of Chlorella S4, a Native Isolate from the Persian Gulf Using Sequential Statistical Designs." *Biofuels* (November). Taylor & Francis: 1–9. doi:10.1080/17597269.2019.1679572.

Ong, Mei Yin, Kit Wayne Chew, Pau Loke Show, and Saifuddin Nomanbhay. 2019. "Optimization and Kinetic Study of Non-Catalytic Transesterification of Palm Oil under Subcritical Condition Using Microwave Technology." *Energy Conversion and Management* 196 (June). Elsevier: 1126–37.

Otondo, Alessandra, Bahareh Kokabian, Savannah Stuart-Dahl, and Veera Gnaneswar Gude. 2018. "Energetic Evaluation of Wastewater Treatment Using Microalgae, Chlorella Vulgaris." *Journal of Environmental Chemical Engineering* 6 (2): 3213–22. https://doi.org/10.1016/j.jece.2018.04.064.

Oyebamiji, Olufunke O., Wiebke J. Boeing, F. Omar Holguin, Olusoji Ilori, and Olukayode Amund. 2019. "Green Microalgae Cultured in Textile Wastewater for Biomass Generation and Biodetoxification of Heavy Metals and Chromogenic Substances." *Bioresource Technology Reports* 7: 100247. https://doi.org/10.1016/j.biteb.2019.100247.

Panesar, Parmjit S., and John F. Kennedy. 2006. "No Title." *International Journal of Biological Macromolecules* 39 (4): 323. https://doi.org/10.1016/j.ijbiomac.2006.03.005.

Pena, Aline C. C., Caroline B. Agustini, Luciane F. Trierweiler, and Mariliz Gutterres. 2020. "Influence of Period Light on Cultivation of Microalgae Consortium for the Treatment of Tannery Wastewaters from Leather Finishing Stage." *Journal of Cleaner Production* 263: 121618. https://doi.org/10.1016/j.jclepro.2020.121618.

Pereira, Alexia Saleme Aona de Paula, Jackeline de Siqueira Castro, Vinícius José Ribeiro, and Maria Lúcia Calijuri. 2021. "Organomineral Fertilizers Pastilles from Microalgae Grown in Wastewater: Ammonia Volatilization and Plant Growth." *Science of The Total Environment* 779: 146205. https://doi.org/10.1016/j.scitotenv.2021.146205.

Peterson, Andrew A., Frédéric Vogel, Russell P. Lachance, Morgan Fröling, Jr. Antal Michael J., and Jefferson W. Tester. 2008. "Thermochemical Biofuel Production in Hydrothermal Media: A Review of Sub- and Supercritical Water Technologies." *Energy & Environmental Science* 1 (1). The Royal Society of Chemistry: 32–65. doi:10.1039/B810100K.

Poveda, Jorge. 2021. "Cyanobacteria in Plant Health: Biological Strategy against Abiotic and Biotic Stresses." *Crop Protection* 141: 105450. https://doi.org/10.1016/j.cropro.2020.105450.

Qu, Wenying, Pau Loke Show, Tomohisa Hasunuma, and Shih-Hsin Ho. 2020. "Optimizing Real Swine Wastewater Treatment Efficiency and Carbohydrate Productivity of Newly Microalga Chlamydomonas Sp. QWY37 Used for Cell-Displayed Bioethanol Production." *Bioresource Technology* 305: 123072. https://doi.org/10.1016/j.biortech.2020.123072.

Rahman, K. M. Atikur, and Dunfu Zhang. 2018. "Effects of Fertilizer Broadcasting on the Excessive Use of Inorganic Fertilizers and Environmental Sustainability." *Sustainability (Switzerland)* 10 (3). doi:10.3390/su10030759.

Ran, Wenyi, Haitao Wang, Yinghui Liu, Man Qi, Qi Xiang, Changhong Yao, Yongkui Zhang, and Xianqiu Lan. 2019. "Storage of Starch and Lipids in Microalgae: Biosynthesis and Manipulation by Nutrients." *Bioresource Technology* 291: 121894. https://doi.org/10.1016/j.biortech.2019.121894.

Rani, Swati, Raja Chowdhury, Wendong Tao, and Asha Srinivasan. 2020. "Tertiary Treatment of Municipal Wastewater Using Isolated Algal Strains: Treatment Efficiency and Value-Added Products Recovery." *Chemistry and Ecology* 36 (1). Taylor & Francis: 48–65. doi:10.1080/02757540.2019.1688307.

Razzak, S. A. 2017. "In Situ Biological CO 2 Fixation and Wastewater Nutrient Removal with Neochloris Oleoabundans in Batch Photobioreactor." 42: 93–105. doi:10.1007/s00449-018-2017-x.

Ribeiro, Lauro A., Patrícia Pereira da Silva, Teresa M. Mata, and António A. Martins. 2015. "Prospects of Using Microalgae for Biofuels Production: Results of a Delphi Study." *Renewable Energy* 75: 799–804. doi:10.1016/j.renene.2014.10.065.

Rizwan, Muhammad, Ghulam Mujtaba, Sheraz Ahmed, Kisay Lee, and Naim Rashid. 2018. "Exploring the Potential of Microalgae for New Biotechnology Applications and beyond : A Review." *Renewable and Sustainable Energy Reviews* 92 (November 2017). Elsevier Ltd: 394–404. doi:10.1016/j.rser.2018.04.034.

Roberti, R., S. Galletti, P. L. Burzi, H. Righini, S. Cetrullo, and C. Perez. 2015. "Induction of Defence Responses in Zucchini (Cucurbita Pepo) by Anabaena Sp. Water Extract." *Biological Control* 82: 61–68. https://doi.org/10.1016/j.biocontrol.2014.12.006.

Rosa, Ana Priscila Centeno da, Lisiane Fernandes Carvalho, Luzia Goldbeck, and Jorge Alberto Vieira Costa. 2011. "Carbon Dioxide Fixation by Microalgae Cultivated in Open Bioreactors." *Energy Conversion and Management* 52 (8): 3071–73. https://doi.org/10.1016/j.enconman.2011.01.008.

Rossi, Federico, Hua Li, Yongding Liu, and Roberto De Philippis. 2017. "Cyanobacterial Inoculation (Cyanobacterisation): Perspectives for the Development of a Standardized Multifunctional Technology for Soil Fertilization and Desertification Reversal." *Earth-Science Reviews* 171: 28–43. https://doi.org/10.1016/j.earscirev.2017.05.006.

Ruiz-Ruiz, Patricia, Adrián Estrada, and Marcia Morales. 2020. "Chapter 8 - Carbon Dioxide Capture and Utilization Using Microalgae." In, edited by Eduardo Jacob-Lopes, Mariana Manzoni Maroneze, Maria Isabel Queiroz, and Leila Queiroz B T—Handbook of Microalgae-Based Processes and Products Zepka, 185–206. Academic Press. https://doi.org/10.1016/B978-0-12-818536-0.00008-7.

Saijo, Yusuke, and Eliza Po-iian Loo. 2020. "Plant Immunity in Signal Integration between Biotic and Abiotic Stress Responses." *New Phytologist* 225 (1). John Wiley & Sons, Ltd: 87–104. https://doi.org/10.1111/nph.15989.

Salih, Fadhil M., and Fadhil M. Salih. 2011. "Microalgae Tolerance to High Concentrations of Carbon Dioxide: A Review." *Journal of Environmental Protection* 2 (5). Scientific Research Publishing: 648–54. doi:10.4236/jep.2011.25074.

Schwarzenbach, René, Beate Escher, Kathrin Fenner, Thomas Hofstetter, Annette Johnson, Urs Gunten, and Bernhard Wehrli. 2006. "El Desafío de Los Microcontaminantes En Los Sistemas Acuáticos." *Science* 313 (August): 1072–77.

Serrà, Albert, Raul Artal, Jaume García-Amorós, Elvira Gómez, and Laetitia Philippe. 2020. "Circular Zero-Residue Process Using Microalgae for Efficient Water Decontamination, Biofuel Production, and Carbon Dioxide Fixation." *Chemical Engineering Journal* 388: 124278. https://doi.org/10.1016/j.cej.2020.124278.

Silambarasan, Sivagnanam, Peter Logeswari, Ramachandran Sivaramakrishnan, Aran Incharoensakdi, Pablo Cornejo, Balu Kamaraj, and Nguyen Thuy Lan Chi. 2021. "Removal of Nutrients from Domestic Wastewater by Microalgae Coupled to Lipid Augmentation for Biodiesel Production and Influence of Deoiled Algal Biomass as Biofertilizer for Solanum Lycopersicum Cultivation." *Chemosphere* 268: 129323. https://doi.org/10.1016/j.chemosphere.2020.129323.

Silva, Maria Rafaele Oliveira Bezerra da, Yanara Alessandra Santana Moura, Attilio Converti, Ana Lúcia Figueiredo Porto, Daniela de Araújo Viana Marques, and Raquel Pedrosa Bezerra. 2021. "Assessment of the Potential of Dunaliella Microalgae for Different Biotechnological Applications: A Systematic Review." *Algal Research* 58: 102396. https://doi.org/10.1016/j.algal.2021.102396.

Silva Vaz, Bruna da, Jorge Alberto Vieira Costa, and Michele Greque de Morais. 2016. "CO2 Biofixation by the Cyanobacterium Spirulina Sp. LEB 18 and the Green Alga Chlorella Fusca LEB 111 Grown Using Gas Effluents and Solid Residues of Thermoelectric Origin." *Applied Biochemistry and Biotechnology* 178 (2): 418–29. doi:10.1007/s12010-015-1876-8.

Singh, Gulab, and S. K. Patidar. 2018. "Microalgae Harvesting Techniques: A Review." *Journal of Environmental Management* 217. Elsevier Ltd: 499–508. doi:10.1016/j.jenvman.2018.04.010.

Skorupskaite, Virginija, Violeta Makareviciene, and Milda Gumbyte. 2016. "Opportunities for Simultaneous Oil Extraction and Transesterification during Biodiesel Fuel Production from Microalgae : A Review." *Fuel Processing Technology* 150. Elsevier B.V.: 78–87. doi:10.1016/j.fuproc.2016.05.002.

Solovchenko, Alexei, Antonie M. Verschoor, Nicolai D. Jablonowski, and Ladislav Nedbal. 2016. "Phosphorus from Wastewater to Crops: An Alternative Path Involving Microalgae." *Biotechnology Advances* 34 (5): 550–64. https://doi.org/10.1016/j.biotechadv.2016.01.002.

Stavi, Ilan, and Rattan Lal. 2015. "Achieving Zero Net Land Degradation: Challenges and Opportunities." *Journal of Arid Environments* 112: 44–51. https://doi.org/10.1016/j.jaridenv.2014.01.016.

Suarez Ruiz, Catalina A., Santiago Zarate Baca, Lambertus A. M. van den Broek, Corjan van den Berg, Rene H. Wijffels, and Michel H. M. Eppink. 2020. "Selective Fractionation of Free Glucose and Starch from Microalgae Using Aqueous Two-Phase Systems." *Algal Research* 46: 101801. https://doi.org/10.1016/j.algal.2020.101801.

Suresh, A., S. Soundararajan, S. Elavarasi, F. Lewis Oscar, and N. Thajuddin. 2019. "Evaluation and Characterization of the Plant Growth Promoting Potentials of Two Heterocystous Cyanobacteria for Improving Food Grains Growth." *Biocatalysis and Agricultural Biotechnology* 17: 647–52. https://doi.org/10.1016/j.bcab.2019.01.002.

Suresh, M., C. P. Jawahar, and Arun Richard. 2018. "A Review on Biodiesel Production, Combustion, Performance, and Emission Characteristics of Non-Edible Oils in Variable Compression Ratio Diesel Engine Using Biodiesel and Its Blends." *Renewable and Sustainable Energy Reviews* 92: 38–49. https://doi.org/10.1016/j.rser.2018.04.048.

Sutherland, Donna L., Jason Park, Peter J. Ralph, and Rupert Craggs. 2021. "Ammonia, PH and Dissolved Inorganic Carbon Supply Drive Whole Pond Metabolism in Full-Scale Wastewater High Rate Algal Ponds." *Algal Research* 58: 102405. https://doi.org/10.1016/j.algal.2021.102405.

Tang, Dahai, Wei Han, Penglin Li, Xiaoling Miao, and Jianjiang Zhong. 2011. "CO2 Biofixation and Fatty Acid Composition of Scenedesmus Obliquus and Chlorella Pyrenoidosa in Response to Different CO2 Levels." *Bioresource Technology* 102 (3): 3071–76. https://doi.org/10.1016/j.biortech.2010.10.047.

Teng, Yihua, and Dongxiao Zhang. 2018. "Long-Term Viability of Carbon Sequestration in Deep-Sea Sediments." *Science Advances* 4 (7): eaao6588. doi:10.1126/sciadv.aao6588.

Teng, Ziyan, Jing Hua, Changsong Wang, and Xiaohua Lu. 2014. "Chapter 4 - Design and Optimization Principles of Biogas Reactors in Large Scale Applications." In, edited by Fan B T—Reactor and Process Design in Sustainable Energy Technology Shi, 99–134. Amsterdam: Elsevier. https://doi.org/10.1016/B978-0-444-59566-9.00004-1.

Thajuddin, Nooruddin, and G Subramanian. 2004. "Cyanobacterial Biodiversity and Potential Applications in Biotechnology." *Current Science* 89 (November).

Tripathi, Shweta, and Krishna Mohan Poluri. 2021. "Heavy Metal Detoxification Mechanisms by Microalgae: Insights from Transcriptomics Analysis." *Environmental Pollution* 285: 117443. https://doi.org/10.1016/j.envpol.2021.117443.

Vieira Salla, Ana Cláudia, Ana Cláudia Margarites, Fábio Ivan Seibel, Luiz Carlos Holz, Vandré Barbosa Brião, Telma Elita Bertolin, Luciane Maria Colla, and Jorge Alberto Vieira Costa. 2016. "Increase in the Carbohydrate Content of the Microalgae Spirulina in Culture by Nutrient Starvation and the Addition of Residues of Whey Protein Concentrate." *Bioresource Technology* 209: 133–41. https://doi.org/10.1016/j.biortech.2016.02.069.

Wan, Chun, Feng-Wu Bai, and Xin-Qing Zhao. 2013. "Effects of Nitrogen Concentration and Media Replacement on Cell Growth and Lipid Production of Oleaginous Marine Microalga Nannochloropsis Oceanica DUT01." *Biochemical Engineering Journal* 78: 32–38. https://doi.org/10.1016/j.bej.2013.04.014.

Wang, Liang, Min Min, Yecong Li, Paul Chen, Yifeng Chen, Yuhuan Liu, Yingkuan Wang, and Roger Ruan. 2009. "Cultivation of Green Algae Chlorella Sp. in Different Wastewaters from Municipal Wastewater Treatment Plant." *Applied Biochemistry and Biotechnology* 162 (4). Springer Science and Business Media LLC: 1174–86. doi:10.1007/s12010-009-8866-7.

Wang, Luyun, Han Xiao, Ning He, Dong Sun, and Shunshan Duan. 2019. "Biosorption and Biodegradation of the Environmental Hormone Nonylphenol By Four Marine Microalgae." *Scientific Reports* 9 (1): 5277. doi:10.1038/s41598-019-41808-8.

Wang, Xiaoxing, and Chunshan Song. 2020. "Carbon Capture From Flue Gas and the Atmosphere: A Perspective." *Frontiers in Energy Research.* 8: 560849. doi:10.3389/fenrg.2020.560849.

Wang, Ya, Chunhua Zhang, Yanheng Zheng, and Ying Ge. 2017. "Bioaccumulation Kinetics of Arsenite and Arsenate in Dunaliella Salina under Different Phosphate Regimes." *Environmental Science and Pollution Research* 24 (26): 21213–21. doi:10.1007/s11356-017-9758-y.

Wastewater and Sewage treatment [n]. 2010. In: Evert KJ., Ballard (deceased) E.B., Elsworth D.J., Oquiñena I., Schmerber JM., Stipe (deceased) R.E. (eds) *Encyclopedic Dictionary of Landscape and Urban Planning*. Springer, Berlin, Heidelberg. https://doi.org/10.1007/978-3-540-76435-9_15980.

Whitton, Rachel, Francesco Ometto, Marc Pidou, Peter Jarvis, Raffaella Villa, and Bruce Jefferson. 2015. "Microalgae for Municipal Wastewater Nutrient Remediation: Mechanisms, Reactors and Outlook for Tertiary Treatment." *Environmental Technology Reviews* 4 (1). Taylor & Francis: 133–48. doi:10.1080/21622515.2015.1105308.

Withers, Paul J. A., Kirsty G. Forber, Christopher Lyon, Shane Rothwell, Donnacha G. Doody, Helen P. Jarvie, Julia Martin-Ortega, et al. 2020. "Towards Resolving the Phosphorus Chaos Created by Food Systems." *Ambio* 49 (5): 1076–89. doi:10.1007/s13280-019-01255-1.

World Economic Forum. 2019. *The Global Risks Report 2019 14th Edition*. World Economic Forum. Geneva, Switzerland. https://www3.weforum.org/docs/WEF_Global_Risks_Report_2019.pdf

Xue, Jiao, Srinivasan Balamurugan, Tong Li, Jia Xi Cai, Ting Ting Chen, Xiang Wang, Wei Dong Yang, and Hong Ye Li. 2021. "Biotechnological Approaches to Enhance Biofuel Producing Potential of Microalgae." *Fuel* 302 (October). Elsevier: 121169. doi:10.1016/J.FUEL.2021.121169.

Yadav, Geetanjali, Sukanta K. Dash, and Ramkrishna Sen. 2019. "A Biorefinery for Valorization of Industrial Waste-Water and Flue Gas by Microalgae for Waste Mitigation, Carbon-Dioxide Sequestration and Algal Biomass Production." *Science of The Total Environment* 688: 129–35. https://doi.org/10.1016/j.scitotenv.2019.06.024.

Yahya, Liyana, Razif Harun, and Luqman Chuah Abdullah. 2020. "Screening of Native Microalgae Species for Carbon Fixation at the Vicinity of Malaysian Coal-Fired Power Plant." *Scientific Reports* 10 (1): 22355. doi:10.1038/s41598-020-79316-9.

Yang, Changyan, Rui Li, Bo Zhang, Qi Qiu, Baowei Wang, Hui Yang, Yigang Ding, and Cunwen Wang. 2019. "Pyrolysis of Microalgae: A Critical Review." *Fuel Processing Technology* 186: 53–72. doi:10.1016/j.fuproc.2018.12.012.

Yap, Jiunn Kwok, Revathy Sankaran, Kit Wayne Chew, Heli Siti Halimatul Munawaroh, Shih-Hsin Ho, J. Rajesh Banu, and Pau Loke Show. 2021. "Advancement of Green Technologies: A Comprehensive Review on the Potential Application of Microalgae Biomass." *Chemosphere* 281: 130886. https://doi.org/10.1016/j.chemosphere.2021.130886.

Yu, Shan, Yanfang Feng, Lihong Xue, Haijun Sun, Lanfang Han, Linzhang Yang, Qingye Sun, and Qingnan Chu. 2019. "Biowaste to Treasure: Application of Microbial-Aged Hydrochar in Rice Paddy Could Improve Nitrogen Use Efficiency and Rice Grain Free Amino Acids." *Journal of Cleaner Production* 240: 118180. https://doi.org/10.1016/j.jclepro.2019.118180.

Yukesh Kannah, R., S. Kavitha, Obulisamy Parthiba Karthikeyan, Eldon R. Rene, Gopalakrishnan Kumar, and J. Rajesh Banu. 2021. "A Review on Anaerobic Digestion of Energy and Cost Effective Microalgae Pretreatment for Biogas Production." *Bioresource Technology* 332: 125055. https://doi.org/10.1016/j.biortech.2021.125055.

Zaidi, Najam W., Manzoor H. Dar, Sudhanshu Singh, and U. S. Singh. 2014. "Chapter 38 - Trichoderma Species as Abiotic Stress Relievers in Plants." In, edited by Vijai K. Gupta, Monika Schmoll, Alfredo Herrera-Estrella, R. S. Upadhyay, Irina Druzhinina, and Maria G B T—Biotechnology and Biology of Trichoderma Tuohy, 515–25. Amsterdam: Elsevier. https://doi.org/10.1016/B978-0-444-59576-8.00038-2.

Zamri, Alif Aiman, Mei Yin Ong, Saifuddin Nomanbhay, and Pau Loke Show. 2021. "Microwave Plasma Technology for Sustainable Energy Production and the Electromagnetic Interaction within the Plasma System: A Review." *Environmental Research* 197: 111204. https://doi.org/10.1016/j.envres.2021.111204.

Zamri, M. F. M. A., Saiful Hasmady, Afifi Akhiar, Fazril Ideris, A. H. Shamsuddin, M. Mofijur, I. M. Rizwanul Fattah, and T. M. I. Mahlia. 2021. "A Comprehensive Review on Anaerobic Digestion of Organic Fraction of Municipal Solid Waste." *Renewable and Sustainable Energy Reviews* 137: 110637. https://doi.org/10.1016/j.rser.2020.110637.

Zhang, Lijie, Libin Zhang, Daoji Wu, Lin Wang, Zhigang Yang, Wenbao Yan, Yan Jin, Feiyong Chen, Yang Song, and Xiaoxiang Cheng. 2021. "Biochemical Wastewater from

Landfill Leachate Pretreated by Microalgae Achieving Algae's Self-Reliant Cultivation in Full Wastewater-Recycling Chain with Desirable Lipid Productivity." *Bioresource Technology* 340: 125640. https://doi.org/10.1016/j.biortech.2021.125640.

Zhao, Bingtao, and Yaxin Su. 2014. "Process Effect of Microalgal-Carbon Dioxide Fixation and Biomass Production: A Review." *Renewable and Sustainable Energy Reviews* 31: 121–32. https://doi.org/10.1016/j.rser.2013.11.054.

Zhao, Bingtao, Yaxin Su, Dunyu Liu, Hang Zhang, Wang Liu, and Guomin Cui. 2016. "SO2/NOx Emissions and Ash Formation from Algae Biomass Combustion: Process Characteristics and Mechanisms." *Energy* 113: 821–30. https://doi.org/10.1016/j.energy.2016.07.107.

Zheng, Mingmin, Xiaowei Ji, Yongjin He, Zhefu Li, Mingzi Wang, Bilian Chen, and Jian Huang. 2020. "Simultaneous Fixation of Carbon Dioxide and Purification of Undiluted Swine Slurry by Culturing Chlorella Vulgaris MBFJNU-1." *Algal Research* 47: 101866. https://doi.org/10.1016/j.algal.2020.101866.

Zhu, Liandong, Y. K. Nugroho, S. R. Shakeel, Zhaohua Li, B. Martinkauppi, and E. Hiltunen. 2017. "Using Microalgae to Produce Liquid Transportation Biodiesel : What Is Next ?" *Renewable and Sustainable Energy Reviews* 78 (May 2016). Elsevier Ltd: 391–400. doi:10.1016/j.rser.2017.04.089.

Zhu, Shunni, Yajie Wang, Wei Huang, Jin Xu, Zhongming Wang, Jingliang Xu, and Zhenhong Yuan. 2014. "Enhanced Accumulation of Carbohydrate and Starch in Chlorella Zofingiensis Induced by Nitrogen Starvation." *Applied Biochemistry and Biotechnology* 174 (7): 2435–45. doi:10.1007/s12010-014-1183-9.

Znad, Hussein, Ahmed M. D. Al Ketife, Simon Judd, Fares AlMomani, and Hari Babu Vuthaluru. 2018. "Bioremediation and Nutrient Removal from Wastewater by Chlorella Vulgaris." *Ecological Engineering* 110: 1–7. https://doi.org/10.1016/j.ecoleng.2017.10.008.

3 Current Issues and Challenges of Applying Microalgae in Environmental Biotechnology

Xiao Gui Xing[1], *Gao Ya Qian*[2], *Sook Sin Chan*[3], *Guo Rui Xin*[4], *Kit Wayne Chew*[5] *and Pau Loke Show*[6]

[1] Pharmaceutical Engineering, Faculty of Engineering, China Pharmaceutical University, Nanjing, China

[2] Biology and Medicine, Faculty of Engineering, China Pharmaceutical University, Nanjing, China

[3] Institute of Biological Sciences, Faculty of Science, University of Malaya, Kuala Lumpur, Malaysia

[4] Faculty of Engineering, China Pharmaceutical University, Nanjing, China

[5] School of Energy and Chemical Engineering, Xiamen University Malaysia, Jalan Sunsuria, Bandar Sunsuria, Sepang, Selangor Darul Ehsan, Malaysia

[6] Department of Chemical and Environmental Engineering, Faculty of Science and Engineering, University of Nottingham Malaysia, Jalan Broga, Semenyih, Selangor Darul Ehsan, Malaysia

CONTENTS

3.1 Introduction 108
3.2 Advancement of Algae-Based Biological Treatment 109
  3.2.1 Microalgal Wastewater Treatment 110
    3.2.1.1 Species and Properties of Common Microalgae 110
    3.2.1.2 Enriched C, N, P Wastewater Treatment 112
    3.2.1.3 Pharmaceutical Wastewater Treatment 114

DOI: 10.1201/9781003202196-3

3.2.1.4 Factors Affecting Microalgae Treatment of Wastewater 115
3.2.1.5 Advantages and Main Defects 117
3.2.2 Activated Sludge Process Wastewater Treatment 118
3.2.2.1 Enriched C, N, P Wastewater Treatment 119
3.2.2.2 Antibiotic Pharmaceuticals Wastewater Treatment 120
3.2.2.3 Factors Affecting the Treatment Effect of Activated Sludge 120
3.2.2.4 Technology and Disadvantages of Activated Sludge Process 122
3.2.3 Algal Symbiotic Wastewater Treatment 123
3.2.3.1 Algal Symbiotic Feasibility Analysis 123
3.2.3.2 Summary of Studies on Algal Symbiotic Wastewater Treatment 124
3.3 Applications and Challenges of Toxicity Evaluation of Microalgae 124
3.3.1 Research Status of Microalgae in Chemical Risk Assessment and Impacts of Respective Source Towards Microalgae 126
3.3.1.1 Nanoparticles 126
3.3.1.2 Metal 127
3.3.1.3 Medical Wastewater 127
3.3.1.4 Antibiotics 127
3.3.1.5 Pesticide 128
3.3.1.6 Bactericide 128
3.3.2 Microalgae Toxicity Test 129
3.3.2.1 Obtainment, Selection, and Culture of Algae Species 129
3.3.2.2 Experimental Methods 130
3.3.2.3 Toxicity Endpoints and Assessment Indices 130
3.3.3 Problems and Challenges of Toxicity Evaluation of Microalgae 132
3.3.3.1 Effects of Environmental Factors on Algae Application 132
3.3.3.2 Limitations of Algae 133
3.3.3.3 Limitations of Research Methods 133
3.4 Summary and Prospects 134
3.5 Conclusions 135
References 135

3.1 INTRODUCTION

With the continuous development and advancement of industrialization, a substantial number of chemical substances such as pollutants from industry, agriculture, medical treatment, and domestic sewage are being discharged into the natural environment. In recent years, increasing attention has been paid to the adverse effects of various toxic substances on the environment and human health. Different techniques have been researched and developed to effectively treat the wastewater produced from the production processes, and the exposure level and risk assessment after treatment are worthy of attention. Although many countries and regions have enforced standards and regulations for wastewater that were discharged after treatment, however, the phenomenon of untreated wastewater discharge still exists. About more than 90% of the water is discharged untreated from ten regions across the globe (Gadipelly et al. 2014). Among the ten regions, the Caribbean, West and Central Africa, Central and East Europe, Caspian Sea, Southern Asia, and East Asia were having serious untreated discharged wastewater issues (Gadipelly et al. 2014).

At present, there are many treatment technologies for pollutants, including physicochemical and biological treatment technologies. Some conventional measurement parameters of treated wastewater are processed and monitored such as pH, total suspended solids (TSS), chemical oxygen demand (COD), and ammonia (sum of ammonium and ammonia, NH_3-N) level. Water quality assessment only aims at general parameters, which often leads to underestimating water quality impact on the environment. The combination of the conventional parameters with biological toxicity tests is an effective method to improve the accuracy of detecting the impact of wastewater exposure, which brings many benefits to wastewater treatment. This process may as well make wastewater meet the discharge requirements of relevant standards. Microalgae can be used not only for the toxicity evaluation of wastewater but also for the risk assessment of emerging chemical substances as well as for the supplementation of the knowledge gap for their potential impact on the ecosystem. Some environmental regulations rely on ecotoxicological data to assess and manage chemical pollutants. Therefore, the obtained toxicological data can provide a reference for the establishment of relevant standards or policies and regulations.

Tremendous studies have reported the microalgae applications in the fields of energy (Taimbú, Martín, and Grossmann 2019), food (Plaza et al. 2009), and environment (Chan, Salsali, and McBean 2013) due to their unique characteristics. The characteristics of fast growth speed, strong nutrient absorption capacity, ability to absorb, and degrade pollutants in sewage and sensitivity to the environment mainly contributed to their application status in pollution treatment and toxicity evaluation which could put forward the problems that need to be further studied in the future. Hence, the following section would be mainly discussing the application of microalgae in environmental technology.

3.2 ADVANCEMENT OF ALGAE-BASED BIOLOGICAL TREATMENT

Water is the source of all things and essential materials for life activities. The progress of society and continuous development of industrialization have led to severe environmental pollution; the pollution degree of water is deteriorating. At the end of 2017, a total of 31.038 billion tons of sewage has been discharged from China's main economic development region, Yangtze River economic belt (Jie et al. 2021). This action has further worsened the situation of ecological environment treatment. Therefore, sewage treatment has attracted extensive attention, and the development of new sewage treatment technology is imminent.

Presently, there are a broad range of common treatment technologies for wastewater treatment, including physical methods, chemical methods, and biological methods. The biological method mainly from the two aspects of microalgal treatment technology that are activated sludge method and algal-bacteria symbiosis technology will be introduced in this chapter. The analysis and summarization of the current research results on wastewater treatment from three perspectives namely the current treatment situation, the treatment principle, and the actual application situation will be providing reference opinions for the current situation and enhancement of wastewater treatment.

3.2.1 Microalgal Wastewater Treatment

Sewage usually has a complex composition of inorganic pollutants (i.e., carbon, nitrogen, and phosphorus), organic pollutants (i.e., pesticides, pharmaceutical), and pathogens (Nsenga Kumwimba et al. 2018; Lim et al. 2021; Chai, Tan et al. 2021). Various methods to treat pollutants in wastewater including biological methods (activated sludge and microalgae) and abiotic methods (oxidation, photolysis, hydrolysis, flocculent precipitation, membrane method) have been invented and applied to reduce the content and toxicity of pollutants in the environment. However, some of the developed methods have a diverse range of disadvantages. For example, activated sludge treatment produces a large amount of waste sludge, as well as easy generation of resistance and higher carbon dioxide (CO_2) emission. With the advantages of easy availability, small size, easy culture, and self-sensitivity to pollutants, microalgae can be used to reduce organic and inorganic pollutants in wastewater, achieve biological activity monitoring, aid in the reduction of carbon emission, and achieve no biological resistance. Based on the aforementioned advantages, the technology of microalgae for wastewater treatment is constantly developing. Microalgae have been studied and used in the treatment of wastewater rich in excessive nutrients namely carbon (C), nitrogen (N), and phosphorus (P), pharmaceutical wastewater (Gadipelly et al. 2014), heavy metal wastewater (Chan, Salsali, and McBean 2013; Chai, Cheun et al. 2021), and municipal wastewater (Abeysiriwardana-Arachchige et al. 2020; Low et al. 2021).

3.2.1.1 Species and Properties of Common Microalgae

Spirulina sp. is fresh water and salt-tolerant microalgae species, which is suitable for growing in high temperatures and alkaline environments. The suitable pH is alkaline (pH 8.0–10.3), and the suitable temperature is within the range of 28–35°C. *Spirulina* sp. also has exorbitant nitrogen and phosphorus recovery efficiency which can reach adsorption of 97% NH_4^+-N, 96.5% total phosphorus, and

98% urea (Chang et al. 2013). At the same time, the CO_2 absorption performance of *Spirulina* sp. is about 10 times higher than that of terrestrial plants (Skjånes, Lindblad, and Muller 2007). This species of microalgae is mainly used for the treatment of high-salt nutrient wastewater with remarkable effect. As reported, *Spirulina* sp. can still maintain the removal of more than 80% of C, N, and P when the salinity is between 2.00% and 3.05% (Zhou et al. 2017). In addition, *Spirulina* sp. can also be used in the treatment of heavy metals. It was proven in one of the studies that the heavy metal removal rates can reach up to 95%, 87%, and 63% when treating aluminium (Al), nickel (Ni), and copper (Cu) wastewater, respectively (Almomani and Bohsale 2021).

Phormidium sp. belongs to Cyanophyta, marine algae, which can survive under different salinity conditions. Under the salinity of 20–60 ppt, its biological growth and various algal proteins increase with the increase of salinity (Hotos 2021). *Phormidium* sp. has a high-efficiency removal of C, N, and P from wastewater. The average removal rate of *Phormidium* sp. is more than 80%, while the removal effect of metal chromium could reach up to 90% (Das et al. 2018). On the other hand, the treatment effect of orthophosphate, nitrate, and ammonia nitrogen is also remarkable, which can be as high as 100%, 87%, and 100%, respectively (Canizaresvillanueva et al. 1994). *Phormidium* sp. could also effectively remove fluoride which is one of the persistent pollutants from wastewater. The study has proven that with an initial fluoride concentration of 3 mg/L, the removal rate can reach 60% (Mittal et al. 2020).

Botryococcus sp. is often used to absorb nitrate in water. However, the presence of nitrate could affect the photosynthesis of microalgae. When nitrate concentration is higher than 100 mg/L, the photosynthetic efficiency of microalgae decreases. *Botryococcus* sp. provides an outstanding output of carbohydrates and bio-oils as energy substances in addition to excellent removal of chemical oxygen demand in wastewater treatment (Gani et al. 2017). However, both temperature and salinity can affect the formation of biofuels. As illustrated in the previous study, when the temperature reaches 32°C, most of the synthesis is inhibited. The increase of salinity can promote oil synthesis in a certain range (HU Zhang-xi 2012)

There are various species of *Scenedesmus* sp., such as *S. obliquus*, *S. acuminatus*, *S. arcuatus*, *S. quadricauda*, and *S. armatus* which are widely distributed and easy to be obtained. These microalgae are mainly distributed in ponds, lakes, ditches, small puddles, various culture tanks, and other water bodies, with wide temperature tolerance (4–40°C) and the ability to directly use organic substances to improve water quality. *Scenedesmus* sp. has relatively excellent nitrogen and phosphorus removal effect (Song et al. 2020; Li, Hu, and Yang 2010) and good oil yield. For example, when *S. quadricauda* is used to treat simulated wastewater, 80% of COD and 78% total nitrogen can be removed. It can be used as an excellent candidate for biodiesel (Silva et al. 2009).

Chlamydomonas reinhardtii, known as photosynthetic yeast, has an excellent photosynthetic ability (Rochaix 1995). Nevertheless, this microalgae species has outstanding nitrogen and phosphorus removal capacity and excellent antioxidant capacity in removing pollutants (Ismaiel, El-Ayouty, and Al-Badwy 2021). It is highly resistant to organophosphorus pesticides and has an obvious removal effect. Previous studies have reported that *Chlamydomonas reinhardtii* could achieve 100% biodegradation of 100 mg/L organophosphorus pesticides (trichlorfon) (Gong et al. 2020).

Chlorella sp. is a very common microalga in sewage treatment, which is widely distributed and easy to obtain and grows rapidly. In terms of pollutant treatment, *Chlorella* sp. can treat as high concentration as 200 mg/L of ammonia nitrogen wastewater which still has a good treatment effect (Lei et al. 2020). It also has an excellent treatment effect on antibiotic wastewater. Under the condition of sodium acetate co-metabolism, the removal rate of sulfamethoxazole is as high as 99.3% (Qian et al. 2020).

Other microalgae include *Microcystis aeruginosa*, and activated carbon powder is immobilized to treat digestive wastewater to achieve more than 90% total nitrogen and total phosphorus removal (Gong et al. 2020). *Ascophyllum nodosum* is also used to treat Cd^{2+} in sewage (Romera et al. 2007) while *Spirogyra* sp. and *Cladophora* sp. can be used to treat Pd^{2+} and Cu^{2+} in sewage (Lee and Chang 2011).

3.2.1.2 Enriched C, N, P Wastewater Treatment

Carbon sources are generally categorized into inorganic carbon and organic carbon. The main forms of inorganic carbon are carbonate and CO_2. Carbonate can enter microalgal cells directly followed by a transformation into CO_2 by carbonate esterase, which is commonly found in microalgae (Picardo et al. 2013). CO_2 in water can diffuse freely into cells and be utilized by photosynthesis. For organic carbon, some microalgae can be used for heterotrophic growth, which is divided into chemical energy heterotrophic growth, light heterotrophic growth, and light-activated heterotrophic growth through the influence of light (Flórez-Miranda et al. 2017). The organic substances available to microalgae are mainly small molecular substances such as acetic acid and its salts, glucose, sucrose, and glycerol. For most macromolecular substances, microalgae cannot use them directly.

Nitrogen-containing wastewater can be used as a source of nitrogen for microalgae growth and metabolism. The main forms of nitrogen elements in nitrogen-containing wastewater are NO_3^-, NO_2^-, and NH_4^+-N, in which NH_4^+-N can be absorbed and utilized by microalgae. This process could provide a theoretical basis for the treatment of nitrogen-containing wastewater, that is, the mechanism of assimilation and absorption. However, according to the actual research, too high NO_3^- concentration will lead to the rise of pH, which is not conducive to the growth of microalgae and eventually reduces the wastewater treatment efficiency. NH_4^+-N can cause a drop in pH and is also detrimental to microalgal growth and metabolism, thereby reducing removal efficiency. The general mechanism of microalgae's absorption and assimilation of nitrogen source is as follows: NO_3^- is reduced to NO_2^- and then reduced to NH_4^+-N. Under the condition of energy consumption, L-glutamine is synthesized for the survival of microalgae, where the inorganic nitrogen sources need to be converted to NH_4^+-N to be absorbed and utilized by microalgae (Cai, Park, and Li 2013; Sanz-Luque et al. 2015). The removal mechanisms of inorganic nitrogen sources including ammonia volatilization, nitrification, and denitrification are virtuous. When pH value is more than 8, ammonia in NH_4^+-N can be converted to NH_3 by volatilization (Basílico et al. 2016). When some microalgae treat toxic substances, microalgae have a tolerance to a nitrogen source. The lack of a nitrogen source will reduce the ability of microalgae to resist the toxicity of pollutants, which provides a new idea for microalgae to treat nitrogen-containing wastewater (Yi et al. 2020).

When phosphorus in phosphorus-containing wastewater is utilized by microalgae, inorganic phosphorus (phosphate, hydrogen phosphate, and phosphide) is mainly used. While organic phosphorus is not able to be directly utilized by microalgae, inorganic phosphorus is mainly used by microalgae as raw material for the preparation of nucleic acid, adenosine triphosphate (ATP), adenosine diphosphate (ADP), and some other enzymes (Cai, Park, and Li 2013). Therefore, it affects the synthesis and metabolism of polysaccharides, lipids, as well as the value-added division of microalgae. The removal mechanism of organic phosphorus is mainly through the biosorption of microalgae and the removal of organic substances secreted by microalgae. For example, carboxylesterase released from microalgae can degrade organic phosphorus into phosphate and remove it through inorganic phosphorus metabolism (Nanda et al. 2019). Studies on microalgae in treating different C, N, and P effluents are presented in Table 3.1.

TABLE 3.1
Treatment Effect of C, N, and P Wastewater By Microalgae Under Different Conditions

Microalgae species	Sewage type	Treatment effect	Reference
Spirulina platensis	High salinity wastewater (salinity 2.24%)	COD: 90.02%; TN: 79.96%; TP: 93.35%	(Zhou et al. 2017)
Chlorella sp. and *Phormidium* sp.	Tannery wastewater	COD: >90%; TN: 91.16%; TP: 88%	(Das et al. 2018)
Botryococcus sp.	Domestic wastewater	COD: 93.9%; TN: 59.9%; TP: 54.5%	(Gani et al. 2017)
Scenedesmus sp.	Simulated wastewater	COD: 73.66%; TN: 75.96%; TP: 95.71%	(Song et al. 2020)
Chlorella sp.	Rice mill effluent	NH_4^+-N: 97.6%; TP: 90.3%	(Singh and Singh 2015)
Scenedesmus sp.	Rice mill effluent	NH_4^+-N: 98.3%; TP: 92%	(Singh and Singh 2015)
Chlorella sp.	High ammonia nitrogen wastewater	NH_4^+-N: 97.83%	(Lei et al. 2020)
Microcystis aeruginosa	Anaerobic digested effluent	TOC: 78.31%; TN: 91.88%; TP: 98.24%	(Gong et al. 2020)

3.2.1.3 Pharmaceutical Wastewater Treatment

Pharmaceutical wastewater contains various drugs that may cause biological effects on humans, animals, and microorganisms such as anti-tumour, regulating hormone secretion, the antibacterial activity of antibiotics, and insecticidal activity of pesticides. Hence, the effluents from pharmaceutical industries and domestic wastewater could contain active ingredients which could be harmful to microorganisms in the environment, including doing significant damage to microalgae cells in the process of biological sewage treatment.

At present, the main types of drugs in pharmaceutical wastewater include antibiotics, hormones, anti-inflammatory and analgesic drugs, anti-tumour drugs, and their biological metabolic degrades. The treatment of pharmaceutical wastewater by microalgae mainly focuses on a few different aspects. One of the aspects is the study of the mechanism of microalgae treatment of pharmaceutical wastewater. The main degradation mechanisms are biosorption, biodegradation, induced photodegradation, and bioaccumulation. In previous studies, the removal rate of carbamazepine-containing wastewater by *Scenedesmus* sp. was up to 28% (Xiong et al. 2016). *Chlorella* sp. was used to treat ketoprofen and salicylic acid, and the removal rates were 63% and 100%, respectively (Ismail et al. 2017). In the study of treating metronidazole with *Chlorella* sp., it is proven that microalgae can produce metabolites such as polysaccharides and amino acids. These metabolites can form hydrogen bonds, ch–π bonds, and π–π dispersion bonds with metronidazole for antibiotic absorption (Hena, Gutierrez, and Croué 2020). *Chlorella vulgaris* was added to the photodegradation experiment of norfloxacin, and the addition of microalgae has demonstrated a photodegradation induction (Junwei Zhang 2012). When *Microcystis aeruginosa* is used to treat tetracycline, the main removal mechanisms involved are the processes of biosorption, bioaccumulation, and biodegradation. In the study conducted, the researcher has illustrated that the biodegradation effect of *Chlorella pyrenoidosa* is much lower than that of *M. aeruginosa*, which also shows that there are great differences in biodegradation levels due to different microalgae (Minmin et al. 2021).

Subsequently, the study of the environmental effects of microalgae treatment of pharmaceutical wastewater (Ting et al. 2021) is also one of the significant aspects of pharmaceutical wastewater treatment. The particularity of drugs as pollutants makes their environmental effects worthy of attention. The toxic effects of spiramycin (SPI), tigecycline (TGC), and amoxicillin (AMX) were studied. Taking *Chlorella pyrenoidosa* and *Anabaena cylindrica* as experimental objects, the concentration of 50% of maximal effect (EC_{50}) of *A. cylindrica* was lower than that of *C. pyrenoidosa*, indicating that *A. cylindrica* had a weaker tolerance of spiramycin (SPI), tigecycline (TGC), and amoxicillin (AMX), and the EC50 value could reach μg/L level, resulting in an obvious damage to microalgae (Xueqing et al. 2021). There are many studies here, and the main evaluative research and technical means will be introduced in the subsequent chapters. A partial study on the treatment of pharmaceutical wastewater by microalgae is presented in Table 3.2.

TABLE 3.2
Effect of Microalgae Treatment on Pharmaceutical Wastewater Under Different Conditions

Microalgae species	Drug	Treatment effect	References
Chlamydomonas sp. Tai-03	Tetracycline	100%	(Xie et al. 2019)
	Sulfamethoxazole	20%	(Jiu-Qiang et al. 2019)
Chlorella PY-ZU1	Tilmicosin	90.2–99.8%	
Chlorella pyrenoidosa	Sulfamethoxazole	99.3%	(Qian et al. 2020)
Green algae, diatom, and cyanobacteria assemblages	Ibuprofen, Oxybenzone, Triclosan, Bisphenol A,N,N-diethyl-3-methylbenzamide (DEET)	70–100%	(Si, Jiahui, and Zhiyou 2021)
Microcystis aeruginosa	Cefradine	85.19%	(Du et al. 2018)
Nannochloris sp.	Amoxicillin	58.20%	(Du et al. 2018)
	Sulfamethoxazole	32%	(Jiu-Qiang et al. 2019)
	Trimethoprim triclosan	70–100%	(Bai and Acharya 2016)
Phaeodactylum tricornutum	Oxytetracycline	97	(Sergio et al. 2016; Jiu-Qiang et al. 2019)
Scenedesmus obliquus	Sulfamethoxazole	1.4–62.3%	(Jiu-Qiang et al. 2019)
Scenedesmus obliquus	Sulfamethazine	27.7–46.8%	(Jiu-Qiang et al. 2019)

Further to that, pharmaceutical wastewater also contains heavy metals (i.e., lead, cadmium and nickel) which are likely to be present in the finished pharmaceutical substance that came from catalysts that are deliberately added to the process (Nessa, Khan, and Abu Shawish 2016). The treatment of microalgae in other wastewater other than enriched C, N, and P wastewater, pharmaceutical wastewater such as heavy metal wastewater, and dyeing textile wastewater have also been studied. The removal of heavy metals from wastewater by microorganisms is done mainly through two ways: extracellular adsorption and intracellular accumulation. Due to the presence of negatively charged groups at the cell wall of microalgae cells, such as carboxyl, nitrate anion, phosphate anion, hydroxyl anion, and alkoxy anion, it can effectively adsorb heavy metals and cations in printing and dyeing wastewater in order to achieve the removal effect. During the adsorption process, the surface potential of microalgae may change (from -20.80 to -27.50 mV) (Lixin et al. 2020) to accelerate the process of coprecipitation. In addition, some ions can accelerate bioaccumulation by stimulating microalgae cells to enhance the membrane permeability of microalgae.

3.2.1.4 Factors Affecting Microalgae Treatment of Wastewater

Many factors affect the treatment of wastewater such as light, temperature, CO_2 concentration, aeration intensity, and water quality. The factors affecting microalgae treatment of wastewater will be discussed in depth in the following section.

3.2.1.4.1 Light

The main energy and material source of microalgae for life activities is photosynthesis, so the light intensity and light wavelength (energy supply) will be affecting the process of photosynthesis. Optimal light intensity is the key to supporting microalgae photosynthesis. The extremely high light intensity could cause light damage to the cell of microalgae. Whereas too weak light will not normally meet the light energy supply of microalgae photosystems, PSI, and PSII, which will inevitably affect the process of photosynthesis. In the previous study, microalgae can grow normally under the light intensity of 5000–25000 lux. Under the light intensity of 8,000 lux, microalgae biomass yield is the highest, and the removal effect of nitrogen source and phosphorus in sewage is the best.

3.2.1.4.2 Temperature

Appropriate temperature is the key to the biological activity of microalgae. If the temperature is too high, microalgae cannot survive. While if the temperature is too low, the enzymes in microalgae cannot function normally, and the metabolism will be affected, hence reducing the rate of pollutants' removal. For example, the most suitable growth temperature range for spirulina SP is 28–35°C (Xiao and Teng 2014).

3.2.1.4.3 CO_2 Concentration

As the material basis of photosynthesis, CO_2 concentration affects the photosynthesis level of microalgae. In the carbon dioxide fixation of *Chlorella vulgaris*, the circulating electron transport, F-type ATPase and Calvin cycle genes of CO_2 tolerant algae strains are up-regulated, which confirms that high CO_2 concentration will change the metabolism and transcription level of microalgae.

3.2.1.4.4 Aeration Intensity

For microalgae, the main purpose of aeration is to increase CO_2 concentration and the contact between microalgae cells and pollutants. However, when the aeration intensity is too strong, for example, above 100 mL/min, the microalgal chlorophyll content is only one-fourth of that without aeration, the growth of microalgae will also be affected; studies have shown that under the condition of appropriate dissolved oxygen (7–8 mg/L), different aeration flow rates have confirmed that aeration affects CO_2 concentration, resulting in the inhibition of microalgae growth. On the other hand, the purpose of agitation is to reduce the alkalization of pH. This is because NH_3 may be generated during the death and degradation of microalgae cells. Also, during the decomposition of nitrogen-containing organic matter produced by microalgae cells, NH_3 could be generated to alkalize the water body hence affecting the biological performance of microalgae.

3.2.1.4.5 Water Quality

Wastewater quality mainly affects the degree of pollutant degradation and microalgal biological properties. The sewage quality mainly includes content and types of pollutants, inorganic salt concentration, and pH. The specific effects are as follows:

Type and content of pollutants: Microalgae have different adaptability to different pollutants. The types of pollutants in sewage (such as nutrients, drugs, and personal care products, microplastics and hormones) and existing forms of compounds (such as organic matter, ionic state, and polymeric state) could make the water quality significantly diverse, resulting in different treatment effects.

Inorganic salt concentration: Previous studies have shown that inorganic salt concentration can inhibit the photosynthetic PSI and PSII systems of microalgae. Zhou et al. (2017) mixed wastewater with different proportions of salinity via different salt concentrations, and the treatment results were significantly different under the treatment of *Spirulina* sp. The total nitrogen removal percentage decreased from 92% to 74% with salinity increasing from 0.93% to 3.04%.

pH: The effect of pH on the treatment effect of microalgae is mainly reflected in two aspects. pH affects the charge on the surface of microalgae, resulting in microalgae producing self-flocculation and different growth level of microalgae due to different pH values. Nonetheless, pH will lead to different forms of some pollutants. For example, under acidic conditions, carboxyl exists in the form of acid while ammonia exists in the form of an ammonium ion. However, under alkaline conditions, carboxyl anion exists, while ammonia exists more in the form of NH_3, which has a greater impact on the adsorption and membrane permeation metabolism of microalgae (Lixin et al. 2020).

3.2.1.5 Advantages and Main Defects

Microalgae have some advantages in wastewater treatment: autotrophy, CO_2 absorption and oxygen production, reducing carbon emission, and low output of waste algae compared with activated sludge. Of course, there are also major defects, which make it difficult to be used on a large scale. The main defects include the following.

3.2.1.5.1 Microalgae Have a Great Impact on Biological Properties

The stability of the biological performance of microalgae is poor: microalgae have a great impact on sewage collection and the environment in the process of sewage treatment, and their biological performance can change. Microalgae are difficult to adapt when the water quality of sewage varies greatly. With the change of light intensity and temperature, there are great changes in the biological activity and biomass increment rate of microalgae in the process of sewage treatment, which makes it difficult to realize industrialization.

3.2.1.5.2 *Difficult to Harvest Microalgae*

At present, the main collection methods of microalgae include centrifugation, filtration, flocculation, immobilization, air flotation, and their combination. However, each method has its disadvantages. Centrifugation is the most effective acquisition method, but it consumes a lot of energy and has low efficiency, so it is not suitable for large-scale use (Japar et al. 2017). The flocculation method requires more flocculants such as chitosan and aluminium chloride. The flocculation effect is obvious, and the operation is simple, but the collection temperature and pH have a great impact, and the treated water will also cause environmental pollution. The filtration method is relatively simple, easy to automate, has low energy consumption and can ensure that the biological activity of microalgae can be well maintained, but there are some problems such as serious membrane pollution and low efficiency. The methods of immobilization and air floatation have high operation costs, and the immobilization has the requirements of algae specificity, which is difficult to be industrialized.

3.2.1.5.3 *Energy Consumption (Light and Aeration)*

Abeysiriwardana-Arachchige et al. (2020) studied the energy demand of the algal-based sewage treatment and resource recovery (STARR) system. The additional energy consumption of dephosphorization operation is about 4.58 kWh/m^3, so it may be necessary to add more than 5% (wt.%) biochar to reduce energy consumption. In addition, according to the standard of 1 mg effluent, the energy consumed by microalgae culture, centrifugation, and hydrothermal liquefaction (HTL) is 1,860 kWh, 429.9 kWh, and 2,811.9 kWh, respectively (Abeysiriwardana-Arachchige et al. 2020). It can also be seen from the aforementioned discussion that microalgae are difficult to settle (recover) and need to consume a lot of energy for centrifugal recovery, which also brings a lot of disadvantages to industrialization.

3.2.1.5.4 *Metabolite Hazard*

Microalgae produce toxins in metabolism, which may lead to a variety of poisonings. In previous studies, microalgae poisoning was divided into six categories according to human poisoning status: paralytic shellfish poisonings (PSP), ciguatera fish poisonings (CFP), diabetic shellfish poisonings (DSP), neurotoxic shellfish poisonings (NSP), amnesic shellfish poisoning (ASP), and azaspiracid poisoning (AZP). In addition, microalgae increase the discharge of macromolecular organic substances, such as polysaccharides, amino acids and some dissolved organic nitrogen and phosphorus, which lead to water eutrophication and water bloom.

3.2.2 Activated Sludge Process Wastewater Treatment

The activated sludge process is the most widely used and mature biological treatment method in sewage treatment technology. Its comprehensive application mainly lies in its characteristics of easy acquisition, high processing efficiency, and high stability. At present, there are many activated sludge technologies for sewage treatment, mainly the activated sludge technology combined with other pretreatment means (such as advanced oxidation method, ultraviolet radiation, and ozone oxidation).

Activated sludge is the general name of microorganisms and their attached organic and inorganic substances. According to the aerobic situation, it is mainly divided into aerobic activated sludge and anaerobic activated sludge. According to its biological characteristics, anaerobic activated sludge is mainly used for the treatment of wastewater with high biochemical indicators, which is intended to carry out anaerobic fermentation degradation and preliminarily destroy the structure of organic substances. Aerobic activated sludge is used for secondary treatment of the initially damaged organic matter. Due to microbial characteristics, the content of dissolved oxygen directly affects the performance of sludge.

The main performance indexes of activated sludge include mixed liquid suspended solids (MLSS), sludge sedimentation ratio (SV), sludge index including sludge volume index (SVI), and sludge density index (SDI). The factors affecting the performance of activated sludge mainly include dissolved oxygen content (2–4 mg/L), temperature (15–25°C), nutrient content and its proportion (usually BOD:N:P = 100:5:1), pH value (6.0–9.0), and other influencing factors.

Common microorganisms in activated sludge mainly include hydrolysis bacteria, fermentation bacteria, hydrogen bacteria and acetic acid bacteria, methanogens, sulphate-reducing bacteria, nitrifying bacteria and denitrifying bacteria, and anaerobic protozoa. Methanogens are the central skeleton of anaerobic activated sludge, but the type and quantity of microorganisms change due to the environment of pollutants.

3.2.2.1 Enriched C, N, P Wastewater Treatment

High-carbon wastewater can be divided into two classes because of its high organic concentration. High biochemical oxygen demand (BOD) is worth organic wastewater, which can be degraded by activated sludge treatment. The other is organic wastewater with low BOD. Although this kind of wastewater has high organic content, it is difficult to be degraded by activated sludge. It needs to be treated with activated sludge again after pretreatment (with processes such as advanced oxidation and photocatalytic degradation). The treatment modes of this kind of wastewater mainly include (1) anaerobic first and then aerobic and (2) anaerobic first and facultative anaerobic then aerobic. In the treatment of high organic wastewater by activated sludge method, the larger organic matter is usually degraded by microorganisms into substances that can be more conducive to conversion and CO_2, but at this time, the COD in the sewage may not be reduced. At this time, aerobic activated sludge needs to be degraded into compounds that can be used by microorganisms through oxidation and hydrolysis.

Inorganic nitrogen sources (NO_3^-, NO_2^-, NH_4^+-N) in high-nitrogen wastewater can be transformed according to microbial conditions. At the same time, the existing form of inorganic nitrogen sources in nitrogen-containing wastewater will also affect the microbial community to produce dominant flora. The transformation mechanisms of inorganic nitrogen sources are mainly assimilation and absorption, ammonia volatilization, nitrification, and denitrification, which are similar to those of microalgae. Nitrification is the oxidation of NH_4^+ to NO_2^- followed by NO_3^- via the ammonia-oxidizing bacteria, ammonia-oxidizing archaea, and NO_2-oxidizing bacteria. Denitrification reduces NO_3^- to NO_2^-, NO, N_2O, and N_2 under the reduction of denitrifying bacteria. The specific conversion is as follows:

(1) Nitrification: $NH_4^+ + O_2 \rightarrow NO_2^-; NO_2^- + O_2 \rightarrow NO_3^-$ (3.1)
(2) Denitrification: $NO_3^- \rightarrow NO_2^- \rightarrow NO \rightarrow N_2O \rightarrow N_2$. (3.2)

Organic nitrogen source is preferentially converted to inorganic nitrogen by nitrifying bacteria (including direct conversion to nitrogen discharge) and then converted by inorganic nitrogen source conversion mechanism.

The removal of inorganic phosphorus from phosphorus-containing wastewater by the activated sludge process is mainly through the growth and utilization of activated sludge and the adsorption of extracellular polymeric substances (EPS) secreted by bacteria. Organic phosphorus needs to be preferentially transformed into inorganic phosphorus by microorganisms and removed by inorganic phosphorus transformation.

3.2.2.2 Antibiotic Pharmaceuticals Wastewater Treatment

Activated sludge can be used to treat antibiotics drugs in pharmaceutical wastewater. When antibiotics are treated by activated sludge, the death of microbial biomass will accelerate, resulting in the imbalance of the microbial community. It can also accumulate in organisms and endanger the ecosystem through the food chain. At the same time, it will also produce resistance, resulting in resistance genes. For example, a neglected genetic material called extracellular DNA (eDNA) can spread extracellular ARGs (eARGs) to environmental bacteria and the environment through horizontal gene transfer (HGT), which has become an important driving factor of multiple antibiotic resistance genes (ARGs). The existence of extracellular polymers (EPSs) can weaken the accumulation of antibiotic resistance genes (ARGs) and play a certain role in environmental protection.

3.2.2.3 Factors Affecting the Treatment Effect of Activated Sludge

3.2.2.3.1 Water Quality (Salinity, BOD Level, pH, Toxicity)

Application of activated sludge in sewage treatment in turn stimulates the growth of activated sludge. Based on this statement, the quality of sewage will directly affect the treatment effect of sewage. The main factors affecting water quality are as follows: (1) Salinity: Generally, most activated sludge is cultured in freshwater. Therefore, the inorganic concentration contained in sewage not only affects the cell osmotic pressure of microorganisms but also causes the deformation of microbial cells, further causing adverse effects to the biological performance of activated sludge and sewage treatment. Salinity increases the extracellular polymeric substances (EPS) content of activated sludge and decreases the content of protein (PN) and loosely bound extracellular polymeric substances (LB-EPS) which leads to better settleability of activated sludge. Moreover, salinity inhibits the dehydrogenase activity (DHA) of activated sludge (Chen et al. 2018). Moreover, the content and type of inorganic salts may directly affect enzyme activity. Most enzymes are composed of protein. The salt concentration can change the stability of enzymes, which affects the effect of sewage treatment. Higher accumulation of H_2O_2 was observed as NaCl increased in the growth medium. The highest amount of H_2O_2 was observed in cells grown in 600-mm NaCl (41.43 μm g^{-1} FW). The activity of ascorbate peroxidase-antioxidative enzyme (APX) in all treated cultures was higher than that of the control. The highest activity (249.76%) was observed in 400 mM NaCl solution (Yun et al. 2019) (2) BOD level: Biodegradation is the main degradation mechanism when organic substances are treated by the activated sludge method, while biochemical oxygen demand can directly determine the degradable level in sewage treatment and affect the process efficiency. (3) pH: pH value mainly affects the growth of sludge. Appropriate pH is conducive to microbial growth and various biochemical properties of microorganisms, such as sedimentation rate, extracellular polymer composition, enzyme activity, microbial flocculation degree, which may also change the structure of activated sludge. In addition, the influence of pH on the pollution itself also exists. For example, cefradine and ceftazidime are easy to be hydrolyzed under weak alkaline conditions. By adjusting the pH from 5 to 9, the hydrolysis rate of cefradine and ceftazidime is 47.42 times and 16.13 times, respectively; (4) Toxicity: Generally, pollutants have certain toxicity to organisms, and the toxicity of pollutants in sewage to activated sludge directly affects the species, quantity, growth and reproduction, and other life activities of microorganisms in activated sludge. Antibiotics are the most prominent here. Due to the nature of antibiotics, when treating antibiotic sewage, the types and content of antibiotics in sewage need to be strictly controlled. Studies have shown that antibiotics can upregulate or downregulate gene expression in activated sludge, produce resistance genes (Li et al. 2021), and can alter microbial communities (Beattie, Skwor, and Hristova 2020).

3.2.2.3.2 Biological Properties of Activated Sludge (Sludge Age)

Sludge age can be regarded as the service cycle of sludge. Generally, activated sludge is reused, but the service life of activated sludge will be different due to the difference in water quality. Therefore, in case of drastic change in water quality, it is necessary to monitor the removal capacity of activated sludge in real-time and update the performance of activated sludge. Sludge age affects the release of fatty acids and phosphorus. In the process of anaerobic acidification, C5 compounds decrease, resulting in the shortening of sludge age, which makes it more suitable for the process of anaerobic acidification.

3.2.2.3.3 Dissolved Oxygen Level

Dissolved oxygen level mainly affects the respiratory and metabolic activities of activated sludge. Whether in an anaerobic system, aerobic system, or facultative anaerobic system, dissolved oxygen level directly affects microbial activity.

It is reported that a dissolved oxygen range from 2 to 4 mg/L would provide a favourable environment for adaptability and sustainability (Tang et al. 2018). DO content below this range may result in an incomplete degradation of organic compounds, change the proportion of O_2 and CO_2 in water, and thus affect the treatment effect of activated sludge.

3.2.2.3.4 Aeration Intensity

Aeration has two functions: it reduces the concentration of CO_2 and N_2 in the water body, so that the dissolved oxygen level increases relatively. On the other hand, the agitation could increase the contact between dissolved oxygen in the water and activated sludge and pollutants and weaken sludge sedimentation. Among them, the concentration of CO_2 sometimes has a greater impact on pollutant removal than the level of dissolved oxygen.

3.2.2.4 Technology and Disadvantages of Activated Sludge Process

3.2.2.4.1 AAO Process

The design concept of this process is to take into account the simultaneous removal of organic matter, nitrogen, and phosphorus. The process is an anaerobic process (A) with the purpose of phosphate secretion. Anoxic process (A) is used for denitrification and denitrification of organic matter, and an aerobic process (O) is used for deep nitrogen and phosphorus removal. The drawback of the process lies in the contradiction between the long mud age of denitrification and the short mud age of phosphorus removal, which is difficult to get a good consideration, usually, as the denitrification effect is limited, and the management is complicated, and the cost is high.

3.2.2.4.2 Oxidation Ditch Process

This process uses rapid large circulation to realize the rapid mixing of sludge and pollutants, which is mainly used for organic matter and ammonia nitrogen removal. The disadvantage of the process is that the reactor covers a large area, the treatment efficiency is low, and the process material has obvious anaerobic and aerobic steps, so the efficiency of nitrogen and phosphorus removal is low.

3.2.2.4.3 Sequencing Batch Reactors (SBR) Process

This process is an intermittent treatment method with time and space separation, which is suitable for the condition of irregular and quantitative sewage discharge. The biggest disadvantage of the process is that each treatment needs to go through the process of water inlet, treatment, sedimentation, and drainage. The operation is complex and time consuming. In addition, it is difficult to deal with the condition when the drainage is too large.

3.2.2.4.4 Membrane Bioreactor (MBR) Process

This process introduces biofilm, which has high organic matter removal efficiency and a high degree of mechanization. However, due to the introduction of biofilm, there is serious membrane pollution and high capital construction cost.

3.2.2.4.5 *Comprehensive Disadvantages*

Energy consumption: Taking the AAO process and 1 mg effluent as an example, the energy consumption of aeration and mixing is 2,677.2 kWh, and the energy consumption of the pump is 950.2 kWh.

Large sludge: The continuous optimization of activated sludge treatment technology also makes the output of waste sludge increase day by day. For example, the daily output of sludge in Zouping County reaches 160,000 tons. Under the condition of various sludge treatment technologies, the treatment cost remains high.

Produce resistance gene: Antibiotic wastewater will induce resistance genes to activated sludge and can persist. Long-term treatment will lead to the decline of treatment effect.

3.2.3 Algal Symbiotic Wastewater Treatment

The activated sludge process is widely used in a variety of sewage treatments. However, based on the disadvantages of high energy consumption (aeration level) and high carbon emission (CO_2 emission), the concept of algae sludge symbiosis is proposed to combine the advantages of microalgae treatment and activated sludge treatment to supplement and alleviate the disadvantages of the two treatment technologies.

During the growth of microalgae, CO_2 is absorbed by the microalgae cells via photosynthesis and transformed into organic substances such as polysaccharides. In the process of sewage treatment, activated sludge needs O_2 supplement through aeration, which consumes a lot of mechanical energy and produces a lot of CO_2. It is reported that the CO_2 emission indirectly related to the sewage treatment plant reaches 54.3 Tg. Synergically, microalgae photosynthesis can make up for O_2 supply and reduce energy consumption caused by aeration. The organic matter produced by microalgae photosynthesis can be used as a nutrient source of activated sludge and reduce secondary pollution caused by microalgae metabolism. At the same time, the CO_2 produced by activated sludge can be used as the carbon source of microalgae photosynthesis to reduce carbon emission, and some small molecules and inorganic substances treated by activated sludge can also be used as the nutrient supply of microalgae. Moreover, microalgae can also strengthen the depth of sewage treatment and reduce the discharge of harmful substances to initially form a closed-loop treatment.

3.2.3.1 Algal Symbiotic Feasibility Analysis

Different species are bound to compete in the same system. Therefore, to maximize the treatment effect, the proportion of biomass needs to be optimized. The main microorganisms of activated sludge are bacteria and fungi, which are pathogenic microorganisms for microalgae themselves and can affect the growth of microalgae. Microalgae compete with microorganisms in the activated sludge system, affecting microbial growth. In the presence of microalgae, because of the pathogenicity of bacteria and fungi and their metabolism, algal toxins will be produced in the water body to resist the properties of activated sludge and reduce the biological properties of activated sludge. The presence of microalgae will affect the particle size and surface potential of activated sludge, thus affecting the flocculation and sedimentation of activated sludge.

Different kinds of pollutants, damage to microalgae, and activated sludge have been different, and selecting a suitable biomass ratio can reduce biological damage appropriately and play a symbiotic role. Different pollutant-removal effects were obtained through different inoculation rates of algae sludge (1:0, 9:1, 3:1, 1:1, 0:1 wt/wt). At the same time, their respective roles under different inoculation rates were explored. The results showed that microalgae played a leading role in biological nitrogen removal, and the assimilation of activated sludge helped to improve the removal rate of COD. It is also important to note that microalgae need photosynthesis. Therefore, the sufficiency of light will directly affect the biological performance of microalgae due to the shading of activated sludge. Therefore, further research is required in optimizing the biomass ratio of microalgae and activated sludge to achieve its best treatment performance.

3.2.3.2 Summary of Studies on Algal Symbiotic Wastewater Treatment

There are many studies on the algae bacteria symbiosis system, but there are few practical application cases at present. In this chapter, some of the current research results are summarized in Table 3.3 mainly from the perspective of different pollution types.

3.3 APPLICATIONS AND CHALLENGES OF TOXICITY EVALUATION OF MICROALGAE

The rapid development of the modern industry meets people's increasing material needs, but it also may produce serious health and environmental safety problems. The continuous advancement of urbanization and industrialization had caused various pollutants to enter the environment in various ways, and it may potentially impact aquatic organisms. The vast majority of chemical toxicity is unknown, so a chemical hazard assessment is needed to provide experimental toxicological data of the aquatic ecosystem.

Microalgae are single-celled photosynthetic organisms that can survive in freshwater, seawater, and land. They can use sunlight and carbon dioxide in the atmosphere for photosynthetic autotrophic growth. As an important contributor to primary productivity in the aquatic ecosystem, microalgae play a very important role in maintaining the stability and balance of an ecosystem (Yao et al. 2018). In addition, microalgae have the characteristics of a small individual, wide distribution, fast reproduction, short growth cycle, and sensitivity to environmental changes and can respond in a short time. Many scholars have studied the *in vivo* and *in vitro* exposure of chemicals at different nutritional levels, such as those of microorganisms, algae, and animals. In the web of science database, the

TABLE 3.3
Effect of Algae Bacteria Symbiosis Wastewater

Microalgae–bacteria consortium	Wastewater	Treatment effect			
		COD (%)	N (%)	P (%)	Other (%)
Chlorella vulgaris + activated sludge	Synthetic wastewater	83.6	89.4	91.4	
Scenedesmus sp. + Bacteria group	Municipal wastewater	92.3	95.7	98.1	
Chlorella. sorokiniana + aerobic sludge	Swine wastewater	62.3	82.7	58.0	
Chlorella sp. & *Scenedesmus* sp. + activated Sludge	Artificial municipal sewage		97 (NH_4^+) 100(NO_3)	100	
C. vulgaris + nitrifying-enriched activated sludge	Synthetic wastewater		100	45	
Wastewater-born algae + activated sludge	Municipal wastewater		100	92	
C. vulgaris + activated sludge	Municipal wastewater		100		
Microalgae and bacteria	Wastewater containing emerging contaminants				Caffeine (99) Ibuprofen (60)
Chlorella sorokiniana + bacteria	WWTP effluents				Sulfamethoxazole (54)
Chlorella vulgaris-Bacillus licheniformi	Synthetic wastewater	86.55	88.95	80.28	
Microcystis aeruginosa–Bacillus licheniformis	Synthetic wastewater	65.62	21.56	70.82	
Microalgae + activated sludge	Synthetic wastewater	84–94			Ibuprofen, naproxen, salicylic acid, triclosan and propylparaben (94 ± 1, 52 ± 43, 98 ± 2, 100 ± 0, 100 ± 0, respectively)

toxicity studies in recent 10 years (2011–2021) were searched and screened with the keywords "pesticide/heavy mental/antibiotic on algae/*Daphnia magna*/rotifer/zebrafish toxicity". Therefore, microalgae are ideal test organisms in toxicity tests to investigate the ecological effects and toxicological mechanism of pollutants in the environment (Zhang et al. 2016).

The adverse effects of substances on algal populations may be harmful to higher trophic organisms with the transmission of the food chain (Pesce, Bouchez, and Montuelle 2011). Alga is an essential research object to evaluate the biological toxicity of environmental pollutants. It has been widely used as a model organism to study the acute and chronic toxicity of substances in an aquatic ecosystem, which will provide valuable information for the environmental risk of organic or inorganic pollutants in water. In recent years, many studies have studied the impact of environmental pollutants on organisms in an aquatic ecosystem from different life levels, such as gene, molecule, cell, organ, individual, population and community, and analysed the damaging effect, impact law, and possible action mechanism of pollutants on algae through various analysis and detection methods (Ahmed and Rodrigues 2013; Al-Jamal et al. 2020; Radix et al. 2000). The results provide theoretical basis and data support for the standard-setting, prevention and control of pollutant discharge in the water environment, environmental protection of the aquatic ecosystem, and ecological risk assessment. At the same time, they also lay a good foundation for the development of aquatic toxicology.

3.3.1 Research Status of Microalgae in Chemical Risk Assessment and Impacts of Respective Source Towards Microalgae

3.3.1.1 Nanoparticles

The nanotechnology industry is developing rapidly, and its applications in electronics, environmental treatment, food, and other fields have brought many benefits to human life (Vance et al. 2015). However, with the increased production and use of nanoparticles (NPs), the increase of their exposure level in the environment also brings corresponding risks. Almost all metal nanoparticles have toxic effects on algae (Mahana, Guliy, and Mehta 2021). Research showed that nanoparticles can react with many aquatic organisms in a short time (Espinasse et al. 2018). Oukarroum et al. (2017) revealed that Nickel Oxide Nanoparticles (NiO-NPs) have some toxic effects on *Chlorella vulgaris*, which could inhibit algal cell division and chlorophyll synthesis, trigger cell oxidative stress, and produce reactive oxygen species (ROS). Furthermore, Li et al. (2017) found that *Chlorella* sp. cells exposed to NiO-NPs had cytoplasmic separation, cell shape shrinkage, plasma membrane rupture, cytoplasmic leakage, and other phenomena, indicating that they had a serious inhibitory effect on the growth of microalgae. According to Zhao et al. (2021), the effects of Ag-NPs on *Chlamydomonas reinhardtii* include chloroplast damage, inhibition of photosynthetic pigment synthesis, an increase of reactive oxygen species (ROS) and malondialdehyde (MDA) content, change of cell membrane permeability, reduction of cell size, and weakening of cell chlorophyll auto-fluorescence. These studies provide new information for a better understanding of the potential toxic risks of NPs in the aquatic environment.

3.3.1.2 Metal

The environmental pollution problem of heavy metals is becoming increasingly prominent; pollution is one of the most serious problems in the world (Danouche et al. 2020). Heavy metals have adverse effects on the metabolism and growth of microalgae. A study showed that Cr(VI) had a negative growth effect on the quantum yield (Fv/Fm) of *Mucidosphaerium pulchellum* and *Micractinium pusillum* when it exceed the concentration range of 100 μg/L (Bashir et al. 2021). By analyzing the changes of algal growth, oxidative damage markers, and antioxidants, Hamed et al. (2017) evaluated the different responses of *Chlorella sorokiniana* and *Scenedesmus acuminatus* under copper stress. Both algae were able to reduce the effects of copper stress and induce an antioxidant defence system to neutralize the oxidative damage caused by copper stress while *Scenedesmus acuminatus* had faster adaptability than *Chlorella sorokiniana* under the long-term exposure of copper. This study provides new data for the copper tolerance and copper removal ability of the two microalgae, making it possible to apply them in the treatment of copper-containing wastewater. Except for the study of single metals on algae, many studies have tried to study the toxicity of binary and mixed metals to microalgae (Koppel et al. 2018). Nevertheless, considering the complexity of natural water, it is difficult to reliably predict the environmental toxicity of algae based on simple mixture toxicity research. Thus, more research is needed to meet the challenge of assessing the risk of complex mixed heavy metal pollutants and study the effects on microalgae.

3.3.1.3 Medical Wastewater

The increasing numbers of humans, animals, and plants demand health protection, and this fact has promoted the development of the pharmaceutical industry. The discharge of pharmaceutical wastewater leads to the release of a large number of drugs and their metabolites into the environment. Cui et al. (2021) evaluated the acute toxicity and ecological risk of 17 kinds of disinfection by-products of medical wastewater to *Scenedesmus obliquus*. Photosystem damage and oxidative stress are the potential reasons of DBPs' toxicity to microalgae; after exposure to disinfection by-products, the levels of reactive oxygen species (ROS), superoxide dismutase (SOD) and malondialdehyde (MDA) in algal cells were increased to varying degrees, indicating that DBPs in wastewater had high ecological risk.

3.3.1.4 Antibiotics

Due to its biological activity, with the increasing use of antibiotics in human and veterinary drugs, most unmetabolized antibiotics are released into the environment, and different types of antibiotics can be observed in an aquatic environment (Yang et al. 2018). Due to the wide distribution and biological activity of antibiotics, their potential risks to aquatic ecosystems have attracted attention.

By investigating the toxicity of 13 different antibiotics to algae, Fu et al. (2017) showed that protein synthesis inhibitors were more toxic to algae than cell wall synthesis inhibitors. Sulfamonomethoxypyrimidine (SMM) is one of the broad-spectrum antibiotics commonly used in aquaculture. Although SMM is widely used in aquaculture and pollutes surface water, there is still insufficient information about the toxicity of SMM to aquatic organisms. Huang et al. (2014) studied the acute and chronic toxicity of SMM to five species of aquatic organism, and SMM was found to be more sensitive to microalgae than *Daphnia magna*, and it had little effect on fish. The effects of SMM in water bodies should also be carefully evaluated to predict the ecological impact of its residues after application and may provide data for reducing its use.

3.3.1.5 Pesticide

Paraquat is a widely used herbicide and has been banned or strictly restricted by some countries due to its high toxicity. Bai et al. (2019) investigated the sensitivity difference of *Microcystis aeruginosa* and *Chlorella aeruginosa* to paraquat by multi endpoint analysis, and the results showed that two kinds of algae had different performances in the photosynthetic process, antioxidant response, oxidative stress, submicroscopic structure change, and growth inhibition. The large-scale destruction of the cell wall and organelle damage of *Microcystis aeruginosa* by paraquat were observed by transmission electron microscopy (TEM). Wan et al. (2020) discussed the toxicity and biodegradation of trichlorfon (TCF) to freshwater algae *Chlamydomonas reinhardtii*. Although the growth of *Chlamydomonas reinhardtii* decreased with the increase of TCF concentration, the analysis of pigment content, chlorophyll fluorescence, and antioxidant enzymes showed that it can produce resistance and adapt to the existence of TCF. Its high tolerance to TCF may have the potential to remove TCF in a natural water environment and treat TCF polluted wastewater. Hence, we can see that the algal toxicity test is also helpful to find microalgae with strong tolerance and potential pollutant treatment potential.

3.3.1.6 Bactericide

Triclosan is commonly used as an antibacterial agent. However, the residue produced from it has potential to harm human health and the environment. Some regions have proposed to prohibit the addition of triclosan to some daily chemicals or control the additional content. There are still no relevant regulations in some regions, so the toxicity test results can provide more systematic and comprehensive evidence for the toxicity of triclosan to the ecosystem and provide a reliable scientific basis for supervision and decision-making. Xin et al. (2019) concluded that triclosan has a long-term impact on the biochemical components such as lipids, proteins, and nucleic acids of five non-target freshwater green algae, which will help to evaluate the toxicity of triclosan systematically and comprehensively in natural water bodies and formulate appropriate risk management strategies.

At present, the research trend of toxicity evaluation varies from the initial toxicity of a single pollutant to the study of mixed toxicities of pollutants, from one test organism to a variety of biological detection chemicals with different nutritional levels. There are also some studies aimed at developing multi-species modelling tools to predict the toxicity of different chemicals in the regulatory toxicology of test species with different nutritional levels (Singh et al. 2014; Sigurnjak Bureš et al. 2021).

3.3.2 Microalgae Toxicity Test

3.3.2.1 Obtainment, Selection, and Culture of Algae Species

3.3.2.1.1 Obtainment of Microalgae

The algae species and culture medium required for the experiment can be purchased from a special algae collection place, and then the microalgae can be activated and cultured according to the instructions. Table 3.4 shows the algal species collections established in some countries; parts of the countries have more than one algal species collection but we only list one in each country.

3.3.2.1.2 Selection of Experimental Algae Species

There are a variety of microalgae in the aquatic environment, which can be divided into marine microalgae and freshwater microalgae. From the research on microalgae toxicity evaluation at home and abroad, freshwater microalgae are mainly used as the test organisms in the toxicity test. Based on the current research, freshwater microalgae commonly used in chemical toxicity detection include *Chlorella* sp. (Liu et al. 2021), *Chlorella pyrenoidosa* (Gao et al. 2020), *Scenedesmus obliquus* (Sun et al. 2020), and *Microcystis aeruginosa* (Feng et al. 2020). Different species of microalgae may have different toxic responses to the same pollutant – that is the toxic effect of pollutants on algae is closely related to the species of tested algae (Dupraz et al. 2016). The selection of tested microalgae is one of the main factors affecting the toxicity evaluation results. Therefore, the selection of tested species is very important in the test. To obtain a fast and effective toxicity test method, the sensitivity of the tested microalgae to pollutants must be considered.

TABLE 3.4
Algae Collection of Some Countries

Country	Algae collection
China	Freshwater Algae Culture Collection at the Institute of Hydrobiology, FACHB
America	Culture collection of algae at the University of Texas at Austin, UTEX
England	Culture collection of algae and protozoa, CCAP
Germany	Culture Collection of Algae at Göttingen University, SAG
Australia	CSIRO Collection of Living Microalgae, CCLM
Japan	National Institute for Environmental Studies, NIES
Canada	Canadian Center for the Culture of Microorganisms, CCCM

3.3.2.1.3 Algae Culture Conditions

The growth state of microalgae is closely related to their surrounding environments, such as nutrients, light intensity, light–dark cycle, temperature, and pH. Light is divided into natural light or artificial light, but for convenience and easy control of the algae culture and experimental conditions, microalgae are mostly cultured using artificial light or in a light incubator. For example, Ouyang et al. (2018) studied the toxicity of nano-colloids to *Chlorella vulgaris*, and the culture condition of *Chlorella vulgaris* was grown in an artificial climate incubator at 25.0°C ± 0.5°C and 80% humidity, the illumination was 10 klux, and the light/dark ratio was 16:8 h. Only under suitable conditions can algae grow well and be suitable for experiments, thus the effects of environmental factors on algae growth are important.

3.3.2.2 Experimental Methods

In the published studies on the toxicity of substances to algae, different research had adopted different experimental methods of algae growth inhibition toxicity. At present, many countries and organizations in the world have developed a series of standard methods for algae toxicity tests, such as the standard methods for algae toxicity test provided by the Organization for Economic Cooperation and Development (OECD), Environmental Protection Agency (EPA), International Organization for Standardization (ISO), which scholars can explore and study based on these standards.

3.3.2.3 Toxicity Endpoints and Assessment Indices

More and more studies have shown that algae are emerging in the understanding of the toxic and inhibitory effects of chemicals. The endpoints of algae toxicity experiments mainly include growth inhibition, cytotoxicity, oxidative stress, genotoxicity, and so on. Based on the results of exposure experiments and combined with chemical analysis, microalgae can provide potential action modes and ecotoxicological effects of pollutants in the aquatic environment. Table 3.5 shows the endpoints used in the study of the toxic effects of some chemicals on algae.

When exposed to exogenous substances, the normal growth and metabolism of algae will be disturbed in many aspects, including algae growth and reproduction, algae cell morphological structure, photosynthesis, and enzyme activity. Many toxicity endpoints can detect the toxicity of pollutants to algae, and some evaluation indexes such as cell density, photosynthetic pigment content, protein content, enzyme activity, or other detection methods are often used to evaluate the effect of chemicals on microalgae.

3.3.2.3.1 Algal Density Determination

In the process of toxicity test, the number of algae may change, resulting in the phenomenon that algae are inhibited or promoted. Algae density is a basic index to evaluate algae biomass, the growth curve of algae can be obtained, and the growth rate or growth inhibition rate can be calculated by using this value. The number of algae can be determined by the absorbance-algae density curve

TABLE 3.5
Summary of a Few Key Toxicity Bioassays Endpoints in Algae

Chemicals	Algae	Endpoints	Reference
Copper	*Chlorella sorokiniana* *Scenedesmus acuminatus*	Algae growth inhibition Oxidative stress	(Hamed et al. 2017)
Herbicides	*Raphidocelissubcapitata*	Algae growth inhibition Algae PSII inhibition	(Glauch and Escher 2020)
Co_3O_4 nanoparticles	*Chlorella minutissima*	Genotoxicity cytotoxic effects oxidative stress	(Sharan and Nara 2020)
Azithromycin	*Raphidocelis subcapitata*	Algae PSII inhibition oxidative stress	(Almeida et al. 2021)
Oxytetracycline	*Chlorella vulgaris* *Phaeodactylum tricornutum* *Microcystis aeruginosa* *Nodularia spumigena*	Algae growth inhibition Algae PSII inhibition	(Siedlewicz et al. 2020)
Microplastic	*Skeletonema costatum*	Algae growth inhibition Algae PSII inhibition	(Zhang et al. 2017)

method, optical microscope, and blood cell counting method or flow cytometry. If the density concentration of algal cells is too high, dilution treatment shall be taken before counting.

3.3.2.3.2 Extraction and Determination of Algal Pigment

Microalgae contain many photosynthetic pigments, which play a central role in plant photosynthesis. The contents of chlorophyll *a*, chlorophyll *b*, and carotenoids are common indicators. The content of algae pigment can be determined by spectrophotometry or flow cytometry. The pigment is insoluble in water but soluble in organic solvents. It has been reported that organic reagents such as the methanol method and dimethylsulphoxide (DMSO) method are used to extract the pigment in algae cells, measure its absorbance at a specific wavelength, and calculate the pigment content according to the formula (Wang et al. 2021; Baruah and Chaurasia 2020). The autofluorescence of algae, compared with spectrophotometry, flow cytometry can directly and quickly detect the pigment content in the FL3 (>650 nm) fluorescence channel (Lou et al. 2013).

3.3.2.3.3 Detection of Nutrients and Enzymes in Algae

The normal operation of algal cells requires a variety of nutrients and enzymes to maintain. From the current research, studies on intracellular nutrients of algae after exposure to pollutants include carbohydrates, lipids, and proteins. The change of conditions makes algae produce oxidative stress response, resulting in the up-regulation or down-regulation of enzyme activity in cells. The research on the changes of enzymes in algae in toxicity experiments mainly focuses on the activities of several enzymes responsible for regulating intracellular redox balance, such as MDA, SOD, LPO, and GSH.

3.3.2.3.4 Observation of Ultrastructural and Molecular Damage

Chemicals can enter algae cells or adhere to the surface of algae cells. Electron microscopy techniques such as SEM and TEM are often used to observe the changes of surface and internal ultrastructure of algae cells after being exposed to pollutants. Single-cell gel electrophoresis (SCGE) is usually used to observe DNA damage at a molecular level, which can qualitatively and quantitatively analyse the extent of DNA damage. With the continuous improvement of this method, it has important applications in toxicology and environmental and ecological sciences. In recent years, many studies have used this method to analyse the genotoxicity of substances to algae (Zhang et al. 2014; Martinez, Di Marzio, and Saenz 2015; Schiavo et al. 2016).

3.3.3 Problems and Challenges of Toxicity Evaluation of Microalgae

The current development of technology has provided various instruments and chemical analysis methods to analyse and determine the impact of pollutants on organisms quickly and sensitively. However, there are still some limitations in the application of toxicity evaluation technology.

3.3.3.1 Effects of Environmental Factors on Algae Application

It has been previously pointed out that algae are sensitive to environmental changes and can respond in a short time, so they can be used for toxicity evaluation. However, the change of environmental factors rather than the exposure of pollutants can also make algae respond, which will affect the correct toxicity evaluation. Considering the effects of environmental factors on algal growth and toxic effects of pollutants, it is still a problem that algae are not applicable under some environmental conditions, which means that the toxicity of pollutants in the real environment may be overestimated or underestimated.

3.3.3.1.1 Temperature

Exceeding the optimum temperature may cause algae death, affecting the cytochemical composition, nutrient absorption, growth rate of algae, and the activity value of enzymes contained in algae. Though some algae can also grow in extreme environments, in general, the optimum growth temperature of common algae is 22–35°C (Singh and Singh 2015).

3.3.3.1.2 Light Intensity

Light intensity and light–dark cycle also play an important role in algae growth and biomass accumulation. Light is the basic condition for the normal growth and reproduction of algae, but an extremely strong or weak light source is not conducive to algae growth (Erickson, Wakao, and Niyogi 2015). In addition, some pollutants have light-shielding (Lou et al. 2013) or light instability and are prone to photolysis. These properties will indirectly affect our evaluation of the toxicity of pollutants in the natural environment.

3.3.3.1.3 pH Value

Different pH values will have a significant impact on the toxic effects of pollutants, and inappropriate pH will inhibit algae growth and photosynthesis. According to a recent study, increasing pH will enhance the toxic effect of metal and inhibit algae growth (Price et al. 2021). It can be seen that environmental factors not only affect the growth of algae but also indirectly lead to the difference of toxic effects of pollutants, which means that the study of the toxicity of some substances to algae may be limited by environmental factors, which will not be conducive to the toxicity study of chemicals. Perhaps new microalgae with a wide range of applications can be developed to solve this problem in the future.

3.3.3.2 Limitations of Algae

Some algae (e.g. *Microcystis aeruginosa*) have microcystins (Schreidah et al. 2020), these toxins pose a significant threat to ecosystems and drinking water quality, which are not suitable for environmental monitoring or environmental treatment. Therefore, the nature of algae species used in environmental biotechnology is also required to be considered. In addition, for some pollutants with low environmental concentration, their impact on algae may not be observed in a short time. For example, Zhang et al. (2017) showed that the growth of microalgae was not affected after 96 hours of exposure to plastic debris. The conclusion that a chemical is non-toxic cannot be drawn only based on the fact that a chemical has no toxic effect on algae. The long-term impact of chemicals on the ecological environment cannot be ignored. It is time-consuming to observe the long-term toxicity of low-concentration pollutants only by prolonging the exposure time. On the other hand, there may be unpredictable problems in the process. Considering that some adverse effects of pollution can only be seen in the environmental after a long time, algae with higher sensitivity than the existing algae may be developed to study the environmental risks of those low-concentration chemicals. Although bioassay is useful in assessing the toxicity of chemicals, there are still some limitations, including, for example, when assessing the impact of pollutants and chemicals, different results may be obtained due to highly specific experimental conditions – that is the repeatability of experiments is limited (Müller et al. 2016).

3.3.3.3 Limitations of Research Methods

Chemicals with complex structures need more accurate toxicity assessment. For estimating and assessing the likelihood of effects on aquatic organisms under chemical exposure, facing main challenges within current risk assessment requires the following: (1) the estimation of combined effects on aquatic organisms exposed to chemicals; (2) the linkage of events in an effect chain that is initiated by interaction and triggers subsequent key events across different biological scales, leading to an adverse outcome at the organism or population level; (3) the current research methods and indicators use a set of systems, and it may not be suitable for future research without developing new methods.

3.4 SUMMARY AND PROSPECTS

Microalgae is one of the most important primary producers on the earth; it fixes a large amount of carbon dioxide for the production of oxygen and various organic carbon compounds which play an important role in providing material and energy basis for the survival and development of other organisms.

On the basic level of wastewater treatment by biological methods, there are some deficiencies, or considerations in terms of feasibility, of single alga, single activated sludge, and algal bacteria symbiosis systems, so it is necessary to develop more rational treatment technologies in using the advantages of microalgae and bacteria, and further research is needed in the following areas:

(1) A more suitable treatment model of alga and bacteria should be established to explore the treatment efficiency and applicability range of the model from different angles.
(2) Screening of microalgal species that are more tolerant to sewage and techniques that do not readily generate resistance gene species and ameliorate this risk.
(3) The establishment of the most basic culture system for algae and bacteria is also of particular importance.
(4) The algal symbiosis system will be an important processing technique worth investigating because it may be effective in favoring the common advantages of the alga.

Harmful substances will affect the life activities, biomass, survival, growth, and reproduction of microalgae. The environmental exposure and ecological risk assessment of algae in chemicals are not only of great significance to the ecological risk research of pollutants but also of important theoretical reference value to the development of pollutant treatment technology based on microalgae. From the current research results, algae have been widely used in the toxicological study of chemicals and environmental pollution. Environmental exposure, impact, and toxicity analysis of pollutants are the basis of hazard assessment and ecological risk assessment. At present, aquatic ecosystem algal toxicology experiment has become one of the standard methods for environmental quality monitoring in many countries. However, only the results of the microalgae biological toxicity test are used to evaluate the safety of substances, which is vulnerable to individual differences and environmental factors. Combined with other tested organisms, research can improve the reliability of aquatic organisms' detection results and evaluate their toxic effects on aquatic organisms more systematically and comprehensively.

Now, the toxic effects of pollutants on microalgae are mainly reflected in the effects of microalgae on interspecific relationship, cell growth and structure, interference with intracellular antioxidant system, photosynthetic process, gene expression, absorption, degradation, and transformation of pollutants. Some chemicals may exist in the water environment in low doses for a long time, and their combined action with other pollutants may cause sustained and far-reaching harm. Attention should be paid to the long-term toxicity of low-concentration pollutants to microalgae, and even the impact on the aquatic ecosystem. In the future, we can strengthen research from the following aspects:

(1) A new type of microalgae should be developed to evaluate the toxicity of pollutants in different conditions due to the impact of environmental factors on their application.
(2) Not only carry out the scene simulation study of microalgae exposed to pollutants but also pay more attention to the actual exposure study. In this process, we should explore the migration and transformation behaviour and occurrence from changes of pollutants in multi-media environments and analyse the actual exposure level of pollutants to microalgae.
(3) Study the toxicity of chemicals at microalgae species or different trophic levels to obtain more reliable results and more accurately evaluate the possible toxic effects of pollutants.

3.5 CONCLUSIONS

Algal toxicity assessment can not only be used for risk assessment of emerging chemicals but also be combined with microalgae pollutant treatment technology to judge whether the treatment technology can effectively reduce the toxicity of wastewater and whether the effluent can meet the discharge requirements, which is conducive to ecological security. There are still some limitations in the application of microalgae. With the deepening of research, new algae species and new treatment methods may be developed in the future to better apply microalgae to environmental technology.

REFERENCES

Abeysiriwardana-Arachchige, I. S. A., S. P. Munasinghe-Arachchige, H. M. K. Delanka-Pedige, and N. Nirmalakhandan. 2020. "Removal and recovery of nutrients from municipal sewage: Algal vs. conventional approaches." *Water Research* no. 175:115709. doi: 10.1016/j.watres.2020.115709.

Ahmed, F., and D. F. Rodrigues. 2013. "Investigation of acute effects of graphene oxide on wastewater microbial community: A case study." *Journal of Hazardous Materials* no. 256–257:33–9. doi: 10.1016/j.jhazmat.2013.03.064.

Al-Jamal, O., H. Al-Jighefee, N. Younes, R. Abdin, M. A. Al-Asmakh, A. B. Radwan, M. H. Sliem, A. F. Majdalawieh, G. Pintus, H. M. Yassine, A. M. Abdullah, S. I. Da'as, and G. K. Nasrallah. 2020. "Organ-specific toxicity evaluation of stearamidopropyl dimethylamine (SAPDMA) surfactant using zebrafish embryos." *Science of The Total Environment* no. 741:140450. doi: 10.1016/j.scitotenv.2020.140450.

Almeida, A. C., T. Gomes, J. A. B. Lomba, and A. Lillicrap. 2021. "Specific toxicity of azithromycin to the freshwater microalga Raphidocelis subcapitata." *Ecotoxicology and Environmental Safety* no. 222:112553. doi: 10.1016/j.ecoenv.2021.112553.

Almomani, Fares, and Rahul R. Bohsale. 2021. "Bio-sorption of toxic metals from industrial wastewater by algae strains Spirulina platensis and Chlorella vulgaris: Application of isotherm, kinetic models and process optimization." *Science of the Total Environment* no. 755:142654. doi: 10.1016/j.scitotenv.2020.142654.

Bai, Fang, Yunlu Jia, Cuiping Yang, Tianli Li, Zhongxing Wu, Jin Liu, and Lirong Song. 2019. "Multiple physiological response analyses aid the understanding of sensitivity variation between Microcystis aeruginosa and Chlorella sp. under paraquat exposures." *Environmental Sciences Europe* no. 31 (1). doi: 10.1186/s12302-019-0255-4.

Bai, Xuelian, and Kumud Acharya. 2016. "Removal of trimethoprim, sulfamethoxazole, and triclosan by the green alga Nannochloris sp." *Journal of Hazardous Materials* no. 315:70–75. doi: 10.1016/j.jhazmat.2016.04.067.

Baruah, P., and N. Chaurasia. 2020. "Ecotoxicological effects of alpha-cypermethrin on freshwater alga Chlorella sp.: Growth inhibition and oxidative stress studies." *Environmental Toxicology and Pharmacology* no. 76:103347. doi: 10.1016/j.etap.2020.103347.

Bashir, Khawaja Muhammad Imran, Hyeon-Jun Lee, Sana Mansoor, Alexander Jahn, and Man-Gi Cho. 2021. "The effect of chromium on photosynthesis and lipid accumulation in two chlorophyte microalgae." *Energies* no. 14 (8). doi: 10.3390/en14082260.

Basílico, Gabriel, Laura de Cabo, Anahí Magdaleno, and Ana Faggi. 2016. "Poultry effluent bio-treatment with spirodela intermedia and periphyton in mesocosms with water recirculation." *Water, Air, & Soil Pollution* no. 227 (6):190. doi: 10.1007/s11270-016-2896-x.

Beattie, Rachelle E., Troy Skwor, and Krassimira R. Hristova. 2020. "Survivor microbial populations in post-chlorinated wastewater are strongly associated with untreated hospital sewage and include ceftazidime and meropenem resistant populations." *Science of the Total Environment* no. 740:140186. doi: 10.1016/j.scitotenv.2020.140186.

Cai, Ting, Stephen Y. Park, and Yebo Li. 2013. "Nutrient recovery from wastewater streams by microalgae: Status and prospects." *Renewable and Sustainable Energy Reviews* no. 19:360–369. doi: 10.1016/j.rser.2012.11.030.

Canizaresvillanueva, R. O., A. Ramos, A. I. Corona, O. Monroy, M. Delatorre, C. Gomezlojero, and L. Travieso. 1994. "Phormidium treatment of anaerobically treated swine wastewater." *Water Research* no. 28 (9):1891–1895. doi: 10.1016/0043-1354(94)90164-3.

Chai, Wai Siong, Jie Ying Cheun, P. Senthil Kumar, Muhammad Mubashir, Zahid Majeed, Fawzi Banat, Shih-Hsin Ho, and Pau Loke Show. 2021. "A review on conventional and novel materials towards heavy metal adsorption in wastewater treatment application." *Journal of Cleaner Production* no. 296:126589. doi: 10.1016/j.jclepro.2021.126589.

Chai, Wai Siong, Wee Gee Tan, Heli Siti Halimatul Munawaroh, Vijai Kumar Gupta, Shih-Hsin Ho, and Pau Loke Show. 2021. "Multifaceted roles of microalgae in the application of wastewater biotreatment: A review." *Environmental Pollution* no. 269:116236. doi: 10.1016/j.envpol.2020.116236.

Chan, Alison, Hamidreza Salsali, and Ed McBean. 2013. "Heavy metal removal (copper and zinc) in secondary effluent from wastewater treatment plants by microalgae." *ACS Sustainable Chemistry & Engineering* no. 2 (2):130–137. doi: 10.1021/sc400289z.

Chang, Yuanyuan, Zucheng Wu, Lei Bian, Daolun Feng, and Dennis Y. C. Leung. 2013. "Cultivation of Spirulina platensis for biomass production and nutrient removal from synthetic human urine." *Applied Energy* no. 102:427–431. doi: 10.1016/j.apenergy.2012.07.024.

Chen, Yujuan, Huijun He, Hongyu Liu, Huiru Li, Guangming Zeng, Xing Xia, and Chunping Yang. 2018. "Effect of salinity on removal performance and activated sludge characteristics in sequencing batch reactors." *Bioresource Technology* no. 249:890–899. doi: 10.1016/j.biortech.2017.10.092.

Cui, H., B. Chen, Y. Jiang, Y. Tao, X. Zhu, and Z. Cai. 2021. "Toxicity of 17 disinfection by-products to different trophic levels of aquatic organisms: Ecological risks and mechanisms." *Environmental Science & Technology* no. 55 (15):10534–10541. doi: 10.1021/acs.est.0c08796.

Danouche, M., N. El Ghachtouli, A. El Baouchi, and H. El Arroussi. 2020. "Heavy metals phycoremediation using tolerant green microalgae: Enzymatic and non-enzymatic antioxidant systems for the management of oxidative stress." *Journal of Environmental Chemical Engineering* no. 8 (5). doi: 10.1016/j.jece.2020.104460.

Das, Cindrella, Nagappa Ramaiah, Elroy Pereira, and K. Naseera. 2018. "Efficient bioremediation of tannery wastewater by monostrains and consortium of marine Chlorella sp and Phormidium sp." *International Journal of Phytoremediation* no. 20 (3):284–292. doi: 10.1080/15226514.2017.1374338.

Du, Yingxiang, Jing Wang, Zhiliang Wang, Oscar Lopez Torres, Ruixin Guo, and Jianqiu Chen. 2018. "Exogenous organic carbon as an artificial enhancement method to assist the algal antibiotic treatment system." *Journal of Cleaner Production* no. 194:624–634. doi: 10.1016/j.jclepro.2018.05.180.

Dupraz, V., N. Coquille, D. Menard, R. Sussarellu, L. Haugarreau, and S. Stachowski-Haberkorn. 2016. "Microalgal sensitivity varies between a diuron-resistant strain and two wild strains when exposed to diuron and irgarol, alone and in mixtures." *Chemosphere* no. 151:241–252. doi: 10.1016/j.chemosphere.2016.02.073.

Erickson, E., S. Wakao, and K. K. Niyogi. 2015. "Light stress and photoprotection in Chlamydomonas reinhardtii." *The Plant Journal* no. 82 (3):449–465. doi: 10.1111/tpj.12825.

Espinasse, B. P., N. K. Geitner, A. Schierz, M. Therezien, C. J. Richardson, G. V. Lowry, L. Ferguson, and M. R. Wiesner. 2018. "Comparative persistence of engineered nanoparticles in a complex aquatic ecosystem." *Environmental Science & Technology* no. 52 (7):4072–4078. doi: 10.1021/acs.est.7b06142.

Feng, L. J., X. D. Sun, F. P. Zhu, Y. Feng, J. L. Duan, F. Xiao, X. Y. Li, Y. Shi, Q. Wang, J. W. Sun, X. Y. Liu, J. Q. Liu, L. L. Zhou, S. G. Wang, Z. Ding, H. Tian, T. S. Galloway, and X. Z. Yuan. 2020. "Nanoplastics promote microcystin synthesis and release from cyanobacterial Microcystis aeruginosa." *Environmental Science & Technology* no. 54 (6):3386–3394. doi: 10.1021/acs.est.9b06085.

Flórez-Miranda, Liliana, Rosa Olivia Cañizares-Villanueva, Orlando Melchy-Antonio, Fernando Martínez-Jerónimo, and Cesar Mateo Flores- Ortíz. 2017. "Two stage heterotrophy/photoinduction culture of Scenedesmus incrassatulus: Potential for lutein production." *Journal of Biotechnology* no. 262.

Fu, L., T. Huang, S. Wang, X. Wang, L. Su, C. Li, and Y. Zhao. 2017. "Toxicity of 13 different antibiotics towards freshwater green algae Pseudokirchneriella subcapitata and their modes of action." *Chemosphere* no. 168:217–222. doi: 10.1016/j.chemosphere.2016.10.043.

Gadipelly, Chandrakanth, Antía Pérez-González, Ganapati D. Yadav, Inmaculada Ortiz, Raquel Ibáñez, Virendra K. Rathod, and Kumudini V. Marathe. 2014. "Pharmaceutical industry wastewater: Review of the technologies for water treatment and reuse." *Industrial & Engineering Chemistry Research* no. 53 (29):11571–11592. doi: 10.1021/ie501210j.

Gani, Paran, Norshuhaila Mohamed Sunar, Hazel Matias-Peralta, Radin Maya Saphira Radin Mohamed, Ab Aziz Abdul Latiff, and Umi Kalthsom Parjo. 2017. "Extraction of hydrocarbons from freshwater green microalgae (Botryococcus sp.) biomass after phycoremediation of domestic wastewater." *International Journal of Phytoremediation* no. 19 (7):679–685. doi: 10.1080/15226514.2017.1284743.

Gao, J., F. Wang, W. Jiang, J. Han, P. Wang, D. Liu, and Z. Zhou. 2020. "Biodegradation of Chiral Flufiprole in Chlorella pyrenoidosa: Kinetics, transformation products, and toxicity evaluation." *Journal of Agricultural and Food Chemistry* no. 68 (7):1966–1973. doi: 10.1021/acs.jafc.9b05860.

Glauch, L., and B. I. Escher. 2020. "The combined algae test for the evaluation of mixture toxicity in environmental samples." *Environmental Toxicology and Chemistry* no. 39 (12):2496–2508. doi: 10.1002/etc.4873.

Gong, Weijia, Yuhui Fan, Binghan Xie, Xiaobin Tang, Tiecheng Guo, Lina Luo, and Heng Liang. 2020. "Immobilizing Microcystis aeruginosa and powdered activated carbon for the anaerobic digestate effluent treatment." *Chemosphere* no. 244.

Hamed, S. M., S. Selim, G. Klock, and H. Abdelgawad. 2017. "Sensitivity of two green microalgae to copper stress: Growth, oxidative and antioxidants analyses." *Ecotoxicology and Environmental Safety* no. 144:19–25. doi: 10.1016/j.ecoenv.2017.05.048.

Hena, Sufia, Leo Gutierrez, and Jean-Philippe Croué. 2020. "Removal of metronidazole from aqueous media by C. vulgaris." *Journal of Hazardous Materials* no. 384.

Hotos, George N. 2021. "Culture growth of the cyanobacterium Phormidium sp. in various salinity and light regimes and their influence on its phycocyanin and other pigments content." *Journal of Marine Science and Engineering* no. 9 (8). doi: 10.3390/jmse9080798.

HU Zhang-xi, XU Ning, DUAN Shun-shan. 2012. "Research progress on energy microalga Botryococcus braunii "*Acta Ecologica Sinica* no. 31 (05):577–584.

Huang, D. J., J. H. Hou, T. F. Kuo, and H. T. Lai. 2014. "Toxicity of the veterinary sulfonamide antibiotic sulfamonomethoxine to five aquatic organisms." *Environmental Toxicology and Pharmacology* no. 38 (3):874–80. doi: 10.1016/j.etap.2014.09.006.

Ismaiel, Mostafa M. S., Yassin M. El-Ayouty, and Asmaa H. Al-Badwy. 2021. "Biosorption of cyanate by two strains of; Chlamydomonas reinhardtii: Evaluation of the removal efficiency and antioxidants activity." *International Journal of Phytoremediation* no. 23 (10).

Ismail, Maha M., Tamer M. Essam, Yasser M. Ragab, Abo El-khair B. El-Sayed, and Fathia E. Mourad. 2017. "Remediation of a mixture of analgesics in a stirred-tank photobioreactor using microalgal-bacterial consortium coupled with attempt to valorise the harvested biomass." *Bioresource Technology* no. 232.

Japar, Azima Syafaini, Nur Mutmainnah Azis, Mohd Sobri Takriff, and Nazlina Haiza Mohd Yasin. 2017. "Application of different techniques to harvest microalgae." *Transactions on Science and Technology* no. 4 (2):98–108.

Jie, CUI, MA Hong – yan, KONG De – cai, WANG Bei – bei, and YANG Zheng – ya. 2021. "Study on the grey relational order between sewage discharge and its influencing factors in the Yangtze river economic belt." *Huaiyin Institute of Technology* no. 30 (3):76–80.

Jiu-Qiang, Xiong, Govindwar Sanjay, B. Kurade Mayur, Paeng Ki-Jung, Roh Hyun-Seog, Khan Moonis Ali, and Jeon Byong-Hun. 2019. "Toxicity of sulfamethazine and sulfamethoxazole and their removal by a green microalga, Scenedesmus obliquus." *Chemosphere* no. 218.

Junwei Zhang, Dafang Fu, and Jilong Wu. 2012. "Photodegradation of Norfloxacin in aqueous solution containing algae." *Journal of Environmental Sciences* no. 24 (4):743–749.

Koppel, D. J., M. S. Adams, C. K. King, and D. F. Jolley. 2018. "Chronic toxicity of an environmentally relevant and equitoxic ratio of five metals to two Antarctic marine microalgae shows complex mixture interactivity." *Environmental Pollution* no. 242 (Pt B):1319–1330. doi: 10.1016/j.envpol.2018.07.110.

Lee, Yi-Chao, and Shui-Ping Chang. 2011. "The biosorption of heavy metals from aqueous solution by Spirogyra and Cladophora filamentous macroalgae." *Bioresource Technology* no. 102 (9):5297–5304. doi: 10.1016/j.biortech.2010.12.103.

Lei, Qin, Gao Mingzhen, Zhang Mengyuan, Feng Lihua, Liu Qiuhua, and Zhang Guoliang. 2020. "Application of encapsulated algae into MBR for high-ammonia nitrogen wastewater treatment and biofouling control." *Water Research* no. 187.

Li, Wang, Yuan Li, Li Zheng-Hao, Zhang Xin, and Sheng Guo-Ping. 2021. "Quantifying the occurrence and transformation potential of extracellular polymeric substances (EPS)-associated antibiotic resistance genes in activated sludge." *Journal of Hazardous Materials* no. 408.

Li, Xin, Hong-Ying Hu, and Jia Yang. 2010. "Lipid accumulation and nutrient removal properties of a newly isolated freshwater microalga, Scenedesmus sp. LX1, growing in secondary effluent." *New Biotechnology* no. 27 (1).

Li, Yongqing, Ran Xiao, Zonglai Liu, Xiujuan Liang, and Wei Feng. 2017. "Cytotoxicity of NiO nanoparticles and its conversion inside Chlorella vulgaris." *Chemical Research in Chinese Universities* no. 33 (1):107–111. doi: 10.1007/s40242-017-6246-3.

Lim, Hooi Ren, Kuan Shiong Khoo, Kit Wayne Chew, Chih-Kai Chang, Heli Siti Halimatul Munawaroh, P. Senthil Kumar, Nguyen Duc Huy, and Pau Loke Show. 2021. "Perspective of Spirulina culture with wastewater into a sustainable circular bioeconomy." *Environmental Pollution* no. 284:117492. https://doi.org/10.1016/j.envpol.2021.117492.

Liu, Shu, Jiayao Li, Seiichi Oshita, Mohammed Kamruzzaman, Minming Cui, and Wenhong Fan. 2021. "Formation of a hydrogen radical in hydrogen nanobubble water and its effect on copper toxicity in chlorella." *ACS Sustainable Chemistry & Engineering* no. 9 (33):11100–11109. doi: 10.1021/acssuschemeng.1c02936.

Lixin, Li, Liu Wanmeng, Liang Taojie, and Ma Fang. 2020. "The adsorption mechanisms of algae-bacteria symbiotic system and its fast formation process." *Bioresource Technology* no. 315.

Lou, L., Q. Yue, F. Liu, F. Chen, B. Hu, and Y. Chen. 2013. "Ecotoxicological analysis of fly ash and rice-straw black carbon on Microcystis aeruginosa using flow cytometry." *Ecotoxicology and Environmental Safety* no. 92:51–56. doi: 10.1016/j.ecoenv.2013.02.014.

Low, Sze Shin, Kien Xiang Bong, Muhammad Mubashir, Chin Kui Cheng, Man Kee Lam, Jun Wei Lim, Yeek Chia Ho, Keat Teong Lee, Heli Siti Halimatul Munawaroh, and Pau Loke Show. 2021. "Microalgae cultivation in palm oil mill effluent (POME) treatment and biofuel production." *Sustainability* no. 13 (6):3247.

Mahana, A., O. I. Guliy, and S. K. Mehta. 2021. "Accumulation and cellular toxicity of engineered metallic nanoparticle in freshwater microalgae: Current status and future challenges." *Ecotoxicology and Environmental Safety* no. 208:111662. doi: 10.1016/j.ecoenv.2020.111662.

Martinez, R. S., W. D. Di Marzio, and M. E. Saenz. 2015. "Genotoxic effects of commercial formulations of Chlorpyrifos and Tebuconazole on green algae." *Ecotoxicology* no. 24 (1):45–54. doi: 10.1007/s10646-014-1353-0.

Minmin, Pan, Lyu Tao, Zhan Lumeng, Matamoros Victor, Angelidaki Irini, Cooper Mick, and Pan Gang. 2021. "Mitigating antibiotic pollution using cyanobacteria: Removal efficiency, pathways and metabolism." *Water Research* no. 190.

Mittal, Yamini, Pratiksha Srivastava, Naresh Kumar, and Asheesh Kumar Yadav. 2020. "Remediation of fluoride contaminated water using encapsulated active growing blue-green algae, Phormidium sp." *Environmental Technology & Innovation* no. 19. doi: 10.1016/j.eti.2020.100855.

Müller, Y., L. Zhu, S. E. Crawford, S. Küppers, S. Schiwy, and H. Hollert. 2016. "The utility of exposure and effect-based analysis in the ecotoxicological assessment of transformation products." In *Assessing Transformation Products of Chemicals by Non-Target and Suspect Screening – Strategies and Workflows*, Vol. 2, 89–109. American Chemical Society.

Nanda, Manisha, Vinod Kumar, Nighat Fatima, Vikas Pruthi, Monu Verma, P. K. Chauhan, Mikhail S. Vlaskin, and Anatoly V. Grigorenko. 2019. "Detoxification mechanism of organophosphorus pesticide via carboxylestrase pathway that triggers de novo TAG biosynthesis in oleaginous microalgae." *Aquatic Toxicology* no. 209: 49–55.

Nessa, Fazilatun, S. A. Khan, and K. Y. I. Abu Shawish. 2016. "Lead, cadmium and nickel contents of some medicinal agents." *Indian Journal of Pharmaceutical Sciences* no. 78 (1):111–119. doi: 10.4103/0250-474X.180260.

Nsenga Kumwimba, Mathieu, Fangang Meng, Oluwayinka Iseyemi, Matthew T. Moore, Bo Zhu, Wang Tao, Tang Jia Liang, and Lunda Ilunga. 2018. "Removal of non-point source pollutants from domestic sewage and agricultural runoff by vegetated drainage ditches (VDDs): Design, mechanism, management strategies, and future directions." *Science of The Total Environment* no. 639:742–759. https://doi.org/10.1016/j.scitotenv.2018.05.184.

Oukarroum, A., W. Zaidi, M. Samadani, and D. Dewez. 2017. "Toxicity of nickel oxide nanoparticles on a freshwater green algal strain of chlorella vulgaris." *BioMed Research International* no. 2017:9528180. doi: 10.1155/2017/9528180.

Ouyang, S., X. Hu, Q. Zhou, X. Li, X. Miao, and R. Zhou. 2018. "Nanocolloids in natural water: Isolation, characterization, and toxicity." *Environmental Science & Technology* no. 52 (8):4850–4860. doi: 10.1021/acs.est.7b05364.

Pesce, S., A. Bouchez, and B. Montuelle. 2011. "Effects of organic herbicides on phototrophic microbial communities in freshwater ecosystems." In *Reviews of Environmental Contamination and Toxicology*, edited by David M. Whitacre, 87–124. Springer Nature.

Picardo, Marta C., José Luiz de Medeiros, Ofélia de Queiroz F. Araújo, and Ricardo Moreira Chaloub. 2013. "Effects of CO2 enrichment and nutrients supply intermittency on batch cultures of Isochrysis galbana." *Bioresource Technology* no. 143.

Plaza, M., M. Herrero, A. Cifuentes, and E. Ibanez. 2009. "Innovative natural functional ingredients from microalgae." *Journal of Agricultural and Food Chemistry* no. 57 (16):7159–7170. doi: 10.1021/jf901070g.

Price, G. A. V., J. L. Stauber, A. Holland, D. J. Koppel, E. J. Van Genderen, A. C. Ryan, and D. F. Jolley. 2021. "The influence of pH on zinc lability and toxicity to a tropical freshwater microalga." *Environmental Toxicology and Chemistry* no. 40 (10):2836–2845. doi: 10.1002/etc.5177.

Qian, Xiong, Liu You-Sheng, Hu Li-Xin, Shi Zhou-Qi, Cai Wen-Wen, He Liang-Ying, and Ying Guang-Guo. 2020. "Co-metabolism of sulfamethoxazole by a freshwater microalga Chlorella pyrenoidosa." *Water Research* (prepublish) no. 175:115656.

Radix, Pascal, Marc Léonard, Christos Papantoniou, Gilles Roman, Erwan Saouter, Sophie Gallotti-Schmitt, Hervé Thiébaud, and Paule Vasseur. 2000. "Comparison of four chronic toxicity tests using algae, bacteria, and invertebrates assessed with sixteen chemicals." *Ecotoxicology and Environmental Safety* no. 47 (2):186–194. doi: 10.1006/eesa.2000.1966.

Rochaix, Jean-David. 1995. "Chlamydomonas reinhardtii as the photosynthetic yeast." *Annual Review of Genetics* no. 29.

Romera, E., F. González, A. Ballester, M. L. Blázquez, and J. A. Muñoz. 2007. "Comparative study of biosorption of heavy metals using different types of algae." *Bioresource Technology* no. 98 (17):3344–3353. doi: 10.1016/j.biortech.2006.09.026.

Sanz-Luque, Emanuel, Alejandro Chamizo-Ampudia, Angel Llamas, Aurora Galvan, and Emilio Fernandez. 2015. "Understanding nitrate assimilation and its regulation in microalgae." *Frontiers in Plant Science* no. 6 (899). doi: 10.3389/fpls.2015.00899.

Schiavo, S., M. Oliviero, M. Miglietta, G. Rametta, and S. Manzo. 2016. "Genotoxic and cytotoxic effects of ZnO nanoparticles for Dunaliella tertiolecta and comparison with SiO2 and TiO2 effects at population growth inhibition levels." *Science of The Total Environment* no. 550:619–627. doi: 10.1016/j.scitotenv.2016.01.135.

Schreidah, C. M., K. Ratnayake, K. Senarath, and A. Karunarathne. 2020. "Microcystins: Biogenesis, toxicity, analysis, and control." *Chemical Research in Toxicology* no. 33 (9):2225–2246. doi: 10.1021/acs.chemrestox.0c00164.

Sergio, Santaeufemia, Torres Enrique, Mera Roi, and Abalde Julio. 2016. "Bioremediation of oxytetracycline in seawater by living and dead biomass of the microalga Phaeodactylum tricornutum." *Journal of Hazardous Materials* no. 320.

Sharan, A., and S. Nara. 2020. "Exposure of synthesized Co3O4 nanoparticles to Chlorella minutissima: An ecotoxic evaluation in freshwater microalgae." *Aquatic Toxicology* no. 224:105498. doi: 10.1016/j.aquatox.2020.105498.

Si, Chen, Xie Jiahui, and Wen Zhiyou. 2021. "Removal of pharmaceutical and personal care products (PPCPs) from waterbody using a revolving algal biofilm (RAB) reactor." *Journal of Hazardous Materials* no. 406.

Siedlewicz, Grzegorz, Adam Żak, Lilianna Sharma, Alicja Kosakowska, and Ksenia Pazdro. 2020. "Effects of oxytetracycline on growth and chlorophyll a fluorescence in green algae (Chlorella vulgaris), diatom (Phaeodactylum tricornutum) and cyanobacteria (Microcystis aeruginosa and Nodularia spumigena)." *Oceanologia* no. 62 (2):214–225. doi: 10.1016/j.oceano.2019.12.002.

Sigurnjak Bureš, M., M. Cvetnić, M. Miloloža, D. Kučić Grgić, M. Markić, H. Kušić, T. Bolanča, M. Rogošić, and Š. Ukić. 2021. "Modeling the toxicity of pollutants mixtures for risk assessment: A review." *Environmental Chemistry Letters* no. 19 (2):1629–1655. doi: 10.1007/s10311-020-01107-5.

Silva, Teresa Lopes, Alberto Reis, Roberto Medeiros, Ana Cristina Oliveira, and Luisa Gouveia. 2009. "Oil production towards biofuel from autotrophic microalgae semicontinuous cultivations monitorized by Flow Cytometry." *Applied Biochemistry and Biotechnology* no. 159 (2).

Singh, K. P., S. Gupta, A. Kumar, and D. Mohan. 2014. "Multispecies QSAR modeling for predicting the aquatic toxicity of diverse organic chemicals for regulatory toxicology." *Chemical Research in Toxicology* no. 27 (5):741–753. doi: 10.1021/tx400371w.

Singh, S. P., and Priyanka Singh. 2015. "Effect of temperature and light on the growth of algae species: A review." *Renewable and Sustainable Energy Reviews* no. 50:431–444. doi: 10.1016/j.rser.2015.05.024.

Skjånes, Kari, Peter Lindblad, and Jiri Muller. 2007. "BioCO 2—A multidisciplinary, biological approach using solar energy to capture CO 2 while producing H 2 and high value products." *Biomolecular Engineering* no. 24 (4).

Song, Chunfeng, Xiaofang Hu, Zhengzheng Liu, Shuhong Li, and Yutaka Kitamura. 2020. "Combination of brewery wastewater purification and CO 2 fixation with potential value-added ingredients production via different microalgae strains cultivation." *Journal of Cleaner Production* no. 268.

Sun, Y., X. Zhang, L. Zhang, Y. Huang, Z. Yang, and D. Montagnes. 2020. "UVB radiation suppresses antigrazer morphological defense in Scenedesmus obliquus by inhibiting algal growth and carbohydrate-regulated gene expression." *Environmental Science & Technology* no. 54 (7):4495–4503. doi: 10.1021/acs.est.0c00104.

Taimbú, Carlos Andrés, Mariano Martín, and Ignacio E. Grossmann. 2019. "Process optimization for the hydrothermal production of algae fuels." *Industrial & Engineering Chemistry Research* no. 58 (51):23276–23283. doi: 10.1021/acs.iecr.9b05176.

Tang, Cong-Cong, Yu Tian, Heng Liang, Wei Zuo, Zhen-Wei Wang, Jun Zhang, and Zhang-Wei He. 2018. "Enhanced nitrogen and phosphorus removal from domestic wastewater via algae-assisted sequencing batch biofilm reactor." *Bioresource Technology* no. 250.

Ting, Zhou, Cao Leipeng, Zhang Qi, Liu Yuhuan, Xiang Shuyu, Liu Tongying, and Ruan Roger. 2021. "Effect of chlortetracycline on the growth and intracellular components of Spirulina platensis and its biodegradation pathway." *Journal of Hazardous Materials* no. 413.

Vance, M. E., T. Kuiken, E. P. Vejerano, S. P. McGinnis, M. F. Hochella, Jr., D. Rejeski, and M. S. Hull. 2015. "Nanotechnology in the real world: Redeveloping the nanomaterial consumer products inventory." *Beilstein Journal of Nanotechnology* no. 6:1769–1780. doi: 10.3762/bjnano.6.181.

Wan, L., Y. Wu, H. Ding, and W. Zhang. 2020. "Toxicity, biodegradation, and metabolic fate of organophosphorus pesticide trichlorfon on the freshwater algae Chlamydomonas reinhardtii." *Journal of Agricultural and Food Chemistry* no. 68 (6):1645–1653. doi: 10.1021/acs.jafc.9b05765.

Wang, X., Y. Li, S. Wei, L. Pan, J. Miao, Y. Lin, and J. Wu. 2021. "Toxicity evaluation of butyl acrylate on the photosynthetic pigments, chlorophyll fluorescence parameters, and oxygen evolution activity of Phaeodactylum tricornutum and Platymonas subcordiformis." *Environmental Science and Pollution Research*. doi: 10.1007/s11356-021-15070-3.

Xiao, Jun, and Jilin Teng. 2014. "Advances in plant Spirulina sp." *Biology Teaching* no. 39 (05):7–8.

Xie, Peng, Shih-Hsin Ho, Jing Peng, Xi-Jun Xu, Chuan Chen, Zi-Feng Zhang, Duu-Jong Lee, and Nan-Qi Ren. 2019. "Dual purpose microalgae-based biorefinery for treating pharmaceuticals and personal care products (PPCPs) residues and biodiesel production." *Science of the Total Environment* no. 688.

Xin, X., G. Huang, C. An, H. Weger, G. Cheng, J. Shen, and S. Rosendahl. 2019. "Analyzing the biochemical alteration of green algae during chronic exposure to triclosan based on synchrotron-based Fourier transform infrared spectromicroscopy." *Analytical Chemistry* no. 91 (12):7798–7806. doi: 10.1021/acs.analchem.9b01417.

Xiong, Jiu-Qiang, Mayur B. Kurade, Reda A. I. Abou-Shanab, Min-Kyu Ji, Jaeyoung Choi, Jong Oh Kim, and Byong-Hun Jeon. 2016. "Biodegradation of carbamazepine using freshwater microalgae Chlamydomonas Mexicana and Scenedesmus obliquus and the determination of its metabolic fate." *Bioresource Technology* no. 205.

Xueqing, Zhong, Zhu Yali, Wang Yujiao, Zhao Quanyu, and Huang He. 2021. "Effects of three antibiotics on growth and antioxidant response of Chlorella pyrenoidosa and Anabaena cylindrica." *Ecotoxicology and Environmental Safety* no. 211.

Yang, Y., W. Song, H. Lin, W. Wang, L. Du, and W. Xing. 2018. "Antibiotics and antibiotic resistance genes in global lakes: A review and meta-analysis." *Environment International* no. 116:60–73. doi: 10.1016/j.envint.2018.04.011.

Yao, K., X. Lv, G. Zheng, Z. Chen, Y. Jiang, X. Zhu, Z. Wang, and Z. Cai. 2018. "Effects of carbon quantum dots on aquatic environments: Comparison of toxicity to organisms at different trophic levels." *Environmental Science & Technology* no. 52 (24):14445–14451. doi: 10.1021/acs.est.8b04235.

Yi, Hu, Meng Fan-Li, Hu Yan-Yun, Habibul Nuzahat, and Sheng Guo-Ping. 2020. "Concentration- and nutrient-dependent cellular responses of microalgae Chlorella pyrenoidosa to perfluorooctanoic acid." *Water Research* no. 185.

Yun, Chol-Jin, Kum-Ok Hwang, Song-Su Han, and Hyong-Guan Ri. 2019. "The effect of salinity stress on the biofuel production potential of freshwater microalgae Chlorella vulgaris YH703." *Biomass and Bioenergy* no. 127.

Zhang, C., X. Chen, J. Wang, and L. Tan. 2017. "Toxic effects of microplastic on marine microalgae Skeletonema costatum: Interactions between microplastic and algae." *Environmental Pollution* no. 220:1282–1288. doi: 10.1016/j.envpol.2016.11.005.

Zhang, Xiaoyue, Xinqing Zhao, Chun Wan, Bailing Chen, and Fengwu Bai. 2016. "Efficient biosorption of cadmium by the self-flocculating microalga Scenedesmus obliquus AS-6–1." *Algal Research* no. 16:427–433. doi: 10.1016/j.algal.2016.04.002.

Zhang, Y. L., B. P. Han, B. Yan, Q. M. Zhou, and Y. Liang. 2014. "Genotoxicity of disinfection by-products (DBPs) upon chlorination of nine different freshwater algal species at variable reaction time." *Journal of Water Supply: Research and Technology-Aqua* no. 63 (1):12–20. doi: 10.2166/aqua.2013.107.

Zhao, Z., L. Xu, Y. Wang, B. Li, W. Zhang, and X. Li. 2021. "Toxicity mechanism of silver nanoparticles to Chlamydomonas reinhardtii: Photosynthesis, oxidative stress, membrane permeability, and ultrastructure analysis." *Environmental Science and Pollution Research* no. 28 (12):15032–15042. doi: 10.1007/s11356-020-11714-y.

Zhou, Weizhi, Yating Li, Yizhan Gao, and Haixia Zhao. 2017. "Nutrients removal and recovery from saline wastewater by Spirulina platensis." *Bioresource Technology* no. 245.

4 How Far Has the Development for Industrial Internet of Things (IoT) in Microalgae?

Vimal Angela Thiviyanathan[1], *Hooi Ren Lim*[2], *Pei En Tham*[2] *and Pau Loke Show*[2]

[1] Institute of Sustainable Energy, Universiti Tenaga Nasional, Kajang, Selangor, Malaysia

[2] Department of Chemical and Environmental Engineering, Faculty of Science and Engineering, University of Nottingham Malaysia, Jalan Broga, Semenyih, Selangor Darul Ehsan, Malaysia

CONTENTS

4.1 Introduction: What Is IoT? 145
4.2 Architecture of IoT 147
4.2.1 Benefits of IoT 153
4.3 The Utilisation of IoT in Microalgae Farming 155
4.3.1 Growth Factors that Affect Microalgae Cultivation 155
4.3.2 Downstream Processing of Microalgae Biomass 156
4.4 Current Application of IoT in Microalgae Cultivation 158
4.4.1 Enhancement of IoT System for Optimal Microalgae Cultivation 160
4.5 Issues and Challenges 161
4.6 Recommendations 163
4.7 Conclusions 163
References 164

4.1 INTRODUCTION: WHAT IS IOT?

Internet of Things (IoT) is mainly a system of interrelated, inter-connected devices, which collects and transfers data over a wireless network without the

DOI: 10.1201/9781003202196-4

intervention of humans (Gupta and Gupta, 2016). The market size of IoT is projected to reach USD 1463.19 billion by 2027 (Fortune Business Insights, 2021). This projection reflects the enormous potential of IoT in various industries. Some of the industries that are currently leveraging the benefits of IoT include the automobile industry (Krasniqi and Hajrizi, 2016; Srinivasan, 2018), manufacturing industry (Singh and Bhanot, 2020), agriculture industry (Pillai and Sivathanu, 2020; Malavade and Akulwar, 2016), and energy and utilities (Hossein Motlagh et al., 2020). The concept of IoT was first introduced by Kevin Ashton in 1999 (Wang et al., 2015) and has evolved as a convenient, cost-effective, and efficient system today.

There are a few key elements that are required to deliver the functionality of the IoT system. These include identification, sensing, communication, computation, services, and schematics. Identification provides detailed information of the objects within a network. The identifiers can be categorised into three major groups that are object identifier, communication identifier, and application identifier. Object identifiers are physical or virtual objects. Communication identifiers differentiate nodes in a network that has communication abilities. Application identifiers are objects and local entities such as Uniform Resource Identifier (URI) and Uniform Resource Locator (URL) (Aftab et al., 2020).

Sensing is the process of obtaining information from objects (Burhan et al., 2018). This information can be collected from a wearable device, appliances, a wall-mounted control, or several commonly found devices. The information can be in the form of biometric data, visual data, audio data, or biological data (Maiman, n.d.). The communication element represents how different devices or objects are connected to communicate with each other. The technologies that are involved in communication include Bluetooth, Radio Frequency Identification (RFID), Wi-Fi, Near Field Communication (NFC), ultra-wide bandwidth (UWB), Z-wave, 3G, 4G, GPS, and Long-Term Evolution-Advanced (LTE-A) (Akpakwu et al., 2017).

The computation process fine-tunes the information obtained from the sensors. Some of the platforms that offer these services include Arduino, Raspberry Pi, Intel Galileo, Tiny OS, Lite OS, and Android (Barik, 2019; Akpakwu et al., 2017). The cloud platform is also a crucial part of IoT as it has enhanced real-time data processing capabilities (Tracy, 2016). The service element can be divided into four groups, which are identity-related services, aggregation services, collaborative services, and ubiquitous services (Mashal et al., 2015). The identity-related services identify the objects that send the request. The aggregation services gather information from the object and process the information. Collaborative services are responsible for decision-making and transmitting suitable responses to the devices. At the end, the ubiquitous service is used to immediately respond to the devices without any time or place restrictions (Burhan et al., 2018). The final element of an IoT system is semantics. This element acts as the brain of IoT by performing human tasks. It makes a decision based on the data received and sends the information to the devices (Al-Osta et al., 2017).

IoT had six impacts on the microalgae biorefinery system. The first is to encourage better behaviour analysis where comprehensive information about the growth, environment, yield, and the interaction between these parameters is collected to predict the next action. The second impact is making accurate predictions by analysing the factors and the impact of these factors on the surroundings. Bringing about better decisions is another impact of IoT on the microalgae biorefinery system. Promoting a highly efficient automation process is also an impact of the IoT system where the flexibility to control the whole operation any time and from anywhere can optimise the microalgae production process. The next IoT impact is the guide to conservative consumption thinking where the decision made from the data collected will be able to reduce wastage of energy and resources. The last impact of IoT is playing a crucial role in industrial production. The application of IoT automates various processes, both simple and complicated, within the production line to allow humans to focus more on the important factors (Wang et al., 2021). These six factors are beneficial to key players as they point to more affordable production of microalgae with a maximised output.

4.2 ARCHITECTURE OF IOT

The most basic and commonly accepted concept of the IoT architecture consists of three layers: the sensing layer, the network layer, and the application layer (Figure 4.1). The sensing layer consists of sensors that can communicate and automatically exchange data among the devices (Gokhale et al., 2018). The main responsibility of the sensing layer is to collect the information from the surrounding. Some examples of the sensing unit include RFID; 2-D barcode; WebCam; smartphone; and common sensors such as pressure, temperature, and humidity sensors (Khattak et al., 2019). The devices in this layer also receive commands from the network layer to initiate an operation (Zhu and Huang, 2017).

These sensors are connected to a hardware component that enables the communication process. These devices can be categorised into two groups. The first group is wearable devices and gadgets, and the second group includes embedded systems (Singh and Kapoor, 2017). The demand for wearable devices and gadgets

FIGURE 4.1 The three-layer architecture of the IoT system.

shows an increasing trend across the years due to their portable and seamless characteristics (Ometov et al., 2021). Apart from that, wearable devices also provide hands-free and quick access to important data such as health condition, the identity of an individual, security, and fitness level of the user. This inference is supported by the analysis published by Cision PR Newswire which reported that the compound annual growth rate (CAGR) for wearable IoT devices is expected to reach $62.82 billion by 2025 (Wood, 2019). However, the advancement of wearable IoT devices is still at an infancy stage where more enhancement is needed to optimise the privacy, data processing, and communication aspects of the devices (Poongodi et al., 2020).

Embedded IoT systems are more commonly used in industries that focus on mass production. Some of the examples include, Beagle Board, Raspberry Pi, Arduino board, and ESP8266 (Payal and Singh, 2021). Each of these hardware devices is suitable for different applications. The following section provides a more comprehensive explanation of these hardware devices. BeagleBone Black was developed by Texas Instruments in 2008. It runs on an advanced RISC machine (ARM) processor (Payal and Singh, 2021) and consists of an Ethernet connection that enables the utilisation of various services such as File Transfer Protocol (FTP), Telnet, and Secure Socket Shell (SSH) (McLaughlin, 2015). Raspberry Pi has been one of the widely used hardware in both industrial and academic fields. The sales of Raspberry Pi reached 30 million in 2019, and this device is being constantly upgraded across the years (Tung, 2019). Arduino Board is also a widely used device as it has the advantage of being able to connect both Ethernet and Wi-Fi (Payal and Singh, 2021). Apart from that, the Arduino Board is developed based on the ATmega2560 AVR microcontroller. This microcontroller has the advantage of low power consumption with a fast start-up. It is also less complex as compared to the 32/64 bit versions (Shawn, 2019). At the end, the ESP8266 is hardware integrated with a 32-bit Tensilica processor with a low-noise amplifier. It also has filters and power management modules (ESPRESSIF, n.d.). Table 4.1 shows a comparison of a few parameters in these devices.

Based on Table 4.1, though Beagle Board has the highest cost, it has a larger number of input and output pins that enable the attachment of more sensors. Apart from that, Beagle Board also has one of the most successful GPUs that enable an efficient system operation as well as a large number of supported interphases (Har-Even, 2021). (Santos and Perestrelo, n.d.). All these advantages justify the pricing for Beagle Board. On the other hand, the advantages of ESP8266 have greatly surpassed its price value. It has a bigger memory as compared to Arduino with an inbuilt Wi-Fi functionality (Mehta, 2015). Its low-energy consumption feature is an important factor that can lower production costs (Patnaik Patnaikuni, 2017).

The network layer gathers and transmits information from the sensing layers to the application layer via various networks (Al Hinai and Singh, 2017). It also connects sensors, network devices, and servers (Burhan et al., 2018). Some of the technologies in this layer comprise wireless networks such as 2G, 3G, 4G, and satellite networks (Khattak et al., 2019). The application layer refers to all applications that use IoT technology. It analyses and processes the information from the

TABLE 4.1
Comparison Between Different IoT Hardware

	Beagle Board	Raspberry Pi	Arduino Board	ESP8266	Reference
Chip	OMAP3530 System	RP2040	ATMega8U2	ESP8266	(Nayyar and Puri, 2016; Shah, 2018)
CPU	Broadcom BCM2835 SoC full HD multimedia applications	TI AM3359	Atmel SAM3X8E ARM Cortex-M3	Single-core RISC-V 32-bit	(Patel and Devaki, 2019; Ishchenko and Nuqui, 2018)
GPU	Arm® Cortex® A8 core PowerVR SGX™ 530 graphics processingunit	Broadcom VideoCore VI (32-bit)	250-MHz, 28-nm PowerVR Series5XE GX5300	-	(He et al., 2018; Sajim et al., 2017; Mittal et al., 2020)
Memory	-512 MB DDR3 —2 GB (4GB with Rev C)	256 MB to 1 GB RAM	32 KB	4MB	(Monk, 2016; Patel and Devaki, 2019; Patnaik Patnaikuni, 2017)
Operating System	Linux	Linux, UNIX, Windows, Solaris, Mac OS X, Novell, NetWare	Windows, Mac OS X or Linux.	Linux/ Mac	(Tutorials, n.d.; Thaker, 2016)
Price	\$63	\$55	\$29.90	\$7.50	(DF Robot, n.d.; Califano, 2019; Upswift, 2021)

sensing layer. Some examples of the software in this layer include ThingSpeak, Google Cloud IoT Core, IBM Watson IoT Platform, G2 Deals, Datadog, AWS IoT Analytics, ThingsBoard, Ubidots, Sensor Cloud, and Blynk. The description of such software is given in Table 4.2.

Based on Table 4.2, each of the software has some advantages over the other and is more suitable for different applications. The best software for different industries can be determined based on few key points such as scalability, data management, adaptability to various sensors, security, connectivity, and usability (Tiempo Development, 2020). The architecture of IoT has been further fabricated to four-layer architecture, five-layer architecture, cloud-based architecture, fog-based architecture, and service-oriented architecture (Al-Qaseemi et al., 2016; Khattab et al., 2016; Hou et al., 2016). As observed in Figure 4.2, four-layer architecture has three basic layers with an additional

TABLE 4.2
Different Analytic Software for IoT Systems

Software	Description	Reference
Microsoft Azure Stream Analytics	— Performs real-time analytics — Deploys AI-powered analytic methods — Runs in the cloud, for large-scale analytics, or in IoT Edge or Azure Stack for ultra-low expectancy analytics.	(Microsoft, 2020; Tolem et al., 2020)
AWS IoT Analytics	— Has pre-built analytical functions with predictive analysis of data — Has tools to clean up data — Has multiple data storage channels to organise the data from multiple devices	(Bastos, 2019; Walker, 2017; Maureira, 2014)
IBM Watson IoT Platform	— Inbuilt security elements to ensure the protection of data — User friendly	(MacGillivray, 2016)
ThingSpeak	— Only open data platform specifically designed for the IoT in 'the cloud' — Uses Phusion Passenger Enterprise, an application server that is supported by Python language	(Maureira, 2014; Nettikadan and Raj, 2018)
Cisco Data Analytics	— Enables quick access to data. — Provides cross-network security and privacy for every aspect of the network	(Cisco, 2015)
Blynk	— Has wide interfaces for data visualisation and analysis — Allows communications between the smartphone and hardware — Can handle thousands of devices and can even be launched on a Raspberry Pi.	(Chavan et al., 2020)

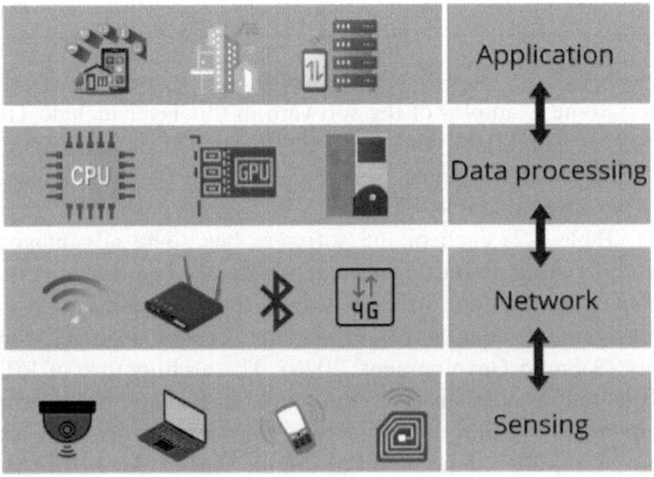

FIGURE 4.2 Four-layer IoT architecture.

layer called data processing (Laubhan et al., 2016). The data processing layer was included to enhance the security of the system. The three-layer architecture is venerable to threats as it sends information directly to the network layer. The data-processing layer in the four-layer architecture confirms the credibility of the data sent by the user and ensures the data is protected from threats. Some of the examples of authentication are using pre-shared keys and passwords (Burhan et al., 2018).

The five-layer architecture has two extra layers from the three-layer architecture (perception layer and application layer). The additional layers are the middleware layer, network layer, and business layer. The network layer is responsible to transmit the data from the perception layer to the middleware layer through several networks such as 3G, LAN, Bluetooth, RFID, and NFC. The advantages and disadvantages of these networks are summarised in Table 4.3.

Based on Table 4.3, though the utilisation of RFID does not require power, it is not compatible with smartphones. Thus, it is crucial to overcome this limitation as a very big percentage of the global population uses smartphones. Catering to the convenience of smartphone users will increase the market potential of RFID. As LAN has better protection against third-party interference and is compatible with smartphones, it has a greater potential to be commercialised. However, more work is required to enhance the connectivity of LAN so that a stable connection can be established. The middleware layer stores evaluate and process large data that is transmitted from the transport layer. The technologies that are available in this layer include cloud computing modules and big data modules (Burhan et al., 2018). The five-layer architecture is able to better interpret the meaning and features of the IoT (Zhong et al., 2015). An illustration of the five-layer IoT architecture is shown in Figure 4.3.

TABLE 4.3
Comparison of the Networks in Transport Layer (Mehl, 2021)

Network	Advantages	Disadvantages
3G	— Stable connection	— High consumption of power
LAN	— Compatible with smartphones — Better protection against hacking	— High-power usage — Unstable connection
Bluetooth	— Widely used technology — Constantly upgraded through new hardware.	— Low battery lifetime — Limited hardware capabilities. Need to be frequently upgraded and improved.
RFID	— Does not require power	— Not well-suited with smartphones — Insecure
NFC	— Requires a low-speed connection with a simple setup — Supports encryption — Can be used to bootstrap more wireless connections	— Short-ranged and may not be feasible in every situation

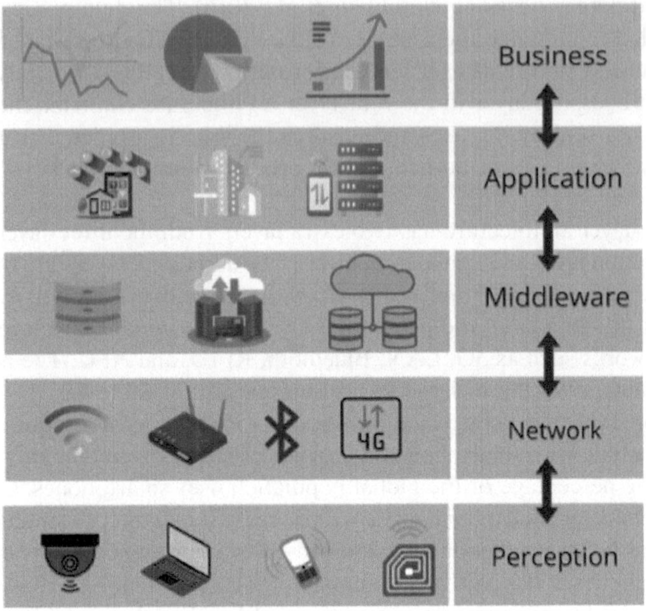

FIGURE 4.3 Five-layer IoT architecture.

The cloud-based architecture consists of the sensing layer, the gateway cloud, and the back application layer (Figure. 4.4) (Khattab et al., 2016). The sensing layer comprises interface for the client that requires access to the cloud computing platforms. Some of the sensing layer interphases include web servers such as Firefox, Chrome, and Internet explorer and mobile devices. The back application layer manages all resources that are required to provide the cloud computing services. This layer is mainly utilised by the service provider. It stores a large amount of data with security and traffic control mechanisms (Melo et al., 2017; Java T Point, n.d.). Cloud-based architecture is one of the effective data-processing platforms as it has high-storage abilities. Nevertheless, the large gap between the amount of data and the network bandwidth has resulted in a bottleneck in this platform (Hu et al., 2017).

The fog architecture consists of the IoT nodes (edge devices), the Fog, and the Cloud (Peralta et al., 2017). The IoT nodes are the closest to the users, consisting of sensors, mobile phones, smart cards, and smart vehicles. The fog layer comprises routers, gateways, access points, and base stations. These fogs are commonly available in cafes, streets, shopping centres, or even in mobile carriers. These fog nodes can transmit and temporarily store the received data. The fog layer is also connected to the cloud data centre using the IP core network (Hu et al., 2017). At the end, the cloud layer consists of computers that provide space for large data storage with high-performance efficiency. It runs computational analysis and permanently stores data for backup purposes. The data centre in

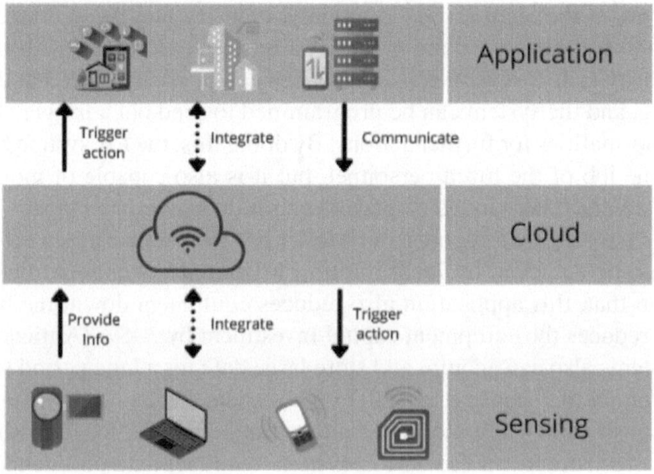

FIGURE 4.4 Cloud IoT architecture.

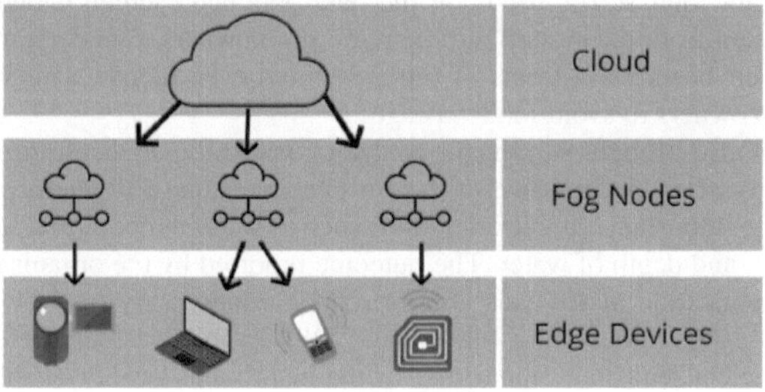

FIGURE 4.5 Fog-computing IoT architecture.

the cloud layer provides the basic features of cloud computing to the users (Naha et al., 2019). Figure 4.5 shows an illustration of the fog computing IoT layer.

4.2.1 Benefits of IoT

One of the major advantages of IoT is operational cost reduction. With IoT technology, the need for expert or human intervention can be greatly reduced. Therefore, industries can save a significant amount of money from monthly wedges. Apart from that the automation nature of the IoT system enables productivity around the clock because, unlike humans, IoT is not limited to the time of operation. For example, in agriculture, constant monitoring is required to ensure that the environment of the plantation, such as temperature, water level, light exposure, and nutrients, are at the ideal level. Commonly, industry players will hire personnel with a monthly wedge to monitor these parameters. However, with the increasing popularity of IoT, several sensing devices can be easily placed at the farm for monitoring, and the system can be programmed to send out a trigger should there be any abnormalities for further actions. By doing this, the IoT system has not only replaced the job of the hired personnel, but it is also capable of monitoring the crops constantly. Thus, industry players can reduce overhead costs. This explanation is supported by a report from McKinsey who stated that a cost reduction of 40% can be achieved by incorporating IoT-based systems (Amicucci, 2019). Apart from that, this application also reduces equipment downtime by 50% and parallelly reduces the equipment capital investment by 3–5% (Amicucci, 2019).

IoT systems also can acquire and store large data for a long period (Kumar and Rajasekaran, 2016; Friansa et al., 2017). These data can be used for further analysis to establish a connection or pattern that can be helpful for the industries. For example, a restaurant can use an IoT system to collect data on kitchen activities. Sensors can be attached to the kitchen utensils, and over a period of time, a correlation can be established between the frequency of the usage of kitchen utensils and the peak hours in the restaurant. This will also allow managers to arrange schedules for their staff's beads on the data obtained (Screen Cloud n.d.). The data obtained from the system will help the shop owners to understand the peak time of their business in terms of time, day, and even month. This information helps shop owners to avoid food wastage.

IoT has also brought significant contributions to the aquaculture sector. The Yield, an Australian company, utilised a Sensing+Aqua IoT platform to collect and analyse important climate conditions such as water temperature, salinity, sea-tide height, and depth of water. The outcome reported by the organisation shows that the application of IoT was able to reduce unnecessary harvest closures of oysters by 30% (Aquaculture Magazine, 2018). In the livestock sector, IoT will enable efficient real-time monitoring of the health and vitality of the livestock. This will help farmers to take quick remedial actions should there be any abnormalities in the health of the livestock. Apart from that, the application of IoT can also prevent the loss of livestock and track the grazing patterns of the livestock. IoT also aids in determining the readiness of the livestock to mate or give birth, optimising breeding practices as well as preventing loss of new births (IoT For All, n.d.; Akhigbe et al., 2021).

The application of IoT is also able to benefit the microalgae industry. Vital factors such as temperature, nutrient concentration, light exposure, and carbon dioxide concentration can be measured remotely. Apart from that, IoT also does data analysis, where growth patterns based on various factors can be identified. This will enable owners to determine the optimum harvesting period. For example, China reported an increase in grape yield by 153% after the application of the IoT system. The system that monitors luminous (light) intensity, soil conditions (carbon dioxide concentration, temperature, humidity) as well as leaf moisture was also helpful in reducing the production cost of the yield by optimising the volume of resources such as water and fertilisers (GSMA, 2020). However, there is very limited information on the utilisation of IoT in microalgae production. Considering the large benefits of this application, more industry players are encouraged to shift to this alternative.

Though the IoT system has significant advantages, there are also several limitations of the system which need further research and improvement. The major limitation with the IoT system is the privacy and the security of the system. Many wearable devices are being utilised by humans, making them a rich source of information (Bertino, 2016; Sadeghi et al., 2015). This information can be easily obtained by third parties who can use it to their advantage. Thus, more attention should be given to this area of study to ensure that the IoT system is not exploited for the wrong objectives.

4.3 THE UTILISATION OF IOT IN MICROALGAE FARMING

The advantages of IoT have been explored in the field of microalgae cultivation. The subsequent subsections elaborate on the crucial factors that encourage the growth of microalgae as well as the application of IoT in ensuring that these criteria are met to achieve optimal production.

4.3.1 Growth Factors that Affect Microalgae Cultivation

As it is well established that microalgae are photosynthetic organisms, it is important to ensure that there is a constant supply of light to facilitate the growth of the microalgae (Wang et al., 2014). The most suitable wavelength for the photosynthesis of microalgae ranges from 400 nm to 700 nm (Vuppaladadiyam et al., 2018). In the growth of microalgae, the light can be supplied using a LED, where a sensor can be used to ensure that the range of light intensity is between 2.5 K to 5 K lux (Erbland et al., 2020). Examples of light sensors include TCS3471 (Krujatz et al., 2016), TSL2561 (Rahmat et al., 2020), and RGB sensors (Wang et al., 2020).

The pH value of the culture medium also plays a major role in the growth kinetics of the microalgae strain (Qiu et al., 2017). Apart from influencing the metabolism rate of the microalgae, the pH value of the culture medium also controls the solubility and the availability of nutrients and carbon dioxide (Qiu et al., 2017). The commonly utilised pH value is from 6 to 8, where optimum growth and lipid accumulation can be achieved. Examples of pH sensors in monitoring the pH of the growth medium are DPD1R1 (Gonzalez-Camejo et al., 2020), SEN0161 (Erbland et al., 2020), and PH3310 (Foladori et al., 2018).

Salinity is also another factor that can influence the growth of microalgae. Culture media that are too saline impose osmotic stress on the culture and jeopardise the growth and metabolism of the microalgae strain. In contrast, optimal salinity encourages growth and lipid production (Ishika et al., 2019). The comprehensive study on the effects of culture medium salinity is mostly reported for marine microalgae strains (Xia et al., 2014; Ho et al., 2014), and very limited work has been reported for freshwater microalgae strains (Minhas et al., 2016).

There are reports of the development of sensors that can measure the salinity of a medium (Severin et al., 2017). However, this development has not been explored in the microalgae cultivation system.

The chemical composition, absorption of nutrients, carbon dioxide and the growth rates of the microalgae also are strongly dependent on the temperature of the culture (Singh and Singh, 2015). A high temperature can cause the deactivation of proteins in the microalgae which may result in the death of the cells (Gonzalez-Camejo et al., 2019). The optimal growth temperature is in the range of 30–37.5°C and varies based on the microalgae strain (Huang et al., 2017; Binnal and Babu, 2017). The optimum temperature also differs based on indoor and outdoor cultivation. The common temperature sensors used in microalgae cultivation are TMP102 (Wishkerman and Wishkerman, 2017), DS18B20, (Rahmat et al., 2020), and WKU-361–00DU (Wolf et al., 2016).

The application of sensors mentioned is more suitable for small-scale cultivation (Wang et al., 2021). More work is needed to better evaluate and improve the performance of the sensors so that they will be applicable for the mass production of microalgae. Apart from that, the development of sensors to detect toxic contaminants in the cultivation is also very limited (Frau et al., 2021; Reverté et al., 2016). This is also an important factor that needs more attention as contaminants can affect the whole production of microalgae.

4.3.2 Downstream Processing of Microalgae Biomass

The downstream production of microalgae consists of harvesting the biomass, extraction of biomolecules, and purification of the biomolecules for commercialisation use (Khanra et al., 2018). It is crucial to understand each of these steps to get a clear picture of where will the application of IoT be practical.

The harvesting techniques include centrifugation, filtration, flotation, flocculation, and electronics based process (Singh and Patidar, 2018). The centrifugation method utilises gravitational force to harvest the suspended cells in the culture (Matter et al., 2019). The filtration process involves passing the microalgae through a membrane with high pressure to obtain a microalgae paste (Mathimani and Mallick, 2018). The flotation method generates gas bubbles that bind to the microalgae cells and causes the cells to float to the top to be harvested (Zhang and Zhang, 2019). Flocculation utilises flocculants to aggregate small particles via surface charge neutralisation, electrostatic patching, and bridging (Ummalyma et al., 2017). Electrical based processes can be categorised into electrocoagulation and electroflotation (Geada et al., 2018). Electrocoagulation solubilises metal electrodes (anode) to coagulate microalgae cells for harvesting (Lucakova et al., 2021). Electroflotation uses an electrolytic method to generate bubbles that attach to the particulates in the culture and lift them to the surface of the culture for harvesting (Luengviriya et al., 2019). These methods have their advantages and disadvantages. For example, the centrifugation method is highly efficient. However, it requires high maintenance costs (Xu et al., 2021). In terms of cost, the filtration method is more preferable to centrifugation as this method requires low energy consumption. However, the operation process of this method is slow (Xu et al., 2021).

The next step in the downstream production is the extraction of the biomolecules from the cells. Microalgae biomolecules such as lipids, polysaccharides, amino acids, pigments, and carotenoids (Kumar et al., 2021) are reported to be useful in many industries (Sydney et al., 2019). Currently, more attention is given to the extraction of lipids from microalgae as lipids have a high potential in the biodiesel and the pharmaceutical industry (Huang et al., 2016; Goh et al., 2019). This statement is well supported by the report published by the *Global Banking and Finance Review* that states that the algae biofuel market is forecasted to reach a value of USD 10193.8 million by 2027 (Global Banking and Finance, 2021). Though the demand for microalgae lipids is showing an increasing trend across the years (Brindhadevi et al., 2021; Enamala et al., 2018), there are still many areas that can be improved to optimise the extraction process. The extraction of microalgae biomolecules begins with cell disruption (Dixon and Wilken, 2018). Cell disruption is a process where the cellular membranes of the cells are disintegrated to facilitate extraction (Corrêa et al., 2021). Some examples of cell disruption methods include bead milling, homogenisation (high or low speed), ultrasonication, irradiation, autoclaving, and using pulsed electric fields.

The bead milling method promotes cell damage by encouraging collision between the cells and the beads (Postma et al., 2017). The homogenisation method uses high-speed stirring to initiate cell disruption. This method is highly efficient for an industrial scale. However, it has the risk of denaturing the protein in the microalgae cell (Corrêa et al., 2021). The ultrasonic method uses acoustic waves to trigger the disruption of the cell membrane. This method is said to be saving more energy as compared to the bead milling method (Skorupskaite et al., 2019). Autoclaving method uses high temperature (121 °C) to rapture the cell membrane of the microalgae (Onumaegbu et al., 2018). The irradiation method uses microwave energy to enhance the efficiency of breaking microalgae cell membranes (Onumaegbu et al., 2018). The pulsed electric field uses high-voltage pulses initiated from two electrodes to displace the charges outside the cells, initiating rapture (Bodenes, 2017).

The last step in the downstream production is the purification of the extracted biomolecules. This step is important because the extracted molecules are usually combined with solvents. Thus, the purification process enables the separation of the desired compound from other substances such as solvents and impurities. The purification process can be done by various methods, such as electrophoresis, membrane separation, and ultracentrifugation and chromatography techniques.

The electrophoresis technique separates the molecules based on their charges with the help of an electric field (Corrêa et al., 2021). Membrane separation requires different materials and pore sizes that target isolating the desired biomolecule (Sarkar et al., 2020). This method is easily tuneable and versatile (Hua et al., 2020). The ultracentrifugation technique is a high-speed dewatering technique that can achieve solid concentration up to 10–20%. However, this method is energy consuming (Al Hattab et al., 2015). Chromatography techniques separate biomolecules using a carrier medium. The biomolecules are separated based on their densities (Vendruscolo et al., 2018). This method has high selectivity but requires tedious procedures (Toraman et al., 2016).

The aforementioned procedures need to be manually done by an expert. Thus, the whole downstream production is limited to human capabilities. This limitation can be easily curbed with the application of IoT systems. For example, the filtration process requires manual adjustment of the pressure to accommodate the size of the cells. This can be automated by using image processing algorithms that can determine the optimum pressure to obtain a high yield.

Apart from that, as the cultivation, harvesting, extraction, and purification process are interconnected, IoT sensors in the cultivation can trigger the harvesting process to start when the prime time for harvesting has arrived. The method of harvesting can also be automatically selected based on the type and volume of the cultivation. All these efforts will be able to reduce human intervention, consequently decreasing the production cost of the microalgae.

Another area that is worth exploring is the utilisation of an electrical relay, an electrically and electrochemically operated switch that opens and closes the circuits based on the signal from an outside source (ElectGo, 2019). It converts small electrical stimuli into larger current (Johnson, 2019). Some examples of relays include electromagnetic relays, non-latching relays, reed relays, and multidimensional relays (Teja, 2021). The recent development of IoT power relays enables better control of the processing in microalgae production. It is carefully designed to be adaptable to microcontrollers such as Arduino and Raspberry Pi (BCRobotics, n.d.). This allows IoT operators to remotely control the opening and closing of the pumps in the microalgae culture. For example, relays can be helpful in remotely initiating harvesting, extraction, and purification processes. Such application of relay has been reported elsewhere (Kothari and Parakh, 2017; Naruka et al., 2016) but has been explored in the microalgae field. Relays also can be used to automate the supply of nutrients to the culture. All these applications will reduce the requirement of manpower in the microalgae production field.

4.4 CURRENT APPLICATION OF IOT IN MICROALGAE CULTIVATION

Italy has reported the application of IoT in their microalgae cultivation plant. Their architecture consists of several IoT sensors that send data to the cloud. The cloud plays an important role by sending the information to a decision support system and subsequently alerts the operator should there be any action to be taken (Esposito et al., 2017). The incorporation of the decision support system was able to increase the productivity of the system by 9% (Esposito et al., 2017).

A research group in Indonesia also reported the application of an IoT system in their microalgae raceways ponds. This work comprises the application of IoT systems for three main objectives. The first objective was to control the speed of the paddlewheel based on sunlight exposure. The second objective was to control the flow of carbon dioxide gases based on the pH value and velocity of the water.

The last objective was to facilitate the harvesting process. These incorporations enabled the optimisation of microalgae production (Hermadi et al., 2021).

Sarawak has also embraced the utilisation of IoT by incorporating sensors that can monitor the growth of microalgae culture in a photobioreactor. The system also automates temperature regulation and gas supply in the photobioreactor. This results in a reduced human intervention in the monitoring of microalgae and therefore reduces operational cost (Sarawak Biodiversity Centre, 2020).

The utilisation of IoT systems was also investigated in closed systems such as photobioreactors. Sensors that detect temperature, light intensity, pH level, and concentration of dissolved oxygen were installed in the photobioreactors, and the data collected were transmitted to the cloud (Carrasquilla-Batista et al., 2017). Apart from that, optical density sensors were also placed in photobioreactors to get real-time data on the condition of the microalgae (Wang et al., 2021). These sensors measure the fluorescence of the culture medium to justify the growth of the cells (Krujatz et al., 2016). All these sensors work together to optimise the production inside the photobioreactor.

A research group in Croatia has exploited the advantages of IoT systems to monitor the growth of microalgae in a closed cultivation system. This resulted in significantly lower consumption of energy in their self-developed photobioreactors (PBRs). The technology enables remote monitoring and control of the system (Sarawak Biodiversity Centre, 2020).

An AGRI-SATT programme in the UK introduced an IoT-incorporated system to monitor the growth of marine microalgae. The system consists of a hybrid of artificial intelligence algorithms, Supervisory Control and Data Acquisition (SCADA)-controlled pond machinery, and IoT system that optimises the nutrient supply to the culture, as well as ensures an optimal environment for constant production of microalgae all around the year. This combined system has significantly increased the production of microalgae as well as reduced the cost of production (Sarawak Biodiversity Centre, 2020).

With the inspiration from Germany, Indonesia has also developed a facade photobioreactor bus shelter. This initiative was taken to accommodate the batch culturing of microalgae in an urban area. The whole system is centred on the application of IoT where various sensors were installed in the facade photobioreactor, and the gathered data were sent to the cloud using ThingSpeak interphase. The data could then be constantly monitored from a remote area (Arbye et al., 2020).

The combined initiative of B-Scada, Inc., and Algae Lab Systems in Colorado has resulted in the development of a microalgae monitoring system that enables the measurement of pH, temperature, DO, density, flow rate, and water level. The data acquired are used to automate the distribution of carbon dioxide gases and nutrients into the culture. The system also automates the pump and valves in the photobioreactor. This collaborative effort can increase microalgae production significantly (ALGAEWORLDNEWS, 2015).

In conclusion, IoT has been significantly improving the efficiency of microalgae production. The monitoring of important parameters such as temperature, light exposure, and water level has been improved. This has resulted in a lower production cost as the intervention of humans can be reduced. Apart from that, many efforts have also been taken up to integrate analytics to further understand the production of microalgae. These analytics help determine the prime time for harvesting as well as the growth rate of microalgae. The application of AI has also been a driving factor for the advancement of microalgae farming. These algorithms are effective in correlating various parameters so that optimal growth of microalgae can be maintained. Most importantly, the application of IoT has enabled remote monitoring of the culture. This results in a more effective monitoring system.

However, due to the low number of industry players who are adapting the application of IoT, the hardware has become more expensive. This is because a commercial price for these devices can be obtained when there is mass demand for these devices. The minimal adaptation of IoT in the microalgae field is mainly due to some of the limitations of the IoT system itself (and this will be discussed in the next section). Nevertheless, with more enhancement of the limitations, the number of industry players who adapt the applications of IoT is expected to increase.

4.4.1 Enhancement of IoT System for Optimal Microalgae Cultivation

Though the application of IoT is showing an increasing trend in microalgae cultivation (Esposito et al., 2017; Giannino et al., 2018; Matos, 2021), the full potential of IoT has not been fully exploited. For example, the production of microalgae largely depends on exposure to sunlight. This becomes challenging for countries with varying weather conditions. This is where IoT can be further applied by using sensors to collect data on climate changes over a period of time. The data obtained can be then analysed to automate the UV light exposure as well as the temperature of the culture medium to ensure optimum production throughout the whole year.

Apart from that, the current process uses electric supply to obtain data from sensor nodes. This can be energy- and cost-consuming. Thus, the utilisation of solar-powered IoT sensors should be encouraged (biz4intellia, 2019). This will combat the limitation of conventionally available sensors as they can enable the acquisition of real-time data that can be used to accurately understand the condition of the culture medium at a lower cost. Nevertheless, the efficiency of solar-powered IoT sensors is limited to tropical countries where sunlight can be obtained through the years.

The IoT system can also be further improved by incorporating the advancement of Artificial Intelligence (AI) (Parhi, 2018; Bang et al., 2020). Currently, IoT application has enabled data sharing and data storage around the globe. However, a comprehensive analysis of these data is still lacking in the IoT system. This is where the contribution of AI can be important. AI has the advantages of recognising patterns and correlations from multiple data and is capable of making accurate predictions based on the analysis (Bang et al., 2020; Makridakis, 2017). These algorithms will be able to reduce the analysis time of the data as well as to aid in decision-making such as the contamination level of the culture, the optimum temperature for growth at different climates, and humidity level. The most attractive feature of AI is the ability to constantly improve the algorithm with more data. Over time, the integration of AI will help key players to optimise their production at a lower cost.

4.5 ISSUES AND CHALLENGES

Though many organisations are utilising IoT in culturing microalgae, there are still some limitations that are yet to be overcome to achieve maximum production potential from microalgae cultivation. These limitations need to be addressed to encourage more industry players to apply IoT systems in their microalgae production. Some of the drawbacks are highlighted in the following subsections.

One of the challenges of utilising an IoT system is ensuring that the sensors used are in good condition to provide reliable data. A faulty sensor will be providing the wrong data to be processed. This will lead to a situation where wrong decisions will be taken. For example, the IoT system may be supplying extensive nutrients or carbon dioxide gases to the culture. This will affect the production of microalgae biomass. Therefore, the condition of the sensors should be constantly monitored to ensure reliable data is obtained.

The most suitable place for mass microalgae cultivation is a rural area. This is because mass cultivation requires a large land, which can be easily found in rural areas at a very reasonable price (Hopulele, 2019). Nevertheless, rural areas have the challenge for obtaining electricity and Internet connection. This limitation needs to be addressed as these are the heart of an IoT system.

Apart from that, the issue of IoT security has also been a pressing topic these days. IBM-X Force reported that the attacks on IoT systems showed a steep increase from October 2019 to June 2020. This rise was 400% higher as compared to the number of attacks for the past 2 years (McMillen, 2021). This shows that the application of IoT has a threat that comes along with the application itself. Thus, there are high possibilities for third parties to hijack and take control of the system. For example, the supply of nutrients can be stopped or the limits within which remedial action should be taken can be altered. This will significantly affect the production of microalgae.

This situation can be curbed by engaging with a good cybersecurity organisation. Some of the well-established cybersecurity organisations include Nuance, Secure Link, KnowBe4, and Lookout. These cyber securities are extremely expensive, owing to the benefits that they offer such as in-depth defence strategy, robust security infrastructure, and upgraded cyber intelligence tools (Garg, 2021). However, there is always a loophole in the security systems (Malik, 2020). Alongside the loopholes, the advancement of hacking tools has made it easier for malicious hackers to easily invade the IoT systems. For example, Facebook uses Facebook Immune System (FIS) as a defence against threats. However, this system was eluded by software created by experts from the University of British Columbia, Canada (Giles, 2011). Thus, the risk of breaching the security of an IoT system is always there. However, cybersecurity organisations are also actively enhancing the strength of their systems.

According to Business Insider, the number of IoT devices are forecasted to reach up to 41 billion by the year 2027 (Gyarmathy, 2020). However, there is a lack of IoT professionals in the current industries. Thus, the efforts of developing IoT devices are getting hampered by shortages of workers with newer analytics skills. This is where the action from the government will come into play. Governments across the countries can encourage professionals in the IT field to expand their knowledge in IoT-related areas such as cloud computing, analytics, software connectivity, and big data. This can be done by engaging current professionals in the IoT field to train IT experts on IoT systems. For example, a government organisation in Malaysia have allocated a total of USD 242.5 million (RM 1 billion) in upskilling current IT experts in areas such as Cybersecurity, Connectivity, IoT, and Digital Talents (International Trade Administration, 2020). Private sectors can also benefit from this allocation by sending their representatives for useful training like these. This effort will result in many certified professionals who can cater to the demand that comes from the development of IoT.

Currently, NPK sensors are widely used in the agriculture field to measure the nutrient concentration in the soil (Na et al., 2016). This sensor enables fast measurement, and the data obtained is highly accurate (Newton, 2021). Therefore, this sensor can also be used to determine the concentration of nutrients in microalgae culture. Apart from NPK sensors, there is also an increasing trend of biochemical sensors in determining the nutrient concentration (Ali et al., 2017; Dubey and Mailapalli, 2016; Chhipa and Joshi, 2016). As compared to NPK sensors, these biochemical sensors have higher accuracy and are more flexible to be used (Lu et al., 2020). However, the cost of these sensors is high, ranging from $9000 to $15000 (Teralytic, n.d.). This will increase the mass production cost of microalgae. Thus, there is a pressing need for cost-effective, convenient, and real-time soil nutrient sensors to optimise the monitoring of the culture condition.

The existing sensors are not sensitive to small changes in nutrient concentration (Burton et al., 2020). Therefore, most industry players prefer to manually measure the nutrient content of the culture using a detector. The process of manually measuring the nutrient concentration is time and cost consuming as an expert is required to perform the measurement regularly.

The exploitation of IoT in the harvesting process of microalgae is not usually reported. The harvesting process, which is currently done manually, is time and cost consuming. Apart from that, the application of IoT in the downstream production of microalgae is also very limited. Thus, though the optimum yield is produced in the microalgae culture, the inefficient harvesting and processing methods can result in a low production of microalgae biomass.

4.6 RECOMMENDATIONS

To date, various sensors have been used to monitor the microalgae culture. However, the utilisation of image-capturing devices in IoT systems is very limited (Yew et al., 2020; Chen et al., 2021; Teng et al., 2020). This is an area that can be exploited further as the technology of image processing has been advancing rapidly. These image-capturing devices will enable the provision of information on the health of the microalgae culture. Various artificial algorithms such as k-Nearest Neighbour, artificial neural networks, and deep learning can be integrated to analyse and establish the patterns between these images. This will result in a more detailed understanding of the growth of the culture. For example, contamination of the culture can be easily identified as remedial action can be immediately taken. Besides that, the analysis of the culture images can provide information on the growth phases of the microalgae culture. This will facilitate the determination of the prime time for harvesting. Further analysis of the images may provide comprehensive information on the biomolecules of the microalgae. Apart from integrating AI, more research can also be conducted to obtain images from the culture *in-situ*. This will make the whole process of monitoring more convenient as well as avoid contamination of the samples, which may jeopardise the analysis process.

One of the major drawbacks is the requirement for constant electricity. As the whole IoT system is fully dependent on electricity, a trip in electric supply may affect the whole microalgae cultivation system. This can be overcome by using IoT devices that are powered by solar energy. Nevertheless, these devices are costly. Thus, more research work should be focused on developing solar power devices that are more reasonable. The advancement of silicon technology can also be applied in this research field.

4.7 CONCLUSIONS

The advancement of industry 4.0 has encouraged the development of enhanced technologies across the globe. One of the most intriguing development is the application of IoT. The application of IoT in a wide range of industries such as agriculture, food and feed, and pharmaceutical fields has been showing an increasing trend due to its enormous benefits and flexibilities. In line with this advancement, the demand for microalgae has also been gaining a lot of attention owing to the benefits of its biomolecules to various industries such as pharmaceutical, biodiesel, food and feed. This review discusses the current application of IoT in the production of microalgae. From the aforementioned comprehensive discussions, it is justified that the combined benefits of microalgae and IoT have not been fully exploited.

In terms of upstream microalgae production, many IoT sensors have been utilised to monitor and optimise the growth of microalgae. However, these sensors are more suitable to be used in small-scale microalgae production. This is the gap that can be enhanced to develop IoT sensors that are applicable for a bigger scale of microalgae production, in terms of durability, sustainability, and maintenance. Sensors to detect contamination in the culture can also be included in the IoT system. The flexibility of these sensors to communicate with various microprocessors should also be taken into account in the research.

Apart from that, data-analysing software in the IoT system can also be improved by increasing the data storage. In addition, the incorporation of AI in analysing the data of various parameters is highly encouraged. This will enable a comprehensive understanding of the factors that influence the growth of microalgae. It is also crucial to augment the protection of the data collected as well as the operation of the sensors in the IoT system.

The application of IoT in the downstream production of microalgae should also be increased to obtain the maximum yield of microalgae biomass and biomolecules. IoT sensors that can aid in determining the prime time of harvesting should be utilised. These sensors include optical density sensors and fluorescence sensors. AI can also be included to determine the method of harvesting based on data such as volume and species of microalgae. For the extraction phase, the benefit of relays can be fully exploited to control the equipment and motors that facilitate the extraction process. Apart from the application of IoT, it is also important to increase the number of experts who are well versed with IoT technologies. This will enable the smooth operation of IoT systems.

This chapter is important to understand the current state of IoT application in microalgae production. This chapter has also shed light on areas that can be further explored to optimise the application of IoT in microalgae production. The content of this chapter serves as a guideline for researchers to come with new, enhanced developments that can fully exploit the benefits of microalgae in both upstream and downstream production. The incorporation of IoT in the microalgae industry will be able to significantly reduce the production cost as well as increase the volume of microalgae yield.

REFERENCES

Aftab, H., Gilani, K., Lee, J., Nkenyereye, L., Jeong, S. and Song, J. (2020) 'Analysis of identifiers in IoT platforms', *Digital Communications and Networks,* 6(3), pp. 333–340.

Akhigbe, B. I., Munir, K., Akinade, O., Akanbi, L. and Oyedele, L. O. (2021) 'IoT technologies for livestock management: A review of present status, opportunities, and future trends', *Big Data and Cognitive Computing,* 5(1), p. 10.

Akpakwu, G. A., Silva, B. J., Hancke, G. P. and Abu-Mahfouz, A. M. (2017) 'A survey on 5G networks for the internet of things: Communication technologies and challenges', *IEEE Access,* 6, pp. 3619–3647.

Al-Osta, M., Ahmed, B. and Abdelouahed, G. (2017) 'A lightweight semantic web-based approach for data annotation on IoT gateways', *Procedia Computer Science,* 113, pp. 186–193.

Al-Qaseemi, S. A., Almulhim, H. A., Almulhim, M. F. and Chaudhry, S. R. (2016) 'IoT architecture challenges and issues: Lack of standardization', in *Future Technologies Conference (FTC).* IEEE, pp. 731–738.

Al Hattab, M., Ghaly, A. and Hammouda, A. (2015) 'Microalgae harvesting methods for industrial production of biodiesel: Critical review and comparative analysis', *Journal of Fundamentals of Renewable Energy and Applications,* 5(2), p. 1000154.

Al Hinai, S. and Singh, A. V. (2017) 'Internet of things: Architecture, security challenges and solutions', in *International Conference on Infocom Technologies and Unmanned Systems (Trends and Future Directions)(ICTUS).* IEEE, pp. 1–4.

ALGAEWORLDNEWS (2015) *B-Scada Provides IOT Solution to Optimize Algae Cultivation in Colorado.* Available at: https://news.algaeworld.org/2015/04/b-scada-provides-iot-solution-to-optimize-algae-cultivation-in-colorado-2/ (Accessed: 12 July 2021).

Ali, M. A., Mondal, K., Wang, Y., Jiang, H., Mahal, N. K., Castellano, M. J., Sharma, A. and Dong, L. (2017) 'In situ integration of graphene foam—titanium nitride based bio-scaffolds and microfluidic structures for soil nutrient sensors', *Lab on a Chip,* 17(2), pp. 274–285.

Amicucci, L. (2019) *How IoT-Based Predictive Maintenance Can Reduce Costs.* Available at: https://blog.nordicsemi.com/getconnected/how-iot-based-predictive-maintenance-can-reduce-costs (Accessed: 5 July 2021).

Aquaculture Magazine (2018) *The Benefits of Aquaculture with IoT Technology.* Available at: https://aquaculturemag.com/2018/06/26/the-benefits-of-aquaculture-with-iot-technology/ (Accessed: 5 August 2021).

Arbye, S., Arianti, R. F., Pradana, Y. S., Suyono, E. A., Koerniawan, M. D., Suwanti, L. T., Siregar, U. J. and Budiman, A. (2020) 'The design of microalgae (Chlorella sp.) photobioreactor as a façade bus shelter building in Indonesia', *American Institute of Physics Conference Series,* 020007.

Bang, G., Barash, G., Bea, R., Cali, J., Castillo-Effen, M., Chen, X., Chhaya, N., Cummings, R., Dhoopar, R. and Dumanci, S. (2020) 'The association for the advancement of artificial intelligence 2020 workshop program', *AI Magazine,* 41(4), pp. 100–114.

Barik, L. (2019) 'IOT based temperature and humidity controlling using Arduino and raspberry pi', *International Journal of Advanced Computer Science Applications* 10(9), pp. 494–502.

Bastos, D. (2019) 'Cloud for IoT—A survey of technologies and security features of public cloud IoT solutions', in *Living in the Internet of Things (IoT 2019).* IET, pp. 1–6.

BCRobotics (n.d.) *IoT Power Relay.* Available at: https://bc-robotics.com/shop/iot-power-relay/ (Accessed: 11 September 2021).

Bertino, E. (2016) 'Data privacy for IoT systems: Concepts, approaches, and research directions', in *IEEE International Conference on Big Data (Big Data).* IEEE, pp. 3645–3647.

Binnal, P. and Babu, P. N. (2017) 'Optimization of environmental factors affecting tertiary treatment of municipal wastewater by Chlorella protothecoides in a lab scale photobioreactor', *Journal of Water Process Engineering,* 17, pp. 290–298.

biz4intellia (2019) *5 Applications of IoT in Agriculture—Making Agriculture Smarter.* Available at: https://www.biz4intellia.com/blog/5-applications-of-iot-in-agriculture/ (Accessed: 25 July 2021).

Bodenes, P. (2017) *Study of the Application of Pulsed Electric Fields (PEF) on Microalgae for the Extraction of Neutral Lipids.* Université Paris-Saclay [Online] Available at: https://tel.archives-ouvertes.fr/tel-01540436/document.

Brindhadevi, K., Mathimani, T., Rene, E. R., Shanmugam, S., Chi, N. T. L. and Pugazhendhi, A. (2021) 'Impact of cultivation conditions on the biomass and lipid in microalgae with an emphasis on biodiesel', *Fuel,* 284, p. 119058.

Burhan, M., Rehman, R. A., Khan, B. and Kim, B.-S. (2018) 'IoT elements, layered architectures and security issues: A comprehensive survey', *Sensors,* 18(9), p. 2796.

Burton, L., Jayachandran, K. and Bhansali, S. (2020) 'The "real-time" revolution for in situ soil nutrient sensing', *Journal of The Electrochemical Society,* 167(3), p. 037569.

Califano, J. (2019) *Raspberry Pi IoT: Is This Tiny Computer Ready for Industrial Applications?* Available at: https://blog.temboo.com/raspberry-pi-iot/ (Accessed: 2 August 2021).

Carrasquilla-Batista, A., Chacón-Rodriguez, A., Murillo-Vega, F., Niiñez-Montero, K., Goomez-Espinoza, O. and Guerrero-Barrantes, M. (2017) 'Characterization of biomass pellets from Chlorella vulgaris microalgal production using industrial wastewater', in *International Conference in Energy and Sustainability in Small Developing Economies (ES2DE).* IEEE, pp. 1–6.

Chavan, R., Kolekar, A. and VPKBIET, B. (2020) 'Solar rooftop power generation system by using IOT (Arduino & Blynk)', *International Journal of Innovative Science and Research Technology,* 5(9), pp. 759–762.

Chen, Z., Cui, S., Zhang, Y. and He, L. (2021) 'Research on statistical algorithm of microalgae growth status based on computer vision', in *IEEE 4th Advanced Information Management, Communicates, Electronic and Automation Control Conference (IMCEC).* IEEE, pp. 1653–1657.

Chhipa, H. and Joshi, P. (2016) 'Nanofertilisers, nanopesticides and nanosensors in agriculture', in *Nanoscience in Food and Agriculture 1.* Springer, pp. 247–282.

Cisco (2015) *Cisco Connected Data and Analytics.* Cisco. Available at: https://www.cisco.com/c/dam/r/en/us/internet-of-everything-ioe/analytics-automation/assets/files/Cisco-Connected_Analytics-AAG.pdf (Accessed: 12 September 2021).

Corrêa, P. S., Morais Júnior, W. G., Martins, A. A., Caetano, N. S. and Mata, T. M. (2021) 'Microalgae biomolecules: Extraction, separation and purification methods', *Processes,* 9(1), pp. 10.

DF Robot (n.d.) *WiDo—An Arduino Compatible IoT (internet of thing) Board.* Available at: https://www.dfrobot.com/product-1159.html (Accessed: 19 July 2021).

Dixon, C. and Wilken, L. R. (2018) 'Green microalgae biomolecule separations and recovery', *Bioresources and Bioprocessing,* 5(1), pp. 1–24.

Dubey, A. and Mailapalli, D. R. (2016) 'Nanofertilisers, nanopesticides, nanosensors of pest and nanotoxicity in agriculture', in *Sustainable Agriculture Reviews.* Springer, pp. 307–330.

ElectGo (2019) *Relay| What Is Relay, Its Function, Types and Relay Wiring.* Available at: https://www.electgo.com/what-is-a-relay/ (Accessed: 8 September 2021).

Enamala, M. K., Enamala, S., Chavali, M., Donepudi, J., Yadavalli, R., Kolapalli, B., Aradhyula, T. V., Velpuri, J. and Kuppam, C. (2018) 'Production of biofuels from microalgae-A review on cultivation, harvesting, lipid extraction, and numerous applications of microalgae', *Renewable and Sustainable Energy Reviews,* 94, pp. 49–68.

Erbland, P., Caron, S., Peterson, M. and Alyokhin, A. (2020) 'Design and performance of a low-cost, automated, large-scale photobioreactor for microalgae production', *Aquacultural Engineering,* 90, p. 102103.

Esposito, S., Cafiero, A., Giannino, F., Mazzoleni, S. and Diano, M. M. (2017) 'A monitoring, modeling and decision support system (DSS) for a microalgae production plant based on internet of things structure', *Procedia Computer Science,* 113, pp. 519–524.

ESPRESSIF (n.d.) *ESP8266.* Available at: https://www.espressif.com/en/products/socs/esp8266 (Accessed: 8 September 2021).

Foladori, P., Petrini, S. and Andreottola, G. (2018) 'Evolution of real municipal wastewater treatment in photobioreactors and microalgae-bacteria consortia using real-time parameters', *Chemical Engineering Journal,* 345, pp. 507–516.

Fortune Business Insights (2021) *Global IoT Market to be Worth USD 1,463.19 Billion by 2027 at 24.9% CAGR; Demand for Real-time Insights to Spur Growth, Says Fortune Business Insights™*. Available at: https://www.globenewswire.com/en/news-release/2021/04/08/2206579/0/en/Global-IoT-Market-to-be-Worth-USD-1-463-19-Billion-by-2027-at-24–9-CAGR-Demand-for-Real-time-Insights-to-Spur-Growth-says-Fortune-Business-Insights.html#:~:text=The%20Internet%20of%20Things%20(IoT,24.9%25%20during%20the%20forecast%20period. (Accessed: 17 September 2021).

Frau, I., Wylie, S., Byrne, P., Onnis, P., Cullen, J., Mason, A. and Korostynska, O. (2021) 'Microwave sensors for in situ monitoring of trace metals in polluted water', *Sensors,* 21(9), p. 3147.

Friansa, K., Haq, I. N., Santi, B. M., Kurniadi, D., Leksono, E. and Yuliarto, B. (2017) 'Development of battery monitoring system in smart microgrid based on internet of things (IoT)', *Procedia Engineering,* 170, pp. 482–487.

Garg, V. (2021) *Covenants without the Sword: Market Incentives for Cybersecurity Investment.* Available at SSRN: https://ssrn.com/abstract=3896578 (Accessed: 16 September 2021).

Geada, P., Rodrigues, R., Loureiro, L., Pereira, R., Fernandes, B., Teixeira, J. A., Vasconcelos, V. and Vicente, A. A. (2018) 'Electrotechnologies applied to microalgal biotech nology—Applications, techniques and future trends', *Renewable and Sustainable Energy Reviews,* 94, pp. 656–668.

Giannino, F., Esposito, S., Diano, M., Cuomo, S. and Toraldo, G. (2018) 'A predictive decision support system (DSS) for a microalgae production plant based on internet of things paradigm', *Concurrency and Computation: Practice Experience,* 30(15), p. e4476.

Giles, J. (2011) *Inside Facebook's Massive Cyber-Security System.* Available at: https://www.newscientist.com/article/dn21095-inside-facebooks-massive-cyber-security-system/ (Accessed: 4 September 2021).

Global Banking and Finance (2021) *North America Algae Biofuel Market Size 2021 to Witness Astonishing Growth with Key Players, Analysis with Impact of COVID-19, Analysis, Demand, Forecast 2027.* Available at: https://www.globalbankingandfinance.com/north-america-algae-biofuel-market-size-2021-to-witness-astonishing-growth-with-key-players-analysis-with-impact-of-covid-19-analysis-demand-forecast-2027/ (Accessed: 13 July 2021).

Goh, B. H. H., Ong, H. C., Cheah, M. Y., Chen, W.-H., Yu, K. L. and Mahlia, T. M. I. (2019) 'Sustainability of direct biodiesel synthesis from microalgae biomass: A critical review', *Renewable and Sustainable Energy Reviews,* 107, pp. 59–74.

Gokhale, P., Bhat, O. and Bhat, S. (2018) 'Introduction to IOT', *International Advanced Research Journal in Science, Engineering Technology,* 5(1), pp. 41–44.

Gonzalez-Camejo, J., Aparicio, S., Ruano, M., Borrás, L., Barat, R. and Ferrer, J. (2019) 'Effect of ambient temperature variations on an indigenous microalgae-nitrifying bacteria culture dominated by Chlorella', *Bioresource Technology,* 290, p. 121788.

Gonzalez-Camejo, J., Robles, A., Seco, A., Ferrer, J. and Ruano, M. (2020) 'On-line monitoring of photosynthetic activity based on pH data to assess microalgae cultivation', *Journal of Environmental Management,* 276, p. 111343.

GSMA (2020) *Improving the Yield and Quality of Grape Production in China with IoT.* Available at: https://www.gsma.com/iot/wp-content/uploads/2020/01/IoT-Grapes-report-final-for-web.pdf (Accessed: 21 August 2021).

Gupta, R. and Gupta, R. (2016) 'ABC of Internet of Things: Advancements, benefits, challenges, enablers and facilities of IoT', in *Symposium on Colossal Data Analysis and Networking (CDAN).* IEEE, pp. 1–5.

Gyarmathy, K. (2020) *Comprehensive Guide to IoT Statistics You Need to Know in 2021.* Available at: https://www.vxchnge.com/blog/iot-statistics (Accessed: 19 August 2021).

Har-Even, B. (2021) *Fun with PowerVR and the BeagleBone Black: Low-Cost Development Made Easy.* Available at: https://www.imaginationtech.com/blog/fun-with-powervr-and-the-beaglebone-black-low-cost-development-made-easy/ (Accessed: 12 September 2021).

He, Q., Weaver, V. and Segee, B. (2018) 'Comparing power and energy usage for scientific calculation with and without GPU acceleration on a raspberry Pi model B+ and 3B'. in *Proceedings on the International Conference on Internet Computing (ICOMP).* The Steering Committee of The World Congress in Computer Science, Computer, pp. 3–9.

Hermadi, I., Setiadianto, I. R., Al Zahran, D. F. I., Simbolon, M. N., Saefurahman, G., Wibawa, D. S. and Arkeman, Y. (2021) 'Development of smart algae pond system for microalgae biomass production'. *IOP Conference Series: Earth and Environmental Science: IOP Publishing*, 012068.

Ho, S.-H., Nakanishi, A., Ye, X., Chang, J.-S., Hara, K., Hasunuma, T. and Kondo, A. (2014) 'Optimizing biodiesel production in marine Chlamydomonas sp. JSC4 through metabolic profiling and an innovative salinity-gradient strategy', *Biotechnology for Biofuels,* 7(1), pp. 1–16.

Hopulele, A. (2019) *The Pros and Cons of Buying a House in a Rural Area.* Available at: https://www.point2homes.com/news/us-real-estate-news/pros-cons-buying-house-rural-area.html (Accessed: 1 September 2021).

Hossein Motlagh, N., Mohammadrezaei, M., Hunt, J. and Zakeri, B. (2020) 'Internet of things (IoT) and the energy sector', *Energies,* 13(2), p. 494.

Hou, L., Zhao, S., Xiong, X., Zheng, K., Chatzimisios, P., Hossain, M. S. and Xiang, W. (2016) 'Internet of things cloud: Architecture and implementation', *IEEE Communications Magazine,* 54(12), pp. 32–39.

Hu, P., Dhelim, S., Ning, H. and Qiu, T. (2017) 'Survey on fog computing: Architecture, key technologies, applications and open issues', *Journal of Network and Computer Applications,* 98, pp. 27–42.

Hua, L., Cao, H., Ma, Q., Shi, X., Zhang, X. and Zhang, W. (2020) 'Microalgae filtration using an electrochemically reactive ceramic membrane: Filtration performances, fouling kinetics, and foulant layer characteristics', *Environmental Science & Technology,* 54(3), pp. 2012–2021.

Huang, Q., Jiang, F., Wang, L. and Yang, C. (2017) 'Design of photobioreactors for mass cultivation of photosynthetic organisms', *Engineering,* 3(3), pp. 318–329.

Huang, Y., Zhang, D., Xue, S., Wang, M. and Cong, W. (2016) 'The potential of microalgae lipids for edible oil production', *Applied Biochemistry and Biotechnology,* 180(3), pp. 438–451.

International Trade Administration (2020) *Malaysia Digital Economy.* Available at: https://www.trade.gov/market-intelligence/malaysia-digital-economy (Accessed: 24 September 2021).

IoT For All (n.d.) *Livestock Management.* Available at: https://www.iotforall.com/use-case/livestock-management (Accessed: 26 July 2021).

Ishchenko, D. and Nuqui, R. (2018) 'Secure communication of intelligent electronic devices in digital substations', in *IEEE/PES Transmission and Distribution Conference and Exposition (T&D).* IEEE, pp. 1–5.

Ishika, T., Moheimani, N. R., Laird, D. W. and Bahri, P. A. (2019) 'Stepwise culture approach optimizes the biomass productivity of microalgae cultivated using an incremental salinity increase strategy', *Biomass & Bioenergy,* 127, p. 105274.

Java T Point. (n.d.) *Cloud Computing Architecture*. Available at: https://www.javatpoint.com/cloud-computing-architecture (Accessed: 14 September 2021).

Johnson, J. (2019) *What is a Relay and Why are They so Important?* Available at: https://amperite.com/blog/relays/ (Accessed: 24 July 2021).

Khanra, S., Mondal, M., Halder, G., Tiwari, O., Gayen, K. and Bhowmick, T. K. (2018) 'Downstream processing of microalgae for pigments, protein and carbohydrate in industrial application: A review', *Food and Bioproducts Processing,* 110, pp. 60–84.

Khattab, H. A., Abdelgawad, A. and Yelmarthi, K. (2016) 'Design and implementation of a cloud-based IoT scheme for precision agriculture', in *28th International Conference on Microelectronics (ICM)*. IEEE, pp. 201–204.

Khattak, H. A., Shah, M. A., Khan, S., Ali, I. and Imran, M. (2019) 'Perception layer security in Internet of Things', *Future Generation Computer Systems,* 100, pp. 144–164.

Kothari, D. and Parakh, A. (2017) 'Application of wireless technologies in agricultural pumps', in *International Conference on Computation of Power, Energy Information and Communication (ICCPEIC)*. IEEE, pp. 75–84.

Krasniqi, X. and Hajrizi, E. (2016) 'Use of IoT technology to drive the automotive industry from connected to full autonomous vehicles', *IFAC-PapersOnLine,* 49(29), pp. 269–274.

Krujatz, F., Fehse, K., Jahnel, M., Gommel, C., Schurig, C., Lindner, F., Bley, T., Weber, J. and Steingroewer, J. (2016) 'MicrOLED-photobioreactor: Design and characterization of a milliliter-scale Flat-Panel-Airlift-photobioreactor with optical process monitoring', *Algal Research,* 18, pp. 225–234.

Kumar, R. and Rajasekaran, M. P. (2016) 'An IoT based patient monitoring system using raspberry Pi', in *International Conference on Computing Technologies and Intelligent Data Engineering (ICCTIDE'16)*. IEEE, pp. 1–4.

Kumar, V., Jaiswal, K. K., Tomar, M. S., Rajput, V., Upadhyay, S., Nanda, M., Vlaskin, M. S., Kumar, S. and Kurbatova, A. (2021) 'Production of high value-added biomolecules by microalgae cultivation in wastewater from anaerobic digestates of food waste: A review', *Biomass Conversion and Biorefinery*, pp. 1–18.

Laubhan, K., Talaat, K., Riehl, S., Morelli, T., Abdelgawad, A. and Yelamarthi, K. (2016) 'A four-layer wireless sensor network framework for IoT applications', in *IEEE 59th International Midwest Symposium on Circuits and Systems (MWSCAS)*. IEEE, pp. 1–4.

Lu, Y., Yang, Q. and Wu, J. (2020) 'Recent advances in biosensor-integrated enrichment methods for preconcentrating and detecting the low-abundant analytes in agriculture and food samples', *TrAC Trends in Analytical Chemistry,* 128, p. 115914.

Lucakova, S., Branyikova, I., Kovacikova, S., Pivokonsky, M., Filipenska, M., Branyik, T. and Ruzicka, M. C. (2021) 'Electrocoagulation reduces harvesting costs for microalgae', *Bioresource Technology,* 323, p. 124606.

Luengviriya, C., Wungmool, P., Rangsi, N., Kumchaiseemak, N. and Sutthiopad, M. (2019) 'Electro-flotation harvesting of microalgae using a combination of electrode types', *The Journal of Applied Science,* 18(1), pp. 1–11.

MacGillivray, C. (2016) 'The platform of platforms in the internet of things', *IBM: White Paper*, pp. 1–7.

Maiman, M. (n.d.) *Internet of Things: The Four Key Elements*. Available at: https://intelligentproduct.solutions/blog/internet-of-things-4-key-elements/ (Accessed: 21 September 2021).

Makridakis, S. (2017) 'The forthcoming artificial intelligence (AI) revolution: Its impact on society and firms', *Futures,* 90, pp. 46–60.

Malavade, V. N. and Akulwar, P. K. (2016) 'Role of IoT in agriculture', *IOSR Journal of Computer Engineering,* 1(13), pp. 56–57.

Malik, J. (2020) 'Making sense of human threats and errors', *Computer Fraud & Security,* 2020(3), pp. 6–10.

Mashal, I., Alsaryrah, O., Chung, T.-Y., Yang, C.-Z., Kuo, W.-H. and Agrawal, D. P. (2015) 'Choices for interaction with things on internet and underlying issues', *Ad Hoc Networks,* 28, pp. 68–90.

Mathimani, T. and Mallick, N. (2018) 'A comprehensive review on harvesting of microalgae for biodiesel—key challenges and future directions', *Renewable and Sustainable Energy Reviews,* 91, pp. 1103–1120.

Matos, Â. P. (2021) 'Advances in microalgal research in Brazil', *Brazilian Archives of Biology and Technology,* 64, p. e21200531.

Matter, I. A., Bui, V. K. H., Jung, M., Seo, J. Y., Kim, Y.-E., Lee, Y.-C. and Oh, Y.-K. (2019) 'Flocculation harvesting techniques for microalgae: A review', *Applied Sciences,* 9(15), p. 3069.

Maureira, M. A. G. (2014) *ThingSpeak—an API and Web Service for the Internet of Things. Available at: https://staas.home.xs4all.nl/t/swtr/documents/wt2014_thingspeak.pdf*

McLaughlin, B. (2015) *The BeagleBone Black Primer.* Que Publishing.

McMillen, D. (2021) *Internet of Threats: IoT Botnets Drive Surge in Network Attacks.* Available at: https://securityintelligence.com/posts/internet-of-threats-iot-botnets-network-attacks/ (Accessed: 29 August 2021).

Mehl, B. B. (2021) *6 Communication Protocols Used by IoT.* Available at: https://www.getkisi.com/blog/internet-of-things-communication-protocols (Accessed: 24 August 2021).

Mehta, M. (2015) 'ESP 8266: A breakthrough in wireless sensor networks and internet of things', *International Journal of Electronics and Communication Engineering and Technology,* 6(8), pp. 7–11.

Melo, R., Bezerra, M. C., Dantas, J., Matos, R., de Melo Filho, I. J., Oliveira, A. S., de Oliveira Feliciano, F. D. and Maciel, P. R. M. (2017) 'Sensitivity analysis techniques applied in cloud computing environments', in *12th Iberian Conference on Information Systems and Technologies (CISTI).* IEEE, pp. 1–7.

Microsoft (2020) *Welcome to Azure Stream Analytics.* Available at: https://docs.microsoft.com/en-us/azure/stream-analytics/stream-analytics-introduction (Accessed: 25 July 2021).

Minhas, A. K., Hodgson, P., Barrow, C. J. and Adholeya, A. (2016) 'A review on the assessment of stress conditions for simultaneous production of microalgal lipids and carotenoids', *Frontiers in Microbiology,* 7, p. 546.

Mittal, R., Pathak, V., Goyal, S. and Mithal, A. (2020) 'A novel approach to localized a robot in a given map with optimization using GP-GPU', in *Recent Trends in Communication and Intelligent Systems Algorithms for Intelligent Systems.* Springer, pp. 157–164.

Monk, S. (2016) *Programming Arduino: Getting Started with Sketches.* McGraw-Hill Education.

Na, A., Isaac, W., Varshney, S. and Khan, E. (2016) 'An IoT based system for remote monitoring of soil characteristics', in *International Conference on Information Technology (InCITe)-The Next Generation IT Summit on the Theme-Internet of Things: Connect your Worlds.* IEEE, pp. 316–320.

Naha, R. K., Garg, S. and Chan, A. (2019) 'Fog computing architecture: Survey and challenges', in Khan, M. U. S., Khan, S. U. and Zomaya, A. Y. (eds.) *Big Data-Enabled Internet of Things.* The Institution of Engineering and Technology, pp. 199–223.

Naruka, T., Singh, A., Janu, A., Gocher, A. and Sharma, A. (2016) 'Automatic regulation of water level through automatic relay switching operation', *International Journal of Engineering and Management Research,* 6(2), pp. 69–72.

Nayyar, A. and Puri, V. (2016) 'A comprehensive review of beaglebone technology: Smart board powered by ARM', *International Journal of Smart Home,* 10(4), pp. 95–108.

Nettikadan, D. and Raj, S. (2018) 'Smart community monitoring system using ThingSpeak IoT platform', *International Journal of Applied Engineering Research,* 13(17), pp. 13402–13408.

Newton, A. (2021) *Measure Soil Nutrient Using Arduino & Soil NPK Sensor.* Available at: https://how2electronics.com/measure-soil-nutrient-using-arduino-soil-npk-sensor/ (Accessed: 23 July 2021).

Ometov, A., Shubina, V., Klus, L., Skibińska, J., Saafi, S., Pascacio, P., Flueratoru, L., Gaibor, D. Q., Chukhno, N. and Chukhno, O. (2021) 'A survey on wearable technology: History, state-of-the-art and current challenges', *Computer Networks,* 193, p. 108074.

Onumaegbu, C., Mooney, J., Alaswad, A. and Olabi, A. (2018) 'Pre-treatment methods for production of biofuel from microalgae biomass', *Renewable and Sustainable Energy Reviews,* 93, pp. 16–26.

Parhi, D. (2018) 'Advancement in navigational path planning of robots using various artificial and computing techniques', *International Robotics & Automation Journal,* 4(2), pp. 133–136.

Patel, A. and Devaki, P. (2019) 'Survey on NodeMCU and Raspberry pi: IoT', *International Research Journal of Engineering and Technology,* 6(4), pp. 5101–5105.

Patnaik Patnaikuni, D. R. (2017) 'A comparative study of Arduino, raspberry Pi and ESP8266 as IoT development board', *International Journal of Advanced Research in Computer Science,* 8(5).

Payal, R. and Singh, A. P. (2021) 'A study on different hardware and cloud based internet of things platforms', *Journal of Physics: Conference Series: IOP Publishing*, 012055.

Peralta, G., Iglesias-Urkia, M., Barcelo, M., Gomez, R., Moran, A. and Bilbao, J. (2017) 'Fog computing based efficient IoT scheme for the industry 4.0', in *IEEE International Workshop of Electronics, Control, Measurement, Signals and Their Application to Mechatronics (ECMSM).* IEEE, pp. 1–6.

Pillai, R. and Sivathanu, B. (2020) 'Adoption of internet of things (IoT) in the agriculture industry deploying the BRT framework', *Benchmarking: An International Journal,* 27(4), pp. 1341–1368.

Poongodi, T., Krishnamurthi, R., Indrakumari, R., Suresh, P. and Balusamy, B. (2020) 'Wearable devices and IoT', in *A Handbook of Internet of Things in Biomedical and Cyber Physical System.* Springer, pp. 245–273.

Postma, P., Suarez-Garcia, E., Safi, C., Yonathan, K., Olivieri, G., Barbosa, M., Wijffels, R. H. and Eppink, M. (2017) 'Energy efficient bead milling of microalgae: Effect of bead size on disintegration and release of proteins and carbohydrates', *Bioresource Technology,* 224, pp. 670–679.

Qiu, R., Gao, S., Lopez, P. A. and Ogden, K. L. (2017) 'Effects of pH on cell growth, lipid production and CO2 addition of microalgae Chlorella sorokiniana', *Algal Research,* 28, pp. 192–199.

Rahmat, A., Jaya, I., Hestirianoto, T., Jusadi, D. and Kawaroe, M. (2020) 'Evaluation of system performance for microalga cultivation in photobioreactor with IOTs (internet of things)', *International Journal of Sciences: Basic and Applied Research,* 49(2), pp. 95–107.

Reverté, L., Prieto-Simón, B. and Campàs, M. (2016) 'New advances in electrochemical biosensors for the detection of toxins: Nanomaterials, magnetic beads and microfluidics systems. A review', *Analytica Chimica Acta,* 908, pp. 8–21.

Sadeghi, A.-R., Wachsmann, C. and Waidner, M. (2015) 'Security and privacy challenges in industrial internet of things', in *52nd ACM/EDAC/IEEE Design Automation Conference (DAC).* IEEE, pp. 1–6.

Sajim, A. S., Rosa, M. R., Wibowo, P. T. and Al-Rahbi, M. (2017) 'Mean shift tracking optimization in beagleboard rev C4', *ResearchGate*, pp. 1–10.

Santos, R. and Perestrelo, L. M. C. (n.d.) *Comparing BeagleBone Black and Raspberry Pi*. Available at: https://www.dummies.com/computers/beaglebone/comparing-beaglebone-black-and-raspberry-pi/ (Accessed: 16 August 2021).

Sarawak Biodiversity Centre (2020) *Smart Monitoring Device*. Available at: https://www.sbc.org.my/programmes/r-d-bioprospecting/algae-research/smart-monitoring-device (Accessed: 20 September 2021).

Sarkar, S., Manna, M. S., Bhowmick, T. K. and Gayen, K. (2020) 'Priority-based multiple products from microalgae: Review on techniques and strategies', *Critical Reviews in Biotechnology,* 40(5), pp. 590–607.

Screen Cloud (n.d.) *5 IoT Solutions for the Connected Restaurant*. Available at: https://screencloud.com/blog/iot-solutions-restaurant (Accessed: 27 August 2021).

Severin, T. S., Plamauer, S., Apel, A. C., Brück, T. and Weuster-Botz, D. (2017) 'Rapid salinity measurements for fluid flow characterisation using minimal invasive sensors', *Chemical Engineering Science,* 166, pp. 161–167.

Shah, H. M. (2018) *Internet of Things—Three Popular Development Boards*. Available at: https://blog.trigent.com/internet-of-things-three-popular-development-boards-trigent/ (Accessed: 22 July 2021).

Shawn (2019) *ATmega2560—Features, Comparisons, and Arduino Mega Review*. Available at: https://www.seeedstudio.com/blog/2019/11/13/atmega2560-features-comparisons-and-arduino-mega-review/ (Accessed: 25 August 2021).

Singh, G. and Patidar, S. (2018) 'Microalgae harvesting techniques: A review', *Journal of environmental management,* 217, pp. 499–508.

Singh, K. J. and Kapoor, D. S. (2017) 'Create your own internet of things: A survey of IoT platforms', *IEEE Consumer Electronics Magazine,* 6(2), pp. 57–68.

Singh, R. and Bhanot, N. (2020) 'An integrated DEMATEL-MMDE-ISM based approach for analysing the barriers of IoT implementation in the manufacturing industry', *International Journal of Production Research,* 58(8), pp. 2454–2476.

Singh, S. and Singh, P. (2015) 'Effect of temperature and light on the growth of algae species: a review', *Renewable and Sustainable Energy Reviews,* 50, pp. 431–444.

Skorupskaite, V., Makareviciene, V., Sendzikiene, E. and Gumbyte, M. (2019) 'Microalgae Chlorella sp. cell disruption efficiency utilising ultrasonication and ultrahomogenisation methods', *Journal of Applied Phycology,* 31(4), pp. 2349–2354.

Srinivasan, A. (2018) 'Iot cloud based real time automobile monitoring system', in *3rd IEEE International Conference on Intelligent Transportation Engineering (ICITE)*. IEEE, pp. 231–235.

Sydney, E. B., Schafranski, K., Barretti, B. R. V., Sydney, A. C. N., Zimmerman, J. F. D. A., Cerri, M. L. and Demiate, I. M. (2019) 'Biomolecules from extremophile microalgae: From genetics to bioprocessing of a new candidate for large-scale production', *Process Biochemistry,* 87, pp. 37–44.

Teja, R. (2021) *Classification of Relays*. Available at: https://www.electronicshub.org/classification-of-relays/ (Accessed: 19 September 2021).

Teng, S. Y., Yew, G. Y., Sukačová, K., Show, P. L., Máša, V. and Chang, J.-S. (2020) 'Microalgae with artificial intelligence: A digitalized perspective on genetics, systems and products', *Biotechnology Advances*, p. 107631.

Teralytic (n.d.) *World's First Wireless NPK Soil Sensor*. Available at: https://order.teralytic.com/ (Accessed: 8 August 2021).

Thaker, T. (2016) 'ESP8266 based implementation of wireless sensor network with Linux based web-server', in *Symposium on Colossal Data Analysis and Networking (CDAN)*. IEEE, pp. 1–5.

Tiempo Development (2020) *Things to Consider When Selecting an IoT Platform.* Available at: https://www.tiempodev.com/blog/things-to-consider-when-selecting-an-iot-platform/ (Accessed: 21 August 2021).

Tolem, S. C., Bogadi, C. R., Korlapati, N. S., Ravichandran, S., Rajendran, R. and Vuppalapati, C. (2020) 'A theoretical study on advances in streaming analytics', in *IEEE Sixth International Conference on Big Data Computing Service and Applications (BigDataService).* IEEE, pp. 41–45.

Toraman, H. E., Franz, K., Ronsse, F., Van Geem, K. M. and Marin, G. B. (2016) 'Quantitative analysis of nitrogen containing compounds in microalgae based bio-oils using comprehensive two-dimensional gas-chromatography coupled to nitrogen chemiluminescence detector and time of flight mass spectrometer', *Journal of Chromatography A,* 1460, pp. 135–146.

Tracy, P. (2016) *These Are the Elements Required to Deploy an IoT Solution.* Available at: https://www.rcrwireless.com/20161007/fundamentals/elements-iot-tag31-tag99 (Accessed: 28 July 2021).

Tung, L. (2019) *Raspberry Pi Has Now Sold 30 Million Tiny Single-Board Computers.* Available at: https://www.zdnet.com/article/raspberry-pi-now-weve-sold-30-million/ (Accessed: 14 August 2021).

Tutorials, R. N. (n.d.) *Installing ESP8266 Board in Arduino IDE (Windows, Mac OS X, Linux).* Available at: https://randomnerdtutorials.com/how-to-install-esp8266-board-arduino-ide/ (Accessed: 14 August 2021).

Ummalyma, S. B., Gnansounou, E., Sukumaran, R. K., Sindhu, R., Pandey, A. and Sahoo, D. (2017) 'Bioflocculation: An alternative strategy for harvesting of microalgae—an overview', *Bioresource Technology,* 242, pp. 227–235.

Upswift (2021) *The New Raspberry Pi 4 VS BeagleBone Black—2021 Comparison.* Available at: https://www.upswift.io/post/raspberry-pi-4-vs-beaglebone-black (Accessed: 12 July 2021).

Vendruscolo, R. G., Facchi, M. M. X., Maroneze, M. M., Fagundes, M. B., Cichoski, A. J., Zepka, L. Q., Barin, J. S., Jacob-Lopes, E. and Wagner, R. (2018) 'Polar and non-polar intracellular compounds from microalgae: Methods of simultaneous extraction, gas chromatography determination and comparative analysis', *Food Research International,* 109, pp. 204–212.

Vuppaladadiyam, A. K., Prinsen, P., Raheem, A., Luque, R. and Zhao, M. (2018) 'Microalgae cultivation and metabolites production: A comprehensive review', *Biofuels, Bioproducts and Biorefining,* 12(2), pp. 304–324.

Walker, T. (2017) *Presenting AWS IoT Analytics: Delivering IoT Analytics at Scale and Faster than Ever Before.* Available at: https://aws.amazon.com/blogs/aws/launch-presenting-aws-iot-analytics/ (Accessed: 6 September 2021).

Wang, B., Wang, Z., Chen, T. and Zhao, X. (2020) 'Development of novel bioreactor control systems based on smart sensors and actuators', *Frontiers in Bioengineering and Biotechnology,* 8, p. 7.

Wang, K., Khoo, K. S., Leong, H. Y., Nagarajan, D., Chew, K. W., Ting, H. Y., Selvarajoo, A., Chang, J.-S. and Show, P. L. (2021) 'How does the internet of things (IoT) help in microalgae biorefinery?' *Biotechnology Advances*, p. 107819.

Wang, P., Valerdi, R., Zhou, S. and Li, L. (2015) 'Introduction: Advances in IoT research and applications', *Information Systems Frontiers,* 17(2), pp. 239–241.

Wang, S. K., Stiles, A. R., Guo, C. and Liu, C. Z. (2014) 'Microalgae cultivation in photobioreactors: An overview of light characteristics', *Engineering in Life Sciences,* 14(6), pp. 550–559.

Wishkerman, A. and Wishkerman, E. (2017) 'Application note: A novel low-cost open-source LED system for microalgae cultivation', *Computers and Electronics in Agriculture*, 132, pp. 56–62.

Wolf, J., Stephens, E., Steinbusch, S., Yarnold, J., Ross, I., Steinweg, C., Doebbe, A., Krolovitsch, C., Müller, S. and Jakob, G. (2016) 'Multifactorial comparison of photobioreactor geometries in parallel microalgae cultivations', *Algal Research,* 15, pp. 187–201.

Wood, L. (2019) *World Market for Wearable Devices, Set to Reach $62.82 Billion by 2025 - Increasing Penetration of IoT & Related Devices Drives Market Growth.* Available at: https://www.prnewswire.com/news-releases/world-market-for-wearable-devices-set-to-reach-62-82-billion-by-2025—increasing-penetration-of-iot—related-devices-drives-market-growth-300974593.html (Accessed: 28 July 2021).

Xia, L., Rong, J., Yang, H., He, Q., Zhang, D. and Hu, C. (2014) 'NaCl as an effective inducer for lipid accumulation in freshwater microalgae Desmodesmus abundans', *Bioresource Technology,* 161, pp. 402–409.

Xu, K., Zou, X., Chang, W., Qu, Y. and Li, Y. (2021) 'Microalgae harvesting technique using ballasted flotation: A review', *Separation and Purification Technology*, p. 119439.

Yew, G. Y., Puah, B. K., Chew, K. W., Teng, S. Y., Show, P. L. and Nguyen, T. H. P. (2020) 'Chlorella vulgaris FSP-E cultivation in waste molasses: Photo-to-property estimation by artificial intelligence', *Chemical Engineering Journal,* 402, pp. 126230.

Zhang, H. and Zhang, X. (2019) 'Microalgal harvesting using foam flotation: A critical review', *Biomass & Bioenergy,* 120, pp. 176–188.

Zhong, C.-L., Zhu, Z. and Huang, R.-G. (2015) 'Study on the IOT architecture and gateway technology', in *14th International Symposium on Distributed Computing and Applications for Business Engineering and Science (DCABES).* IEEE, pp. 196–199.

Zhu, Z. and Huang, R.-G. (2017) 'Study on the IoT architecture and access technology', in *16th International Symposium on Distributed Computing and Applications to Business, Engineering and Science (DCABES).* IEEE, pp. 113–116.

5 Understanding Environmental Biotechnology 4.0

*Wai Yan Cheah[1], Tengku Nilam Baizura Tengku Ibrahim[2], Nurul Syahirah Mat Aron[3], Wai Siong Chai[4] and Pau Loke Show[3],**

[1] Centre of Research in Development, Social and Environment (SEEDS), Faculty of Social Sciences and Humanities, Universiti Kebangsaan Malaysia, 43600 Bangi, Selangor, Malaysia

[2] Department of Environmental Health, Faculty of Health Sciences, Universiti Teknologi MARA, Cawangan Pulau Pinang, Kampus Bertam, Kepala Batas, Penang, Malaysia

[3] Department of Chemical and Environmental Engineering, Faculty of Science and Engineering, University of Nottingham Malaysia, Jalan Broga, Semenyih, Selangor Darul Ehsan, Malaysia

[4] School of Mechanical Engineering and Automation, Harbin Institute of Technology, Shenzhen, Guangdong, China

CONTENTS

5.1 Introduction ... 176
5.2 Understanding Industrial Revolution 4.0 ... 176
5.3 The Advancement of Environmental Biotechnology in Industrial Revolution 4.0 ... 179
5.4 Biomass to Biofuel Conversion Technologies ... 181
5.5 Conventional Pretreatment: Physical, Chemical, and Biological Pretreatment ... 181
5.5.1 Physical Pretreatment ... 181
5.5.2 Chemical Pretreatment ... 182
5.5.3 Physico-Chemical Pretreatment ... 183
5.5.4 Biological Pretreatment ... 184
5.5.5 Novel and Emerging Technologies: Ionic Liquid and Hydrothermal Pretreatment ... 184

DOI: 10.1201/9781003202196-5

5.6 Biofuel Conversion Technologies........185
5.6.1 Biochemical Conversion........185
5.6.2 Thermochemical Conversion........187
5.7 Pollution Control, Prevention, and Cleaner Production........188
5.7.1 Wastewater Treatment........188
5.7.2 Organic Pollutant Removal........189
5.7.3 Heavy Metal Removal........192
5.7.4 Micropollutant Removal........193
5.7.5 Air Pollution Control........196
References........197

5.1 INTRODUCTION

Industrial revolution had brought tremendous positive effects and challenges to all countries and human race. The transformation has been led by Great Britain with the invention of the commercial steam engine. The invention has brought industrial developments due to the improvement in communication and transportation. First industrial revolution was followed by second industrial revolution which was led by United States, with the invention of telephone, revolutionizing the art of communication during that era. The development has evolved further to third industrial revolution, with the invention of Internet communication. With the public infrastructure built, the Internet has transformed the economic landscape in all sectors. All these industrial revolutions, undeniably, have brought impactful economic growth, increase in productivity and quality goods as well as services, improved social development, and the overall advancement in the well-being of the society. Conversely, there are also challenges arisen, with the most significant one including environmental degradation. Resources' depletion, climate change, and global warming have resulted from rapid economic growth and industrialization, with excessive utilization of fossil fuels. Thus, moving towards sustainability in overall, pillars of social equity, economic viability and the environmental protection are vital to reach the 'balance'.

Industrial revolution 4.0 is not like an exemption to the previous industrial revolution. It is also predicted to bring both benefits and challenges to the society. Industrial revolution 4.0 is the marriage between production environment with technologies like big data, Internet of Things (IoT), cyber-physical system (CPS), 3D printing, autonomous vehicles, automated systems, artificial intelligence, and new materials. The following section describes industrial revolution 4.0 and the advancement of environmental biotechnology in industrial revolution 4.0.

5.2 UNDERSTANDING INDUSTRIAL REVOLUTION 4.0

The term 'revolution' refers to the concept of radical change. The first industrial revolution began with the steam engine invention in seventeenth and eighteenth centuries. This was the key for mass production in these eras. The textile industry has changed from having to rely on man-made products into having to rely on machinery. Equipment are used to increase the productivity. The United States, England, and Japan have shown that industrialization has improved the standard of living of their citizens. Between nineteenth and twentieth centuries, second industrial revolution occurred after infrastructure advancement, for instance, electricity utilization, mass production, and method of division of labor. It applied technology innovation to reshape industries and societies. Steel (upgraded from iron), electricity, light bulb, telephone, aeroplane, automobiles, antibiotics and mass production are in use and in place. Third industrial revolution emerged in 1970s with digital revolution, information technologies, and automation of production. It brought us advanced communication capabilities, with invention of television, radio, Internet, personal computer, smartphone, and satellites (Morrar, Arman, and Mousa 2017).

Fourth industrial revolution, based on the concept called smart factory, is the one we are currently heading to. It was created in Germany in 2011 by Institute Fraunhofer-Gesellschaft and the German Federal Government as a collective term that defines the application of information flow technologies, automation, and manufacturing, as a high technology strategy for 2020. Germany, indeed, is the global leader for manufacturing equipment. The initiation of industrial revolution 4.0 can be seen as an action to sustain the position as one of the most influential countries in machinery and automotive manufacturing (Rojko 2017). It is also the strategy to mitigate the increasing competition worldwide and to differentiate German and European Union industries from the international market (Rojko 2017; Morrar, Arman, and Mousa 2017). German government wanted to apply intelligent monitoring in the manufacturing processes, via digitalization and exploitation of new technologies to aid decision-making and cost reduction, thereby increasing the competitiveness (Rojko 2017). Since then, parties including politicians, academics, and entrepreneurs have come out with ideas in improving the industrial processes from planning, operation, engineering to logistics for the whole product life cycle. The integration could be categorized into (1) vertical integration which refers to the implementation of manufacturing system with flexibility and reconfigurability characteristics; (2) horizontal integration which creates collaborations between companies in which physical, intellectual, energy, and financial resources can flow between companies; and (3) end-to-end engineering which incorporates these two goals to build power chain of software tools for mass production and communications (Da Silva and Massabni 2019; Morrar, Arman, and Mousa 2017).

Industrial revolution 4.0 embraces new technologies such as development in advanced robotics, artificial intelligence, nanotechnology, 3D printing, and biotechnology. The impacts of industrial revolution 4.0 are expected to be more profound, irreversible, and more rapid than the previous three industrial revolutions. Smart systems, smart production, skills, and humans are the focus of this in industrial revolution 4.0 (Da Silva and Massabni 2019). Industrial revolution 4.0 emphasizes data interconnection, integration, and innovation. The following are the core areas that are required to be advanced for a successful transformation in productions (Da Silva and Massabni 2019).

i. Cloud computing: Cloud computing refers to the practice of using interconnected remote servers hosted on the Internet for information storage, management, and processing. The delivery of computer system includes networking, software, analytics, and intelligence over the Internet for flexible resources.
ii. Internet of Things (IoT): a concept that refers to connections between physical objects like sensors or machines and the Internet. Connected devices could also be a car, sensor, building or machinery to aid the internal operations via cloud environment where data is stored. Equipment and operation can be optimized by leveraging the insights of others using the same equipment. This will allow small enterprises to access the technology they wouldn't able to afford on their own. It was estimated that 46% of the global economy can benefited by industrial Internet (Rojko 2017).
iii. Cybersecurity: a protect management system to ensure data confidentiality, integrity, and availability; risk assessment to identify the ever-changing and growing threat landscape and, therefore, placement of measures to mitigate the possible damage.
iv. Augmented reality: a combination of real and virtual worlds (computer-generated). A real image is captured on video, while that real-world image is "augmented" with layers of digital information.
v. Big data analytics: large sets of structured or unstructured data are generated by sensors and devices. The data should be collected, integrated, stored, organized, processed, and analyzed to reveal patterns, trends, associations, and opportunities.
vi. Autonomous robots: robots with autonomous capacity used in production lines to perform complex activities without explicit human control. The potential of autonomous robots includes increased productivity and efficiency, decreased cost due to lower risk and error, and improved safety in high-risk environment.
vii. Additive manufacturing/3D printing: this technology has improved rapidly in the last decade, which included processes from prototyping to production; advancement in metal additive manufacturing to create more possibilities for production.
viii. Simulation: the technique of applying real or hypothetic model of system and process to analyze and predict the behavior/outcome of the system and process. This contributes to decision-making, reducing risk and loss, shortening the development cycle, enhancing product quality, and improving performance and efficiency (Rodič 2017).
ix. System integration: a combination of computer system with software packages to create a larger system. This enables the multiple systems to work cohesively in a unified and coordinated manner.

Industrial revolution 4.0 is important as it is estimated to make companies more flexible and responsive to the business trend. With the current situation of increasing market volatility, shorter product lifecycles, higher product complexity, and global supply chains, companies have to be more flexible and able to adapt to the changing trend (Morrar, Arman, and Mousa 2017). Industrial production is also expected to adapt to this changes fast, with ever-changing market demand (Rojko 2017). For instance, the digitization of product lifecycle allows companies to use data from production, service, and social media. This will result in the improvement of products at a faster rate, including the changes applied on operation and so on. Second, industrial revolution 4.0 will lead to an innovation economy. The new business model could be transformed. For instance, business model could be changed from selling products like engines to providing services. Third, the consumer today demands customization. With the advancement in digitization and crowd sourcing, design process could be speeded up. Industrial revolution 4.0, indeed, would eliminate workers with operating simple manual tasks as machines are smarter. Conversely, industrial revolution 4.0 is also putting humans in the centre of production as humans are required to manage and coordinate more complex projects. This will result in more flexibility to employees rather than routine work. The focus will be more towards problem solving and self-organization. Next, industrial revolution 4.0 could contribute to reduce energy consumption. Industrial revolution 4.0 is expected to find ways to cope with constraints on energy; resources; environmental, social, and economic impacts. Rojko (Rojko 2017) claimed that production cost can be reduced by 10–30%, followed by another 10–30% of logistic cost and another 10–20% of quality management cost, via industrial revolution.

For industrial revolution 4.0 to be realized, three stages that are digitization, automation and integration have to take place (Oláh et al. 2020). The concept of 'Industrie du futur' was introduced as the core principle for French industrial policy. The concept stressed on cooperation of industry and science, with technologies including virtual plant, Internet of Things, and augmented reality (Rojko 2017). In China, initiative of 'Made in China 2025' was introduced, with the goal of upgrading Chinese industry using the inspiration of Germany's industrial revolution 4.0 and adapting to the China's market demands. China has prioritized the sectors related to information technology, robotics, and automated machinery. Their long-term goals are to reform China's manufacturing industry by producing high-quality products and to take over the dominance of Germany and Japan in manufacturing until 2035 (Rojko 2017).

5.3 THE ADVANCEMENT OF ENVIRONMENTAL BIOTECHNOLOGY IN INDUSTRIAL REVOLUTION 4.0

Industrialization, undoubtedly, has brought undesirable environmental impacts such as climate change, global warming, and air pollution due to excessive consumption of fossil fuels. In contrast, as industrial revolution 4.0 emphasizes on innovation and new technology, this is also contributing to addressing environmental security (Herweijer and Waughray 2018). Great innovation and transformation are estimated for the field of environmental biotechnology due to industrial biotechnology 4.0. Advanced biotechnology should ideally include multidisciplinary group with applications for health and manufacturing industries like fermentation, agriculture, livestock, and mining (Da Silva and Massabni 2019). Artificial intelligence is expected to address earth challenge areas including climate change for clean power, sustainable land use, smart cities and homes; biodiversity and conservation for habitat protection and restoration, and pollution control; healthy oceans for fishing sustainably and pollution control; water security; clean air and weather and disaster resilience (Herweijer and Waughray 2018).

Biotechnology is estimated to substitute large number of processes in the future and create innovative and sophisticated solutions for problems. The environmental biotechnology field that would be transformed following industrial revolution 4.0 includes new biological therapies, new energy sources' discovery, structuring of analytical tools, nanobiotechnology, proliferation of transgenic technology, bioinformatics, biomass to biofuel conversion technologies, and research based on sustainability. With the integration of digital system and technologies as mentioned in Section 5.2, biotechnology companies need to evolve for the automation of research laboratories for greater productivity and reproducibility and research accuracy. Network and robotics could be incorporated to monitor experiments and therefore producing more accurate data of experimental results, predictions, simulations, and modeling. This allows the integration of physical system with virtual system, benefiting biotechnology companies, manufacturing and service laboratories, and academic research laboratories. German national innovation network is working on the integration between intelligent laboratories and industrial revolution 4.0. It aims to drive the development of innovation technologies and work toward simplified process flow, greater process reliability, process efficiency, and outcomes with better quality. Innovation center for laboratory automation is developed in Stuttgart, Germany. The aims are to create intelligent laboratory for intelligent tracking, automatic documentation, and analyzing hand movement using 3D images. The system is expected to capture and record all processes, so as to reduce time spent and workload and for a better outcome. Japan is inventing two-arm robots for pharmaceutical laboratories. The project is led by the company named Yaksawa and the National Institute of Advanced Industrial Science and Technology of Japan. This robot named 'Mahoro', which is able to perform laboratory work, works faster, accurately, and efficiently. The robot is suitable to be applied for working with biohazards and conducting clinical trials. The researchers from University of Technology in Vienna, Austria, are currently working on high-resolution 3D printing process for live cells generation. Technology using special 'biological ink' is applied, which allows cells to be incorporated into a 3D matrix (Da Silva and Massabni 2019). Smart agriculture is expected to produce high yield with automated data collection and decision-making at farm level, for instance, to spray, harvest, or to plant and to allow early detection of crop disease. Sensors will be monitoring the soil moisture and temperature by sending signal to adding moisture (Herweijer and Waughray 2018). Biotechnology innovation provides vital contribution to the transition from current unsustainable economic practice to renewable industrial system. It could also be known as the circular and bio-based economy.

There are undeniably challenges that have arisen with the technological advancement and innovation: for instance, the change in labor market, skills demands, environmental sustainability, globalization, population aging, and so on. According to the United Nation, 40% of the jobs existing today will be eliminated in 2035 (Da Silva and Massabni 2019). Looking at the positive side, significant employment opportunities could also be created in the environmental biotechnology field, such as analyst and data scientist, specialist in artificial intelligence, specialist in machine and automation, environmental engineering, waste consultant, and scientific consultant. Demand for professionals with skills is emerging due to industrial revolution 4.0.

5.4 BIOMASS TO BIOFUEL CONVERSION TECHNOLOGIES

Lignocellulosic feedstock such as forest woody and herbaceous biomass, agricultural residue, and microalgae are commonly applied as feedstocks for biofuel production (Aron et al. 2020). The lignocellulosic feedstock is complex in structure, with components of lignin, hemicellulose, and cellulose. Therefore, pretreatment of biomass feedstock is an essential step to break down these structures, releasing glucose for bioethanol and for biogas production. Pretreatment is important to (1) increase amorphous region to ease hydrolysis, (2) enhance the porosity of matrix to ease chemical and enzymatic hydrolysis, and (3) release cellulose from lignin and hemicellulose (Cheah et al. 2020). The pretreatment methods which are currently in use include physical, chemical, physico-chemical, biological pretreatment, as well as the recent ionic liquid and hydrothermal pretreatment. Effective pretreatment method should be able to

i. increase glucose yield for downstream processing,
ii. pretreat all types of lignocellulosic feedstocks,
iii. assist in lignin recovery for later on combustion,
iv. lessen the formation of inhibitors,
v. be effective in lowering cost and energy, and
vi. regenerate valuable lignin coproducts. In fact, all the mentioned pretreatment technologies have their advantages and drawbacks.

The choice of pretreatment technologies is usually subjected to economical factor, the lignocellulosic feedstock used, and the environmental impact.

5.5 CONVENTIONAL PRETREATMENT: PHYSICAL, CHEMICAL, AND BIOLOGICAL PRETREATMENT

5.5.1 Physical Pretreatment

Physical pretreatment includes milling, extrusion, freezing, ultrasound and microwave, and these are the common methods applied on lignocellulosic biomass feedstock. Milling reduces the size of the biomass and provides larger surface area for bioethanol and biogas production. There are various types of milling processes, including rod milling, ball milling, two-roll milling, hammer milling, and wet-disk milling (Baruah et al. 2018). Gu et al. (2018) stated that reported on planetary ball milling of post-harvest forest residues, yielding the maximal glucose and xylose/mannose with low energy input ranging at duration of 7 to 30 min of milling. While Baruah et al. (2018) claimed that both rod and hammer milling are effective to pretreat wheat straw in reduction of size prior to pyrolysis at duration of 60 min, but decrease in crystallinity in using rod milling. Nevertheless, the drawbacks of milling pretreatment are cost of operation, maintenance of the equipment, and mostly the high-energy consumption. Wet-disk milling is known to be lower in energy consumption (Baruah et al. 2018). Zakaria et al. (2015) reported that wet-disk milling has resulted in the defibrillation of oil palm mesocarp fiber, and SEM analysis showed an increase in surface area for cellulose-to-glucose conversion (Zakaria et al. 2015). Yield of 98.1% was attained using a combination of hot compressed water (200°C, 20 min) and wet-disk milling, with an energy consumption of only 9.6 MJ kg^{-1} of substrate. Four cycles of freezing at -18°C and thawing at 22°C resulted in the highest glucose yield (Rooni, Raud, and Kikas 2017). Microwave-assisted pretreatment has also been found to be very effective in pretreating switchgrass and miscanthus (Jędrzejczyk et al. 2019). *Panicum* sp. and *Miscanthus* sp. were pretreated using microwave at optimal temperatures of 60°C and 120°C, respectively. About 7–10% of pretreated substrates were found to be of higher solubility in subcritical water, as compared to non-pretreated substrates (Baruah et al. 2018). Microwave pretreatment of *Hyacinthus* sp. was found able to increase the methane yield via anaerobic digestion (Zhao et al. 2017). Ultrasound pretreatment of soft wood biomass was found able to increase bio-oil yield via conventional pyrolysis (Cherpozat, Loranger, and Daneault 2019). There was an enhanced biomethane production from grape pomace pretreated by using ultrasound at a frequency of 50 kHz and temperature of less than 25°C with a residence time of 40–70 minutes (El Achkar et al. 2018). Extrusion is another pretreatment method in which raw materials are passed through the barrel under high temperature (>300°C). Screw design, screw speed, and barrel temperature are controlling factors for an effective extrusion in pretreatment of biomass. Extrusion pretreatment alone of olive tree pruning has produced 69% yield over combined extrusion and alkaline pretreatment (Baruah et al. 2018). Hubenov et al. (Hubenov et al. 2020) has reported that vacuum freeze drying can enhance the pore sizes of biomass poplar, switch grass, and wheat straw through water–ice transformation, as compared to conventional drying for biochar production.

5.5.2 Chemical Pretreatment

Chemical pretreatment is commonly applied at a commercial scale. Acid, alkali, organic acids, and ionic liquids (green solvent) are chemicals used for pretreatment. Sulfuric acid, nitric acid, phosphoric acid, and hydrochloric acid are commonly applied, while the hydroxide salts are used for alkali pretreatment. Biogas generated from corn cob was two times higher than untreated corn cob, after lime pretreatment to remove lignin (Gu et al. 2018). The lime applied was calcium hydroxide, which is cost-effective and safe to handle. Rice straw pretreated with sodium hydroxide of 1% at room temperature at 3 hours has been found to be effective in reducing hemicellulose and cellulose content, therefore increase biomethane production by 34% (Law et al. 2022). Acid pretreatment can be applied using concentrated acids (30–70%) with low temperature (<100°C) or using diluted acids (0.1–10%) at higher temperature of 100–250°C, respectively (Gelosia et al. 2020). Kundu and Samudrala (Zhou et al. 2020) have reported thermally assisted peracetic acid pretreatment on biomass (hardwood and softwood), at 90°C at 5 hours, which has brought effective lignin removal with only negligible carbohydrate loss. Organosolv is a mixture of organic solvents which are also able to solubilize hemicellulose and extract lignin. The organic solvents used include methanol, ethanol, and acetone with organic acids (Baruah et al. 2018). Gelosia and Bertini (Gelosia et al. 2020) have studied organosolv pretreatment with a slight amount of acid to pretreat *Cynara cardunculus L.* for glucose production. Yield of 80% glucose was obtained from wheat straw, pretreated at 190°C for 60 minutes, using 25% of ethanol with 1% H_2SO_4 as catalyst. Corn stover was pretreated using 60% of ethanol and n-propylamine as catalyst at 140°C for 40 min. Total yields of 87.1% glucose and 75.4% xylose were attained (Foong et al. 2021).

5.5.3 Physico-Chemical Pretreatment

Apart from the physical pretreatment or chemical pretreatment method mentioned before, there is physico-chemical pretreatment method as well that could be applied on lignocellulosic biomass. Physico-chemical pretreatment refers to a combination of physical pretreatment with chemical pretreatment in order to enhance the efficiency. Pretreatment using milling and alkali on corn stover has found an improvement of up to 110% of enzymatic hydrolysis for glucose production (Kundu et al. 2021). The physico-chemical pretreatment methods also include steam explosion, ammonia fiber explosion, carbon dioxide (CO_2) explosion, microwave/chemical pretreatment, and using liquid hot water. Combined pretreatment using microwave with ionic liquid applied on Crotalaria juncea at temperature of 160°C for 46 min had produced 78.7% of glucose yield (Pecha and Garcia-Perez 2020). Liquid hot water pretreatment using water at the temperature of 160–240°C, with high pressure, was able to pretreat up to 80% of hemicellulose, without using any chemicals (Kundu et al. 2021). Ammonia to biomass loading of 5:1, 70% of moisture content, and a temperature of 170°C are optimal parameters to enhance the enzymatic digestibility of corn stover (Gelosia et al. 2020). da silva et al. (2016) reported that an amount of 5.2 L of bioethanol per kg of dried biomass was obtained via liquid hot water pretreatment on corn stover, whereas 5.4 L of bioethanol was attained in corn stover pretreated with ammonia fiber explosion (da Silva et al., 2016). Liquid hot water offers advantages of low cost, non-toxicity, and no catalyst requirement with no inhibitor formation and also better yield in the study mentioned. CO_2 explosion method is the combination method in which steam and supercritical CO_2 are used. In high-pressure condition, CO_2 is explosively released through nozzle, resulting in biomass cell to break. Hydrolysis of sugarcane bagasse using supercritical CO_2 has yielded 60% of fermentable sugars (Gelosia et al. 2020). Due to the consumption of CO_2, this pretreatment method is considerably an environment-friendly method. This pretreatment method has been applied on corn stalk, corn cob, and rice straw at 80–160°C and on a range of pressure (5–20 MPa) for 15–60 min (Kundu et al. 2021).

5.5.4 Biological Pretreatment

Biological pretreatments using fungi and bacteria are applied to biodegradable biomass using their lignin-degrading enzymes. These enzymes include phenol oxidase, lignin peroxidase, manganese peroxidase, laccases, and versatile peroxidase. Biological pretreatment offers low operating cost, low energy consumption, no chemical application, and environmental-friendly process as advantages. Fungi such as white-rot, brown-rot, and soft-rot fungi are applied to degrade lignin in biomass feedstock. Among all types of fungi, white-rot fungus is mostly applied for pretreatment as it usually provides high sugar yield. Biological pretreatment of bamboo clums using *Ceriporiopsis subvermispora* showed lignin degradation efficiency of 50% (Gelosia et al. 2020). Paddy straw showed an enhanced sugar recovery followed by enzymatic saccharification using Trametes hirsuta (Gelosia et al. 2020). Aside from lignin degradation, some fungi can as well remove specific components such as antimicrobial substances. Aside from whole cell pretreatment, pretreatment of lignocellulosic biomass can also be carried out using enzymes. The enzymatic pretreatment is generally performed under mild temperature ranging about 40–50°C at the pH values of 4.5–5 (Vasić, Knez, and Leitgeb 2021). Vasic and Knez (Vasić, Knez, and Leitgeb 2021) stated that 91% of bioethanol yield was obtained from sugarcane bagasse, pretreated with low-temperature aqueous ammonia soaking and commercial enzyme named Cellic CTec2 Cellulase. Around 39% lignin was removed from wheat straw using laccase and peroxidase from polyporus brumalis (Gelosia et al. 2020). Nevertheless, biological pretreatment is subjected to microbial activity which is governed by factors of temperature, moisture, incubation period, subject size, pH, nutrients, and the performance of the species itself.

5.5.5 Novel and Emerging Technologies: Ionic Liquid and Hydrothermal Pretreatment

The application of ionic liquid (IL) for lignocellulosic pretreatment has gained attention due to its green properties, high thermal stabilities, and promising solvating potential. The commonly used ionic liquids are imidazolium-, pyridinium-, [Emim]Ac, [Emim]Cl, and [Bmim]Cl. These ionic liquids are found to be efficient due to their ability to form strong hydrogen bonds with hydroxyl groups. The industrial scale application is still limited as ionic liquid is expensive. Hu and Cheng (Hu et al. 2018) reported on adding 50% (w/w) water into [Bmim] BF4 IL, and enzymatic hydrolysis of cornstalk has improved up to about 82%. Ionic liquid and co-solvent combination have also been found effective. Xu and Zhang (Xu et al. 2013) found enhanced cellulose dissolution by using DMSO and 1-butyl-3-methylimidazolium acetate ([Bmim]Ac). Hydrothermal pretreatment applies high-temperature subcritical water to destruct plant cell wall, degrade hemicellulose, and convert lignin into glucose. Hydrothermal pretreatment provides the advantages of high energy conversion, low corrosion, and no catalyst. This technique enables improved cellulose accessibility to enzyme by increasing biomass surface area and by decreasing the cellulose crystallinity. There are three main stages of hydrothermal pretreatments, which include carbonization, liquefaction, and gasification. Hydrothermal pretreatment is applied at varied temperatures for all these stages depending on the desired output. Hydrothermal carbonization is performed at 200 and 270°C, and the product is a carbon-rich solid char. Hydrothermal liquefaction is performed between 250°C and 400°C, producing bio-oil, water-soluble constituents, char, and CO_2. Hydrothermal gasification works at temperatures higher than 400°C, and the product is fuel gases (Toor, Rosendahl, and Rudolf 2011; Chai, Chew et al. 2021). Nevertheless, the constraints of hydrothermal pretreatment are high cost, and the research on the effectiveness of it on contributing to biofuel production is still very limited at large scale.

5.6 BIOFUEL CONVERSION TECHNOLOGIES

5.6.1 Biochemical Conversion

After the biomass has undergone pretreatment, the subsequent process could be classified under biochemical conversion and thermochemical-conversion, depending on the desired biofuels to be attained. Alcoholic fermentation is the process of converting cellulose, starch, and glucose into fermentative biofuels using bacteria, yeast, or fungi under anaerobic digestion. Bioethanol, today's largest volume of biofuel, is produced via biochemical conversion of corn starch and sugarcane (Vimmerstedt et al. 2014). Saccharomyces cerevisiae is the commonly applied strain for fermentation, at commercial scale. It is the fermenting yeast that assimilates glucose into pyruvate at the glycolytic pathway, subsequently forms acetaldehyde from pyruvate, and finally acetaldehydes are reduced to generate ethanol. All these processes require close monitoring and regulations, as they are subjected to various parameters, which will greatly affect the microbial activity of the fermenting microorganisms. Distillation is the final step to remove the CO_2 and water, producing high-purity biofuel. The bioalcohol will be subsequently condensed for storage and for engine usage. (Cheah et al. 2016) reported that, theoretically, maximal yield of 0.5 kg of ethanol could be produced from every kg of glucose. Biofuel feedstock and microalgae are of great concern over the past decades (Low et al. 2021). *Chlorella vulgaris and Chlamydomonas* sp. are well-applied microalgae species for bioethanol production via starch fermentation. For every 1 g of biomass of *Chlamydomonas reinhardtii*, an amount of 0.235 g of bioethanol could be produced via separated hydrolysis and fermentation method. In fact, there are two types of fermentation processes which include separated hydrolysis and fermentation method, meanwhile the second type is known as simultaneous hydrolysis and fermentation method. The former provides an optimal performance in separated unit to yield high content of glucose prior to fermentation. However, the increase in glucose and cellodextrin content might inhibit the cellulose activity. In contrast, the single step of the simultaneous hydrolysis and fermentation method is with no inhibition constraint, offers shorter reaction time and lower equipment cost and lower risk of contamination as fewer steps are involved. Acetone-butanol-ethanol (ABE) fermentation is applied aiming to produce biobutanol. *Clostridium* sp. is commonly applied for ABE fermentation, producing biobuthanol. Biobutanol is produced at ratio of 3:6:1 via ABE fermentation.

Gaseous biofuels, mainly biohydrogen and biomethane, are generated from an anaerobic digestion of carbohydrate-rich biomass. Conventionally, hydrogen is generated from energy-intensive processes such as reverse water gas shift reaction, water electrolysis, steam methane reforming, and gasification. Besides, biohydrogen can be generated via photofermentation and dark fermentation of biomass, using photosynthetic microorganisms like *Rhodobacter* sp. to convert glucose into CO_2 and hydrogen. Dark fermentation uses acidogenic bacteria like *Basillus* sp. and *Clostridium* sp. to ferment glucose without sunlight (Cheah et al. 2016). Microalgae species of *Dunaliella tertiolecta* and *Chlamydomonas reinhardtii* have produced biohydrogen of 61% and 52%, respectively, via photofermentation. Biomass like rice, corn, and wheat residue can be well applied for dark fermentation, yielding range of 12–7017 mL H_2 L^{-1} via dark fermentation (Soares et al. 2020). The lowest yield of 12 mL H_2 L^{-1} was obtained from cashew apple bagasse, whereas the highest yield of 7017 mL H_2 L^{-1} was attained from sugarcane bagasse. The former has lowest H_2 yield as it contains high amount of lignin (35.26%). Mesophilic temperature, pH of neutral value, and hydraulic retention time of 72 hours are commonly applied parameters of dark biohydrogen parameters (Soares et al. 2020). Biomethane is produced, in the final stage named as methanogenesis, after hydrolysis and fermentation. Methanogens are producing gases of mainly methane and CO_2. Types of substrates, retention time in the digester, temperature, pH values are governing factors to biomethane production output. Shetty and Kshirsagar (Shetty et al. 2017) reported that maximal methane yield of 514 L kg^{-1} VS (with about 59% of methane) was attained from milled rice straw pretreated with 1% of sodium hydroxide at an ambient temperature for 180 min. Hubenov and Carcioch (Hubenov et al. 2020) has stated that biomethane of yields 1116 cm^3 L^{-1}, 1350.5 cm^3 L^{-1} and 1293.25 cm^3 L^{-1} were attained from non-pretreated, pretreated using ultrasonic and microwaved maize stalks, respectively.

5.6.2 Thermochemical Conversion

Thermochemical conversion is as well a process of converting lignocellulosic biomass into carbon-rich biochar (solid), syngas, and bio-oil (ElFar et al. 2020). Torrefaction, pyrolysis, gasification, and hydrothermal carbonization are the main thermochemical conversion techniques involved in the decomposition of the biomass using heat energy. Thermochemical conversion of agricultural biomass to energy is promising, with high productivity yield (Verma et al. 2012). Biochemical conversion relies microbes and enzymes and thus lacks robustness at industrial scale, as compared to the thermochemical conversion which is independent from environmental conditions (Verma et al. 2012). ElFar and Chang (ElFar et al. 2020) claimed that thermochemical technology which initiates bond breaking and reforming of organic matter is applied to produce biochar from microalgae. Hydrothermal gasification thermochemical process can be applied to convert wet biomass into biogas at high pressure 22.1 MPa and temperature 374°C (ElFar et al. 2020). Law and Cheah (Mat Aron et al. 2021; Law et al. 2022) have also recently reported that the slow pyrolysis of microalgae *Synechococcus* sp., *Dunaliella tertiolecta*, and *Chaetocerous muelleri* at temperature 500°C yielded biochar of 44–63%. Fast pyrolysis at short duration and high heating rate produced relatively lower microalgae biochar yield, except for *Chlorella protothecoldes*, which had produced higher yield of 54%. Foong and Chan (Foong et al. 2021) have also recently reported on advanced pyrolysis which includes catalytic pyrolysis, vacuum pyrolysis, microwave pyrolysis, co-pyrolysis, CO_2-pyrolysis, and solar pyrolysis (Foong et al. 2021). Zhou and Zhou (Zhou et al. 2020) reported that microwave-assisted pyrolysis biomass of wood pellet generated 67% of syngas at lower heating rate (18 MJ Nm-3) and, therefore, has reduced the overall energy consumption from 7.2 to 3.45 MJ kg^{-1} of substrate. The selection of feedstock, feedstock size, reactor temperature, pressure, heating rate, residence time, catalyst, carrier gas flow rate, and supply of oxygen are the key factors to govern the production of the targeted products (Foong et al. 2021). Solar pyrolysis of rice husk has produced 13 mol% of H_2, 12 mol% of CH_4, and 25 mol% of CO_2 at heating rate of 800°C, whereas 42 mol% of H_2, 6 mol% of CH_4, and 6 mol% of CO_2 at heating rate of 1400°C (Weldekidan et al. 2020). Solar pyrolysis of willow wood was found to produce 10.8–47.0% of gas yield, with composition of syngas with 0.16–12.1 mol kg^{-1} of H_2, 1.9–12.9 mol kg^{-1} of CO, 0.2–1.7 mol kg^{-1} of CH_4, and 0–0.58 mol kg^{-1} of C_2H_6 (Foong et al. 2021). Fast pyrolysis is one of the pathways to produce bio-oil. Particle size (<2 mm), fast heating rate (>100°C per seconds), and temperatures of 400°C and 650°C are commonly applied parameters for fast pyrolysis. Bio-oil yield of 60–75%, biochar yield of 15–25% and syngas yield of 10% are as well commonly obtained from the fast pyrolysis of biomass (Pecha and Garcia-Perez 2020). Ansari and Arora (Ansari et al. 2019) have worked on thin-film pyrolysis of biopolymers which includes cellulose, xylan, and lignin. Dominant of products are bio-oil with 55.35%, followed by 20.08% of biochar and 4.69% of gases, when thin-film pyrolysis was performed at a temperature of 550°C. The bio-oil produced had the composition of 45.17% of low-molecular weight phenols, 25.10% of phenolic aldehydes/ketones, 20.23% of methoxyphenols, and 9.5% of light oxygenates (Ansari et al. 2019).

5.7 POLLUTION CONTROL, PREVENTION, AND CLEANER PRODUCTION

5.7.1 Wastewater Treatment

Natural resources such as water are extremely valuable. In contrast, the number of developing countries without having any access to clean water currently poses a serious threat to the quality of national water supplies in industrialized countries. Several organic and inorganic pollutants contaminate municipal, agricultural, and industrial wastewater, including microplastics, xenobiotics, heavy metals, nitrate (NO_3^-), phosphates (PO_4^{3-}), and carbon compounds and therefore endangering the food chain and also human health (Ajibade et al. 2021; Chai, Tan et al. 2021). The water treatment problem cannot be solved by any one technology due to the vast differences in scale, type of contaminants, and regional conditions (Rawat et al. 2016; Taheran et al. 2018; Yang et al. 2019). Biodegradation prevents eutrophication of downstream waters, such as rivers and lakes, by breaking down organic molecules and inorganic compounds. The use of conventional technologies is insufficient to degrade heavy metals, extremely high nutrients, and xenobiotic substances, resulting in an increase in their concentration in the groundwater supply (Alcántara et al. 2015; Rawat et al. 2016).

Because microalgae have a wide range of metabolic abilities, including photoautotrophic, mixotrophic, and heterotrophic metabolism, they can be used as biological systems for treating a variety of wastewaters. Depending on local regulations on algal reuse, algae biomass generated via wastewater streams might be used to make sustainable bioproducts (e.g., proteins, fatty acids, pigments, biofertilizers/biochar, and animal feed). Algae-based WWT technologies have in fact been researched since the 1950s (Wollmann et al. 2019; Nirmalakhandan et al. 2019; Chye et al. 2017).

The history of the commercial use of algal cultures spans about 75 years with application to wastewater treatment and mass production of different strains such as Chlorella and Dunaliella, which accumulate large volumes of lipids and starch, and fix N and Pcompounds in very efficient ways (Abdel-Raouf, Al-Homaidan, and Ibraheem 2012; Wollmann et al. 2019; Aron et al. 2020). Presently, significant international interest has been developed in countries like Australia, the United States, Thailand, Taiwan, and Mexico. The biologists in these nations possess a wealth of knowledge and experience, enabling them to design and operate high-margin algal cultures for producing high-value products such as pharmaceuticals and biotechnology, each of which have a critical role in algae farming. There are also antibacterial products, antivirals, anticancer products, antitumor products, and antihistamines (Nur and Buma 2019; Abdel-Raouf, Al-Homaidan, and Ibraheem 2012).

Because they use natural carbon as well as inorganic nitrogen and phosphorus plentiful in wastewater, a variety of microalgae may grow there. In addition, the use of microalgae is still really restricted in wastewater production. Microalgae growth is used worldwide, albeit typically on a modest scale, for wastewater treatment. High-rate microalgae lakes, for example, have been proven to be extremely beneficial when treating wastewater with oxidation ponds or regular oxidation (adjusting) lakes or the more established suspension microalgal lakes (Daguerre-Martini et al. 2018; Kamyab et al. 2019; Ganesan et al. 2020). The requirement for treatment is remarkable for the expulsion of large concentrations of additives, particularly P and N, which can typically lead to eutrophication risks if these additives are accumulated in rivers and ponds. In wastewater treatment, microalgae can effectively remove phosphorus, nitrogen, and toxic metals from the water and would thereby contribute significantly to the remediation process, especially during the last (tertiary) stage. The possible saving costs and the lower-level innovation are the remarkable advantages of algal treatment of wastewater over the normal treatment methods based on substances (Kamyab et al. 2019).

5.7.2 Organic Pollutant Removal

The chemicals and microorganisms in wastewater which might cause watercourse contamination can be identified, as already indicated. In addition to the microbiological contents, wastewater pollution may occur in three wide categories: organic materials, inorganic pollutants, and microbial contents. Wastewater contains a substantial quantity of organic molecules containing at least one carbon atom. Both chemical and biological reactions can result in CO_2 being produced from these carbon atoms. The test of biological oxidation is known as the Biochemical Oxygen Demand Test (BOD), while the tests of chemical oxidation are known as the Chemical Oxygen Demand Tests (CODs). BOD, or biodegradable organic matter, involves the use of molecular oxygen as an oxidizing agent to change organic matter to CO_2 and water. This means that biochemical oxygen demand is indeed an assessment of the respiration needs of bacteria in wastewater, which decompose organic material. Excess BOD should be removed from wastewater because it can cause fish deaths and anaerobiosis by lowering the dissolved oxygen concentration of receiving water. Algae have been studied for their potential to treat biological wastewater. The results showed that BOD and COD were eliminated from domestic wastewater by 68.4% and 67.2%, respectively (Madima et al. 2020; Sonawane, Ezugwu, and Ghosh 2020; Abdel-Raouf, Al-Homaidan, and Ibraheem 2012).

There has been over 50 years of research on algae bio-treatment, which is beneficial for removing nutrients such as nitrogen and phosphorus from water and for providing oxygen to aerobic bacteria. Since then, a number of laboratory and pilot investigations have been carried out, and multiple sewage treatment facilities have been built using different variants of these systems. N is mostly produced by the metabolism of the extracted chemicals in sewage effluent, whereas 50% and more percentage of P are produced by synthetic detergents. They are most commonly found in wastewater as NH_3 (ammonia), NO_2^- (nitrite), NO_3^-, and PO_4^{3-}. The term 'nutrient stripping' refers to the process for removing these two components jointly as nutrition and nutrient (Mohsenpour et al. 2021).

Wastewater is mostly processed by aerobic or anaerobic organic decomposition, although the treated water still includes inorganic chemicals such as NO_3^-, NH^{4+} (ammonium), and PO_4^{3-} ions. NO_3^- and NH^{4+} are to be the key of eutrophication. The water environment must thus be treated further in order to prevent eutrophication. Nutrient enrichment has the potential to create eutrophication in waterbodies because it promotes the growth of invasive plants like algae and macrophytes. Non-ionized ammonia toxicity in fish and other aquatic animals, disinfection when free chlorine residues are required, and influencing methemoglobinemia owing to high NO_3^- levels (above 45 mg/lgm^{-3}) in drinking water are some of the additional effects of N compounds in wastewater effluent (Yan et al. 2021; Rahimi, Modin, and Mijakovic 2020).

Microalgal cultures can significantly reduce wastewater nutrient levels (tertiary wastewater treatment). It is possible to grow mass cultures of microalgae in outdoor solar bioreactors because of their high capacity to absorb inorganic nutrients. It appears that the biological processes perform better than chemical and physical processes, which tend to be too expensive in most places, and can cause second-generation pollution. In addition to being able to use inorganic nutrients and P for growth, microalgae can also remove heavy metals, making them a good alternative to tertiary and quaternary treatments (Li et al. 2019; Paździor, Bilińska, and Ledakowicz 2019).

Chlorella vulgaris was recently shown to be able to remove nutrients from water in a study. Results indicated a nutrient removal efficiency of 86% for inorganic N and 70% for inorganic P. The reduction of N (50.2%) and P (85.7%) reported by the study was achieved by a treatment of industrial wastewater, while the reduction of P (97.8%) was achieved by a treatment of domestic wastewater using algae. It is thought that conventional treatment processes are hampered by some major disadvantages, which is why microalgal cultures have gained such a great deal of interest: (a) It depends on the nutrient to be removed when determining its efficiency; (b) operating costs are high; (c) secondary pollution is often caused by chemical processes; and (d) nutritional value is lost (N, P). This last disadvantage is particularly significant, as conventional treatment processes consume too much of the environment's resources (Solimeno, Gómez-Serrano, and Acién 2019).

Numerous studies have demonstrated that algae can remove nutrients from wastewater that contains high levels of N and P compounds. Abdel-Raouf (Abdel-Raouf, Al-Homaidan, and Ibraheem 2012) observed that *Scenedesmus* sp. occurs in almost all types of fresh water bodies, main primary producers and contribute to the purification of eutrophic water. A study found that certain species of *Scenedesmus* can be used to determine the quality of water for environmental and economic reasons, and it is critically important to recycle the algae cells or the biomass generated, in order to avoid recycling nutrients and to prevent excess nutrients from being retained in receiving waters. Many of the experiments undertaken before have used planktonic and unicellular microalgae, which are difficult to collect (Abdel-Raouf, Al-Homaidan, and Ibraheem 2012).

Renewability and use of solar energies provide benefits for microalgae or plants to remove inorganic compounds. Cyanobacteria may develop at rates that are greater than those of plant species in suitable circumstances so that cyanobacteria may look to be having substantial potential in inorganic nutrient removal systems. With regard to N pollution, the greatest concern arising from the possibility of health hazards that can be attributed to NO_2^- as either directly causing or indirectly causing methemoglobinemia is the source of N resulting from the increasing use of inorganic N fertilizers and waste from humans or animals (Camargo and Alonso 2006).

Also, as a preventer of N-nitroso compounds, NO_2^- themselves are significant, particularly nitrosamines which, because of their potential carcinogenic, teratogenic, and mutagenic characteristics, have attracted considerable attention. Because the conventional water treatment does not considerably reduce NO_3^-, considerable research focuses on developing novel NO_3^- reduction strategies in water tolerability, that is, less than 50 mg L^{-1}. Biological N removal is typically a legitimate alternative and offers several benefits in relation to tertiary chemical and physical–chemical treatments. *Phormidium bohneri* were cultured to remove nitrates from the effluents obtained after anaerobic digestion of swine manure. The removal of N from contaminated waters by means of the thermophilic cyanobacterium *Phormidium laminosum* bench-top bioreactors has also been described. As thermophilic cyanobacteria are tolerant of high temperatures and can be treated at high temperatures, the use of these organisms in wastewater treatment is beneficial, as contamination may be avoided (45°C) (Brienza et al. 2020; Ruiz-Marin, Mendoza-Espinosa, and Stephenson 2010).

Phormidium sp. cells were fed to flakes of chitosan and utilized in urban secondary effluents in the removal of N (NH^{4+}, NO_3^-, NO_2^-) and orthophosphates. Although P removal from municipal and industrial wastewater does not appear to constitute an issue for human health, it is necessary to safeguard water from eutrophication. During the past three decades, the elimination of biological P has been substantial. NO_3^- absorption and reductions are photosynthetic-driven processes that probably involve the product(s) of NH^{4+} in the carbon skeletons in N-sufficient cells of cyanobacteria. NO_3^- assimilation is also regulated by a range of environmental and nutritional variables, including availability of light, temperature, pH, and carbon supply. The metabolism of P and N is strongly connected as P is not very common in abundance if N is not available and vice versa. The absorption of cyanobacteria by phosphate is a hyperbolic feature of the external phosphate concentration and has already been described in a number of strains. The cleaned water may be decanted and the cyanobacteria can be collected once the cyanobacteria have absorbed nutrients into the effluents. The commercially useful pigments are extracted from possible end applications of the collected biomass (Das et al. 2018).

5.7.3 Heavy Metal Removal

Availability of freshwater is one of the most important problems that mankind has in respect to natural resources. It is still available to humankind. Approximately 1.8 billion people have no access to freshwater according to a United Nations World Water Assessment Programme's (WWAP) study. The rise in the freshwater use and the demand on that natural resource are connected to population expansion and industrial development. In addition to increasing demand for freshwater, its supply has been reduced owing to human pollution activity. Lakes and rivers have received many waste products from the various primary (agricultural and animal) and secondary (industrial) sectors, including municipal wastewater and effluent that has been left untreated or are just partially treated. Farming and cattle effluents high in nitrogen and phosphate include noxious compounds—for instance, pesticides and fertilizers—contaminating the surface and ground water. They're also made up of heavy metal wastes like copper and mercury, which are made by fertilizer and livestock feed. Hazardous organic components such as hydrocarbons and high amounts of heavy metals are found in industrial effluents, which vary depending on the activity (Parmar and Thakur 2013).

The heavy metal industries are the primary industries that create liquid effluents of high-pollutant concentrate and are involved in mining, metalworking, and electroplating. If not adequately managed, they can lead to significant environmental concerns, which can pose great risk to living creatures. Some of the elements may also be derived from minerals in rock and soil (natural sources), in quantities (in the water) which are not hazardous to living organisms, which includes copper (Cu), zinc (Zn), etc. Nevertheless, industrial effluents can increase the concentration of these compounds (as well as others like mercury and lead) in water bodies, growing at a rate where exposure of these compounds to living organisms can lead to toxic effects. To treat wastewater with a high concentration of heavy metals, chemical precipitation, ion exchange techniques, electrochemical treatment, and activated carbon adsorption are utilized. Unfortunately, these are highly costly methods that are usually unsuccessful. Furthermore, the removal of significant amounts of hazardous sludge generated by chemical precipitation or coagulation/flocculation procedures is a serious concern. As a result, developing a more cost-effective and ecologically friendly way to treat heavy metal-polluted wastewater is vital (Raikova et al. 2019; Parmar and Thakur 2013; Chai, Cheun et al. 2021).

Bioremediation using microalgae is a low-cost option for addressing this environmental issue. Bioadsorption and bioaccumulation are the two phases of bioremediation. Adsorption takes place in two phases: a solid phase (bioadsorbents are in this instance microalgae) and liquid phase (solvent) that contains the adsorbing species. Metal penetrates the cell wall and accumulates inside the cells during the bioaccumulation phase. Microalgae can also remove nutrients and CO_2 from the environment. They can also be used to produce biofuels as well as for heavy metal removal. In wastewater treatment, microalgae have been extensively studied for the removal of nutrients, especially nitrogen and phosphorus, as these are the two nutrients that they require to grow. Produced sludge can be used for fertilization because of its high nutritional value, enabling the reuse of nutrients. Microalgae may grow in these effluents due to the presence of heavy metals. This can also affect the ability of the algae to remove nitrogen and phosphorus. Microalgae can help with heavy metal removal in low concentrations, as micronutrients can be used in some culture media for microalgae growth but not for high concentrations of heavy metals. The large surface area and strong affinity of microalgae make them excellent for the removal of heavy metals from wastewater. The buildup of heavy metals in biomass, on the other hand, restricts its utility as a fertilizer, and its deposition requires cremation (Sibi 2019; Suresh Kumar et al. 2015).

Effects of heavy metals on microalgae growth and nitrogen and phosphorus removal efficiency from effluents are still being researched. Wastewater is typically treated at a wastewater treatment plant (WWTP) after being filtered in the manufacturing facility. In this instance, discharge of effluent containing a high quantity of heavy metals without pretreatment can lead to high concentrations of such metals affecting the biological treatment process at WWTPs and decreasing the effectiveness of treatment, so high concentrations are dangerous. Thus, studies into microalgal growth in heavy metal polluted environments are highly relevant, and some have already been conducted. Microalgae have been proven to be capable of removing heavy metals through research currently being conducted. Further research is needed in wastewater treatment, though. This has resulted in a lack of studies on the removal of nutrients such as nitrogen and phosphorus when heavy metals are present. Considering the need to remove nitrogen and phosphorus from wastewater, a future study should address the effects of heavy metals on microalgae (Travieso et al. 1999).

5.7.4 Micropollutant Removal

Micropollutants (MPs) are chemicals that have been anthropogenically introduced into waterbodies and whose concentration stays at trace amounts, up to few micrograms per liter. A majority of these organic pollutants come from pharmaceuticals, personal-care products, industrial chemicals, pesticides, and polyaromatic hydrocarbons as well as new compounds. MPs are only used in this context to refer to organic MPs. These contaminants have been linked to a variety of negative consequences, including endocrine disruption, acute and chronic toxicities in various species, and the development of antibiotic resistance in microorganisms (Liu et al. 2021). The continual release of MPs into the environment, though at relatively low levels, may cause anomalies in sensitive species' development and reproduction. Surfactants, industrial chemicals, and pesticides are among the MPs, which have been regulated in just a few industrialized nations; medicines, typically, personal care products (PCPs), and steroid hormones are excluded. Additionally, these processes are not able to efficiently remove bulk contaminants and MPs simultaneously. Despite the fact that wastewater treatment methods can lower the concentrations of several MPs, they are not meant to eliminate them entirely. MPs are hard to quantify and analyze during therapy due to their low concentration and variety.

There are four main processes in microalgae PBRs that remove contaminants: biodegradation, photodegradation, volatilization, and sorption to biomass. Biodegradation and photodegradation appear to be the most important routes for the elimination of chemicals, whereas sorption and volatilization appear to be crucial primarily for hydrophobic compounds and pollutants with a relatively high Henry's law constant, respectively. Adsorption is a typical method for removing micropollutants and is also a key component of microalgae-based systems. There were 21 micropollutants that were specifically found to be reduced by biosorption. A study looked at how well *Chlorella vulgaris* removed ten chemicals. According to a study, biosorption accounted for more than 80% of total pesticide elimination in 1 hour (Ding, Yang et al. 2017). Biosorption is a dependable route since it is unaffected by bioactivity or lighting conditions. The biosorption process might take over some of the roles of other pathways to remove micropollutants in dark conditions when other modes of removal become inefficient or incapable. As well as biosorption, biodegradation is also entangled with biosorption, eradicating micropollutants together. For instance, a research on nonylphenol removal effectiveness and processes by three distinct Chlorella species discovered that *Chlorella vulgaris* was able to remove about 73 percent of nonylphenol after a 12-hour exposure owing to adsorption. *Chlorella vulgaris* then gradually and fully biodegraded the adsorbed nonylphenol. Furthermore, in microalgae-based systems, comparable removal patterns for ceftazidime, levofloxacin, and certain hormone acting compounds have been described, which might be a benefit.

By definition, biodegradation is the breakdown of organic chemicals catalyzed by microorganisms. Metabolic degradation and co-metabolism are two possible processes. Organic molecules are the only carbon and energy sources employed in metabolic breakdown. MP breakdown is dependent on non-specific enzymes found in the environment, which catalyze the metabolism of many other compounds in co-metabolism. As MP concentrations in wastewater are typically in the ng L^{-1} to mg L^{-1} range, metabolic degradation may not be enough to keep microalgae growing. As a result, co-metabolism might be the primary mechanism causing MP degradation. For example, a research conducted by Ding et al. (Ding, Yang et al. 2017) compared the biodegradation routes of several acidic medicines in the wastewater treatment process with and without the addition of an additional source of carbon. Just co-metabolic degradation of bezafibrate, naproxen, and ibuprofen was discovered.

As a method for eliminating organic MPs from aqueous phase, biodegradation is widely regarded as being the most effective. There are two types of degradation: intracellular and extracellular. A microalgae's catalytic activity results in the degradation of complex parent compounds. This permits simpler molecules to be formed. The microalgae have an extensive enzyme system that performs multiple biodegradation reactions with the MPs, that is hydroxylation, carboxylation, oxidation, hydrogenation, demethylation, and ring cleavage (Naveed et al. 2019; Ding, Yang et al. 2017). A hydrated biofilm matrix can also be formed by microalgae excreting various polymeric extracellular substances. In this way, the extracellular enzymes help trigger the cells to metabolize organic compounds by staying close to them (Costa, Raaijmakers, and Kuramae 2018). Biofilms can sequester, accumulate, and utilize organic compounds (dissolved, colloidal, or solid) imported through the water phase of the matrix, as extracellular enzymes are retained within the matrix (Liu et al. 2021).

Researchers in the past have conducted laboratory batch experiments using microalgae to remove MPs. In addition to removing estrogens, antibiotics, anti-epileptics, and antibacterials, microalgae also can remove nonsteroidal anti-inflammatory drugs and antibiotics. Biodegradation was the major means of removing most of the contaminants examined.

The degradation of diclofenac by microalgae was recognized as being the major method for the removal this nonsteriol anti-inflammatory drug from water bodies, because *Scenedesmus obliquus* has the highest removal efficiency at 99% (Mustafa et al. 2021). Ibuprofen, caffeine, carbamazepine, and tris(2-chloroethyl) phosphate were all biodegraded by microalgae in urban or synthetic wastewater in previous studies (Ding, Lin et al. 2017). In other studies, microalgae were able to biodegrade several pharmaceutically active compounds in urban or synthetic wastewater, including ibuprofen, caffeine, carbamazepine, and tris (2-chloroethyl) phosphate (Wang et al. 2019). For instance, Jalilian (Jalilian, Najafpour, and Khajouei 2020) looked into acetaminophen micropollutant occurrences, toxicity, removal methods, and transformation routes. Over 20 acetaminophen by-products and intermediates have been discovered, each with its own distinct level of toxicity. Many of those, such as N-acetyl-p-benzoquinone imine, seem to be more harmful than the parent form. As a result, they concluded that treated wastewater was not completely devoid of acetaminophen metabolites' harmful effects. As a conclusion, the impacts of the final effluent on live creatures must be investigated in order to fully assess the microalgae-based treatment's effectiveness.

The photodegradation mechanism is influenced by the structure of micropollutant molecules as well as environmental factors. The photodegradation of seven micropollutants was investigated. Sulfamethoxazole and triclosan were shown to be sensitive to sunlight exposure, but carbamazepine, diuron, simazine, caffeine, and 2,4-dichlorophenoxyacetic acid were not. Ketoprofen, diclofenac, fenofibric acid, and metronidazole were identified to be fast-photodegradable micropollutants. Acebutolol, theophylline, sulfamethoxazole, sotalol, and isoproturon were identified to be medium-photodegradable micropollutants. Diazepam, atrazine, dimethoate, diuron, simazine, and cyclophosphamide were among the six slow-photodegradable micropollutants. The rate of photodegradation of a compound is also impacted by changes in solar irradiation intensity as a function of latitude and season. The inadequate degradation of micropollutants and also the high reliance upon the chemical characteristics of pollutants could be drawbacks of this removal process. Since certain intermediary molecules are more hazardous than the parent chemicals, the transformation products must be identified. For example, acridine, a carbamazepine photodegradation product, is mutagenic and carcinogenic. Naproxen's photodegradation metabolites are likewise more hazardous than the original molecule. Furthermore, not every micropollutant is photodegradable, but this phenomenon happens mostly in the upper layer of the water column. Photodegradation may not be the primary removal mechanism for micropollutants in some cases, but it may be used in conjunction with other methods.

Volatilization is used to extract just galaxolide (or tonalide) and tributyl phosphate. This is similar to what happens in activated sludge systems. Compounds having a molecular weight less than 200 g mol^{-1} and a Henry's law constant higher than 10^{-3} (atm-m^3) mol^{-1} are considered volatile, whereas those with a Henry's law constant less than 5×10^{-5} (atm-m^3) mol^{-1} are considered soluble and continue to stay in water. These two micropollutants have Henry's law constants of 1.1×10^{-4} and 1.4×10^{-6} (atm-m^3) mol^{-1} and molecular weights of 258.41 and 266.32 g mol^{-1}, respectively—that is when micropollutants were transferred from liquid to air-by-air striping. Nevertheless, because volatilization somehow doesn't remove the structure of micropollutant molecules, it may result in air pollution if no off-gas collection is present.

5.7.5 Air Pollution Control

Changes in climate have been called one of the most critical human concerns of the last several decades, according to some. CO_2 accounts for roughly 68 percent of total emissions, and it is the major cause. There are major sources of CO_2 emissions from the transportation system as well as industries such as power plants. Furthermore, GHGs affect human life and the climate in addition to causing global warming (GW). As more and more CO_2 is emitted from human activity every year, the oceans are absorbing the majority of the gas—which is why, alongside with increasing levels of GHG in the atmosphere, the liquid pH in the seas is progressively changing into an acidic mode. A decline in pH can be traced to many causes, for instance, the expeditious habitat destruction in aquatic and coral reef ecosystems, which have subtle implications for ocean organisms and, ultimately, for other organisms on earth (Jalilian, Najafpour, and Khajouei 2020).

Because global warming affects so many elements of human existence and the global environment, it necessitates not just one but a varied collection of measures to be handled. The reduction of CO_2 has been the subject of numerous research and development projects in recent years. Chemical, physical, and biological methods were employed in the processes studied. Other methods are much less trustworthy than the biological technique. CO_2 biofixation is thought to be most effective when algae are involved in the photosynthesis process. They are chosen primarily based on their photosynthetic capacities in CO_2 biofixation, the production of high amounts of lipids and biomass, and the production of non-fuel co-products. Photosynthetic or autotrophic microalgae fix CO_2 more effectively than terrestrial plants and seem to be 10–50 times more efficient than terrestrial plants. Terrestrial plants are responsible for just 3–6% of global CO_2 emissions. Microalgae's growth depends on carbon more than nitrogen and phosphorus. Microalgae may be grown with CO_2 by injecting flue gas from power plants into the culture media or growing it in a raceway pond to reduce CO_2 levels in the atmosphere (Molazadeh et al. 2019; Jalilian, Najafpour, and Khajouei 2020).

The light-absorbing pigment cyclic tetrapyrrolen has a chemical structure similar to that of hemoglobin's heme group. The metal ion Mg^{2+}, which is found in the core of chlorophyll, activates the electron transport mechanism. While some oxidation and reduction may occur, a variety of enzymatic processes should also occur in light-harvesting pigments. CO_2 is transformed into organic carbon-like lipids and hydrocarbons through this process, which reflects the meaning of the term 'CO_2 fixation'. In addition to cultivation systems, biomass production and CO_2 fixation affect the microalgae species characteristics in addition to physicochemical processes. Among the organisms studied was *Neochloris oleoabundans*, which produces biodiesel under different alkaline concentrations and saline concentrations. The biomass productivity was found to be highest at a CO_2 concentration of 2% and nitrogen restriction. During the photosynthetic process, this form of microalga is capable of converting organic phosphates and nitrates in the medium into complex lipid molecules ideal for biodiesel generation (Lim et al. 2021). Generally, algal types should exhibit high sinking capacities and high tolerances of various toxic pollutants, CO_2 concentrations, nutrient restrictions, pH levels, and temperature variations. The most promising microalgae species for carbon capture and biofuel production are *Botryococcus braunii*, *Scenedesmus obliquus*, *Nannochloropsis oculate*, and *Chlorella vulgaris*. CO_2 concentrations that result in high biomass productivity and high tolerance to CO_2 are reported. Microalgae can fix roughly 1.83 kg of CO_2 every kilogram of biomass produced, indicating that they are key actors in CO_2 bio sequestration. Certain blue green algae, however, including *Anabaena*, *Nostoc*, *Aulosira*, and *Tolypothrix*, may fix ambient nitrogen that can be used for plant development in addition to CO_2 fixation (Gupta and Dubey 2019).

REFERENCES

Abdel-Raouf, N., A. A. Al-Homaidan, and I. B. M. Ibraheem. 2012. "Microalgae and Wastewater Treatment." *Saudi Journal of Biological Sciences* 19 (3). King Saud University: 257–75. doi:10.1016/j.sjbs.2012.04.005.

Achkar, Jean H. El, Thomas Lendormi, Dominique Salameh, Nicolas Louka, Richard G. Maroun, Jean Louis Lanoisellé, and Zeina Hobaika. 2018. "Influence of Pretreatment Conditions on Lignocellulosic Fractions and Methane Production from Grape Pomace." *Bioresource Technology* 247: 881–9. doi:10.1016/j.biortech.2017.09.182.

Ajibade, Fidelis O., Bashir Adelodun, Kayode H. Lasisi, Oluniyi O. Fadare, Temitope F. Ajibade, Nathaniel A. Nwogwu, Ishaq D. Sulaymon, Adamu Y. Ugya, Hong Cheng Wang, and Aijie Wang. 2021. "Environmental Pollution and Their Socioeconomic Impacts." In *Microbe Mediated Remediation of Environmental Contaminants*, edited by Ajay Kumar, Vipin Kumar Singh, Pardeep Singh, and Virendra Kumar Mishra, 1st ed., 321–54. Cambridge: Woodhead Publishing.

Alcántara, Cynthia, Esther Posadas, Benoit Guieysse, and Raúl Muñoz. 2015. "Microalgae-Based Wastewater Treatment." In *Handbook of Marine Microalgae*, edited by Se-Kwon Kim, 1st ed., 439–55. Cambridge: Academic Press. doi:10.1016/B978-0-12-800776-1.00029-7.

Ansari, Khursheed B., Jyotsna S. Arora, Jia Wei Chew, Paul J. Dauenhauer, and Samir H. Mushrif. 2019. "Fast Pyrolysis of Cellulose, Hemicellulose, and Lignin: Effect of Operating Temperature on Bio-Oil Yield and Composition and Insights into the Intrinsic Pyrolysis Chemistry." *Industrial and Engineering Chemistry Research* 58 (35): 15838–52. doi:10.1021/acs.iecr.9b00920.

Baruah, Julie, Bikash Kar Nath, Ritika Sharma, Sachin Kumar, Ramesh Chandra Deka, Deben Chandra Baruah, and Eeshan Kalita. 2018. "Recent Trends in the Pretreatment of Lignocellulosic Biomass for Value-Added Products." *Frontiers in Energy Research* 6: 1–19. doi:10.3389/fenrg.2018.00141.

Brienza, Monica, Rayana Manasfi, Andrés Sauvêtre, and Serge Chiron. 2020. "Nitric Oxide Reactivity Accounts for N-Nitroso-Ciprofloxacin Formation under Nitrate-Reducing Conditions." *Water Research* 185: 116293. doi:10.1016/j.watres.2020.116293.

Camargo, Julio A., and Álvaro Alonso. 2006. "Ecological and Toxicological Effects of Inorganic Nitrogen Pollution in Aquatic Ecosystems: A Global Assessment." *Environment International* 32 (6): 831–49. doi:10.1016/j.envint.2006.05.002.

Chai, Wai Siong, Jie Ying Cheun, P. Senthil Kumar, Muhammad Mubashir, Zahid Majeed, Fawzi Banat, Shih-Hsin Ho, and Pau Loke Show. 2021. "A Review on Conventional and Novel Materials towards Heavy Metal Adsorption in Wastewater Treatment Application." *Journal of Cleaner Production* 296: 126589. doi:10.1016/j.jclepro.2021.126589.

Chai, Wai Siong, Chee Hong Chew, Heli Siti Halimatul Munawaroh, Veeramuthu Ashokkumar, Chin Kui Cheng, Young-Kwon Park, and Pau-Loke Show. 2021. "Microalgae and Ammonia: A Review on Inter-Relationship." *Fuel* 303: 121303. doi:10.1016/j.fuel.2021.121303.

Chai, Wai Siong, Wee Gee Tan, Heli Siti Halimatul Munawaroh, Vijai Kumar Gupta, Shih-Hsin Ho, and Pau Loke Show. 2021. "Multifaceted Roles of Microalgae in the Application of Wastewater Biotreatment: A Review." *Environmental Pollution* 269: 116236. doi:10.1016/j.envpol.2020.116236.

Cheah, Wai Yan, Tau Chuan Ling, Pau Loke Show, Joon Ching Juan, Jo Shu Chang, and Duu Jong Lee. 2016. "Cultivation in Wastewaters for Energy: A Microalgae Platform." *Applied Energy* 179: 609–25. doi:10.1016/j.apenergy.2016.07.015.

Cheah, Wai Yan, Revathy Sankaran, Pau Loke Show, Tg. Nilam Baizura Tg. Ibrahim, Kit Wayne Chew, Alvin Culaba, and Jo-Shu Chang. 2020. "Pretreatment Methods for Lignocellulosic Biofuels Production: Current Advances, Challenges and Future Prospects." *Biofuel Research Journal* 7 (1). Department of Environmental Health, Faculty of Health Sciences, MAHSA University, 42610 Jenjarom, Selangor, Malaysia.: 1115–27. doi:10.18331/BRJ2020.7.1.4.

Cherpozat, Lucie, Eric Loranger, and Claude Daneault. 2019. "Ultrasonic Pretreatment of Soft Wood Biomass Prior to Conventional Pyrolysis: Scale-up Effects and Limitations." *Biomass and Bioenergy* 124: 54–63. doi:10.1016/j.biombioe.2019.03.009.

Chye, Jonah Teo Teck, Lau Yien Jun, Lau Sie Yon, Sharadwata Pan, and Michael K. Danquah. 2017. "Biofuel Production from Algal Biomass." In *Bioenergy and Biofuels*, edited by Ozcan Konur, 1st ed., 84–116. Boca Raton: CRC Press. doi:10.1201/9781351228138-3.

Costa, Ohana Y. A., Jos M. Raaijmakers, and Eiko E. Kuramae. 2018. "Microbial Extracellular Polymeric Substances: Ecological Function and Impact on Soil Aggregation." *Frontiers in Microbiology* 9 (Jul): 1–14. doi:10.3389/fmicb.2018.01636.

da Silva, A. R. G., C. E. Torres Ortega, and B. -G. Rong. 2016. "Techno-Economic Analysis of Different Pretreatment Processes for Lignocellulosic-Based Bioethanol Production." *Bioresource Technology* 218: 561–570. doi: 10.1016/j.biortech.2016.07.007.

Daguerre-Martini, S., M. B. Vanotti, M. Rodriguez-Pastor, A. Rosal, and R. Moral. 2018. "Nitrogen Recovery from Wastewater Using Gas-Permeable Membranes: Impact of Inorganic Carbon Content and Natural Organic Matter." *Water Research* 137: 201–10. doi:10.1016/j.watres.2018.03.013.

Das, Cindrella, Nagappa Ramaiah, Elroy Pereira, and K. Naseera. 2018. "Efficient Bioremediation of Tannery Wastewater by Monostrains and Consortium of Marine Chlorella Sp. and Phormidium Sp." *International Journal of Phytoremediation* 20 (3): 284–92. doi:10.1080/15226514.2017.1374338.

Ding, Tengda, Kunde Lin, Bo Yang, Mengting Yang, Juying Li, Wenying Li, and Jay Gan. 2017. "Biodegradation of Naproxen by Freshwater Algae Cymbella Sp. and Scenedesmus Quadricauda and the Comparative Toxicity." *Bioresource Technology* 238: 164–73. doi:10.1016/j.biortech.2017.04.018.

Ding, Tengda, Mengting Yang, Junmin Zhang, Bo Yang, Kunde Lin, Juying Li, and Jay Gan. 2017. "Toxicity, Degradation and Metabolic Fate of Ibuprofen on Freshwater Diatom Navicula Sp." *Journal of Hazardous Materials* 330: 127–34. doi:10.1016/j.jhazmat.2017.02.004.

ElFar, Omar Ashraf, Chih Kai Chang, Hui Yi Leong, Angela Paul Peter, Kit Wayne Chew, and Pau Loke Show. 2020. "Prospects of Industry 5.0 in Algae: Customization of Production and New Advance Technology for Clean Bioenergy Generation." *Energy Conversion and Management: X* 10: 100048. doi:10.1016/j.ecmx.2020.100048.

Foong, Shin Ying, Yi Herng Chan, Wai Yan Cheah, Noor Haziqah Kamaludin, Tengku Nilam Baizura Tengku Ibrahim, Christian Sonne, Wanxi Peng, Pau Loke Show, and Su Shiung Lam. 2021. "Progress in Waste Valorization Using Advanced Pyrolysis Techniques for Hydrogen and Gaseous Fuel Production." *Bioresource Technology* 320 (Part A): 124299. doi:10.1016/j.biortech.2020.124299.

Ganesan, Ramya, S. Manigandan, Melvin S. Samuel, Rajasree Shanmuganathan, Kathirvel Brindhadevi, Nguyen Thuy Lan Chi, Pham Anh Duc, and Arivalagan Pugazhendhi. 2020. "A Review on Prospective Production of Biofuel from Microalgae." *Biotechnology Reports* 27: e00509. https://doi.org/10.1016/j.btre.2020.e00509.

Gelosia, Mattia, Alessandro Bertini, Marco Barbanera, Tommaso Giannoni, Andrea Nicolini, Franco Cotana, and Gianluca Cavalaglio. 2020. "Acid-Assisted Organosolv Pre-Treatment and Enzymatic Hydrolysis of Cynara Cardunculus L. For Glucose Production." *Energies* 13 (16): 4195. doi:10.3390/en13164195.

Gu, Bon Jae, Jinwu Wang, Michael P. Wolcott, and Girish M. Ganjyal. 2018. "Increased Sugar Yield from Pre-Milled Douglas-Fir Forest Residuals with Lower Energy Consumption by Using Planetary Ball Milling." *Bioresource Technology* 251: 93–8. doi:10.1016/j.biortech.2017.11.103.

Gupta, Vaishali, and Jaishree Dubey. 2019. "Cyanobacterial Biomass—A Tool for Sustainable Management of Environment." In *Emerging Energy Alternatives for Sustainable Environment*, edited by D. P. Singh, Richa Kothari, and V. V. Tyagi, 1st ed., 193–210. FL: CRC Press. doi:10.1201/9780429058271-11.

Herweijer, Celine, and Dominic Waughray. 2018. "Fourth Industrial Revolution for the Earth Harnessing Artificial Intelligence for the Earth." *A report of Pricewaterhouse-Coopers (PwC)*. Switzerland: World Economic Forum.

Hu, Xiaohui, Li Cheng, Zhengbiao Gu, Yan Hong, Zhaofeng Li, and Caiming Li. 2018. "Effects of Ionic Liquid/Water Mixture Pretreatment on the Composition, the Structure and the Enzymatic Hydrolysis of Corn Stalk." *Industrial Crops and Products* 122: 142–7. doi:10.1016/j.indcrop.2018.05.056.

Hubenov, Venelin, Ramiro Ariel Carcioch, Juliana Ivanova, Ivanina Vasileva, Krasimir Dimitrov, Ivan Simeonov, and Lyudmila Kabaivanova. 2020. "Biomethane

Production Using Ultrasound Pre-Treated Maize Stalks with Subsequent Microalgae Cultivation." *Biotechnology and Biotechnological Equipment* 34 (1). Taylor & Francis: 800–9. doi:10.1080/13102818.2020.1806108.

Jalilian, Neda, Ghasem D. Najafpour, and Mohammad Khajouei. 2020. "Macro and Micro Algae in Pollution Control and Biofuel Production—A Review." *ChemBioEng Reviews* 7 (1): 18–33. doi:10.1002/cben.201900014.

Jędrzejczyk, Marcin, Emilia Soszka, Martyna Czapnik, Agnieszka M. Ruppert, and Jacek Grams. 2019. "Physical and Chemical Pretreatment of Lignocellulosic Biomass." In *Second and Third Generation of Feedstocks: The Evolution of Biofuels*, edited by Angelo Basile and Francesco Dalena, 1st ed. Netherlands: Elsevier. doi:10.1016/B978-0-12-815162-4.00006-9.

Kamyab, Hesam, Chew Tin Lee, Shreeshivadasan Chelliapan, Tayebeh Khademi, Amirreza Talaiekhozani, and Shahabaldin Rezania. 2019. "Role of Microalgal Biotechnology in Environmental Sustainability-a Mini Review." *Chemical Engineering Transactions* 72: 451–6. doi:10.3303/CET1972076.

Kundu, Chandan, Shanthi Priya Samudrala, Mahmud Arman Kibria, and Sankar Bhattacharya. 2021. "One-Step Peracetic Acid Pretreatment of Hardwood and Softwood Biomass for Platform Chemicals Production." *Scientific Reports* 11 (1). Nature Publishing Group UK: 1–11. doi:10.1038/s41598-021-90667-9.

la Noüe, J. de, and A. Bassères. 1989. "Biotreatment of Anaerobically Digested Swine Manure with Microalgae." *Biological Wastes* 29 (1): 17–31. doi:10.1016/0269-7483(89)90100-6.

Law, Xin Ni, Wai Yan Cheah, Kit Wayne Chew, Mohamad Faizal Ibrahim, Young Kwon Park, Shih Hsin Ho, and Pau Loke Show. 2022. "Microalgal-Based Biochar in Wastewater Remediation: Its Synthesis, Characterization and Applications." *Environmental Research* 204 (PA): 111966. doi:10.1016/j.envres.2021.111966.

Li, Kun, Qiang Liu, Fan Fang, Ruihuan Luo, Qian Lu, Wenguang Zhou, Shuhao Huo, et al. 2019. "Microalgae-Based Wastewater Treatment for Nutrients Recovery: A Review." *Bioresource Technology* 291. Elsevier BV: 121934. doi:10.1016/j.biortech.2019.121934.

Lim, Yi An, Meng Nan Chong, Su Chern Foo, and I. M. S. K. Ilankoon. 2021. "Analysis of Direct and Indirect Quantification Methods of CO2 Fixation via Microalgae Cultivation in Photobioreactors: A Critical Review." *Renewable and Sustainable Energy Reviews* 137: 110579. doi:10.1016/j.rser.2020.110579.

Liu, Ranbin, Siqi Li, Yingfan Tu, and Xiaodi Hao. 2021. "Capabilities and Mechanisms of Microalgae on Removing Micropollutants from Wastewater: A Review." *Journal of Environmental Management* 285: 112149. doi:10.1016/j.jenvman.2021.112149.

Low, Sze Shin, Kien Xiang Bong, Muhammad Mubashir, Chin Kui Cheng, Man Kee Lam, Jun Wei Lim, Yeek Chia Ho, Keat Teong Lee, Heli Siti Halimatul Munawaroh, and Pau Loke Show. 2021. "Microalgae Cultivation in Palm Oil Mill Effluent (POME) Treatment and Biofuel Production." *Sustainability* 13 (6): 3247. doi:10.3390/su13063247.

Madima, N., S. B. Mishra, I. Inamuddin, and A. K. Mishra. 2020. "Carbon-Based Nanomaterials for Remediation of Organic and Inorganic Pollutants from Wastewater. A Review." *Environmental Chemistry Letters* 18: 1169–91. doi:10.1007/s10311-020-01001-0.

Mat Aron, Nurul Syahirah, Kuan Shiong Khoo, Kit Wayne Chew, Ashokkumar Veeramuthu, Jo-Shu Chang, and Pau Loke Show. 2021. "Microalgae Cultivation in Wastewater and Potential Processing Strategies Using Solvent and Membrane Separation Technologies." *Journal of Water Process Engineering* 39: 101701. doi:10.1016/j.jwpe.2020.101701.

Mat Aron, Nurul Syahirah, Kuan Shiong Khoo, Kit Wayne Chew, Pau Loke Show, Wei-Hsin Chen, and The Hong Phong Nguyen. 2020. "Sustainability of the Four Generations of Biofuels—A Review." *International Journal of Energy Research* 44 (12): 9266–82. doi:10.1002/ER.5557.

Mohsenpour, Seyedeh Fatemeh, Sebastian Hennige, Nicholas Willoughby, Adebayo Adeloye, and Tony Gutierrez. 2021. "Integrating Micro-Algae into Wastewater Treatment: A Review." *Science of the Total Environment* 752: 142168. doi:10.1016/j.scitotenv.2020.142168.

Molazadeh, Marziyeh, Hossein Ahmadzadeh, Hamid R. Pourianfar, Stephen Lyon, and Pabulo Henrique Rampelotto. 2019. "The Use of Microalgae for Coupling Wastewater Treatment with CO2 Biofixation." *Frontiers in Bioengineering and Biotechnology* 7: 42. doi:10.3389/fbioe.2019.00042.

Morrar, Rabeh, Husam Arman, and Saeed Mousa. 2017. "The Fourth Industrial Revolution (Industry 4.0): A Social Innovation Perspective." *Technology Innovation Management Review* 7 (23): 12–20. doi:10.25073/0866-773x/97.

Mustafa, Shazia, Haq Nawaz Bhatti, Munazza Maqbool, and Munawar Iqbal. 2021. "Microalgae Biosorption, Bioaccumulation and Biodegradation Efficiency for the Remediation of Wastewater and Carbon Dioxide Mitigation: Prospects, Challenges and Opportunities." *Journal of Water Process Engineering* 41: 102009. doi:10.1016/j.jwpe.2021.102009.

Naveed, Sadiq, Chonghua Li, Xinda Lu, Shuangshuang Chen, Bin Yin, Chunhua Zhang, and Ying Ge. 2019. "Microalgal Extracellular Polymeric Substances and Their Interactions with Metal(Loid)s: A Review." *Critical Reviews in Environmental Science and Technology* 49 (19): 1769–802. doi:10.1080/10643389.2019.1583052.

Nirmalakhandan, N., T. Selvaratnam, S. M. Henkanatte-Gedera, D. Tchinda, I. S. A. Abeysiriwardana-Arachchige, H. M. K. Delanka-Pedige, S. P. Munasinghe-Arachchige, Y. Zhang, F. O. Holguin, and P. J. Lammers. 2019. "Algal Wastewater Treatment: Photoautotrophic vs. Mixotrophic Processes." *Algal Research* 41: 101569. doi:10.1016/j.algal.2019.101569.

Nur, Muhamad Maulana Azimatun, and Anita G. J. Buma. 2019. "Opportunities and Challenges of Microalgal Cultivation on Wastewater, with Special Focus on Palm Oil Mill Effluent and the Production of High Value Compounds." *Waste and Biomass Valorization* 10 (8). Springer Netherlands: 2079–97. doi:10.1007/s12649-018-0256-3.

Oláh, Judit, Nemer Aburumman, József Popp, Muhammad Asif Khan, Hossam Haddad, and Nicodemus Kitukutha. 2020. "Impact of Industry 4.0 on Environmental Sustainability." *Sustainability* 12 (11): 4674. doi:10.3390/su12114674.

Parmar, M., and L. S. Thakur. 2013. "Heavy Metal Cu, Ni and Zn: Toxicity, Health Hazards and Their Removal." *International Journal of Plant Sciences* 3 (3): 143–57.

Paździor, Katarzyna, Lucyna Bilińska, and Stanisław Ledakowicz. 2019. "A Review of the Existing and Emerging Technologies in the Combination of AOPs and Biological Processes in Industrial Textile Wastewater Treatment." *Chemical Engineering Journal* 376: 120597. doi:10.1016/j.cej.2018.12.057.

Pecha, M. Brennan, and Manuel Garcia-Perez. 2020. *Pyrolysis of Lignocellulosic Biomass: Oil, Char, and Gas. Bioenergy*, 2nd ed. doi:10.1016/b978-0-12-815497-7.00029-4.

Rahimi, Shadi, Oskar Modin, and Ivan Mijakovic. 2020. "Technologies for Biological Removal and Recovery of Nitrogen from Wastewater." *Biotechnology Advances* 43: 107570. doi:10.1016/j.biotechadv.2020.107570.

Raikova, Sofia, Marco Piccini, Matthew K. Surman, Michael J. Allen, and Christopher J. Chuck. 2019. "Making Light Work of Heavy Metal Contamination: The Potential for Coupling Bioremediation with Bioenergy Production." *Journal of Chemical Technology and Biotechnology* 94 (10): 3064–72. doi:10.1002/jctb.6133.

Rawat, Ismail, Sanjay K. Gupta, Amritanshu Shriwastav, Poonam Singh, Sheena Kumari, and Faizal Bux. 2016. "Microalgae Applications in Wastewater Treatment." In *Algae Biotechnology Products and Processes*, edited by Faizal Bux and Yusuf Chisti, 1st ed., 249–68. Switzerland: Springer Nature. doi:10.1007/978-3-319-12334-9_13.

Rodič, Blaž. 2017. "Industry 4.0 and the New Simulation Modelling Paradigm." *Organizacija* 50 (3): 193–207. doi:10.1515/orga-2017-0017.

Rojko, Andreja. 2017. "Industry 4.0 Concept: Background and Overview." *International Journal of Interactive Mobile Technologies* 11 (5): 77–90. doi:10.3991/ijim.v11i5.7072.

Rooni, Vahur, Merlin Raud, and Timo Kikas. 2017. "The Freezing Pre-Treatment of Lignocellulosic Material: A Cheap Alternative for Nordic Countries." *Energy* 139: 1–7. doi:10.1016/j.energy.2017.07.146.

Ruiz-Marin, Alejandro, Leopoldo G. Mendoza-Espinosa, and Tom Stephenson. 2010. "Growth and Nutrient Removal in Free and Immobilized Green Algae in Batch and Semi-Continuous Cultures Treating Real Wastewater." *Bioresource Technology* 101 (1): 58–64. doi:10.1016/j.biortech.2009.02.076.

Shetty, Deepa J., Pranav Kshirsagar, Sneha Tapadia-Maheshwari, Vikram Lanjekar, Sanjay K. Singh, and Prashant K. Dhakephalkar. 2017. "Alkali Pretreatment at Ambient Temperature: A Promising Method to Enhance Biomethanation of Rice Straw." *Bioresource Technology* 226: 80–8. doi:10.1016/j.biortech.2016.12.003.

Sibi, G. 2019. "Factors Influencing Heavy Metal Removal by Microalgae-a Review." *Journal of Critical Reviews* 6. Innovare Academics Sciences Pvt. Ltd: 29–32. doi:10.22159/jcr.2019v6i6.35600.

Silva, Gilson José Da, and Antonio Carlos Massabni. 2019. "Biotechnology and Industry 4.0: The Professionals of the Future." *International Journal of Advances in Medical Biotechnology* 2 (2): 45–53. doi:10.25061/2595-3931/ijamb/2019.v2i2.39.

Soares, Juliana Ferreira, Tássia Carla Confortin, Izelmar Todero, Flávio Dias Mayer, and Marcio Antonio Mazutti. 2020. "Dark Fermentative Biohydrogen Production from Lignocellulosic Biomass: Technological Challenges and Future Prospects." *Renewable and Sustainable Energy Reviews* 117: 109484. doi:10.1016/j.rser.2019.109484.

Solimeno, Alessandro, Cintia Gómez-Serrano, and Francisco Gabriel Acién. 2019. "BIO_ALGAE 2: Improved Model of Microalgae and Bacteria Consortia for Wastewater Treatment." *Environmental Science and Pollution Research* 26: 25855–68. doi:10.1007/s11356-019-05824-5.

Sonawane, Jayesh M., Chizoba I. Ezugwu, and Prakash C. Ghosh. 2020. "Microbial Fuel Cell-Based Biological Oxygen Demand Sensors for Monitoring Wastewater: State-of-the-Art and Practical Applications." *ACS Sensors* 5 (8): 2297–316. doi:10.1021/acssensors.0c01299.

Suresh Kumar, K., Hans Uwe Dahms, Eun Ji Won, Jae Seong Lee, and Kyung Hoon Shin. 2015. "Microalgae—A Promising Tool for Heavy Metal Remediation." *Ecotoxicology and Environmental Safety* 113: 329–52. doi:10.1016/j.ecoenv.2014.12.019.

Taheran, Mehrdad, Mitra Naghdi, Satinder K. Brar, Mausam Verma, and R. Y. Surampalli. 2018. "Emerging Contaminants: Here Today, There Tomorrow!" *Environmental Nanotechnology, Monitoring and Management* 10: 122–6. doi:10.1016/j.enmm.2018.05.010.

Toor, Saqib Sohail, Lasse Rosendahl, and Andreas Rudolf. 2011. "Hydrothermal Liquefaction of Biomass: A Review of Subcritical Water Technologies." *Energy* 36 (5): 2328–42. doi:10.1016/j.energy.2011.03.013.

Travieso, L., R. O. Cañizares, R. Borja, F. Benítez, A. R. Domínguez, R. Dupeyrón, and Y. V. Valiente. 1999. "Heavy Metal Removal by Microalgae." *Bulletin of Environmental Contamination and Toxicology* 62 (2): 144–51. doi:10.1007/s001289900853.

Vasić, Katja, Željko Knez, and Maja Leitgeb. 2021. "Bioethanol Production by Enzymatic Hydrolysis from Different Lignocellulosic Sources." *Molecules* 26 (3): 753. doi:10.3390/molecules26030753.

Verma, M., S. Godbout, S. K. Brar, O. Solomatnikova, S. P. Lemay, and J. P. Larouche. 2012. "Biofuels Production from Biomass by Thermochemical Conversion Technologies." *International Journal of Chemical Engineering* 2012: 542426. doi:10.1155/2012/542426.

Vimmerstedt, Laura J., Brian W. Bush, Dave D. Hsu, Daniel Inman, and Steven O. Peterson. 2014. "Maturation of Biomass-to-Biofuels Conversion Technology Pathways for Rapid Expansion of Biofuels Production: A System Dynamics Perspective." *Biofuels, Bioproducts, Biorefining* 9: 158–76. doi:10.1002/bbb.

Wang, Luyun, Han Xiao, Ning He, Dong Sun, and Shunshan Duan. 2019. "Biosorption and Biodegradation of the Environmental Hormone Nonylphenol By Four Marine Microalgae." *Scientific Reports* 9 (1): 5277. doi:10.1038/s41598-019-41808-8.

Weldekidan, Haftom, Vladimir Strezov, Rui Li, Tao Kan, Graham Town, Ravinder Kumar, Jing He, and Gilles Flamant. 2020. "Distribution of Solar Pyrolysis Products and Product Gas Composition Produced from Agricultural Residues and Animal Wastes at Different Operating Parameters." *Renewable Energy* 151: 1102–9. doi:10.1016/j.renene.2019.11.107.

Wollmann, Felix, Stefan Dietze, Jörg Uwe Ackermann, Thomas Bley, Thomas Walther, Juliane Steingroewer, and Felix Krujatz. 2019. "Microalgae Wastewater Treatment: Biological and Technological Approaches." *Engineering in Life Sciences* 19 (12): 860–71. doi:10.1002/elsc.201900071.

Xu, Airong, Yajuan Zhang, Yang Zhao, and Jianji Wang. 2013. "Cellulose Dissolution at Ambient Temperature: Role of Preferential Solvation of Cations of Ionic Liquids by a Cosolvent." *Carbohydrate Polymers* 92 (1): 540–4. doi:10.1016/j.carbpol.2012.09.028.

Yan, Zhiwei, Qiuyue Wang, Yang Li, Ling Wu, Junnan Wang, Bin Xing, Dan Yu, Ligong Wang, and Chunhua Liu. 2021. "Combined Effects of Warming and Nutrient Enrichment on Water Properties, Growth, Reproductive Strategies and Nutrient Stoichiometry of Potamogeton Crispus." *Environmental and Experimental Botany* 190: 104572. doi:10.1016/j.envexpbot.2021.104572.

Yang, Qi, Bo Wu, Fubing Yao, Li He, Fei Chen, Yinghao Ma, Xiaoyu Shu, Kunjie Hou, Dongbo Wang, and Xiaoming Li. 2019. "Biogas Production from Anaerobic Co-Digestion of Waste Activated Sludge: Co-Substrates and Influencing Parameters." *Reviews in Environmental Science and Biotechnology* 18 (4): 771–93. doi:10.1007/s11157-019-09515-y.

Zakaria, Mohd Rafein, Mohd Nor Faiz Norrrahim, Satoshi Hirata, and Mohd Ali Hassan. 2015. "Hydrothermal and Wet Disk Milling Pretreatment for High Conversion of Biosugars from Oil Palm Mesocarp Fiber." *Bioresource Technology* 181: 263–9. doi:10.1016/j.biortech.2015.01.072.

Zhao, Bai Hang, Jie Chen, Han Qing Yu, Zhen Hu Hu, Zheng Bo Yue, and Jun Li. 2017. "Optimization of Microwave Pretreatment of Lignocellulosic Waste for Enhancing Methane Production: Hyacinth as an Example." *Frontiers of Environmental Science and Engineering* 11 (6): 1–9. doi:10.1007/s11783-017-0965-z.

Zhou, Nan, Junwen Zhou, Leilei Dai, Feiqiang Guo, Yunpu Wang, Hui Li, Wenyi Deng, et al. 2020. "Syngas Production from Biomass Pyrolysis in a Continuous Microwave Assisted Pyrolysis System." *Bioresource Technology* 314: 123756. doi:10.1016/j.biortech.2020.123756.

6 What Are Smart Microalgae?

Nur Azalina Suzianti Feisal[1], Noor Haziqah Kamaludin[2], Dingling Zhuang[3], Kit Wayne Chew[4] and Pau Loke Show[5]

[1] Department of Environmental Health, Faculty of Health Sciences, MAHSA University, Bandar Saujana Putra, Jenjarom, Selangor, Malaysia

[2] Center of Environmental Health & Safety, Faculty of Health Sciences, Universiti Teknologi MARA, Puncak Alam, Selangor, Malaysia

[3] Institute of Biological Sciences, Faculty of Science, Universiti Malaya, Kuala Lumpur, Malaysia

[4] School of Energy and Chemical Engineering, Xiamen University Malaysia, Jalan Sunsuria, Bandar Sunsuria, Sepang, Selangor Darul Ehsan, Malaysia

[5] Department of Chemical and Environmental Engineering, Faculty of Science and Engineering, University of Nottingham Malaysia, Jalan Broga, Semenyih, Selangor Darul Ehsan, Malaysia

CONTENTS

6.1 Introduction 206
6.2 Mechanisms of Smart Microalgae 207
6.2.1 Genetic Engineering 207
6.2.2 Internet of Things (IoT) 211
6.2.3 Intelligent Biomanufacturing 212
6.2.3.1 Big-Data Software 212
6.2.3.2 Artificial Intelligence and Artificial Neural Network 214
6.3 Advantage and Disadvantage of Smart Microalgae 215
6.4 Conclusions 218
References 219

DOI: 10.1201/9781003202196-6

6.1 INTRODUCTION

It has been shown that algae are eukaryotic organisms with diverse ecologies, which can occur in fresh-water and saltwater environments. Their capacities to generate food and energy have been critical in meeting the sustainable development goals (SDGs). Microalgae are one of the promising and sustainable protein sources due to the content of high protein and the presence of essential amino acids in its composition, with the requirement of minimal resources for growth. Smart microalgae are defined on the basis of their increasing demand for food and energy where the microalgae have demonstrated their best capacity to contribute to global energy and food security via their technology processes. Microalgae are a group of unicellular microscopic photosynthetic plant-like organisms that are rich source of carbon compounds (Khan et al., 2018), which grow through the autotrophs photosynthesis that have roots or leafy shoots and lack of vascular tissues such as *Chlamydomonas reinhardtii*, *Dunaliella salina* and various *Chlorella* species *(Botryococcus braunii)* photosynthesis takes place of turning solar radiation into chemical energy (Debowski et al., 2020). It is a promising source of superior biomass as alternative to sustainable food, high value products and biofuel commodities (Godvin et al., 2021; Putri et al., 2019; Hermadi et al., 2021). Microalgae feed on nitrogen that comes from nitrates, phosphorus from phosphates, and carbon dioxide. Accessing microalgal genomes to discover hidden genomes and oncogenes in metabolic pathways can greatly enhance biofuel production.

Nowadays, microalgae are widely used in most of the applications and industries such as agar, alginate, antioxidants, astaxanthin, beta-carotene and carotenoids, bioenergy and biofuels, biochar, biorefinery, biosorbents, carragen or carrageenan, catalysts, chemicals, conditioner, digester residue, extraction and production, wastewater treatment, gas treatment, waste management, leachate treatment, and biogas upgrading (Debowski et al., 2020; Chai et al., 2021). Laterally, microalgae have become an emerged area and getting more attention as a source in biofuels' generation to substitute the conventional method which is produced from an edible plant where the fatty acids produced can be extracted and used and the residual can be fermented to produce ethanol or methane (Luisa, 2014; Ravindran et al., 2016; Jeffrey et al., 2020). The smart microalgae applied in biotechnology industry involves metabolic engineering or genetic modification of microalgae species (Lu et al., 2011). Recently, the biotechnology industry using microalgae has been expanding rapidly and currently getting attention in order to meet the growing demand for bioactive ingredients, food, and feedstock with low resource consumption, microalgae being one of the most effective sources of renewable energy production (Mohamed Hussian, 2018; Lakatos and Strieth, 2017; Mutanda et al., 2020). Based on the life cycle analysis, microalgae are one of the biofuels that have been identified as being major renewable energy sources for sustainable development to replace fossil-based fuels.

In line with industrial revolution 4.0, microalgae have an approach that no matter how the biomass is processed in an optimized biorefinery, they allow for the production of the greatest numbers of goods, maintaining the co-products and the lowest residual quantity, downstream capital in optimal returns where they can automate the growth of algae and harvesting system to reduce the operational cost and enable to track the growth operators and output of the microalgae in real time by creating the replica or digital duplicate of the device from the sensor data of microalgae production (Omar et al., 2020). Biomanufacturing involves the use of biological systems such as living microorganisms, cells, tissues, or enzymes for product manufacturing or commercializing. The revolution of biomanufacturing started in the early nineteenth century where it was more focused on the production of metabolites and then into the production of protein and enzyme. A new paradigm was evolving in the early 2000s with the advent of synthetic biology. Microorganisms, and even higher life systems, became agents for industrial-scale conversions of fossil feeds into valuable products, completely changing the game. Biomanufacturing is moving forward in parallel to the fourth industrial revolution that emphasizes big data production by using smart software, artificial intelligence, novel modelling, robotics, the intervention of three-dimensional (3D) printing, and the Internet of Things (IoT) (Hossain et al., 2019).

Microalgae cultivation will be performed on a global scale in order to contribute meaningfully to the future promotion of sustainable industry in terms of supplying biomass with higher-value low-cost products. The successful usage of technologies of culturing microalgae is needed to create enormous amounts of biomass to make the use of foods stuffs for bioethanol production. The characteristics that contribute to the successful use of smart microalgae are light, temperature, nutrients' content, mixing, pH and salinity, and mixotrophic cultivation (Khan et al., 2018). As such, the commercialization of technology of smart microalgae, which need extensive research, conceptual study, operational analysis, and marketing, is one of the essential elements regarding the prospect of microalgae in industry 4.0. In this chapter, mechanism of smart microalgae, which is divided into several aspects such as genetic engineering technologies, Internet of Things (IoT) and intelligence biomonitoring, is explained which could contribute greatly to the novelty and relevance in the field of smart microalgae to develop more smarter, safer, cleaner, greener, and economically efficient techniques for energy recovery in the future.

6.2 MECHANISMS OF SMART MICROALGAE

6.2.1 Genetic Engineering

Microalgae are one of the natural sources which contain many bioactive compounds (proteins, lipids, vitamins, enzymes, sterols, and other high-value compounds) known to produce different types of biocompounds where they can be obtained from biomass or released extracellularly (Paniagua-Michel, 2015). Microalgae naturally can tolerate diverse abiotic stress conditions of temperature, light intensity, pH, salinity, mineral reduction, phytohormones, and UV radiation (Elfar et al., 2020; Skjanes et al., 2013) to accumulate lipids, carbohydrates, protein, and secondary metabolites (Salama et al., 2019; Gulheneuf et al., 2016; Low et al., 2021). There are four main groups of microalgae that frequently are used: *Cyanophyceae* which are blue–green algae, *Chlorophyceae* which are green algae, *Bacillariophyceae* that include diatoms, and *Chrysophyceas* that include golden algae (Rajkumar et al., 2016). Some microalgae species are able to switch between photoautotrophic and heterotrophic growth where the process of microalgae cultivation utilizes the organic substrates through aerobic respiration where it generates the energy without source of light while some species will be in mixotrophic growth where microalgae are processed using inorganic and organic carbon sources in the presence of light (Muhamad and Anita, 2019).

One of the major limiting factors of microalgae cultivation is light intensity where it affects the photosynthesis of microalgae and influences the biochemical composition of microalgae and biomass where the light intensity varies with the culture and reduces the culture depth where it can be considered for modelling of bioreactor or open pond system (Khan et al., 2018). Additionally, microalgae are flexible in terms of metabolism and have a short doubling time where photoautotrophs account for the vast majority of microalgal species where develop to heterotroph based on the conditions and environmental factors (Low et al., 2021). Developing a novel microalgal strain with multiple characteristics, such as higher biomass production, lipid accumulation, CO_2 utilization, and pollution removal, would be beneficial (Mishra et al., 2019). The organic carbon plays a vital role in the growth of microalgae under mixotrophic and heterotrophic cultivation conditions (Muhamad and Anita, 2019).

In smart microalgae, some technologies have been applied to increase efficiency and effectiveness. Cultivation, harvesting, drying, cell destruction, and fossil fuel extraction are already used to make microalgae biodiesel (lipid extraction). Cell disruption is the most crucial process to ensure the best quality and greatest quantity of biofuel production for lipid extraction (Lee et al., 2010; Ravindran et al., 2016). In order to facilitate the interest of the genetic engineering process, the knowledge on the microalgae genome is needed where the strain of microalgae can be used as a model organism tool at the moment. The societal implications of genetically modified (GM) microalgae where they convert the carbon dioxide into organic biomass and harvest the light from the sun are naturally and rightfully having the strong green credentials. Microalgae have the good survival potential where the mass cultivation of microalgae inevitably leads to the low-level release of the microorganisms which can survive in the natural environment away from the controlled industrial cultivation site.

Transgene evasion techniques as shown in Figure 6.1 will be useful in optimizing the production of protein in metabolically modified microalgae (Gulheneuf et al., 2016). An understanding of microalgae physiology is necessary in order to improve strain performance, especially in relation to carbon allocation and regulatory mechanisms for the distribution of desired metabolites, particularly in response to environmental factors (Rai et al., 2016). The primary goal of omics is to detect comprehensively genes (genomics), mRNA (transcriptomics), proteins (proteomics), and metabolites (metabolomics) in a microalgae sample (Horgan

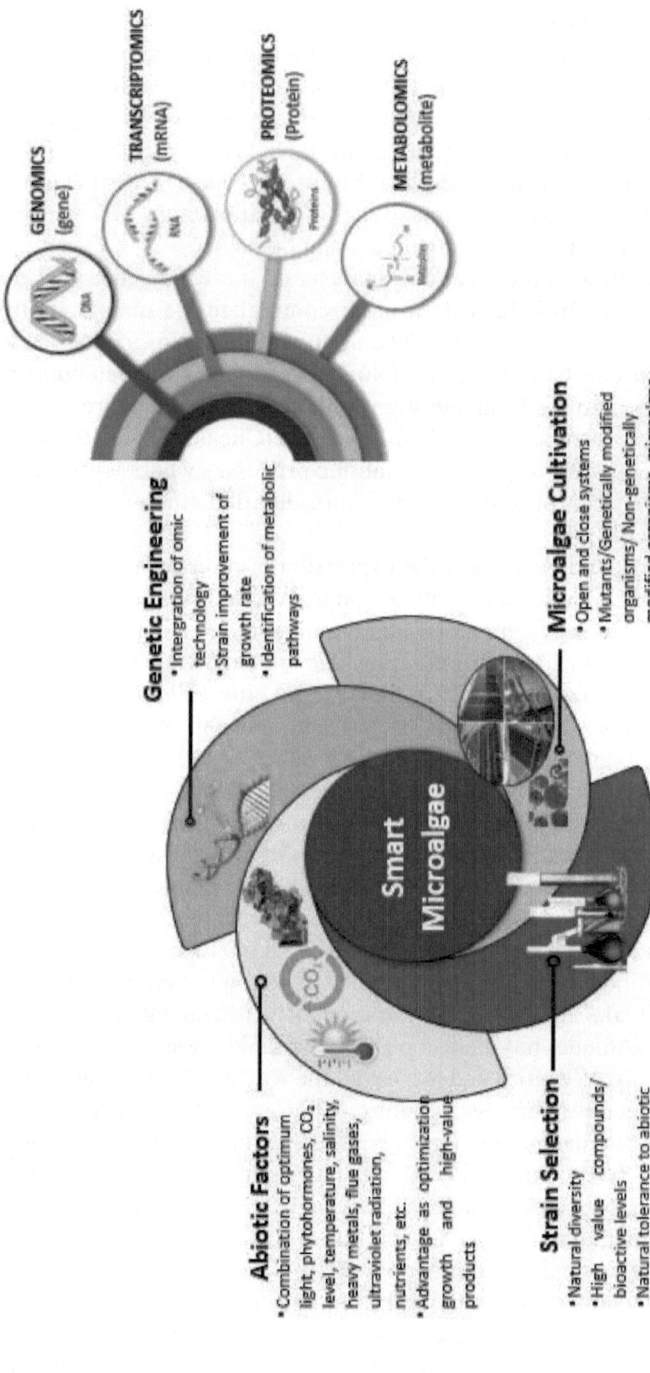

FIGURE 6.1 Integrated smart microalgae using genetic engineering.

and Kenny, 2011). Rather than investigating genes, proteins, and metabolic pathways, integrated omics platforms were used to optimize bio-component synthesis in microalgae (Salama et al., 2019).

The branch of genetics called genomics focuses on generating and analyzing the entire genome transcripts and collections of cDNA (Dal'Molin and Nielsen, 2016; Salama et al., 2019). In advanced microalgae biotechnology, genome-scale modeling can be used to build, analyze, and apply metabolic models at the genome level to study metabolism (Dal'Molin and Nielsen, 2016). It is a computational method that analyzes all metabolic information at the genome scale for an organism and converts it into a mathematical model.

Transcriptomics is biochemical analysis of the occurrence and quality of RNA transcripts which provide a more comprehensive understanding of the functions of cells than genomics. The application of transcriptomics to microalgal research can help identify biological processes or metabolic pathways and can be used to study metabolic regulation and gene expression based on carbon sources (such as glycerol and glucose) (Chen et al., 2016). It provides a broad overview of the important metabolic processes where it efficiently builds hypotheses that can be guided into future-detailed studies on improving the accumulation of lipid.

A proteomic study considers the expression of proteins and their interactions with other proteins associated with genomics and transcriptomics which provide extra insights into gene expression and regulation (Salama et al., 2019). By using proteomics, we can identify post-translational modifications of proteins, how proteins are organized, and where they exist within cells. It is now known that proteomics clarifies how the expression of a protein can be modulated both under variable conditions that affect a specific product as well as during cell survival (Anand et al., 2019). As the result of genetic and environmental changes inside a cell, metabolomics analyzes the amounts of these released low-molecular weight biomolecules. Most of the metabolic research in microalgae focuses on identifying and quantifying secondary metabolites with economic potential in the food, pharmaceutical, and public health sectors (Salama et al., 2019).

Microalgae are moving toward using artificial intelligence analytics in genome sequencing to help with optimization, detection, and alterations in gene expression (Teng et al., 2020). The application of artificial intelligence to genome sequencing techniques has made it possible for rapid retrieval of relevant information on microalgal genetics and has paved the way for efficient gene editing as an important aspect of microalgae's future (Elfar et al., 2020; Teng et al., 2020). In addition, this can make microalgae genes more compatible with mutagenesis and recombinant DNA technology, which enhances microalgae nutrient compounds or physiology (Ryan Georgianna and Mayfield, 2012). By integrating gene expression with artificial intelligence algorithms, the identification process is assisted, improved, and automated (Teng et al., 2020; Chuai et al., 2018; Lin and Wong, 2018). The integrated omics modification and technologies would improve the genetic engineering technologies that can approach precisely a specific target and genetic diversity of molecular system. The big data on the comprehension of microalgal cellular dynamics would be enriched with the help of integrated omics and advanced techniques.

Microalgae cultivation systems can be optimized through research on transcriptomes, proteomes, and metabolomes under a variety of stressful conditions to produce high-quality microalgae biofuel. There are insufficient genome and transcriptome data limiting the microalgae genetic manipulation, as well as efficient tools. Microalgae transformation systems have to be updated and redeveloped, which include methods for introducing transgenic DNA, identifying suitable promoters, distinct selectable markers, and expression vectors as well as for optimization of codon usage (Gulheneuf et al., 2016). Previous studies found that microalgal strains or biomass produce lab-scale microalgae cultivation system with different efficiencies where *Chlamydomonas* sp. do not have any algal growth or biomass production where *Chorella* sp. are able to produce algal growth and biomass productivity (Debowski et al., 2020; Zewdie and Ali, 2020). The microalgae-based bioproducts point toward a promising future, providing novel characteristics and one of the populations needs and growth in this current situation, where the algae-based products are considered to supply clean sustainable energy and food resources.

6.2.2 Internet of Things (IoT)

Internet of Things (IoT) can be used in order to monitor and control the cultivation of microalgae along with biological modelling for Decision Support System (DSS). It is a system that is related to computing devices, mechanical and digital machines, objects, animals or people that contain a unique identifying ability to transfer the data over the network without any need of human-to-human or human-to-computer interactions. There are several functions of prototype IoT with the application of modelling on the microalgae cultivation information system such that the real-time microalgae biomass concentration can be sensed and parameters such as atmospheric conditions in greenhouse, algae growth in cultivation ponds, greenhouse atmospheric conditions and real time to remote IoT platform with wireless sensor of network and Internet connection can be sent. The second function is that the change of trend of biomass concentration can be forecast and the cultivation pond status can be based on real-time data, and the prediction of modelling using a photosynthetic microalgae model can be done using this application. The third function is that the DSS operation can improve the production of biomass, and the health of cultivation could be preserved. Fourth function is that DSS can be used to prevent the difficulties during microalgae growth and culture (Serena et al., 2017).

Due to the development of our global economy that needs large supply of energy, many countries used microalgae as an alternative of raw material for biodiesel. Therefore, a photobioreactor with the IoT System was developed for microalgae cultivation (Rahmat et al., 2020) where the high-density cultivation process with two approaches such as metabolism control and cultivation system in a bioreactor was used. Meanwhile, in the biomanufacturing industry (Carina et al., 2020), the production process was upgraded using the latest industrial revolution where it monitored and controlled the cultivation of microorganisms using microalgae. At this point, the smart microalgae have improved the efficiency of smart biomanufacturing in both academia and industrial development.

By the year 2050, our world population is estimated to be more than 10 billion people, and our agriculture industry faces several challenges relating to climate change and urban expansion (Fabris et al., 2020). The uses of microalgae in agriculture industry together with the application of IoT will help to understand the biological of algal and established the algal-based bioeconomy. Currently, the usage of algae in industrial applications is relatively small whereby highlighting the IoT application and principles, it will produce the algal culture and optimize the operation to meet the standard demand. With IoT's ability to respond immediately to changes, algal culture and harvest operations can be automated, which reduce expenditures and enable IoT sensors to analyze algae cultivation in real-time.

Biotechnological approaches that use IoTs provide new applications where microalgae production through photobioreactors can be used as an architecture component or a home air treatment system since they produce biofuel which yields a variety of beneficial compounds, including the fixation of carbon dioxide. Next, it showed that smart microalgae can undergo water treatment during rain and thaw where the photobioreactors could be installed in the city's water distribution systems (Pavel, 2020). Thus, the integration between microalgae and IoTs is one of the focuses in terms of new possibilities and issues that pertain in most of the biological systems and industrial systems for the better future.

6.2.3 Intelligent Biomanufacturing

6.2.3.1 Big-Data Software

The intelligence of biomanufacturing can be classified into the smart software used for microalgae synthesis and artificial intelligence of microalgae genetics. The concept of biomanufacturing 4.0 is beyond constructing a simulation, digitalization, and remote data sensing. Using simulations, operating practices can be adjusted to meet customer demands and reduce waste in real time. The development of mathematical models is of great value, especially when it comes to optimizing and controlling bioprocesses. In order to ensure that the operations have effective monitoring that is capable of reproducing bulky and persistent processes, it is important to build and execute a modelling strategy, monitor, and then optimise the processes in real-time before controlling them accordingly.

The mathematical modeling and statistical software methods are examples of a tremendously integrated approach that combines all data types concurrently and produces a single integrated clustered sample (Mo et al., 2013; Yu and Zeng, 2018). These theories of smart software are to elucidate and interpret the complexity of the system with the investigation exploration of the data integration. This technology would preserve data resources and can be used as analysis tool for the prevalent and prospective designation for big-data formation.

Phenomics databases are currently not searchable, which limit the process of microalgal phenomics. In silico experiments on the plant and yeast can be designed and even performed by using The Arabidopsis Information Resource (TAIR) database (Lamesch et al., 2011). Meanwhile, Fernandez-Ricaud et al. (2016) is using the PROPHECY database for microbial growth phenomics tool for automated extraction and visualization exploration components. Researchers can use these tools to analyze how different genes are related through shared phenotypes and to distinguish between the functions of duplicated genes. Automated digitization of morphological phenotypes is also possible via machine learning (ML) approaches using support vector machines (SVNs) and convolutional neural networks (CNNs) (Mohanty et al., 2016; Sladojevic et al. (2016).

It is crucial for biomanufacturing companies to automatically and appropriately control bioprocesses in their optimal state by reducing or maintaining production costs and increasing yields, while maintaining uniform product quality. Smart sampling method and automated analysis technique may significantly save time in the process of observation and governance by using the technological readiness level (TRL) which is used as an assessment of a new technology's capability toward economic viability. The TRL technologies are an exploration of microalgae for easier multi-omics translation.

In the pharmaceutical industry, one of the analytical technologies tools that have been recommended by the Food and Drug Administration (FDA) to ensure the quality of the product is by using Process Analytical Technology (PAT) (Food and Drug Administration (FDA), 2004). PAT is a critical tool that verifies the quality and performance attributes of raw and processed materials are analyzed and controlled to ensure final product quality. The goal is to reduce overprocessing, ensure efficiency, and minimize waste through the elimination or reduction of unnecessary elements in the process. The software package devoted to PAT application tools includes sensor technologies, multivariate data analysis, experimental designation, and bioprocess modelling (Gargalo et al., 2020). Recently, PAT has been used as a tool for integrated omics and smart process analyzer in the pharmaceutical industry (Streefland et al., 2013). The application of PAT is general in nature; thus, it will be an evolution in any biomanufacturing process. PAT will be an advanced technology that will be applied to microalgae on a computerized basis with automatic information-driven to boost productivity.

Analytical instrumentation including biosensors comprising biochemical signals is transformed into measurable electrical signals by an immobilized sensing component in close proximity to an appropriate transducer. A standard biosensor consists of three parts: an immobilized biological sensor component on a condition transducer section that is intensified by a signal conversion unit. The biocomponents identify the analytes in high sensitivity and selectivity approach, whether by a catalytic mechanism such as a process involving enzymatic cells, tissues, and organelles or via interaction of protein channels, antibodies, and nucleic acids (Gargalo et al., 2020). Thus, the transducer component of the biosensor can synergize to analyte by adapting the visualisation concept, electrochemical, calorimetrical or piezoelectrical as the detection unit. These sensors provide rapid, affordable, and reliable analytical results (Prasad et al., 2009). Biosensors' conjunction with an increasing computation power is also a promising solution for enabling smart manufacturing.

The application of spectroscopic sensors is done for monitoring and managing bioprocesses assuring an immediate accessibility with real-time data in optimum condition control. The application of spectroscopic sensor can separate different types of genomic data with the principle used of NAD(P)H or fluorophores such as GFP and YFP (Samorski et al., 2005; Kensy et al., 2009). BioLector® is one of the efficient systems applying 2-D fluorescence spectroscopy in biomanufacturing fermentation (Gargalo et al., 2020). Hence, this will promise a fast procedure in process selection and giving a real-time response and monitoring.

In order to capture continuous real-time data of bioprocess variables, spectroscopic sensing generates enormous datasets, necessitating chemometrics or multivariate data analysis (MVDA). Machine learning techniques such as MVDA analyze many variables at the same time and thus decrease complexity of data sets, and MVDA is employed for the purpose of extracting information from spectra (Gargalo et al., 2020). This can be accomplished by preprocessing data to reveal useful information within it. The preprocessed spectral data is calibrated to extract both qualitative and quantitative information. The Principal Component Analysis (PCA) is often used in MVDA to analyze the structure, variance, and/or distribution of a dataset and to find outliers (Glassey, 2013). By using PCA, spectral data is used for process supervision, batch categorization, and determining the process state (Pomerantsev and Ye, 2012). The process goal line, or trajectory, commonly referred to as a golden batch, can be determined from similar and perfect process runs (Henriques et al., 2010). By utilizing these data, it is possible to predict multiple variables online, allowing comprehensive access to a process and enabling automated control and rapid problem detection via closed-loop controls.

6.2.3.2 Artificial Intelligence and Artificial Neural Network

Artificial intelligence (AI) is one of the platforms or recent gene editing tools and a platform where microalgae can be precisely converted for a system design and integration. AI can stimulate the human brain activities which are applied through different techniques of computer science such as heuristic algorithms, machine learning, and fuzzy logic (Xing et al., 2021). Due to their high oil content and productivity such as lipid and carbohydrate, microalgae are one of the best methods used for biofuels' production. Artificial intelligence or AI is another optional tool used in order to develop the modelling techniques for various complex systems (Mayol et al., 2020). The alteration of microalgae using artificial intelligence can modify the metabolic pathway of microalgae to increase yields production of lipids, biomass, and other components (Teng et al., 2020). AI may be crucial in order to get a more efficient use of some resource but it also may be used for some data analysis, production management of microalgae, and knowledge gap in higher productivity and competences where it is proved to be much more efficient as compared to processing the microalgae manually.

Nutraceuticals are often produced from species of microalgae, for example *Spirulina*, *Chlorella*, and *Klamath*, which perform photosynthesis and provide a wide range of nutrients to other organisms in the food chain to illustrate some important relationships between preventive medicine using AI techniques and patient outcomes (Archana et al., 2019). The application of AI to microalgae will accelerate the production of microalgae through algorithms such as active learning, semi-supervised learning, and meta learning (Teng et al., 2020). AI is one of the applications playing an important role and implementing the microalgae source to increase the production of healthy and nutritive food products in order to achieve the rapid demands of the human population.

Artificial neural networks (ANNs) are one part of artificial intelligence where they are part of algorithms which are characterized by the development of data and are able to perform without explicating the programmed. In this network, two types of microalgae have been used, such as Chlorella vulgaris and Scenedesmus almeriensis, whose nodes are in the previous layer (Otalora et al., 2021). Recently, these predictions have got popular where they can predict the bloom and the trend of the microalga biomass where it solves the complex time series in forecasting problems. Because of the higher rate of CO_2 fixation, ANNs have a great deal of significance in determining the potential of biomass and possibilities for industrial purposes (Kumar et al., 2010).

6.3 ADVANTAGE AND DISADVANTAGE OF SMART MICROALGAE

The enhancement of microalgae technologies contributes to the global industrial revolution. These smart microalgae have been widely used in many industries as tools of environment-protecting technologies, product modifications, and life quality improvements. The exploration of microalgae as environment-protecting technologies has been applied in gases' treatment, wastewater treatment, leachate treatment, waste management, and biogas upgrading for nutrient removal where the smart microalgae treatments are one of the well-established technologies that compared to typical wastewater treatment systems have reduced capital and running costs as well as a better degree of efficiency (Debowski et al., 2020; Fabris et al., 2020; Simone et al., 2021). Undisputed involvement of new technologies on microalgae will boost up the product modification in agriculture, chemical, pharmaceuticals, and personal care (Barsanti and Gualtieri, 2018; Bhalamurugan et al., 2018; Fabris et al., 2020).

In fact, the food cycle of many people globally will turn out to be significantly more sustainable if microalgae are used as they are rich in proteins, carbohydrates, and high-density lipoproteins (Bhalamurugan et al., 2018). As compared to animal-based proteins, the microalgae protein requires less land (Caporgno and Mathys, 2018). Continuous research proves that smart microalgae can aid in getting a better life quality. Various health benefits have been attributed to microalgae as they are good sources of vitamins and antioxidants and having anti-inflammatory properties and cancer-fighting abilities (Bhalamurugan et al., 2018; Liu et al., 2000). Microalgae provide new energy avenues of biofuels pertaining to biogas, biohydrogen, bioethanol, and biodiesel production, as well as the production of biochar and bioethanol (Kumar et al., 2020; Debowski et al., 2020). This could be one of the advantages in the use of smart microalgae where in the developed countries the local population consumes high-calorie foods due to their busy schedule and modern lifestyle. Some examples of microalgae species in the food industry are *Spirulina plantesis, Chorella sp., Dunaliella terticola, Dunaliela saline, and Aphanizomenon flos-aquae* which are rich in high-protein contents (Sathasivam et al., 2019).

It is estimated that toward 2040, the worldwide energy usage demand will dramatically increase to 85% of energy production (Parsaeimehr et al., 2015). Thus, the development of biofuel energy as an alternative source of renewable energy to accommodate the global demand is encouraged (Ravindran et al., 2016). The smart microalgae used as a prospective source to increase biofuel production. Towards achieving industrial revolution 4.0, smart microalgae refineries have been optimized and integrated to produce an extensive yield, reduce the production of residual, and strengthen profits (Elfar et al., 2020) where the microalgae lipids can be extracted to get some of the yield oil similar to the land-based oilseed crops where the use of microalgae in energy generation involves low-cost production and an efficient provision of the scalable reactor design (Melinda et al., 2014).

Figure 6.2 presents the use of smart microalgae, capable of the production of sustainable and clean energy. In Indonesia, microalgae are used for CO_2

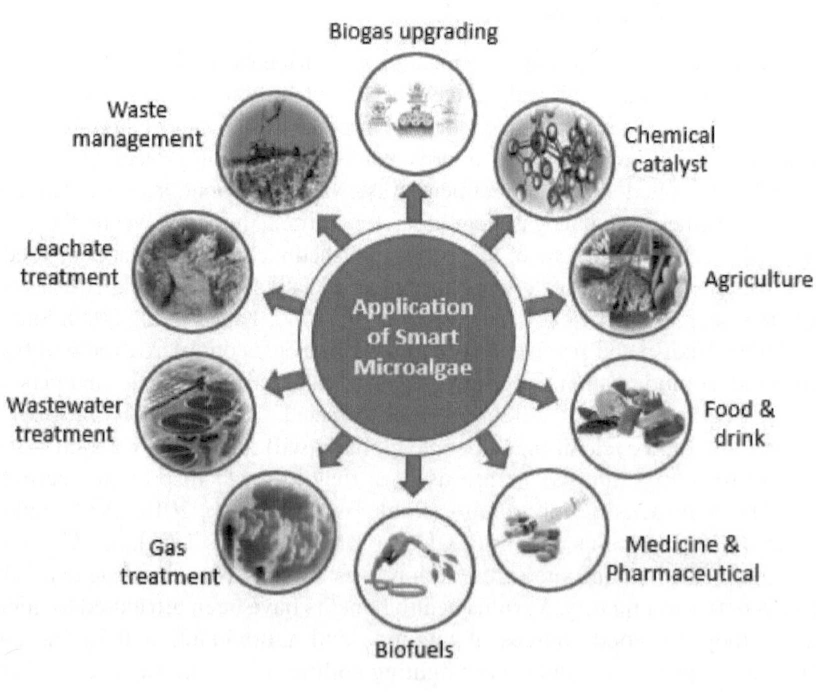

FIGURE 6.2 Application of smart microalgae.

absorption and O_2 production using photobioreactors in solving the issues of high CO_2 emissions which trigger the global warming and influence climate changes (Putri et al., 2019; Bermejo et al., 2021). While the transportation sector contributes to emission of more than 19% of CO_2 in Malaysia (Mathieu et al., 2020). Hence, the use of microalgae in biodiesel production to replace the biomass is a preeminent solution. Besides that, as microalgae are one of the natural methods in waste stabilization, their use can form one of the advantageous methods in order to treat the rubber mill effluent (Jayakumar et al., 2017).

One of the first existing and potential applications of microalgae was in the industrial sector of aquaculture in the developed economies that cultivated the microalgae and used directly in formulated feeds for larval and juvenile fish (Carrasco et al., 2018). The second application was in the preparation of biofuel that is a form of renewable energy to sustain human population growth and progress by enhancing their feasibility, effectiveness, and benefits, such as easy and effective maintenance, high biomass concentration, water saving capability, two-route cultivation, rapid growth of algae, and synthesis of valuable metabolites, resulting in superior energy recovery and carbon footprint reduction. (Culaba et al., 2020; Marcin et al., 2020).

A study attempted to determine the success of using IoTs (Pavel, 2020) in smart microalgae application in photobioreactors so that the data can be efficiently exchanged through farm systems and city data centers to assist in the development of predictive mathematical models from all installed systems. The wireless sensor networks (WSN) in biofuel cell application, power supply is expanded without using battery charging or any other energy source supply systems. While in novel biosensors they can enhance number of different sensing molecules and are able to analyze the preliminary data where different parameters can be measured and more reliable data in terms of predictive models can be received. Thus, it can be concluded that using smart microalgae in terms of IoTs can be one of the integrated modern biotechnologies that will invite significant positive feedback for communities and city developments. Microalgae using smart microalgae technology produce high yields which are photosynthetic efficiencies and are environmentally manipulatable, as they use low-cost saline water and low-quality nonarable land, and the water loss is by evaporation only. Nutrients, especially fertilizer, can be recycled, and most of the biomass and genetic enhancement can be used.

Artificial intelligence (AI) technology makes it cheaper and easier to monitor microalgae, as it is capable of identifying and quantifying different types of microalgae. The time of operational AI system uses software in the combination of microscope are one of the advantages where it needs about one to two hours including the confirmation of the result by a human analyst. Third advantage is that it provides an early warning of problems where testing could be done more quickly and frequently. Moreover, using AI technologies does not involve any revolutionary changes in the legacy system to produce policy evidence where a more practical approach adopts new analytics and predictive approaches where it can also enhance the policymakers' decisions when facing the problems in terms of responding it promptly with a limited data. The undeniable strength of using these technologies of microalgae is well-established due to the characteristics of high photosynthetic efficiency where it can convert solar energy at 4% to chemical energy at 10% where the final productivity can yield to reach 36 tonnes of dry microalgal.

The first disadvantage of smart microalgae is their indigenous habitat where some strains of microalgae are difficult to adapt to laboratory growth conditions in artificial media and some strains are non-culturable in artificial media. Nevertheless, sourcing of the strains from the collection sites can be time consuming where they occur in mixed cultures and need purification and optimization which can be long and tedious exercise and expensive. The second disadvantage is that the genomic data for most microalgal strain is limited where the modified strains may become invasive in local habitats and may not be easily propagated under outdoor and open ponds (Mutanda et al., 2020). The third disadvantage of using smart microalgae in biofuels are low production of biomass, low contents of lipids in the cells, and the small size of the cell, which makes the harvesting process costly (Srikanth et al., 2015). Although using smart microalgae had some limitations and challenges, the technology of industry needs to be more upgraded from pilot-phase to industrial level where the crucial part of using microalgae is to enhance the microalgae rate and to synthesize the products.

However, in terms of using smart microalgae in IoTs, there are several issues that need to be discussed where the first issues are related to the complexity of the biological systems in which the biological objects were sometimes small element in the system like a cell in a cell-based biosensors, and the processes were overly complex inside the small cell and were influenced by lots of factors that need to be addressed in this system. The second issue is the dependence on environment where microalgae usually can survive in a narrow range of climate parameters, and they can be even narrower for some microalgae species where the common weather usually changes throughout the year. The third issue is that sometimes it is difficult to receive the data concerning internal processes taking place inside the objects without destruction whereas the synthetic biology needs additional sensors such as optical and biochemical to solve the issues.

However, there are certain limits to applying artificial intelligence (AI) in some microalgae species, such as cyanobacteria, due to inaccurate prediction based on faulty data. Furthermore, the adjustment of higher accuracy is needed in order to obtain the optimal values of each region (Lee and Lee, 2018).

6.4 CONCLUSIONS

Microalgae are tiny little factories: renewable, sustainable, and economical sources for biofuels; bioactive compounds of medicinal products; and sources of production of food ingredients. Smart microalgae form a large-scale industry that provides an alternative and sustainable by-product in the form of microalgae biomass which can contribute in more than one industrial sector. Microalgae industry can be one of the best and great benefits to our environment and to our quality of life, providing a better living standard of society. To pursue sustainable development goals, microalgae biofuels can be a thorough source for fossil fuels, which can increase energy security and environmental sustainability. In terms of food production, it is clearly proved that microalgae production can meet the demands of human needs. In the future, it can be developed into algal-based food industry, focusing on innovative functional food products and improving sustainable development.

A promising source of biofuels is microalgae oil technologies, which is being endorsed as an alternative to fossil fuels. Nevertheless, the microalgae could also be potentially modified to synthesize other types of fuels such as ethanol, butanol, isopropanol, and hydrocarbons or in terms of downstream processing of microalgae which can be modified to transform any biomass into energy containing fuels via thermal processes. Even so, more needs to be done to achieve improved productivity and lower prices. A major component of advancing economic viability is metabolic engineering for increasing yields of microalgae-derived biofuels focusing on the lipids in microalgae. The strains of microalgae that are relevant for biofuel production could be developed by using such omics technologies. Novel gene and gene product discoveries within metabolic pathways are imperative for optimum biofuel production using microalgal genomes. It might be possible to explore system-emergent traits that may assist microalgae farming, microalgae biotechnology, and biofuel production by combining different 'omics' datasets with knowledge concerning the functions of genes related with carbon acquisition and accumulation.

To improve the world economic competitiveness of microalgae products, the efficiency of the extraction methods with the genetic engineering of microalgae has an immense potential to improve the world economics although it is limited due the genetic information. In the future, purpose-specific robust bioengineered strains for high photosynthetic efficiency, high carbon dioxide (CO_2) fixation, and high productivities of biomass need to be achieved and developed. The advantages and potentials of microalgae make them a suitable candidate to solve the CO_2 and energy issues. The main insights discussed in this chapter demonstrate that smart microalgae highlight the most important aspects of technology to point out the requirement to face contemporary challenges more effectively and make use of technology that can bring up the smart microalgae where it can be expanded to a broader concept.

REFERENCES

Anand, V., Kashyap, M., Samadhiya, K., and Kiran, B. (2019). Strategies to unlock lipid production improvement in algae. *International Journal of Environmental Science Technology*, 16, pp. 31829–31838. https://doi.org/10.1007/s13762-018-2098-8.

Archana, P., Lala, B. S., Debabrata, P., and Vinita, S. (2019). Artificial intelligence and virtual environment for microalgal source for production of nutraceuticals. *Biomedical Journal of Scientific and Technical Research,* 13(5), pp. 10239–10243. https://doi.org/10.26717/ BJSTR.2019.13.002459.

Barsanti, L., and Gualtieri, P. (2018). Is exploitation of microalgae economically and energetically sustainable? *Algal Research*, 31, pp. 107–115. https://doi.org/10.1016/j.algal.2018.02.001.

Bermejo, E., Filali, R., and Taidi, B. (2021). Microalgae culture quality indicators: A review. *Critical Reviews in Biotechnology*, 41(4), pp. 457–473. https://doi.org/10.1080/07388551.2020.1854672.

Bhalamurugan, G. L., Valerie, O., and Mark, L. (2018). Valuable bioproducts obtained from microalgal biomass and their commercial applications: A review. *Environmental Engineering Research*, 23(3), pp. 229–241. https://doi.org/10.4491/eer.2017.220.

Caporgno, M. P., and Mathys. A. (2018). Trends in microalgae incorporation into innovative food products with potential health benefits. *Frontiers in Nutrition*. 5(58), pp. 1–10. https://doi.org/10.3389/fnut.2018.00058.

Carina, L. G., Isuru, U., Katrin, P., Pau, C. L., Rasmus, F. N., Aliyeh, H., Seyed, S. M., Christoph, B., Helena, J., and Krist, V. G. (2020). Towards smart biomanufacturing: A perspective on recent developments in industrial measurement and monitoring technologies for bio-based production processes. *Journal of Industrial Microbiology and Biotechnology*. 47, pp. 947–964. https://doi.org/10.1007/s10295-020-02308-1.

Carrasco, R., Fajardo, C., Guarnizo, P., Vallejo, R. A., and Fernandez-Acero, F. J. (2018). Biotechnology applications of microalgae in the context of EU 'blue growth' initiatives. *Journal of Microbiology and Genetics*, 1, pp. 1–14. https://doi.org/10.29011/2574-7371.

Chai, W. S., Tan, W. G., Munawaroh, H. S. M., Gupta, V. K., Ho, S. H., and Show, P. L. (2021). Multifaceted roles of microalgae in the application of wastewater biotreatment: A review. *Environmental Pollution*. 269, p. 116236. doi: 10.1016/j.envpol.2020.116236.

Chen, W., Zhou, P. P., Zhang, M., Zhu, Y., Wang, X. P., Luo, X. A., Bao, Z. D., and Yu, L. J. (2016). Transcriptome analysis reveals that upregulation of the fatty acid synthase gene promotes the accumulation of docosahexaenoic acid in Schizochytrium sp S056 when glycerol is used. *Algal Research*, 15, pp. 83–92. https://doi.org/10.1016/j.algal.2016.02.007.

Chuai, G., Ma, H., Yan, J., Chen, M., Hong, N., Xue, D., Zhou, C., Zhu, C., Chen, K., Duan, B., Gu, F., Qu, S., Huang, D., Wei, J., and Liu, Q., (2018). Deep CRISPR: Optimized CRISPR guide RNA design by deep learning. *Genome Biology*, 19(80), pp. 1–18. https://doi.org/10.1186/s13059-018-1459-4.

Culaba, A. B., Ubando, A. T., Ching, P. M. L., Chen, W. H., and Chang, J. S. (2020). Biofuel from microalgae: Sustainable pathways. *Energy Development for Sustainability*, 12(8009), pp. 1–19. doi: https://doi.org/10.3390/su12198009.

Dal'Molin, C. G., and Nielsen, L. K. (2016). Algae genome-scale reconstruction, modelling and applications. *The Physiology of Microalgae*, pp. 591–598. https://doi.org/10.1007/978-3-319-24945-2_22.

Debowski, M., Zielinski, M., Kazimierowicz, J., Kujawska, N., and Talbierz, S. (2020). Microalgae cultivation technologies as an opportunity for bioenergetic system development—advantages and limitations. *Sustainability*, 12(9980), pp. 1–37. https://doi.org/10.3390/su12239980.

Elfar, O. A., Chang, C. K., Leong, H. Y., Peter, A. P., Chew, K. W., and Show, P. L. (2020). Prospects of Industry 5.0 in algae: Customization of production and new advance technology for clean bioenergy generation. *Energy Conversion and Management*, X(10), pp. 1–10. https://doi.org/10.1016/j.ecmx.2020.100048.

Fabris, M., Abbriano, R. M., Pernice, M., Sutheland, D. L., Commault, A. S., Hall, C. C., Labeeuw, L., McCauley, J. I., Kuzhiuparambil, U., Ray, P., Kahlke, T., and Ralph,

P. J. (2020). Emerging technologies in algal biotechnology: Toward the establishment of a sustainable, algae-based bioeconomy. *Frontiers in Plant Science*, 11(279), pp. 1–22. https://doi.org/10.3389/fpls.2020.00279.

Fernandez-Ricaud, L., Kourtchenko, O., Zackrisson, M., Warringer, J., and Blomberg, A. (2016). PRECOG: A tool for automated extraction and visualization of fitness components in microbial growth phenomics. *BMC Bioinformatics,* 17, p. 249. https://doi.org/10.1186/s12859-016-1134-2.

Food and Drug Administration (FDA). (2004). Guidance for industry, PAT-A framework for innovative pharmaceutical development, manufacturing and quality assurance. Retrieved from https://www.fda.gov/media/71012/download.

Gargalo, C. L., Udugama, I., Pontius, K., Lopez, P. C., Nielsen, R. F., Hasanzadeh, A., Mansouri, S. S., Bayer, C., Junicke, H., and Gernaey, K. V. (2020). Towards smart biomanufacturing: A perspective on recent developments in industrial measurement and monitoring technologies for bio-based production processes. *Journal of Industrial Microbiology & Biotechnology*, 47, pp. 947–964. https://doi.org/10.1007/s10295-020-02308-1.

Glassey, J. (2013). Multivariate data analysis for advancing the interpretation of bioprocess measurement and monitoring data. *Advance Biochemistry Engineering Biotechnology*, 132, pp. 167–191. https://doi.org/10.1007/10_2012_171.

Godvin S. V., Dinesh K. M., Arulazhagan P., Amit K. B., Poornachander G., and Banu R. J. (2021). Biofuel production from Macroalgae: Present scenario and future scope. *Bioengineered,* 12(2), pp. 9216–9238. https://doi.org/10.1080/21655979.2021.1996019.

Gulheneuf, F., Khan, A., and Tran, L. S. P. (2016). Genetic engineering: A promising tool to engender physiological, biochemical, and molecular stress resilience in green microalgae. *Frontiers in Plant Science,* 7(400), pp. 1–8. https://doi.org/10.3389/fpls.2016.00400.

Henriques, J. G., Buziol, S., Stocker, E., Voogd, A., and Menezes, J. C. (2010). Monitoring mammalian cell cultivations for monoclonal antibody production using near-infrared spectroscopy. *Advance Biochemical Engineering Biotechnology*. https://doi.org/10.1007/10_2009_11.

Hermadi, I., Setiadianto, I. R., Al Zahran, D. F., Simbolon, M. N., Saefurahman, G., Wibawa, D. S., and Arkeman, Y. (2021). Development of smart algae pond system for microalgae biomass production. *IOP Conference Series: Earth and Environmental Science*, 749, pp. 1–8. https://doi.org/10.1088/1755-1315/749/1/012068.

Horgan, R. P., and Kenny, L. C. (2011). "Omic" technologies: Genomics, transcriptomics, proteomics and metabolomics. *The Obstetrician & Gynaecologist*, 13(3), pp. 189–195. https://doi.org/10.1576/toag.13.3.189.27672.

Hossain, K., Rahman, M., and Roy, S. (2019). IoT data compression and optimization techniques in cloud storage: Current prospects and future directions. *International Journal of Cloud Applications and Computing,* 9(2), pp. 43–59. https://doi.org/10.4018/IJCAC.2019040103.

Jayakumar, S., M. Yusoff, M., Ab. Rahim, M. H., Maniam, G. P., and Govindan, N. (2017). The prospect of microalgal biodiesel using agro-industrial and industrial wastes in Malaysia. *Renewable and Sustainable Energy Reviews,* 72, pp. 33–47. https://doi.org/10.1016/j.rser.2017.01.002.

Jeffrey, H., Hideharu, M., Hiroshi, K., Takuro, I., and Keisuke, G. (2020). Accurate Classification of microalgae by intelligent frequency-division-multiplexed fluorescence imaging flow cytometry. *Osa Continuum,* 3(3), pp. 430–440. https://doi.org/10.1364/OSAC.387523.

Kensy, F., Zang, E., Faulhammer, C., Tan, R. K., and Büchs, J. (2009). Validation of a high-throughput fermentation system based on online monitoring of biomass and fluorescence in continuously shaken microtiter plates. *Microbiology Cell Fact*, 8(1), p. 31. https://doi.org/10.1186/1475-2859-8-31.

Khan, M. I., Shin, J. H., and Kim, J. D. (2018). The promising of microalgae: Current status, challenge, and optimization of a sustainable and renewable industry for biofuels, feed, and other products. *Microbial Cell Factories,* 17(36), pp. 1–21. https://doi.org/10.1186/s12934-018-0879-x.

Kumar, A., Ergas, S., Yuan, X., Sahu, A., Zhang, Q., Dewulf, J., Malcata, F., and Van Langenhove, H. (2010). Enhanced CO_2 fixation and biofuel production via microalgae: Recent developments and future directions. *Trends in Biotechnology,* 28(7), pp. 371–380. https://doi.org/10.1016/j.tibtech.2010.04.004.

Kumar, M., Sun, Y., Rathour, R., Pandey, A., Thakur, I. S., and Tsang, D. C. W. (2020). Algae as potential feedstock for the production of biofuels and value-added products: Opportunities and challenges. *Science of the Total Environment*, 716, pp. 1–17. https://doi.org/10.1016/j.scitotenv.2020.137116.

Lakatos, M. and Strieth, D. (2017) 'Terrestrial Microalgae: Novel Concepts for Biotechnology and Applications', in Cánovas, F. M., Lüttge, U., Matyssek, R. (Eds.), *Progress in Botany Volume* 79. Cham, Switzerland: Springer International Publishing. pp. 269–312. https://doi.org/10.1007/124_2017_10.

Lamesch, P., Berardini, T. Z., Li, D., Swarbreck, D., Wilks, C., Sasidharan, R., Muller, R., Dreher, K., Alexander, D. L., Garcia-Hernandez, M., Karthikeyan, A. S., Lee, C. H., Nelson, W. D., Ploetz, L., Singh, S., Wensel, A., and Huala E. (2011). The Arabidopsis Information Resource (TAIR): Improved gene annotation and new tools. *Nucleic Acids Research,* 40, pp. D1202–D1210. https://doi.org/10.1093/nar/gkr1090.

Lee, J. Y., Yoo, C., Jun, S. Y., Ahn, C. Y., and Oh, H. M. (2010). Comparison of several methods for effective lipid extraction from microalgae. *Bioresource Technology*, 101, pp. 75–77. https://doi.org/10.1016/j.biortech.2009.03.058.

Lee, S., and Lee, D. (2018). Four major south Korea's river using deep learning models. *International Journal of Environmental Research and Public Health*, 15(7), 1322. https://doi.org/10/3390/ijerph15071322.

Lin, J., and Wong, K. C., (2018). Off-target predictions in CRISPR-Cas9 gene editing using deep learning. *Bioinformatics,* 34(17), pp. 656–663. https://doi.org/10.1093/bioinformatics/bty554.

Liu, Y., Xu, L., Cheng, N., Lin, L., and Zhang, C. (2000). Inhibitory effect of phycocyanin from Spirulina platensis on the growth of human leukemia K562 cells. *Journal of Applied Phycology*, 12, pp. 125–130. https://doi.org/10.1023/A:1008132210772.

Low, S. S., Bong, K. X., Mubashir, M., Cheng, C. K., Lam, M. K., Lim, J. W., Ho, Y. C., Lee, K. T., Munawaroh, H. S. H., and Show, P. L. (2021). Microalgae cultivation in palm oil mill effluent (POME) treatment and biofuel production. *Sustainability,* 13(6), pp. 1–17. https://doi.org/10.3390/su13063247.

Lu, J., Sheahan, C., and Fu, P. C. (2011). Metabolic engineering of algae for fourth generation biofuels production. *Energy Environmental Science*, 4, pp. 2451–2466. doi: https://doi.org/10.1039/C0EE00593B.

Luisa, G. (2014). From tiny microalgae to huge biorefineries. *Oceanography*, 2(120), pp. 55–76. https://doi.org/10.1201/b18525-5.

Marcin, D., Marcin, Z., Joanna, K., Natalia, K., and Szymon, T. (2020). Microalgae cultivation technologies as an opportunity for bioenergetic system development—advantages and limitations. *Sustainability,* 12, pp. 1–37. https://doi.org/10.3390/su12239980.

Mathieu, D., Ales, G., Javier, G., Silke, H., and Frank, H. (2020). Smart specialisation and blue biotechnology in Europe. *Joint Research Centre Science for Policy Report,* 30521. https://doi.org/10.2760/19274.

Mayol, A. P., San Juan, J. L. G., Sybingco, E., Bandala, A., Dadios, E., Ubando, E. T., Culaba, A. B., Chen, W. H., and Chang, J. S. (2020). Environmental impact prediction of microalgae to biofuels chains using artificial intelligence: A life cycle perspective. *International Conference on Sustainable Energy and Green Technology,* 463, pp. 012011. http://dx.doi.org/10.1088/1755-1315/463/1/012011.

Melinda, J., G., Reay, G., D., Christine, R., and Susan, T. L. H. (2014). Advantages and challenges of microalgae as a source of oil for biodiesel. *Biodiesel Feedstocks and Processing Technologies.* http://dx.doi.org/10.5772/30085.

Mishra, A., Medhi, K., Malaviya, P., and Thakur, I. S. (2019). Omics approaches for microalgal applications: Prospects and challenges. *Bioresource Technology*, 291(121890), pp. 1–12. https://doi.org/10.1016/j.biortech.2019.121890.

Mo, Q., Wang, S., Seshan, V. E., Olshen, A. B., Schultz, N., Sander, C., Powers, R. S., Ladanyi, M., and Shen, R. (2013). Pattern discovery and cancer gene identification in integrated cancer genomic data. *National Academy of Sciences USA*, 110(11), pp. 4245–4250. https://doi.org/10.1073/pnas.1208949110.

Mohamed Hussian, A. E. (2018). The role of microalgae in renewable energy production: Challenges and opportunities. *Marine Ecology—Biotic and Abiotic Interactions*, 12, pp. 257–283. https://dx.doi.org/10.5772/intechopen.73573.

Mohanty, S. P., Hughes, D. P., and Salathé, M. (2016). Using deep learning for image-based plant disease detection. *Frontiers in Plant Science,* 7, pp. 1419. https://doi.org/10.3389/fpls.2016.01419.

Muhamad, M. A. N., and Anita, G. J. B. (2019). Opportunities and challenges of microalgal cultivation on wastewater, with special focus on palm oil mill effluent and the production of high value compounds. *Waste and Biomass Valorization,* 10, pp. 2079–2097. https://doi.org/10.1007/s12649-018-0256-3.

Mutanda, T., Naidoo, D., Bwapwa, J. K., and Anandraj, A. (2020). Biotechnological applications of microalgal oleaginous compounds: Current trends on microalgal bioprocessing of products. *Frontiers in Energy Research,* 8, pp. 598803. https://doi.org/10.3389/fenrg.2020.598803.

Omar A. E., Chang, C. K., Leong, H. Y., Angela, P. P., Kit. W. C., and Pau, L. S. (2020). Prospects of industry 5.0 in algae: Customization of production and new advance technology for clean bioenergy generation, energy conservation and management: X. *ScienceDirect*, 10, 100048. https://doi.org/10/1016/j.ecmx.2020.100048.

Otalora, P., Guzman, J. L., Acien, F. G., Berengue, M., and Reul, A. (2021). Microalgae classification based on machine learning techniques. *Algal Research,* 55. https://doi.org/10.106/j.algal.2021.102256.

Paniagua-Michel, J. (2015) 'Chapter 16-Microalgal Nutraceuticals', in Se-Kwon, K. (Ed), *Handbook of Marine Microalgae: Biotechnology Advances.* London, UK: Elsevier, pp. 255–267. https://doi.org/10.1016/B978-0-12-800776-1.00016-9.

Parsaeimehr, A., Sun, Z., Dou, X., and Chen, Y. F. (2015). Simultaneous improvement in production of microalgal biodiesel and high-value alpha-linolenic acid by a single regulator acetylcholine. *Biotechnology for Biofuels*, 8(11), pp. 1–10. https://doi.org/10.1186/s13068-015-0196-0.

Pavel, G. (2020). How IoT can integrate biotechnological approaches for city applications—review of recent advancements, issues and perspectives. *Applied Science,* 10, p. 3990. https://doi.org/10.3390/app10113990.

Pomerantsev, A. L. and Ye, O. (2012). Rodionova, "process analytical technology: A critical view of the chemometricians". *Journal of Chemometrics*, 26(6), pp. 299–310. https://doi.org/10.1002/cem.2445.

Prasad, K., Ranjan, R. K., Lutfi, Z., and Pandey, H. (2009). Biosensors: Applications and overview in industrial automation. *International Journal on Applied Bioengineering*, 3(1), pp. 66–70. https://doi.org/10.18000/ijabeg.10041.

Putri, A., Pratama, A. O., Rohdiana, A., and Saraswati, R. R. (2019). Smart microalgae photobioreactor helped by solar energy as eco-green technology to reduce CO_2 emissions in Jakarta, Indonesia. *Asean Youth Conference*, 2018, pp. 2599–2643. https://doi.org/10.5281/zenodo.2541282.

Rahmat, A., Jaya, I., Hestirianoto, T., Jusaidi, D., and Kawaroe, M. (2020). Design a photobioreactor for microalgae cultivation with the IOTs (internet of things) system. *Omni Akuatika,* 16(1), pp. 53–61. http://dx.doi.org/10.20884/1.oa.2020.16.1.791.

Rai, V., Karthikaichamy, A., Das, D., Noronha, S., Wangikar, P. P., and Srivastava, S. (2016). Multi-omics frontiers in algal research: Techniques and progress to explore biofuels in the postgenomics world. *OMICS: A Journal of Integrative Biology,* 20(7), pp. 387–399. https://doi.org/10.1089/omi.2016.0065.

Rajkumar, R., and Takriff, M. S. (2016). Prospects of algae and their environmental applications in Malaysia: A case study. *Journal of Bioremediation and Biodegradation,* 7, pp. 321. http://dx.doi.org/10.4172/2155-6199.1000321.

Ravindran, B., Gupta, S. K., Cho, W. M., Kim, J. K., Lee, S. R., Jeong, K. H., Lee, D. J., and Choi, H. C. (2016). Microalgae potential and multiple roles—current progress and future prospects—an overview. *Sustainability*, 8(1215), pp. 1–16. https://doi.org/10.3390/su8121215.

Ryan Georgianna, D. and Mayfield, S. P. (2012). Exploiting diversity and synthetic biology for the production of algal biofuels. *Nature*, 488, pp. 329–335. https://doi.org/10.1038/nature11479.

Salama, E. S., Govindwar, S. P., Khandare, R. V., Roh, H. S., Jeon, B. H., and Li, X. (2019). Can omics approaches improve microalgal biofuels under abiotic stress? *Trends in Plant Science,* 24(7), pp. 611–624. https://doi.org/10.1016/j.tplants.2019.04.001.

Samorski, M., Mueller-Newen, G., and Buechs, J. (2005). Quasi-continuous combined scattered light and fluorescence measurements: A novel measurement technique for shaken microtiter plates. *Biotechnology Bioengineering*, 92(1), pp. 61–68. https://doi.org/10.1002/bit.20573.

Sathasivam, R., Radhakrishnan, R., Hashem, A., and Abd_Allah, E. F. (2019). Microalgae metabolites: A rich source for food and medicine. *Saudi Journal of Biological Sciences,* 26, pp. 709–722. https://doi.org/10.1016/j.sjbs.2017.11.003.

Serena, E., Antonio, C., Francesco, G., Stefano, M., Marcello, M. D. (2017). A monitoring, modelling and decision support system (DSS) for a microalgae production plant based on internet of things structure. *Procedia Computer Science*, 113, pp. 519–524. https://doi.org/10.1016/j.procs.2017.08.316.

Simone, M., Xavier, G. D., Angela, P. T., and Sergio, R. (2021). Marine microalgae contribution to sustainable development. *Water,* 13, p. 1373. https://doi.org/10.3390/w13101373.

Skjånes K., Robours, C., and Lindblad, P. (2013). Potential for green microalgae to produce hydrogen, pharmaceuticals and other high value products in a combined process. *Critical Reviews in Biotechnology,* 33(2), pp. 172–215. https://doi.org/10.3109/07388551.2012.681625.

Sladojevic, S., Arsenovic, M., Anderla, A., Culibrk, D., and Stefanovic, D. (2016). Deep neural networks based recognition of plant diseases by leaf image classification. *Computational Intelligence and Neuroscience*. https://doi.org/10.1155/2016/3289801.

Srikanth, R. M., Fatimah, M. Y., Sanjoy, B., and Shariff, M. (2015). Microalgae as sustainable renewable energy feedstock for biofuel production. *Biomed Research International,* 519513. https://doi.org/10.1155/2015/519513.

Streefland, M., Martens, D. E., Beuvery, E. C., and Wiffels, R. H. (2013). Process analytical technology (PAT) tools for the cultivation step in biopharmaceutical production. *Engineering in Life Sciences*, 13, pp. 212–223. https://doi.org/10.1002/elsc.201200025.

Teng, S. Y., Yew, G. Y., Sukačovác, K., Show, P. L., Mášaa, V., and Chang, J. S. (2020). Microalgae with artificial intelligence: A digitalized perspective on genetics, systems and products. *Biotechnology Advances*, 44(107631), pp. 1–17. https://doi.org/10.1016/j.biotechadv.2020.107631.

Xing, Y., Zheng, Z., Sun, Y., and Alikhani, M. A. (2021). A review on machine learning application in biodiesel production studies. *International Journal of Chemical Engineering,* 2154258. https://doi.org/10/1155/2021/2154258.

Yu, X. T., and Zeng, T. (2018). Computational systems biology: Methods and protocols, methods in molecular biology; Chapter 7: Integrative analysis of omics big data. *Springer Nature*, 1754, pp. 109–135. https://doi.org/10.1007/978-1-4939-7717-8_7.

Zewdie, D. T., and Ali, A. Y. (2020). Cultivation of microalgae for biofuel production: Coupling with sugarcane-processing factories. *Energy, Sustainability and Society,* 10, pp. 27. https://doi.org/10.1186/s13705-020-00262-5.

7 Towards the Digitalization of Environmental Biotechnology

Viggy Tan Wee Gee[1], *Kuan Shiong Khoo*[2] *and Pau Loke Show*[1]

[1] Department of Chemical Engineering, Faculty of Science and Engineering, University of Nottingham Malaysia, Jalan Broga, Semenyih, Selangor Darul Ehsan, Malaysia

[2] Faculty of Applied Sciences, UCSI University, Cheras, Kuala Lumpur, Malaysia

CONTENTS

7.1 Introduction 228
7.2 Data Mining and Profiling of Microalgal Cell Dynamics 229
7.2.1 Connection and Data-to-Information Conversion Level 230
7.2.2 Big-Data Analysis 231
7.2.3 Cloud Computing 233
7.2.4 Imaging of Microalgal Cell Line Development with Advanced Microscopy 233
7.2.5 Monitoring System for Microalgal Culture Conditions in Bioreactor 234
7.3 Advanced Quantification and Classification of Microalgal Trajectories 235
7.3.1 Microfluidic Technique 235
7.3.2 Multiplexed Imaging Cytometry with Machine Learning 237
7.3.3 RGB Colour Analysis 238
7.3.4 Convolutional Neural Network (CNN) 238
7.3.5 Matrix-Assisted Laser Desorption/ionization (MALDI)-Based Spectrometer 238
7.4 Bioinformatics and Computational Tools for Microalgal Discovery 238
7.4.1 Whole Genome Sequencing 240
7.4.2 Gene Editing 241
7.4.3 Database 244

DOI: 10.1201/9781003202196-7

7.5 Statistical Tools for the Optimization of Microalgal Biofuel Production 246
7.5.1 Central Composite Design (CCD) 247
7.5.2 Artificial Neural Network (ANN) 247
7.5.3 Predictive Modelling 250
7.6 Conclusions 251
References 251

7.1 INTRODUCTION

Modelling of microalgal cell cultivation of biotechnological scrutiny remains to be an imperative subject manner. The software tools for microalgal biorefineries comprise simulation, system design, optimization, and modulation of bioprocesses (Eriksen 2008). Photo-physiological instruments are capable to assess the growth and biochemical traits of microalgae in the culture medium. There are various arithmetical function approaches that express microalgal cellular dynamics with ordinary differential equations (ODE), partial differential equation (PDE), and individual-based model (IBM) (Carreño-Zagarra et al. 2019).

The functions of differential calculations and individual-based modeling are integrated in a intersectional modeling structure to demonstrate and illustrate complicated biology eco-systems. Recent developments of digitalization model regarding the theoretical approaches of biosystem dynamics are gaining attention. The model has been distinguishingly accomplished for its cultivated cell variations modeling, and the primary attribute of the model is the phenomenon of growth inhibition as a response to the build-up of toxic components in the medium. Conceptual digitalized modelling mechanism indicates that the cumulation of certain toxic components is capable to elucidate the acclivity of species variety and spatial compartmentalization of greenery and ornamentation (Olaniyan and Adetunji 2021).

The shortcomings and inefficacy of conventional control and statistical technique in microalgal study in the prospects of determination of growth rate of microalgae, productivity, and extraction of biomass can be tackled by coupling of digitalization system. Moreover, the selectiveness in algal growth conditions is highly dependent on the concentration of nutrients, pH, interaction with other elements, ionic strength, fluid dynamics, and so on. Therefore, the study of microalgal biotechnology requires precise model predictiveness, metaheuristic evaluation, and computation modeling.

Microalgal biorefineries, culture conditions, and operating parameters require optimization of the organism and environment to support algal growth in terms of maximizing yield, especially algal biodiesel that could be economically contentious with conventional fossil fuels. The significance of genomic scale reconstruction designs and kinetic models is demonstrated in this chapter to optimize the metabolic pathway by comprehending the intricacy of algal morphology profile (Goatley and Bellwood 2011).

Various general implications of digital techniques in algal study such as data analysis methods, data collection methods, data description, algal ecosystem monitoring, with their respective advantages and disadvantages are presented. Genomic data retrieved from algal species can be utilized to determine and modify the native genetic elements and expressions and localization of particular sites of genome in microalgae for further approach of genetic engineering and organism transformation.

7.2 DATA MINING AND PROFILING OF MICROALGAL CELL DYNAMICS

Digitalization is a software-driven implementation that encompasses the facility of digital technology to retrieve digitized data, conjecture inclinations, and develop a more constructive business model. Interpretation of data acquired by web-connected instruments enhances decision-making process in the attainment of the most favourable outcome thus generating new revenue streams. The transformation of the reporting process from analogue to digital enables a real-time data collection which initiates a work order in the system and prompts users to mitigate flagging issues by tracking sources of error which comprise technical, machinery, and environmental in a more efficient and organized manner (Cavalcanti et al. 2022).

Microalgal study constitutes upstream, midstream, and downstream processing (Daneshvar et al. 2021). Upstream processing involves biomass productivity and cultivation technologies of microalgae. Midstream processing focuses on microalgal harvesting from culture media, collection of biomasses, and pretreatment of microalgal cell before extraction. Downstream processing comprises the extraction and refinement of value-added bioproducts from microalgal biomass (Liberato et al. 2019). The adaptation of digitalization into microalgal biotechnology elevates the efficiency of midstream processing, attains an energy balance and net positive cost from the bioproducts extraction in downstream processing, and enhances the quality and quantity of the generated biomass in upstream processing (Karemore and Sen 2016). These advances will maximize the exploitation of renewable resources and bioproducts which is an imperative approach towards the zero-waste purpose (Krüger et al. 2020).

It is crucial for users to refer primarily on data instead of their surmise during the decision-making process to avoid incorrect estimation. The necessity of big data analytics for manufacturing industry is conspicuous for creation of value for Industry 4.0 (Figure 7.1). The main objective would be to construct a system that is more inclined towards self-adaptation and effective automation. The system can be known as Cyber-Physical Production Systems (CPPS) that is capable to detect and adjust according to its environment. The system involves the collaboration of sensors and actuators in complex big data applications. The baseline of the adaptable and automated systems is attaining predictive functionalities by enhancing decision-making process from data analytics (Penas et al. 2017). The industry 4.0

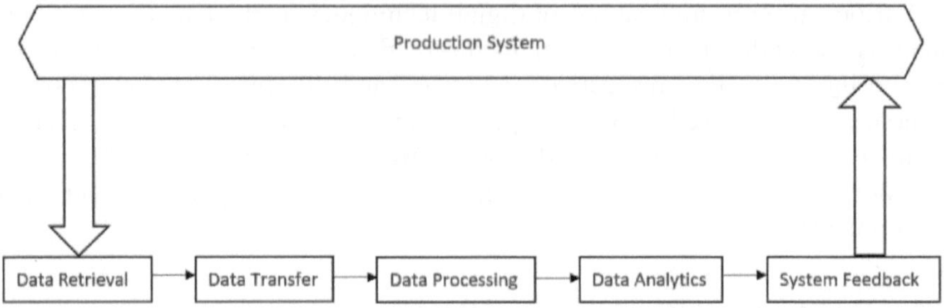

FIGURE 7.1 Data value chain.

index designs a 5C architecture system that is constituted of five segregated levels in order to achieve predictive functionalities. These levels include connection, conversion, cyber, cognition, and configuration (Åkerman et al. 2018).

7.2.1 Connection and Data-to-Information Conversion Level

Connection level denotes the connectivity of the real-time database (Åkerman et al. 2018). IoT paradigm is built on communicative and connected devices and software tools, information exchange, data processing, and data generation. Therefore, IoT can be utilized to monitor and control the cultivation of microalgae. The complex culture environment of microalgae is more prone to fluctuations of their growth parameters such as irradiation, nutrients concentration, pH level, fluid dynamics, and such. Failure to control the parameters critically will result in concession of biomass productivity. Hence, the forecast and control of biochemical dynamics require data collection and processing to avoid the undesirable outcomes (Villa-Henriksen et al. 2020). Decision support systems (DSSs) are data-driven models that incorporate methods and processes to extract important information from physical data. IoT utilizes the integration of differential equations and individual-based techniques to detect the biochemical characteristics of microalgae. The arithmetical model which is System Dynamics (SD) is a main tool to narrate the cultivation of microalgae (Hu et al. 2012).

The system involves the monitoring of microalgae culture medium by IoT sensors, delivering of signals generated from the input data to Central IoT System (CS), and involvement between IoT system and the consumers supported by DSS to enable optimizing-related decision-making in terms of improving microalgal growth rate based on model outputs provided (Rybkina and Reissell 2017). The IoT sensors have a monitoring tool, which collects and transmits information to the IoT Cloud. The IoT cloud distributes data to the modelling structure. The modelling system utilizes the retrieved data as input variable for predictive simulations and delivers output variable to the IoT cloud (Giannino et al. 2018). The Internet-connected IoT sensors that can be accessed from the IoT cloud are able to monitor and identify temperature, pH level, biomass concentration, and irradiance in the culture medium (Antony et al. 2020).

The arithmetical model of IoT system simulates the biomass productivity in culture medium by incorporating the equations of the photo-physiological processes such as photosynthesis, photo-inhibitors, photo-acclimatization, nutrient uptake, and respiration of microalgae (Nižetić et al. 2019). The intensity of irradiance provided to microalgal cells is estimated by the IoT light sensors that are controlled by the surface area of culture ponds and the photoinhibition effect resulted from overgrowing cell concentration which is detected by the IoT biomass sensors. The photosynthesis process is modulated by the ambient temperature detected by sensors and irradiation calculated as elaborated earlier (Anyakora et al. 2017).

The temperature-based empirical correspondence is denoted in Eqs. (7.1) and (7.2). The carbon-specified utmost photosynthesis rate P_{max} correlates the nutrient assimilations (Q) and ambient temperature (T) with a sigmoidal formula where the asymptotic state relates to high irradiance values [Eq. (7.1)] (Giannino et al. 2018).

$$P_{max} = P_{ref} \times \frac{Q - Q_{min}}{Q_{max} - Q_{min}} \times \frac{1}{\left(1 + e^{a-bT}\right)} \tag{7.1}$$

Moreover, respiration losses (resp) are dependent on the energy required for cell development and nutrient consumption (V_N) and varies exponentially with ambient temperature (T) Eq. (7.2) (Giannino et al. 2018).

$$resp = V_N \times N_{cost} \times \frac{e^{cT}}{e^{cT_{ref}}} \tag{7.2}$$

Hence, the integration of IoT and DSS is highly effective in analysing complex microbiology systems and optimizing microalgae cultivation with the automated prediction and control of negative growth culture medium (Lawal and Rafsanjani 2022)

7.2.2 Big-Data Analysis

The big data collected with the assistance of integrated omics paired with state-of-the-art technologies like confocal microscopy sets a great approach to the analysis of microalgal cellular dynamics (Raj and Raman 2017). The assay of phylogenomic and localization of bioactive metabolites is critical in determining the absent linkages in the biosynthesis of a targeted metabolite, which facilitates the identification process of any on/off transcription hubs thus enhancing the growth of microalgae (Viegas et al. 2016). Hence, a collateral method of establishing the genome-scale metabolic models (GEMs) and algorithm design with multi-omics and engineering techniques would improve the upregulation of microalgal metabolism (Rahmat et al. 2020).

There is an essential requirement to overcome the bottlenecks of processing huge data sets in the form of genomic snippets in the field of microalgal biotechnology. Big-data analysis is an arduous work for the biologists, with the involvement of varying techniques such as phylogenomics, subcellular localization, and so forth (Ashwaniy, Perumalsamy, and Pandian 2020). Sensitivity analysis is also critical as the nutrient consumption found in algal biomass substrates is subject to variations caused by the source of sewage or microalgal culture medium. Monte-Carlo simulation was executed in terms of determining the sensitivity of the microalgae to the various growth conditions of different system designs (Esakkimuthu et al. 2019).

The unpredictability of the process was estimated referring to the pedigree-matrix approach. The predictive simulation of each design in comparison to the reference scheme is estimated to assess the system biology of the respective culture mediums. The Monte-Carlo simulation outcome displays a confidence interval of 95% for further testing (Kumar et al. 2020). The test results that were based on the responsiveness to nutrient compositions establish a threshold limit of nutrient composition synopsis identified for respective substrate, with reference of the fluctuation scopes displayed in Table 7.1.

For the assessment of availability of substrate, the spatially segregated ecosystems and yearly biogas production of the targeted substrates were utilized in terms of progressing a graphical instrument to evaluate the ideal consolidated microalgal biomass generation. Thereafter, the standardized annual biogas potential (NABP) was derived from Eq. (7.3), where $UWWTP_{Capis}$ represents the UWWTPs dimensions in unit per capita loading (p.e.), P and C represent the pig stock per km^2 and mean livestock, respectively (Dayton and Foust 2019).

$$NABP = UWWTP_{Cap} \times 7.2 \frac{m^3}{p.e.} + C \times 398.6 \frac{m^3}{animal} + P \times 34.45 \frac{m^3}{animal} \tag{7.3}$$

TABLE 7.1
Minimum and Maximum Nutrient Concentration from Monte-Carlo Simulation

	Minimum nutrient concentration (g/kg digestate)			Maximum nutrient concentration (g/kg digestate)			
Wastewater	**N**	**P**	**K**	**N**	**P**	**K**	**References**
Biowaste	0.0168	0.0052	0.0164	0.0678	0.0240	0.073	(Kugler et al. 2019)
Sewage sludge	0.0000	0.0970	0.0260	0.9500	3.9440	0.5065	(Sommer, Karsten, and Glaser 2020)
Livestock manure	1.05	0.35	2.08	5.24	2.66	9.38	(X. H. Chen et al. 2007)
Pig manure	1.83	0.35	1.49	4.46	3.49	6.89	(Thiébaut 2006)

In order to determine the microalgae cultivation potential, geospatial analysis is recommended as it helps localizing an ideal terrain that has greater substrate availability for microalgal cultivation to decrease the environmental burden. A study conducted by (Bussa, Zollfrank, and Röder 2020) performed the geospatial evaluation in the Bavarian-Czech division. The geospatial analysis demonstrated a maximal amount of 0.21 million tons dry mass of microalgal biomass can be cultivated in the region involving all the accessible substrates (Bussa, Zollfrank, and Röder 2020; Sörenson et al. 2019). The subarea of UWWTP in the border of Cesky Krumlov possesses 2,300 p.e., equivalent to 16,560 m^3 biogas annually (Ozyurt 2004).

7.2.3 Cloud Computing

Harmful Algal Blooms (HABs) occur when toxic microalgae increase out of control and affect waterbodies with negative environmental and socioeconomic impacts such as red tides, marine life, and the resource of safe drinking water (Nwankwegu et al. 2019); (Davies and Ogidiaka 2019). The management of algal blooms is a challenge for governments, environmental organizations, and the society that depends on clean water bodies for their living (Rose et al. 2001); (Davidson et al. 2014).

Albeit the investment in waste management and monitoring tools, conventional systems and methods is still unfavourable (Sanseverino et al. 2016; Mishra et al. 2020). CyberHAB is a platform powered by cloud computing to process and analyze paramount volumes of ecological data for the management of harmful algal blooms (Dong et al. 2011). This cloud service can manage harmful algal blooms (HABs) by extracting information from data monitoring and converting variations and parameters into visualizations to improve decision-making (Tian and Huang 2019). The predictive models require calibration. The computing demands can be fulfilled by EOSChub. The CyberHAB model is a prototype involving the Jupyter Notebooks interface that integrates software components in an open manner and supplies direct access to cloud computing resources.

End-users can utilize CyberHAB via the Internet or mobile interfaces (Wang et al. 2014). Basic functionalities of raw data, models to forecast scenarios, or methods to generate additional data for a given case are provided (Kim, Jonoski, and Solomatine 2022).

7.2.4 Imaging of Microalgal Cell Line Development with Advanced Microscopy

Coelastrella vacuolata MACC-549 cells cultivated under sterile conditions in a tris-acetate-phosphate medium for 7 days were dispersed in an 8 microlitre volume on a silicon disc covered with 0.01 percent poly-L-lysine for electron microscopy (Barbano et al. 2015). The cells were then immobilized overnight at 4°C with 2.5 percent (v/v) glutaraldehyde and 0.05 M cacodylate buffer (pH 7.2 in PBS). The discs were then gently rinsed three times with PBS before being dehydrated

FIGURE 7.2 SEM image of C. vacuolata MACC-549 cells (Shetty et al. 2021).

using a graded ethanol series (30 percent, 50 percent, 70 percent, 80 percent, and 100 percent of ethyl alcohol, each left for overnight at 4 degrees Celsius) (Wirth et al. 2012). The samples were rinsed once more with 100 percent ethanol before being dehydrated using a critical point heater and then coated with a 12 nm gold coating before being examined under a field emission scanning electron microscope (Figure 7.2). For the characterization of C. vacuolata's morphology, the cell walls were smooth and ridge-free, and the cell surface morphology was remarkably similar to that of known C. vacuolata isolates. On C. vacuolata MACC-549 cells, no flagella of any type were seen. With two to eight autospore cells, asexual reproduction could be identified (Shetty et al. 2021).

7.2.5 Monitoring System for Microalgal Culture Conditions in Bioreactor

Exclusively closed PBRs may be built and adjusted to give the optimal growing conditions for individual microbe strains and therefore enhance biomass production efficiency (Cecchin et al. 2020). Furthermore, the amount of biomass used in the synthesis of certain added-value metabolites is not necessarily the most important factor. The capacity to alter and accurately regulate the operational conditions of the culture process (temperature, light intensity and wavelength range, and medium pH) is a far more important feature (Yuan et al. 2016). The most essential factors here are light and temperature, as they will be critical for microalgae development and biomass generation. They are deemed to be regarded jointly when optimizing, as they do in environment. Because it is impossible to provide ideal lighting for all of the grown cells, a trade-off must be made. This is due to the fact that the surface layer of cultures capture the majority of the light (López-Rosales et al. 2015). Light-emitting diodes (LEDs) are one of the most common luminaires. This is due to the fact that they are both effective and inexpensive (He, Subramanian, and Tang 2012).

Glass surface resistance sensors (STPt-100; hydrometer) were employed to detect the temperature in the incubation module, allowing operation in the range of temperature of 0–100°C. An RS-485 interface and the Modbus RTU network protocols were used to connect the sensors to the temperature sensors by AlgaeLabs (Husselmann and Hawick 2013). DS18B20 temperature sensors coated in stainless steel were used to take temperature readings within AMAPh-S's spawning chambers. An STR-5321-D (Ultima) temperature exchanger was utilized to connect with the sensors, allowing simultaneous collaboration with up to 16 DS18B20 sensors over a 1-Wire bus (Pires, Alvim-Ferraz, and Martins 2017). The findings were communicated to the master computer through the RS-485 interface and the Modbus RTU protocol via the STR-5321-D converter. Glass surface ERH- 13X2 sensors (Hydromet) were utilized to monitor pH in all cultivation chambers. The pH sensors in the inoculated modules worked in tandem with pH monitoring sensors with an RS-485 interface and the Modbus RTU protocol. LED panels were used to produce artificial lighting in the AMAPh-S growing chambers (Takouridis et al. 2015). Lumileds light-emitting diodes were used in all of the lighting panels, which were powered by 24 V DC. At 250 mol/m^2/s, the lighting system's output was adjusted to guarantee uniform illumination across the incubation chambers for each light hue (Borowiak et al. 2021) (Figure 7.3).

7.3 ADVANCED QUANTIFICATION AND CLASSIFICATION OF MICROALGAL TRAJECTORIES

The implementation of advanced methods for quantifying microalgal species will result in lower research costs and enhance microalgal bioproduct research in underdeveloped countries. Moreover, microalgal research also employs biomass measurement as a metric. Direct biomass weighing, flow cytometry, automated cell quantifying, molar concentration, and spectrophotometric assimilation are some of the contemporary microalgal biomass quantification techniques (Sunoj et al. 2021). Microalgae have a variety of carotenoids that assimilate and emit light in diverse ways and give the algae their distinct colour features and identities. Complexion and formation studies have recently been used to determine the amount of microalgal biomass productivity directly (Sarrafzadeh et al. 2015).

7.3.1 Microfluidic Technique

The microfluidic techniques are emerging to become one of the highly optimized methods to produce microalgae than the commercial one due to its lower production cost and high yield. Thus, the microfluidic techniques consist of various

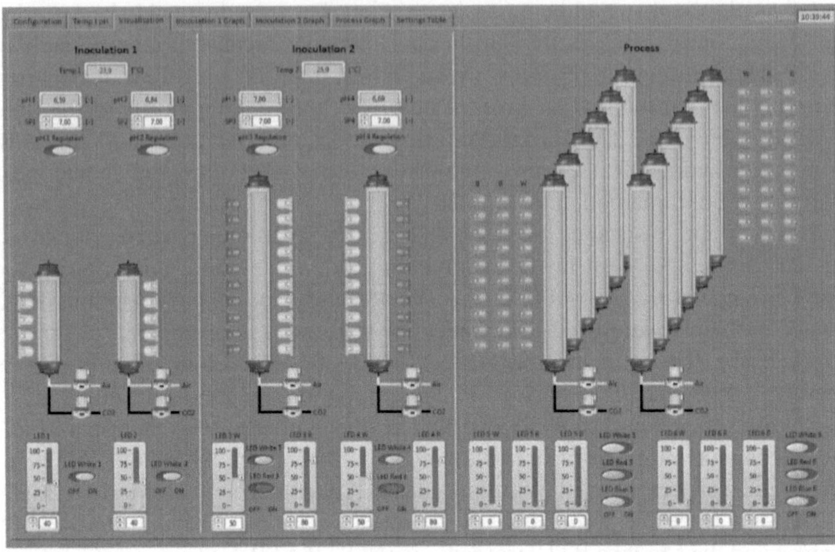

FIGURE 7.3 The AMAPh-S user interface in LabVIEW 2014 (Borowiak et al. 2021).

microbial applications like detection of pathogenic microorganisms (Foudeh et al. 2012), processes where prolific yeast strains are being screened while guided evolution is taking place (K. Chen, Crane, and Kaeberlein 2017), as well as the miniature microbial fuel cells (Bodénès et al. 2019).

Implementing the microfluidic techniques can potentially enhance the microalgal fuel and the biorefinery industry. In the microfluidic technique studies, the identification and selection of the microalgae strains are important to elevate the productivity of microalgae. Thus, the identification of microalgae is also crucial to establish a microscale bioreactor for the microalgae cultivation.

The microscale cultivation of microalgae is widely used due to its ability to optimize large-scaled operations which eases the screening and control of treatment conditions and microalgal strains. With the microfluidic techniques, the conditions of fluid, nutrient, and light content can be regulated to optimize the growth of microalgae. Apart from that, microfluidic studies can be used to determine the growth kinetics and enhance the production of lipids from the microalgae cell strains to increase the yield and productivity (Udayan et al. 2022).

Mechanical traps, droplets, and microchambers are three different types of microfluidic bioreactors. Mechanical traps would be used to examine single cells that are immobilized in a liquid that is constantly flowing. The microalgae cells injected into the microfluidic device can be captured by the traps. Nonetheless, when the held microalgae cells overflow during cell division or are washed out, cross-contamination may develop in between catches. Other than that, microchannel or microchamber is also used instead of the physical traps with the height lower than the cell diameter. Compressions are used to immobilize the cells during cultivation for more convenient cell monitoring. But it may impact the cell structure and metabolism due to the mechanical stress exerted from the compressions. Thus, it is hard to retain the cell if its morphology even though with slight influence.

Furthermore, microfluidic droplets were also utilized to surround single and multiplex colonies in their own habitat, simulating growth conditions. The continuous flow emulsion and electrowetting are two key methods for producing microfluidic droplets (Bodénès et al. 2019). In comparison to electrowetting-based droplets, the production of microfluidic droplets related to continuous flow dispersion is less common in microfabrication and surface treatment. Droplets can be produced indefinitely, making it difficult to identify particular cells. Aside from that, nutrients in the droplet could be eaten, therefore protracted lipid deposition studies may be limited.

Moreover, microchamber is also a downscaled photobioreactor. The culture scale of the microchambers is larger than the microfluidic devices. Thus, the microchamber can be used to analyse the biomass and the bulk culture conditions of the microalgae. Microchamber devices typically have a capacity of 40 to 400 microliter and are tailored to accommodate into a commercial plate analyser for direct inspection. Microcolumns linked in series autonomous microcolumns with singular entrance and one exit and microcolumns with numerous orifices for a variety of stress assessments are all examples of microchamber designs. (Kim et al. 2016).

In the microfluidic bioreactors, the growth rates of the microalgae cells are largely studied on through the cell count and optical density measurement. Deeper mechanical traps can produce a lower growth rate compared to the low-depth traps (Bae et al. 2013). Other than that, the single-cell cultivation in microdroplets can allow to achieve single-cell trapping-like growth rates (Kim et al. 2016). The microchambers can produce slower growth rates (Karimi 2017). In general, a single-cell monitoring with mechanical traps may be the most practical way to keep track of cells in ideal development circumstances. Hence, as compared to earlier gadgets, there would be no restrictions on nourishment or radiation. The single-cell trapping also allow more higher accuracy of monitoring on the various cells with different structure and content from the same strains of microalgae.

7.3.2 Multiplexed Imaging Cytometry with Machine Learning

Characterization and classification of microalgal cells are highly important to determine the feasibility of the microalgae in water treatment, biofuel application, and others. Employing frequency-division-multiplexed image cytometry and machine learning, precise identification and spherical microalgal populations are discovered (Foudeh et al. 2012). Monitoring the microalgae populations is important to determine the changes in ecosystems. Microalgal structural similarity and species diversity can make characterization more complicated, resulting in longer analysis times. By implementing the imaging flow cytometry, it is believed that the analysis time can be reduced. By collecting pictures with a singular pixel photodetector, the device substantially increases the performance of fluorescence optical sensing cell lines (Villa-Henriksen et al. 2020). The automation of the study and the numerous biological properties collected with the fluorescence optical sensing flow cytometer can solve the issue with comparable features that formerly restricted microalgal morphology categorization (Bradford et al. 2020).

7.3.3 RGB Colour Analysis

The concentration of the microalgae is studied by applying the red-green-blue (RGB) colour analysis that is utilized to analyse digital images (Coltelli et al. 2017). The colouration of microalgae, which was influenced by factors such as chlorophyll pigment content, provides a considerable promise for biomass estimation utilizing colour modelling approach. In the study, colour is represented as an effective indicator to identify the concentration of the suspension (Sarrafzadeh et al. 2015). The processing of macroscopic suspension colour may be utilized to build a straightforward and low-cost-microalgae-culture measuring technique.

7.3.4 Convolutional Neural Network (CNN)

The convolutional neural network is used to correctly categorize limited microalgae pictures with minimal bioinformatics (Narayanan et al. 2020). The deep learning models have been widely popular due to the fact that their visual characteristics from hierarchical data can be captured. CSS has long been a framework for categorizing visual information. These systems are feed-forward artificial neural networks that are bio-inspired on the visual cortex and have a layer-based design that takes a particular data volume and transforms it into a corresponding output sequence (Huang 2011).

7.3.5 Matrix-Assisted Laser Desorption/ionization (MALDI)-Based Spectrometer

The matrix-assisted laser desorption/ionization (MALDI) is used for study so that microalgae can be distinguished at the class and variant stages, and a basic microalgal combination may be characterized. Microalgae have different growth efficiency and can be affected by environment interference like temperature and nutrient supply (Andrade et al. 2015). The MALDI is highly effective in characterizing bacteria at the genus rapidly. The application is made possible by correlating mass spectrometry obtained from unidentified species' crude protein extraction with precedent spectra stored in a database (Barbano et al. 2015). Apart from that, the MALDI mass spectra methods have better taxonomic resolution compared to the conventional molecular techniques (Wirth et al. 2012).

7.4 BIOINFORMATICS AND COMPUTATIONAL TOOLS FOR MICROALGAL DISCOVERY

Light, phosphorus, nitrogen, carbon dioxide, and water are imperative parameters for microalgal cell growth. They are also the main source of energy for cell development through the processes of assimilation and conversion (Singh and Pandey 2018). Automated nutrient screening system optimizes algal growth conditions. Remote monitoring and control of algal culture which indicate temperature, pH, blending, radiation absorption, gas injection, and fluid dynamics are all factors to consider (Oey et al. 2013).

For system biology, 13C metabolic flux statistics and a following metabolic mapping produced from oleaginous algae can be used to identify targets involved in lipid metabolism. These flux statistics show that enzymes and pathways in regulation are rate controlling and have a major impact on overall metabolism (Jagadevan et al. 2018). Flux analysis provides monitoring the carbon flow during lipid formation and carbon fixation, and growth as a whole is an efficient way of predicting the future (Dal'Molin et al. 2011). Information on the enzyme flow control coefficients of Calvin cycle enzymes would be used to identify enzymes that could be aimed to improve development and carbon sequestration (Sirikhachornkit et al. 2018).

Microalgae have high storage of lipids, which can be processed to make biodiesel. The screening of emerging strains of accessible species is imperative for optimal output in commercialization of biodiesel production (Sharma et al. 2020). The outlook of biofuel production has entailed thorough multi-omics research in microalgae. Microalgae genomics and transcriptomics have generated informative and fundamental knowledge regarding lipid production. Proteomics and metabolomics are progressively being utilized to supplement algal omics, providing meticulous practical discernment into the associated static and dynamic physiological settings (Rai et al. 2021). The field of multi-omics has advanced from shotgun to targeted tactics. Various multi-omics methods and technologies can be utilized to delve further into microalgal physiology. As a result, it is intended to present different high-throughput biotechnologies and omics applications in microalgae, as well as integrating a compilation of cutting-edge literature. Emerging omics studies have discovered transcriptional, post-transcriptional, and translational pathways that regulate lipid accumulation in microalgae. The incorporation of data streams from multi-omics technologies can offer a thorough perspective on microalgal cell-line characterization (Costello and Martin 2018).

Transcriptomics focuses on the kind, shape, and the role of transcripts generated by a species under specific circumstances (X. Liu et al. 2021). Genomic data may be sufficient for design in prokaryotic cyanobacteria. Using new Bag2D-workflow codes, non-excessive transcripts after de novo synthesis have been discovered using high-throughput sequencing in microalgae models *Chlamydomonas reinhardtii*, *Coccomyxa subellipsoidea C-169*, *Ostreococcus lucimarinus*, *Volvox carteri*, and *Chlorella variabilis NC64A* (Gargouri et al. 2015). There are 481,381 contigs in the 181 Mb of contigs gathered, encompassing 10,185 genes. According to bioinformatics study, the route from inositol phosphate breakdown to lipid synthesis has been most strongly connected to enhanced fatty acid output inside this variant (Rismani-Yazdi et al. 2012).

An analysis of algal omics would be incomplete without proteomics, as it provides a notion of the exact functional group in both dynamic and static biochemical and physiological properties (Gargouri et al. 2015). There haven't been any specific or focused proteomics researches on microalgae yet. Proteomics has emerged to be a more effective tool for deciphering the complexities of biological systems. This may be extensively utilized to acquire algal oil and convert it to biodiesel via biochemical conversion (Anand et al. 2017). The fast advancement of biotechnology necessitated sophisticated gene modification as well as multi-omics research, resulting in the birth of different terminologies including synthetic biology, metabolic engineering, and systems biology. Synthetic biology aims to find the most common biopart in terms of creating a quick and innovative biological solution. The biopart of structural proteins has indeed been retrieved from a number of animals via genome sequencing and evaluation (Bogen et al. 2013).

7.4.1 Whole Genome Sequencing

Microalgae have a wide range of species due to their diverse biological pathway. The great diversity of lipids of various classifications and of unusual fatty acids observed in distinct algal strains, even within the same division, suggests varying lipid metabolism within microalgae. This can be exemplified by the contrast shown by the fatty acid composition of diverse microalgal species such as *Bacillariophyceae*, *Chlorophyceae*, *Cyanophyceae*, *Haptophyceae*, *Prymnesiophyceae*, and *Raphidophyceae* determined by mass spectrometry (Lang et al. 2011). Scores of systematic screenings, aimed at identifying microalgal species with high level of lipid content and been undertaken in recent times, stemmed from this information. With these methods for identifying interesting strains, the demand for comprehensive genome studies using generation sequencing, annotation, and reconstruction of lipid metabolic pathways is not trivial. These techniques provide further information about the phototrophic microalgal lipid metabolism (Bogen et al. 2013).

The generic process workflow of next generation sequencing is demonstrated in Figure 7.4. The process begins from the extraction of nucleic acid from the algal sample. The library preparation process produces a population of DNA fragments with specified lengths and oligomer sequences on both ends that are compatible with the sequencing method being used. Following that, the actual sequencing run on the relevant system occurs. In order to acquire the necessary information, the sequencing data is processed via a bioinformatics analysis pipeline (Hess et al. 2020).

As a prerequisite for metabolic network reconstruction, characterization and annotation of a draft genome are being performed on the organisms. Next-generation sequencing techniques were used to analyse the genome of *M. neglectum* in order to comprehend its metabolic processes and establish the groundwork for a thorough genetic study (Yao et al. 2015). The Illumina MiSeq technique was used to generate the genome sequence, which resulted in pair-end reads of 2 × 250 bases in length. Nearly 8 million pair-end reads were constructed, yielding around 6,700 scaffolds, with 857 of them exceeding 20 kb in length (Bogen et al. 2013) (Table 7.2).

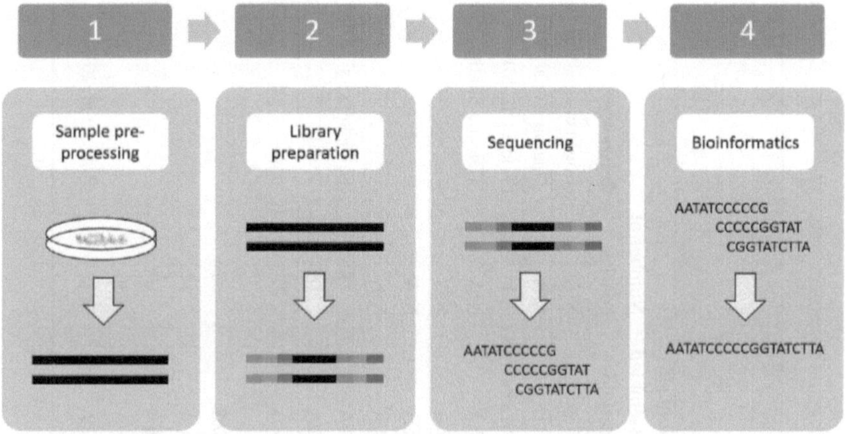

FIGURE 7.4 Next generation sequencing process in general.

7.4.2 Gene Editing

Metabolic engineering seeks to build a stable host for high-level synthesis of a value-added chemical by modifying the genome and designing genetic circuits to readdress carbon flux, as well as including proteomics and metabolomics in some cases (Reijnders et al. 2014). Creating a thorough map of biosynthetic pathways for rational metabolic engineering is one of the key challenges in algae research (Gimpel, Henríquez, and Mayfield 2015). To do this, multi-omics data was used to build genomic scale metabolic models (GSMs), which track metabolic processes linked to the genomes and enzymes involved (Tibocha-Bonilla et al. 2018). Enhanced gene sequence of particular endogenic genes, emphasis of heterological genes that could strengthen enzymatic modifications, resulting in greater production of secondary metabolites; modification of promoters, resulting in increased gene transcription in metabolite biosynthetic pathways; deactivation, down legislation, or suppressing of genes that encode enzymes are all examples of re-engineering (Kang et al. 2017). Figure 7.5 depicts algal cell genetic engineering for enhanced metabolite synthesis.

Molecular tools *for C. reinhardtii*, *P. tricornutum*, *T. pseudonana*, and *Nannochloropsis* species are becoming increasingly advanced (Fu et al. 2016). These comprise a multitude of vectors, promoters, and targeted sequences for transgene expression, as well as TALEN and CRISPR-Cas9-based genome editing. In order to discover new regulatory sequences, genome sequence and transcriptome data are progressively being analysed, and this technique also offers data that may be used to regulate other algae (Shin et al. 2016). Simultaneously, synthetic biology techniques based on engineering design concepts are being progressively implemented on algae. Various permutations may be evaluated for the optimum gene transcription of concern using the Design–Build–Test–Learn

TABLE 7.2
Genome Assembly Statistics for Nuclear Draft Genome of Various Microalgal Species

Species	Sequencing method	Genome size (Mb)	Genomic G+C content (%)	Genome Coverage	Aligned reads	Number of assembled scaffolds	Number of contigs	References
M. neglectum	Illumina MiSeq	68	64.74	49.30	16194053	6739	N/A	(Bogen et al. 2013)
C. vulgaris UTEX 395	Illumina HiSeq 2000	37.34	61.5	N/A	165874962	113	N/A	(Guarnieri et al. 2018)
Tetraselmis striata	PacBio	300	57.9	55.00	227954216	3613 contigs	3613	(Tyler et al. 2019)
P. piriformis	Illumina RNA	N/A	N/A	55.50	69700000	N/A	N/A	(Tian and Smith 2016)
P. parva		N/A	N/A	67.60	52900000	N/A	N/A	
P. capuana		N/A	N/A	99.20	93100000	N/A	N/A	
P. magna		N/A	N/A	95.10	260000000	N/A	N/A	
Picochlorum NBRC 102739	Illumina	22.76	44.3	N/A	N/A	N/A	728	(Foflonker et al. 2018)
P. oculata	Illumina	14.52	44.3	N/A	N/A	N/A	218	
P. oklahomensis primary	Pacbio	13.36	46.1	N/A	N/A	N/A	13	
P. oklahomensis haplotig	Pacbio	12.05	46.1	N/A	N/A	N/A	80	
Picochlorum SE3 v2 primary	Pacbio	13.47	46.2	N/A	N/A	N/A	14	
Picochlorum SE3 v2 haplotig	Pacbio	0.46	46.8	N/A	N/A	N/A	14	
P. soleocismus DOE 101 v1 (9)	N/A	15.14	44.4	N/A	N/A	N/A	36	
C. subellipsoidea C-169	Sanger	48.8	53	N/A	28322	29	N/A	(Blanc et al. 2012)

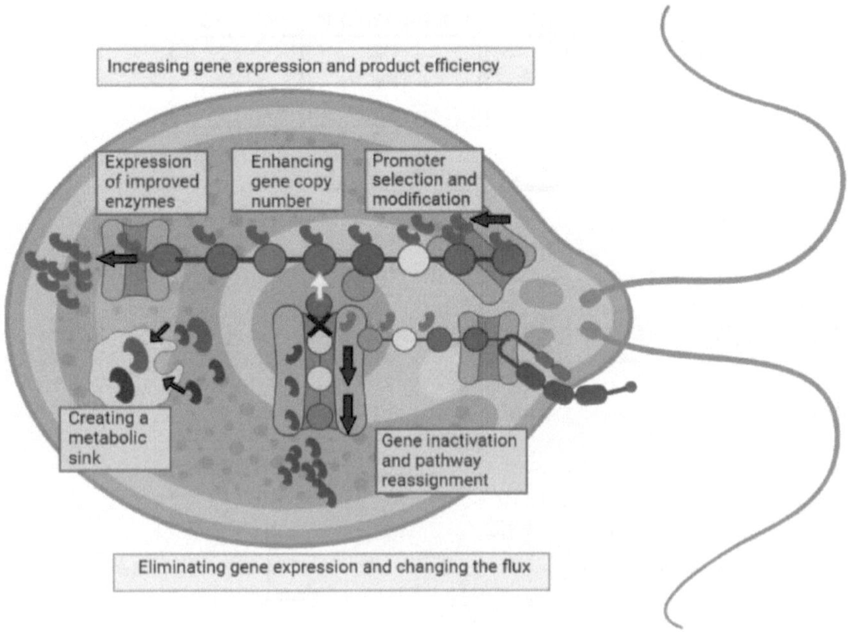

FIGURE 7.5 Steps of a metabolic pathway of algae call genetic engineering.

cycle (Figure 7.6) utilizing specified pieces like promoters and terminators with compatible behaviour and a standardized manner to assemble them (Brodie et al. 2017).

Recombinant nucleases are used in genome editing to detect and cleave specific regions in the genome, culminating in double-strand breaks (DSBs). Non-homologous end joining (NHEJ), a homology-independent and error-prone DNA repair process, is used to repair DSBs, leading to alterations at the cleavage site (Jeon et al. 2017).

Cas9 has endonuclease sequences that are well-structured which could be amended to make nickase and perhaps as well as dCas9 mutants, which may be utilized for novel and interesting approaches. dCas9, in example, may attach to a target site without cleaving DNA, which might disrupt biological functions such as transcription (Poliner et al. 2018). By fusing domains engaged in transcriptional activation (CRISPRa), repression (CRISPRi), and epigenetic structure, dCas9 can be reprogrammed for various activities (Nymark et al. 2016). It is worth noting that CRISPRi is utilized for a variety of purposes. By fusing CRISPRa and CRISPRi which are modules engaged in transcriptional activation, repression, and agamogenetic control, dCas9 can be repurposed for various activities (Chuai et al. 2018). It's worth noting that CRISPRi is utilized for both basic intervention with no linked proteins and effective interference with repressor domains, both of which might be handled in the future (Baek et al. 2016). In

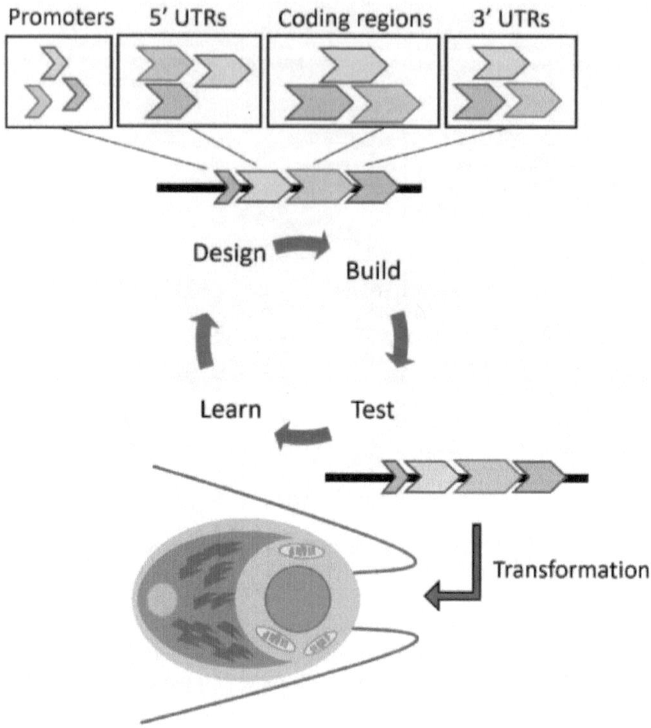

FIGURE 7.6 Design–Build–Test–Learn cycle.

eukaryotic cell systems, the numerous repetitions of the hepatitis simplex VP16 activation domain (VP64) and the nuclear factor-B activating component activation domain (p65AD) have been widely utilized as CRISPRa activator domains (Costello and Martin 2018).

7.4.3 Database

Several gene co-expression databases are collected through model organisms such as humans, mice, Arabidopsis, and rice. The functional classification of flora and fauna genes has been made easier in these databases by offering smart interface design and services that deliver sustained and reinforced information through constant renewal, validity analysis, and certification of data and offering extensive data elucidation, such as interspecies analogies and network analyses (Andrade et al. 2015). In contrary, there has never been a genetic co-expression data source of microalgae, which meets these functional criteria. Albeit the Algae Path database has gene co-expression data for a portion of green algae; it focuses on expression outline analysis of single genomes and does not include complete co-expression data. To address this shortcoming, comprehensive gene co-expression databases for microalgal species are created (Aoki et al. 2016). Numerous contemporary algae genera have official diagnoses in AlgaeBase. AlgaeBase also contains a thorough Glossary with 28,160 words already defined, each with a detailed breakdown and, where applicable, extensive comments on the origin of each phrase. Details on the number of species and other taxonomies in AlgaeBase, as well as forecasts of the quantity of yet-to-be-included known taxonomies and unidentifiable species are known; currently, 40,000 algal species have been included (Guiry et al. 2014).

To deepen our understanding of given algal capabilities and to offer a foundation for biotechnological processes, breakthroughs in algal annotations will be compulsory to join forces alongside system modelling of metabolism and regulatory networks (Thiriet-Rupert et al. 2018). Figure 7.7 depicts the relationship between systemic biology and bioinformatics modelling phases. Novel microalgal genomic, transcriptomic, and proteomic data are gathered (step 1), enabling genes and proteins to be identified (step 2). Following the use of first-generation greater cognitive identification (step 3), a refinement phase using second-generation

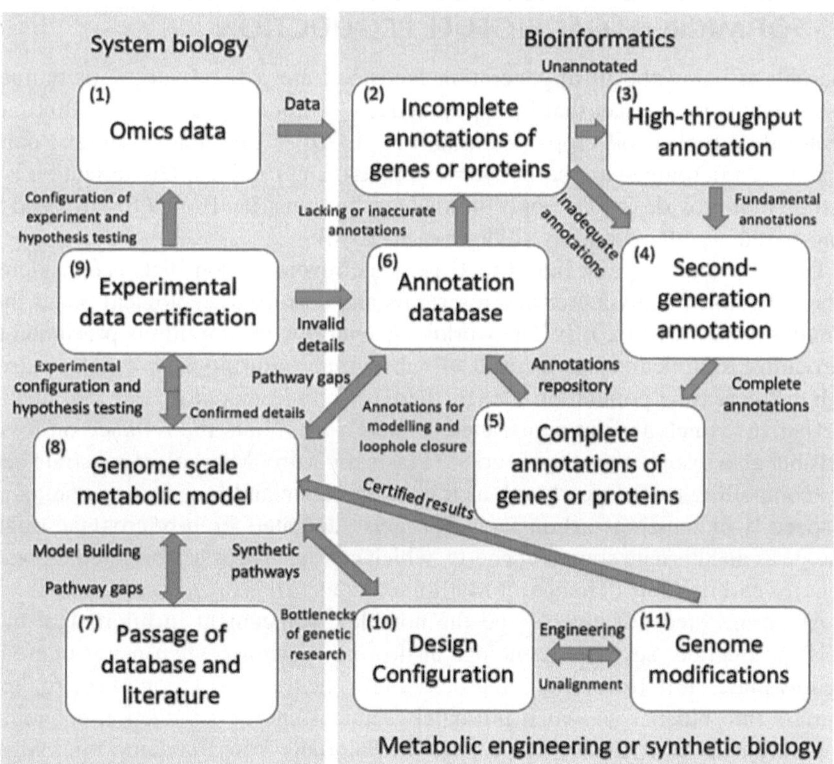

FIGURE 7.7 Interconnection between system biology, bioinformatics, and metabolic engineering.

functional annotation methodologies (step 4) is completed. In bioinformatics, gene and protein annotation is a gradual process that continues until they are deemed acceptable (step 5). Systems biology modelling uses these descriptors (step 6) and information from online databank as well as research (step 7) to reverse-engineer a genome-scale metabolic model GSMM (step 8) to investigate metabolism interconnections in a range of circumstances. The metabolic modelling that is validated experimentally (step 9) is prompted to be undertaken after achieving a GSMM to confirm predictive performance or identify errors and research gaps (Loira et al. 2017).

Supplemental omics data or annotation modifications may be necessary as a result of these findings. Due to the scarcity of experimentally verified algal proteomics, the feedback loop derived from algal modelling to gene/protein function prognostication is critical for increasing the base of knowledge, which will eventually underlie effective algae genomic engineering for commercial product manufacture (Leow et al. 2015). Once an algae GSMM has been created, it should be publicly released in a repository and literature.

7.5 STATISTICAL TOOLS FOR THE OPTIMIZATION OF MICROALGAL BIOFUEL PRODUCTION

Microalgae being the third-generation feedstock are a good supply of biofuel. This stems from the fact that lipid productivity of microalgae is 15 to 300 times greater than typical oil crops (Zullaikah et al. 2019). The main chemical components of microalgae are carbohydrates, lipids, and proteins. Digitalization is a method that provides broad opportunities for the transition from a fossil-based to a bio-based circular economy (Krüger et al. 2020).

The continual usage of liquid fuels generated from carbon fuels is considered to be untenable due to decreasing reservoirs and correlated ecological issues that remain severe (S. Liu 2010). The worldwide energy consumption is presumed to aggrandize significantly as a result of substantial anthropogenic events, paired with the increasing population. Hence, there is a requirement for more sustainable alternatives which are not constricted to merely eliminate the reliance on fossil fuels but also resolve the climate crisis (Ptasinski 2016). Microalgal substrate acts as a compelling biofuel feedstock as it is regenerative and long-lasting compared to speedily exhausting carbon fuels. Biofuels generated from microalgae might emerge as an alternate source of energy which can survive as a protracted replacement for carbon fuels (Thangavel Mathimani et al. 2017).

Albeit assorted advantages and the notable advancement in microalgal biofuels' production, several technological downfalls persist when using microalgae raw material in the extraction processes for the transformation of microalgae biomass into bioenergy, which ultimately reduces energy production economic growth (Sarkar et al. 2020). The recovery of biofuels from the damp microalgae biomass has posed to be a pragmatic problem for the microalgal bioenergy sector (Ozturk et al. 2017). Aside from cultivation, the downstream processes account for nearly 70% of the sum of biodiesel production cost. In general, adjourn microalgal cultivation entails substantial bottlenecks of biomass extraction or drying that can contribute to approximately 20–40% of the total production cost (Abomohra et al. 2020).

7.5.1 Central Composite Design (CCD)

Central composite design (CDD) is pivoted on the modeling and improvement of microalgal growth, and bio-fixation rates of microalgae regarding the linkage between temperature of photobioreactor, carbon dioxide concentration, nutrients' assimilation, gaseous flow rate, commencing incubation concentration, and the degree of irradiance in terms of optimizing microalgal growth (Wang et al. 2014). The energy ratio (ER) is identified for individual experiment and evaluated for the option of ideal functioning culture medium. The process optimization system was utilized according to CDD utilizing ANOVA statistical method.

Prism graph pad program is employed to initiate the central composite investigational design statistical approach (CCD-SA) (Ido et al. 2018). The CCD-SA is implemented with the identification of central outcome of respective shifting of process-inclined variable. Several culture conditions comprising TPBR, Itot, CCO2, Q gas, NUreq (TIN, TP and TC), and INden have been collectively enlisted in the computations (Raheem et al. 2015). The integration of the parameters has a repercussion of 28 testings, each repeated three times, and was used in calculations (Derakhshandeh and Tezcan Un 2019). A linear (L), two-factor interaction (2FI), quadratic (QU), and cubic four-level factorial design (CFLF4k) were assessed (T. Mathimani, Beena Nair, and Ranjith kumar 2016).

The outcomes of each of the culture parameters evaluated in the middle level (0) at the central point in the between levels of the lowest (–1) and the highest (+1) of the standardized readings are incorporated into computations (Kiran et al. 2016). The acquired cultivation data have been suited to the replica, and the magnitude of the replica has been assessed utilizing ANOVA computational tool to determine the relation between the growth conditions and growth rate as well as carbon dioxide biofixation (Ellison, Overa, and Boldor 2019).

For all experiments, it was obvious that growth rate attained plateau or static phase after 20 to 28 days (Thangavel Mathimani, Uma, and Prabaharan 2018). The reported values of growth rate and carbon dioxide bio-fixation were thoroughly contingent on the functioning parameters with a prevailing tendency displaying an enhancement of growth rate and carbon dioxide bio-fixation values by increasing the initial carbon dioxide concentration in the gaseous streamlet as great as 20% (v/v), and a subsequent reduction in growth rate as well as carbon dioxide bio-fixation values for Chlorella vulgaris grown at higher carbon dioxide concentration (Dragone et al. 2011; Thangavel Mathimani et al. 2021; Almomani 2020).

7.5.2 Artificial Neural Network (ANN)

Artificial neural network (ANN) has been renowned as a calculation technique to demonstrate non-linear correlations and can be utilized for predictive modeling linked to microalgae bioproducts and biomass productivity (Sewsynker-Sukai, Faloye, and Kana 2016). Table 7.3 demonstrates several ANN models that have been studied to estimate the storage of bio-active components and the biomass productivity of microalgae. For example, Richmond and Hu (2013) developed an ANN to identify the fluctuations in the rate of lutein amassing in a culture of *Desmodesmus* sp. On the contrary, Muthuraj et al. (2014) employed an ANN operated with feed forward practice in order to enhance the growth media for optimizing the generation of triglycerides in *Chlorella vulgaris* where Li et al. (2013) matured an ANN model to estimate the biohydrogen productivity through the aerobically assimilated biomass of *Chlorella vulgaris*.

Additionally, ANN has also been employed to foretell about the culture of microalgae involving *Spirulina platensis* which has garnered interest acting as a feedstock of food (Ravishankar and Ambati 2019). In the experiment conducted by (Susanna et al. 2019), the biomass productivity of *Sprirulina platensis* in an outdoor environment was prognosticated utilizing an ANN model, whereas Liyanaarachchi et al. (2020) developed an ANN to enhance the production of the algal species in photobioreactors. Suresh, Srivastava, and Mishra (2009) utilized ANN to replicate the various nutrient reciprocations in the culture medium of *Protoceratium reticulatum* (López-Rosales et al. 2015) and *Karlodinium veneficum* (Yun et al. 2016).

In contrary to the research aiming on an individual microalgal species, Reynolds, Davey, and Aldridge (2019) developed ANN to calculate the growth rate of polyculture microalgae in an outdoor raceway medium having the intent of manufacturing microalgal biomass as a compromising bioenergy feedstock. Table 7.3 demonstrates the results of denoted models that are restricted as a response to the diminishing impact of micronutrients. Hence, in order to make an ANN model prosper with more efficient predictive functionality, the aspects of ambient conditions and impact of micronutrients and macronutrients are important to be included in experimental parameter settings.

The ANN is utilized to fit empirical patterns in the observed data set, whereas the software program of Diesel-RK offers numerical solutions at defined operational parameters at a specific computing cost. As a result, the network is able to anticipate behavior over the whole spectrum of operations with which it was programmed. The corresponding combustion, efficiency, and emission characteristics were predicted utilizing the structure in accordance to the process variables of load, blending, and fuel injection pressure.

The engine was simulated for 80 various operating circumstances, including four distinct loading rates, five separate blending ratios, and five distinct fuel injection pressures. Seventy percent (56 cases) of the 80 circumstances were utilized to upskill the matrix that adjusts the web weights as well as biases based on the projected output error. Furthermore, 15% (12 cases) were utilized to validate for networking generalization measurement and to halt development when generalization stopped expanding. Remaining 15% (12 cases) seemed to have no influence on training and have been used to assess the network's functionality after programming. The candidates were divided into three groups with the help of the MATLAB function divider.

TABLE 7.3
Models Utilizing ANN to Estimate Biomass Productivity, Cell Growth and Quality of Microalgae Bioproducts

Approach	Neural network design	Manipulated variables	Measured variables	References
Growth simulation of microalga *Chlorella vulgaris*	Multilayer perceptron	• Macronutrients. • Micronutrients. • Environmental conditions.	Cell concentration in timely manner.	(Liyanaarachchi et al. 2020)
Growth monitoring of *Spirulina platensis* in PBR.	Multilayer perceptron	• Macronutrients. • Environmental conditions.	Productivity of biomass daily.	(Susanna et al. 2019)
Growth modelling of polyculture microalgae.	Multilayer perceptron	• Macronutrients. • Environmental conditions.	Biomass productivity.	(Supriyanto et al. 2019)
Illustration of lutein photo-production dynamic process of *Desmodesmus* sp.	Multilayer perceptron	• Macronutrients. • Environmental conditions.	Concentration of nutrients, biomass, and lutein production	(Rio-Chanona et al. 2017)
Estimation of chlorophyll variations.	Multilayer perceptron	• Environmental conditions.	Chlorophyll-*a*	(Ferreira and Sant'Anna 2016)
Culture pond enhancement to optimize lipid and biomass productivity of *Chlorella vulgaris*.	Multilayer perceptron	• Macronutrients.	Biomass and lipid productivity.	(Liyanaarachchi et al. 2020)
Forecasting of the hydrogen production through anaerobic digestion of *Chlorella vulgaris*.	Multilayer perceptron	• Macronutrients. • Environmental conditions.	Concentration of hydrogen	(Maulana, Nur, and Hadiyanto 2015)
Representation of various nutrient interactivities to the cultivation of *Protoceratium reticulatum*	Feed-forward back-propagation	• Macronutrients. • Micronutrients.	Cell growth rate	(López-Rosales et al. 2013)
Estimating the growth rate of *Karlodinium veneficum*.	Feed-forward back-propagation	• Macronutrients. • Micronutrients.	Cell concentration	(García-Camacho et al. 2016)
Duplication of phytoplankton variations based on the fluctuations of substrate quality.	Multilayer perceptron	• Macronutrients. • Environmental conditions.	Chlorophyll-a	(Ghafar Channar et al. 2014)
Evaluating the biomass production of *Sprirulina platensis*.	Multilayer perceptron	• Macronutrients. • Environmental conditions.	Chlorophyll-a	(Sharon Mano Pappu, Vijayakumar, and Ramamurthy 2013)
Progression of a pre-alerting system for harmful algal blooms	Feed-forward back-propagation	• Macronutrients. • Environmental conditions.	Chlorophyll-a	(Wei, Sugiura, and Maekawa 2001)
Demonstration of phytoplankton bloom	Multilayer perceptron	• Macronutrients. • Environmental conditions.	Chlorophyll-a	

7.5.3 Predictive Modelling

Figure 7.8 highlights the research technique used in a case study which is a predictive modelling that characterizes microalgal biodiesel behaviour (Salam and Verma 2019). The simulation was performed in the program Diesel-RK after the process parameters were determined using design of experiments and/or domain knowledge. Prior to actually supplying the dataset to the ANN, the different output parameters of the engine operation were standardized. This included utilizing Eq. (7.4) to normalize both input and output data on a variable-by-variable basis, where X_n is the value obtained through normalization of process parameters of value X_i, while X_{min} and X_{max} are minimum and maximum values of the variable in the dataset correspondingly, and the maximum and minimum limits of the normalized values are a and b, respectively (Salam and Verma 2019).

$$X_n = (b-a)\frac{X_i - X_{min}}{X_{imax} - X_{min}} + a \tag{7.4}$$

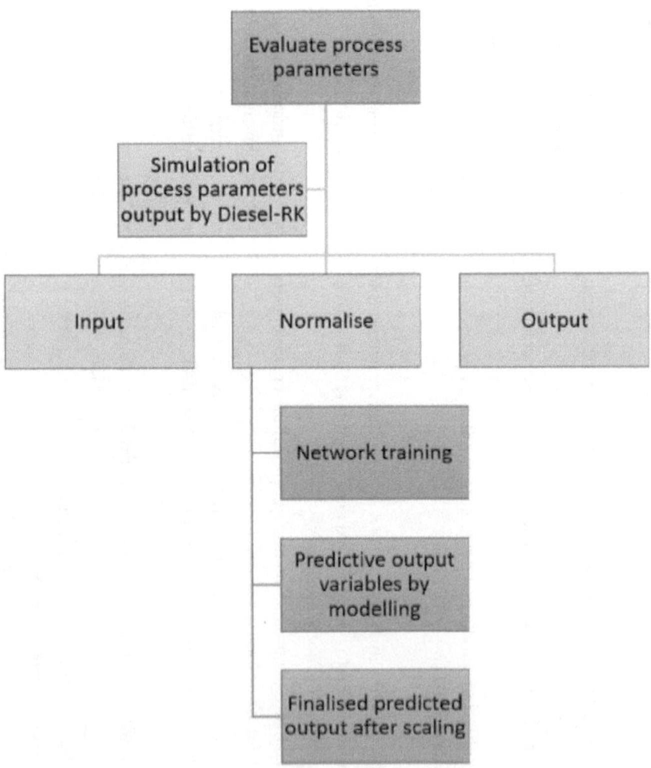

FIGURE 7.8 Predictive behavior of process output flowchart.

Following that, the system is trained using the program settings. The network's estimated results are then scaled in the direction opposite to the standardization performed before with Eq. (7.5) (Salam and Verma 2019).

$$X_i = X_{min} + \frac{(X_{max} - X_{min}) - (X_n - a)}{(b - a)} \tag{7.5}$$

7.6 CONCLUSIONS

The algal omics requires further study in the future to constitute technological, economic, and societal consequences in the post-genomics age. The generation of biofuels from oil-rich algae is an essential and promising research topic for the environment. High-throughput multi-omics techniques have made significant contributions to microalgal genetic transformation, including increased lipid and pigment synthesis. The importance of proteomic and metabolomic technologies in the research of metabolic processes is highlighted by the fact that eukaryotic microalgae have metabolic functions distinct from other species and frequently assert protein variants with distinct activities. The difficulties lie in the creation of particular microalgal databank, data amalgamation technologies, as well as the advanced comprehensive genetic engineering instruments which could allow focused metabolic pathway construction to fully exploit the data generated by those emerging novel methods. In addition, research emphasis needs to be placed on integrating microalgae datasets while considering the multi-omics pathway as well as commercialization concerns. This paves the way for microalgal biotechnologies and commodities to have a more digitally enhanced and sustainable development.

REFERENCES

Abomohra, Abd El Fatah, Mahdy Elsayed, Sivakumar Esakkimuthu, Mostafa El-Sheekh, and Dieter Hanelt. 2020. "Potential of Fat, Oil and Grease (FOG) for Biodiesel Production: A Critical Review on the Recent Progress and Future Perspectives." *Progress in Energy and Combustion Science* 81: 100868. doi:10.1016/J.PECS.2020.100868.

Åkerman, Magnus, Camilla Lundgren, Maja Bärring, Mats Folkesson, Viktor Berggren, Johan Stahre, Ulrika Engström, and Martin Friis. 2018. "Challenges Building a Data Value Chain to Enable Data-Driven Decisions: A Predictive Maintenance Case in 5G-Enabled Manufacturing." *Procedia Manufacturing* 17: 411–18. doi:10.1016/J.PROMFG.2018.10.064.

Almomani, Fares. 2020. "Kinetic Modeling of Microalgae Growth and CO2 Bio-Fixation Using Central Composite Design Statistical Approach." *Science of The Total Environment* 720: 137594. doi:10.1016/J.SCITOTENV.2020.137594.

Anand, Vishal, Puneet Kumar Singh, Chiranjib Banerjee, and Pratyoosh Shukla. 2017. "Proteomic Approaches in Microalgae: Perspectives and Applications." *3 Biotech* 7: 197. doi:10.1007/S13205-017-0831-5.

Andrade, L. M., M. A. Mendes, P. Kowalski, and C. A. O. Nascimento. 2015. "Comparative Study of Different Matrix/Solvent Systems for the Analysis of Crude Lyophilized Microalgal Preparations Using Matrix-Assisted Laser Desorption/Ionization Time-of-Flight Mass Spectrometry." *Rapid Communications in Mass Spectrometry* 29 (3): 295–303. doi:10.1002/RCM.7110.

Antony, Anish Paul, Kendra Leith, Craig Jolley, Jennifer Lu, and Daniel J. Sweeney. 2020. "A Review of Practice and Implementation of the Internet of Things (IoT) for Smallholder Agriculture." *Sustainability* 12 (9). doi:10.3390/SU12093750.

Anyakora, Anthony N., Oladiran K. Abubakre, Edeki Mudiare, and M. A. T. Suleiman. 2017. "Effect of Fibre Loading and Treatment on Porosity and Water Absorption Correlated with Tensile Behaviour of Oil Palm Empty Fruit Bunch Fibre Reinforced Composites." *Advances in Materials Research* 6 (4): 329–41. doi:10.12989/amr.2017.6.4.329.

Aoki, Y., Y. Okamura, H. Ohta, K. Kinoshita, and T. Obayashi. 2016. "ALCOdb: Gene Coexpression Database for Microalgae." *Plant & Cell Physiology* 57 (1). Plant Cell Physiol: e3. doi:10.1093/PCP/PCV190.

Ashwaniy, V. R. V., Muthiah Perumalsamy, and Sivakumar Pandian. 2020. "Enhancing the Synergistic Interaction of Microalgae and Bacteria for the Reduction of Organic Compounds in Petroleum Refinery Effluent." *Environmental Technology & Innovation* 19: 100926. doi:10.1016/J.ETI.2020.100926.

Bae, S., C. W. Kim, J. S. Choi, J. W. Yang, and T. S. Seo. 2013. "An Integrated Microfluidic Device for the High-Throughput Screening of Microalgal Cell Culture Conditions That Induce High Growth Rate and Lipid Content." *Analytical and Bioanalytical Chemistry* 405 (29). Anal Bioanal Chem: 9365–74. doi:10.1007/S00216-013-7389-9.

Baek, Kwangryul, Duk Hyoung Kim, Jooyeon Jeong, Sang Jun Sim, Anastasios Melis, Jin-Soo Kim, EonSeon Jin, and Sangsu Bae. 2016. "DNA-Free Two-Gene Knockout in Chlamydomonas Reinhardtii via CRISPR-Cas9 Ribonucleoproteins." *Scientific Reports* 6 (1): 1–7. doi:10.1038/srep30620.

Barbano, Duane, Regina Diaz, Lin Zhang, Todd Sandrin, Henri Gerken, and Thomas Dempster. 2015. "Rapid Characterization of Microalgae and Microalgae Mixtures Using Matrix-Assisted Laser Desorption Ionization Time-Of-Flight Mass Spectrometry (MALDI-TOF MS)." *PLoS One* 10 (8). doi:10.1371/JOURNAL.PONE.0135337.

Blanc, G., I. Agarkova, J. Grimwood, A. Kuo, A. Brueggeman, D. D. Dunigan, J. Gurnon, et al. 2012. "The Genome of the Polar Eukaryotic Microalga Coccomyxa Subellipsoidea Reveals Traits of Cold Adaptation." *Genome Biology* 13 (5). doi:10.1186/GB-2012-13-5-R39.

Bodénès, Pierre, Hsiang Yu Wang, Tsung Hua Lee, Hung Yu Chen, and Chun Yen Wang. 2019. "Microfluidic Techniques for Enhancing Biofuel and Biorefinery Industry Based on Microalgae." *Biotechnology for Biofuels* 12 (1). doi:10.1186/S13068-019-1369-Z.

Bogen, C., A. Al-Dilaimi, A. Albersmeier, J. Wichmann, M. Grundmann, O. Rupp, K. J. Lauersen, et al. 2013. "Reconstruction of the Lipid Metabolism for the Microalga Monoraphidium Neglectum from Its Genome Sequence Reveals Characteristics Suitable for Biofuel Production." *BMC Genomics* 14 (1). doi:10.1186/1471-2164-14-926.

Borowiak, Daniel, Paweł Lenartowicz, Michał Grzebyk, Maciej Wiśniewski, Jacek Lipok, and Paweł Kafarski. 2021. "Novel, Automated, Semi-Industrial Modular Photobioreactor System for Cultivation of Demanding Microalgae That Produce Fine Chemicals—The next Story of H. Pluvialis and Astaxanthin." *Algal Research* 53: 102151. https://doi.org/10.1016/j.algal.2020.102151.

Bradford, Eric, Lars Imsland, Dongda Zhang, and Ehecatl Antonio del Rio Chanona. 2020. "Stochastic Data-Driven Model Predictive Control Using Gaussian Processes." *Computers & Chemical Engineering* 139: 106844. doi:10.1016/J.COMPCHEMENG.2020.106844.

Brodie, J., C. X. Chan, O. De Clerck, J. M. Cock, S. M. Coelho, C. Gachon, A. R. Grossman, et al. 2017. "The Algal Revolution." *Trends in Plant Science* 22 (8): 726–38. doi:10.1016/J.TPLANTS.2017.05.005.

Bussa, Maresa, Cordt Zollfrank, and Hubert Röder. 2020. "Life-Cycle Assessment and Geospatial Analysis of Integrating Microalgae Cultivation into a Regional Economy." *Journal of Cleaner Production* 243: 118630. doi:10.1016/J.JCLEPRO.2019.118630.

Carreño-Zagarra, J. J., J. L. Guzmán, J. C. Moreno, and R. Villamizar. 2019. "Linear Active Disturbance Rejection Control for a Raceway Photobioreactor." *Control Engineering Practice* 85: 271–79. doi:10.1016/J.CONENGPRAC.2019.02.007.

Cavalcanti, D. R., T. Oliveira, and F. de Oliveira Santini. 2022. "Drivers of Digital Transformation Adoption: A Weight and Meta-Analysis." *Heliyon* 8 (2): e08911. doi:10.1016/J.HELIYON.2022.E08911.

Cecchin, Michela, Silvia Berteotti, Stefania Paltrinieri, Ivano Vigliante, Barbara Iadarola, Barbara Giovannone, Massimo E. Maffei, Massimo Delledonne, and Matteo Ballottari. 2020. "Improved Lipid Productivity in Nannochloropsis Gaditana in Nitrogen-Replete Conditions by Selection of Pale Green Mutants." *Biotechnology for Biofuels* 13 (1): 1–14. doi:10.1186/S13068-020-01718-8.

Chen, K. L., M. M. Crane, and M. Kaeberlein. 2017. "Microfluidic Technologies for Yeast Replicative Lifespan Studies." *Mechanisms of Ageing and Development* 161 (Pt B): 262–69. doi:10.1016/J.MAD.2016.03.009.

Chen, Xiao Hua, Alexandra Koumoutsi, Romy Scholz, Andreas Eisenreich, Kathrin Schneider, Isabelle Heinemeyer, Burkhard Morgenstern, et al. 2007. "Comparative Analysis of the Complete Genome Sequence of the Plant Growth—Promoting Bacterium Bacillus Amyloliquefaciens FZB42." *Nature Biotechnology* 25 (9): 1007–14. doi:10.1038/nbt1325.

Chuai, G., H. Ma, J. Yan, M. Chen, N. Hong, D. Xue, C. Zhou, et al. 2018. "DeepCRISPR: Optimized CRISPR Guide RNA Design by Deep Learning." *Genome Biology* 19 (1). doi:10.1186/S13059-018-1459-4.

Coltelli, Primo, Laura Barsanti, Valter Evangelista, and Paolo Gualtieri. 2017. "Algae through the Looking Glass." *Microscopy Research and Technique* 80 (5): 486–94. doi:10.1002/JEMT.22820.

Costello, Zak, and Hector Garcia Martin. 2018. "A Machine Learning Approach to Predict Metabolic Pathway Dynamics from Time-Series Multiomics Data." *Npj Systems Biology and Applications 2018 4:1* 4 (1): 1–14. doi:10.1038/s41540-018-0054-3.

Dal'Molin, C. G., L. E. Quek, R. W. Palfreyman, and L. K. Nielsen. 2011. "AlgaGEM—a Genome-Scale Metabolic Reconstruction of Algae Based on the Chlamydomonas Reinhardtii Genome." *BMC Genomics* 12: S5. doi:10.1186/1471-2164-12-S4-S5.

Daneshvar, Ehsan, Yong Sik Ok, Samad Tavakoli, Binoy Sarkar, Sabry M. Shaheen, Hui Hong, Yongkang Luo, Jörg Rinklebe, Hocheol Song, and Amit Bhatnagar. 2021. "Insights into Upstream Processing of Microalgae: A Review." *Bioresource Technology* 329: 124870. https://doi.org/10.1016/j.biortech.2021.124870.

Davidson, Keith, Richard J. Gowen, Paul J. Harrison, Lora E. Fleming, Porter Hoagland, and Grigorios Moschonas. 2014. "Anthropogenic Nutrients and Harmful Algae in Coastal Waters." *Journal of Environmental Management* 146: 206–16. doi:10.1016/J.JENVMAN.2014.07.002.

Davies, O. A., and E. Ogidiaka. 2019. "Science Arena Publications Specialty Journal of Biological Sciences Harmful Algal Blooms (HABs) in Nigerian Inland and Coastal Waters." *Specialty Journal of Biological Sciences* 5 (4): 14–24.

Dayton, David C., and Thomas D. Foust. 2019. *Analytical Methods for Biomass Characterization and Conversion*. Elsevier. doi:10.1016/C2017-0-03467-5.

Derakhshandeh, Masoud, and Umran Tezcan Un. 2019. "Optimization of Microalgae Scenedesmus SP. Growth Rate Using a Central Composite Design Statistical Approach." *Biomass and Bioenergy* 122: 211–20. doi:10.1016/J.BIOMBIOE.2019.01.022.

Dong, Li, Ze Zhao, Li Cui, He Zhu, Le Zhang, Zhaoliang Zhang, and Yi Wang. 2011. "A Cyber Physical Networking System for Monitoring and Cleaning up Blue-Green Algae Blooms with Agile Sensor and Actuator Control Mechanism on Lake Tai." 2011 IEEE Conference on Computer Communications Workshops, INFOCOM WKSHPS 2011, 732–37. doi:10.1109/INFCOMW.2011.5928908.

Dragone, Giuliano, Bruno D. Fernandes, Ana P. Abreu, António A. Vicente, and José A. Teixeira. 2011. "Nutrient Limitation as a Strategy for Increasing Starch Accumulation in Microalgae." *Applied Energy* 88 (10): 3331–35. doi:10.1016/J.APENERGY.2011.03.012.

Ellison, Candice R., Sean Overa, and Dorin Boldor. 2019. "Central Composite Design Parameterization of Microalgae/Cyanobacteria Co-Culture Pretreatment for Enhanced Lipid Extraction Using an External Clamp-on Ultrasonic Transducer." *Ultrasonics Sonochemistry* 51: 496–503. doi:10.1016/J.ULTSONCH.2018.05.006.

Eriksen, Niels T. 2008. "The Technology of Microalgal Culturing." *Biotechnology Letters* 30 (9): 1525–36. doi:10.1007/S10529-008-9740-3.

Esakkimuthu, Sivakumar, Venkatesan Krishnamurthy, Shuang Wang, Abd El-Fatah Abomohra, Sabarathinam Shanmugam, Sankar Ganesh Ramakrishnan, Sadhasivam Subrmaniam, and Swaminathan K. 2019. "Simultaneous Induction of Biomass and Lipid Production in Tetradesmus Obliquus BPL16 through Polysorbate Supplementation." *Renewable Energy* 140: 807–15. doi:10.1016/J.RENENE.2019.03.104.

Ferreira, Veronica da Silva, and Celso Sant'Anna. 2016. "Impact of Culture Conditions on the Chlorophyll Content of Microalgae for Biotechnological Applications." *World Journal of Microbiology and Biotechnology* 33 (1): 1–8. doi:10.1007/S11274-016-2181-6.

Foflonker, F., D. Mollegard, M. Ong, H. S. Yoon, and D. Bhattacharya. 2018. "Genomic Analysis of Picochlorum Species Reveals How Microalgae May Adapt to Variable Environments." *Molecular Biology and Evolution* 35 (11): 2702–11. doi:10.1093/MOLBEV/MSY167.

Foudeh, Amir M., Tohid Fatanat Didar, Teodor Veres, and Maryam Tabrizian. 2012. "Microfluidic Designs and Techniques Using Lab-on-a-Chip Devices for Pathogen Detection for Point-of-Care Diagnostics." *Lab on a Chip* 12 (18): 3249–66. doi:10.1039/C2LC40630F.

Fu, Weiqi, Amphun Chaiboonchoe, Basel Khraiwesh, David R. Nelson, Dina Al-Khairy, Alexandra Mystikou, Amnah Alzahmi, and Kourosh Salehi-Ashtiani. 2016. "Algal Cell Factories: Approaches, Applications, and Potentials." *Marine Drugs* 14 (12). doi:10.3390/MD14120225.

García-Camacho, F., L. López-Rosales, A. Sánchez-Mirón, E. H. Belarbi, Yusuf Chisti, and E. Molina-Grima. 2016. "Artificial Neural Network Modeling for Predicting the Growth of the Microalga Karlodinium Veneficum." *Algal Research* 14: 58–64. doi:10.1016/J.ALGAL.2016.01.002.

Gargouri, M., J. J. Park, F. O. Holguin, M. J. Kim, H. Wang, R. R. Deshpande, Y. Shachar-Hill, L. M. Hicks, and D. R. Gang. 2015. "Identification of Regulatory Network Hubs That Control Lipid Metabolism in Chlamydomonas Reinhardtii." *Journal of Experimental Botany* 66 (15): 4551–66. doi:10.1093/JXB/ERV217.

Ghafar Channar, Abdul, Ali Muhammad Rind, Ghulam Murtaza Mastoi, Khalida Faryal Almani, Khalid Hussain Lashari, Muhammad Ameen Qurishi, and Nasrullah Mahar. 2014. "Comparative Study of Water Quality of Manchhar Lake with Drinking Water Quality Standard of World Health Organization." *American Journal of Environmental Protection* 3 (2): 68–72. doi:10.11648/j.ajep.20140302.15.

Giannino, Francesco, Serena Esposito, Marcello Diano, Salvatore Cuomo, and Gerardo Toraldo. 2018. "A Predictive Decision Support System (DSS) for a Microalgae Production Plant Based on Internet of Things Paradigm." *Concurrency Computation* 30 (15). doi:10.1002/cpe.4476.

Gimpel, J. A., V. Henríquez, and S. P. Mayfield. 2015. "In Metabolic Engineering of Eukaryotic Microalgae: Potential and Challenges Come with Great Diversity." *Frontiers in Microbiology* 6: 1376. doi:10.3389/FMICB.2015.01376.

Goatley, Christopher H. R., and David R. Bellwood. 2011. "The Roles of Dimensionality, Canopies and Complexity in Ecosystem Monitoring." *PLoS One* 6 (11): e27307. doi:10.1371/JOURNAL.PONE.0027307.

Guarnieri, Michael T., Jennifer Levering, Calvin A. Henard, Jeffrey L. Boore, Michael J. Betenbaugh, Karsten Zengler, and Eric P. Knoshaug. 2018. "Genome Sequence of the Oleaginous Green Alga, Chlorella Vulgaris UTEX 395." *Frontiers in Bioengineering and Biotechnology* 37 (April). doi:10.3389/FBIOE.2018.00037.

Guiry, Michael D., Gwendoline M. Guiry, Liam Morrison, Fabio Rindi, Salvador Valenzuela Miranda, Arthur C. Mathieson, Bruce C. Parker, et al. 2014. "AlgaeBase: An on-Line Resource for Algae." *Cryptogamie, Algologie* 35 (2): 105–15. doi:10.7872/crya.v35.iss2.2014.105.

He, Lian, Venkat R. Subramanian, and Yinjie J. Tang. 2012. "Experimental Analysis and Model-Based Optimization of Microalgae Growth in Photo-Bioreactors Using Flue Gas." *Biomass and Bioenergy* 41: 131–38. doi:10.1016/J.BIOMBIOE.2012.02.025.

Hess, J. F., T. A. Kohl, M. Kotrová, K. Rönsch, T. Paprotka, V. Mohr, T. Hutzenlaub, et al. 2020. "Library Preparation for Next Generation Sequencing: A Review of Automation Strategies." *Biotechnology Advances* 41: 107537. doi:10.1016/J.BIOTECHADV.2020.107537.

Hu, Dawei, Rui Zhou, Yi Sun, Ling Tong, Ming Li, and Houkai Zhang. 2012. "Construction of Closed Integrative System for Gases Robust Stabilization Employing Microalgae Peculiarity and Computer Experiment." *Ecological Engineering* 44: 78–87. doi:10.1016/J.ECOLENG.2012.04.001.

Huang, Feng Yuan. 2011. "Synthesis and Properties of Cellulose Stearate." *Advanced Materials Research* 228–229: 919–24. doi:10.4028/www.scientific.net/AMR.228-229.919.

Husselmann, A. V., and K. A. Hawick. 2013. "Simulating Growth Kinetics in a Data-Parallel 3d Lattice Photobioreactor." *Modelling and Simulation in Engineering* 2013. doi:10.1155/2013/153241.

Ido, Alexander L., Mark Daniel G. de Luna, Sergio C. Capareda, Amado L. Maglinao, and Hyungseok Nam. 2018. "Application of Central Composite Design in the Optimization of Lipid Yield from Scenedesmus Obliquus Microalgae by Ultrasound-Assisted Solvent Extraction." *Energy* 157: 949–56. doi:10.1016/J.ENERGY.2018.04.171.

Jagadevan, Sheeja, Avik Banerjee, Chiranjib Banerjee, Chandan Guria, Rameshwar Tiwari, Mehak Baweja, and Pratyoosh Shukla. 2018. "Recent Developments in Synthetic Biology and Metabolic Engineering in Microalgae towards Biofuel Production." *Biotechnology for Biofuels* 11 (1): 185. doi:10.1186/S13068-018-1181-1.

Jeon, Seungjib, Jong-Min Lim, Hyung-Gwan Lee, Sung-Eun Shin, Nam Kyu Kang, Youn-Il Park, Hee-Mock Oh, Won-Joong Jeong, Byeong-Ryool Jeong, and Yong Keun Chang. 2017. "Current Status and Perspectives of Genome Editing Technology for Microalgae." *Biotechnology for Biofuels* 10: 267. doi:10.1186/s13068-017-0957-z.

Kang, Nam Kyu, Eun Kyung Kim, Young Uk Kim, Bongsoo Lee, Won-Joong Jeong, Byeong-Ryool Jeong, and Yong Keun Chang. 2017. "Increased Lipid Production by Heterologous Expression of AtWRI1 Transcription Factor in Nannochloropsis Salina." *Biotechnology for Biofuels* 10: 231. doi:10.1186/s13068-017-0919-5.

Karemore, Ankush, and Ramkrishna Sen. 2016. "Downstream Processing of Microalgal Feedstock for Lipid and Carbohydrate in a Biorefinery Concept: A Holistic Approach for Biofuel Applications." *RSC Advances* 6 (35): 29486–96. doi:10.1039/c6ra01477a.

Karimi, Mahmoud. 2017. "Exergy-Based Optimization of Direct Conversion of Microalgae Biomass to Biodiesel." *Journal of Cleaner Production* 141: 50–55. doi:10.1016/J.JCLEPRO.2016.09.032.

Kim, Jaoon Young Hwan, Ho Seok Kwak, Young Joon Sung, Hong Il Choi, Min Eui Hong, Hyun Seok Lim, Jae-Hyeok Lee, Sang Yup Lee, and Sang Jun Sim. 2016. "Microfluidic High-Throughput Selection of Microalgal Strains with Superior Photosynthetic Productivity Using Competitive Phototaxis." *Scientific Reports* 6 (1): 1–11. doi:10.1038/srep21155.

Kim, J., A. Jonoski, and D. P. Solomatine. 2022. "A Classification-Based Machine Learning Approach to the Prediction of Cyanobacterial Blooms in Chilgok Weir, South Korea." *Water* 14 (4): 542. doi: 10.3390/W14040542.

Kiran, Bala, Kratika Pathak, Ritunesh Kumar, and Devendra Deshmukh. 2016. "Statistical Optimization Using Central Composite Design for Biomass and Lipid Productivity of Microalga: A Step towards Enhanced Biodiesel Production." *Ecological Engineering* 92: 73–81. doi:10.1016/J.ECOLENG.2016.03.026.

Krüger, Anna, Christian Schäfers, Philip Busch, and Garabed Antranikian. 2020. "Digitalization in Microbiology—Paving the Path to Sustainable Circular Bioeconomy." *New Biotechnology* 59 (June): 88–96. doi:10.1016/j.nbt.2020.06.004.

Kugler, Amit, Boris Zorin, Shoshana Didi-Cohen, Maria Sibiryak, Olga Gorelova, Tatiana Ismagulova, Kamilya Kokabi, et al. 2019. "Long-Chain Polyunsaturated Fatty Acids in the Green Microalga Lobosphaera Incisa Contribute to Tolerance to Abiotic Stresses." *Plant and Cell Physiology* 60 (6): 1205–23. doi:10.1093/PCP/PCZ013.

Kumar, Pushpendar, Arghya Bhattacharya, Sanjeev Kumar Prajapati, Anushree Malik, and Virendra Kumar Vijay. 2020. "Anaerobic Co-Digestion of Waste Microalgal Biomass with Cattle Dung in a Pilot-Scale Reactor: Effect of Seasonal Variations and Long-Term Stability Assessment." *Biomass Conversion and Biorefinery*: 1–13. doi:10.1007/S13399-020-00778-Y.

Lang, I., L. Hodac, T. Friedl, and I. Feussner. 2011. "Fatty Acid Profiles and Their Distribution Patterns in Microalgae: A Comprehensive Analysis of More than 2000 Strains from the SAG Culture Collection." *BMC Plant Biology* 11. doi:10.1186/1471-2229-11-124.

Lawal, K., and H. N. Rafsanjani. 2022. "Trends, Benefits, Risks, and Challenges of IoT Implementation in Residential and Commercial Buildings." *Energy and Built Environment* 3 (3): 251–266. doi: 10.1016/J.ENBENV.2021.01.009.

Leow, Shijie, John R. Witter, Derek R. Vardon, Brajendra K. Sharma, Jeremy S. Guest, and Timothy J. Strathmann. 2015. "Prediction of Microalgae Hydrothermal Liquefaction Products from Feedstock Biochemical Composition." *Green Chemistry* 17 (6): 3584–99. doi:10.1039/C5GC00574D.

Li, Changling, Hailin Yang, Xiaole Xia, Yuji Li, Luping Chen, Meng Zhang, Ling Zhang, and Wu Wang. 2013. "High Efficient Treatment of Citric Acid Effluent by Chlorella Vulgaris and Potential Biomass Utilization." *Bioresource Technology* 127: 248–55. doi:10.1016/J.BIORTECH.2012.08.074.

Liberato, Vanessa, Carolina Benevenuti, Fabiana Coelho, Alanna Botelho, Priscilla Amaral, Nei Pereira, and Tatiana Ferreira. 2019. "Clostridium Sp. as Bio-Catalyst for Fuels and Chemicals Production in a Biorefinery Context." *Catalysts* 9 (11): 962. doi:10.3390/CATAL9110962.

Liu, Shijie. 2010. "Woody Biomass: Niche Position as a Source of Sustainable Renewable Chemicals and Energy and Kinetics of Hot-Water Extraction/Hydrolysis." *Biotechnology Advances* 28 (5): 563–82. doi:10.1016/J.BIOTECHADV.2010.05.006.

Liu, Xiao, Dan Zhang, Jianhui Zhang, Yuhong Chen, Xiuli Liu, Chengming Fan, Richard R-C. Wang, Yongyue Hou, and Zanmin Hu. 2021. "Overexpression of the Transcription Factor AtLEC1 Significantly Improved the Lipid Content of Chlorella Ellipsoidea." *Frontiers in Bioengineering and Biotechnology* 113. doi:10.3389/FBIOE.2021.626162.

Liyanaarachchi, Vinoj Chamilka, Gannoru Kankanamalage Sanuji Hasara Nishshanka, Pemaththu Hewa Viraj Nimarshana, Thilini Udayangani Ariyadasa, and Rahula Anura Attalage. 2020. "Development of an Artificial Neural Network Model to Simulate the Growth of Microalga Chlorella Vulgaris Incorporating the Effect of Micronutrients." *Journal of Biotechnology* 312: 44–55. doi:10.1016/J.JBIOTEC.2020.02.010.

Loira, Nicolás, Sebastian Mendoza, María Paz Cortés, Natalia Rojas, Dante Travisany, Alex Di Genova, Natalia Gajardo, Nicole Ehrenfeld, and Alejandro Maass. 2017. "Reconstruction of the Microalga Nannochloropsis Salina Genome-Scale Metabolic Model with Applications to Lipid Production." *BMC Systems Biology* 11 (1). BioMed Central: 1–17. doi:10.1186/S12918-017-0441-1.

López-Rosales, L., J. J. Gallardo-Rodríguez, A. Sánchez-Mirón, A. Contreras-Gómez, F. García-Camacho, and E. Molina-Grima. 2013. "Modelling of Multi-Nutrient Interactions in Growth of the Dinoflagellate Microalga Protoceratium Reticulatum Using Artificial Neural Networks." *Bioresource Technology* 146: 682–88. doi:10.1016/J.BIORTECH.2013.07.141.

López-Rosales, L., F. García-Camacho, A. Sánchez-Mirón, and Yusuf Chisti. 2015. "An Optimal Culture Medium for Growing Karlodinium Veneficum: Progress towards a Microalgal Dinoflagellate-Based Bioprocess." *Algal Research* 10: 177–82. doi:10.1016/J.ALGAL.2015.05.006.

Mathimani, T., B. Beena Nair, and R. Ranjith kumar. 2016. "Evaluation of Microalga for Biodiesel Using Lipid and Fatty Acid as a Marker—A Central Composite Design Approach." *Journal of the Energy Institute* 89 (3): 436–46. doi:10.1016/J.JOEI.2015.02.010.

Mathimani, Thangavel, Manigandan Sekar, Sabarathinam Shanmugam, Jamal S. M. Sabir, Nguyen Thuy Lan Chi, and Arivalagan Pugazhendhi. 2021. "Relative Abundance of Lipid Types among Chlorella Sp. and Scenedesmus Sp. and Ameliorating Homogeneous Acid Catalytic Conditions Using Central Composite Design (CCD) for Maximizing Fatty Acid Methyl Ester Yield." *Science of The Total Environment* 771: 144700. doi:10.1016/J.SCITOTENV.2020.144700.

Mathimani, Thangavel, Tamilkolundu Senthil Kumar, Murugesan Chandrasekar, Lakshmanan Uma, and Dharmar Prabaharan. 2017. "Assessment of Fuel Properties, Engine Performance and Emission Characteristics of Outdoor Grown Marine Chlorella Vulgaris BDUG 91771 Biodiesel." *Renewable Energy* 105: 637–46. doi:10.1016/J.RENENE.2016.12.090.

Mathimani, Thangavel, Lakshmanan Uma, and Dharmar Prabaharan. 2018. "Formulation of Low-Cost Seawater Medium for High Cell Density and High Lipid Content of Chlorella Vulgaris BDUG 91771 Using Central Composite Design in Biodiesel Perspective." *Journal of Cleaner Production* 198: 575–86. doi:10.1016/J.JCLEPRO.2018.06.303.

Maulana, Muhamad, Azimatun Nur, and H Hadiyanto. 2015. "Enhancement of Chlorella Vulgaris Biomass Cultivated in POME Medium as Biofuel Feedstock under Mixotrophic Conditions." *Article in Journal of Engineering and Technological Sciences*. doi:10.5614/j.eng.technol.sci.2015.47.5.2.

Mishra, Deepak R., Abhishek Kumar, Lakshmish Ramaswamy, Vinay K. Boddula, Moumita C. Das, Benjamin P. Page, and Samuel J. Weber. 2020. "CyanoTRACKER: A Cloud-Based Integrated Multi-Platform Architecture for Global Observation of Cyanobacterial Harmful Algal Blooms." *Harmful Algae* 96: 101828. doi:10.1016/J.HAL.2020.101828.

Muthuraj, Muthusivaramapandian, Niharika Chandra, Basavaraj Palabhanvi, Vikram Kumar, and Debasish Das. 2014. "Process Engineering for High-Cell-Density Cultivation of Lipid Rich Microalgal Biomass of Chlorella Sp. FC2 IITG." *BioEnergy Research* 8 (2): 726–39. doi:10.1007/S12155-014-9552-3.

Narayanan, Barath Narayanan, Russell C. Hardie, Manawaduge Supun De Silva, and Nathaniel K. Kueterman. 2020. "Hybrid Machine Learning Architecture for Automated Detection and Grading of Retinal Images for Diabetic Retinopathy." *Journal of Medical Imaging* 7 (3): 034501. doi:10.1117/1.JMI.7.3.034501.

Nižetić, Sandro, Nedjib Djilali, Agis Papadopoulos, and Joel J. P. C. Rodrigues. 2019. "Smart Technologies for Promotion of Energy Efficiency, Utilization of Sustainable Resources and Waste Management." *Journal of Cleaner Production* 231: 565–91. doi:10.1016/J.JCLEPRO.2019.04.397.

Nwankwegu, Amechi S., Yiping Li, Yanan Huang, Jin Wei, Eyram Norgbey, Linda Sarpong, Qiuying Lai, and Kai Wang. 2019. "Harmful Algal Blooms under Changing Climate and Constantly Increasing Anthropogenic Actions: The Review of Management Implications." *3 Biotech* 9 (12): 1–19. doi:10.1007/S13205-019-1976-1.

Nymark, M., A. K. Sharma, T. Sparstad, A. M. Bones, and P. Winge. 2016. "A CRISPR/Cas9 System Adapted for Gene Editing in Marine Algae." *Scientific Reports* 6. doi:10.1038/SREP24951.

Oey, M., I. L. Ross, E. Stephens, J. Steinbeck, J. Wolf, K. A. Radzun, J. Kügler, A. K. Ringsmuth, O. Kruse, and B. Hankamer. 2013. "RNAi Knock-down of LHCBM1, 2 and 3 Increases Photosynthetic H2 Production Efficiency of the Green Alga Chlamydomonas Reinhardtii." *Plos One* 8 (4): e61375. doi:10.1371/JOURNAL.PONE.0061375.

Olaniyan, Olugbemi Tope, and Charles Oluwaseun Adetunji. 2021. "Biological, Biochemical, and Biodiversity of Biomolecules from Marine-Based Beneficial Microorganisms: Industrial Perspective." In *Microbial Rejuvenation of Polluted Environment*, edited by Charles Oluwaseun Adetunji, Deepak G. Panpatte, and Yogeshvari K. Jhala, 57–81. Springer. doi:10.1007/978-981-15-7459-7_4.

Ozturk, Munir, Naheed Saba, Volkan Altay, Rizwan Iqbal, Khalid Rehman Hakeem, Mohammad Jawaid, and Faridah Hanum Ibrahim. 2017. "Biomass and Bioenergy: An Overview of the Development Potential in Turkey and Malaysia." *Renewable and Sustainable Energy Reviews* 79: 1285–1302. doi:10.1016/J.RSER.2017.05.111.

Ozyurt, Mustafa. 2004. "Conversion of Agricultural And Industrial Wastes for Single-Cell Protein Production and Pollution Potential Reduction: A Review." *Fresenius Environmental Bulletin* 13 (8): 693–96.

Penas, Olivia, Régis Plateaux, Stanislao Patalano, and Moncef Hammadi. 2017. "Multi-Scale Approach from Mechatronic to Cyber-Physical Systems for the Design of Manufacturing Systems." *Computers in Industry* 86: 52–69. doi:10.1016/J.COMPIND.2016.12.001.

Pires, José C. M., Maria C. M. Alvim-Ferraz, and Fernando G. Martins. 2017. "Photobioreactor Design for Microalgae Production through Computational Fluid Dynamics: A Review." *Renewable and Sustainable Energy Reviews* 79: 248–54. doi:10.1016/J.RSER.2017.05.064.

Poliner, E., T. Takeuchi, Z. Y. Du, C. Benning, and E. M. Farré. 2018. "Nontransgenic Marker-Free Gene Disruption by an Episomal CRISPR System in the Oleaginous Microalga, Nannochloropsis Oceanica CCMP1779." *ACS Synthetic Biology* 7 (4): 962–68. doi:10.1021/ACSSYNBIO.7B00362.

Ptasinski, K. J. 2016. *Efficiency of Biomass Energy: An Energy Approach to Biofuels, Power, and Biorefineries* (K. J. Ptasinski (ed.)). John Wiley & Sons, Inc. Hoboken, New Jersey.

Raheem, Abdul, Wan Azlina W. A. K. G., Y. H. Taufiq Yap, M. K. Danquah, and Razif Harun. 2015. "Optimization of the Microalgae Chlorella Vulgaris for Syngas Production Using Central Composite Design." *RSC Advances* 5 (88): 71805–15. doi:10.1039/C5RA10503J.

Rahmat, Ayi, Indra Jaya, Totok Hestirianoto, Dedi Jusadi, and Mujizat Kawaroe. 2020. "Evaluation of System Performance for Microalga Cultivation in Photobioreactor with IOTs (Internet of Things)." *International Journal of Sciences: Basic and Applied Research (IJSBAR) International Journal of Sciences: Basic and Applied Research* 49 (2): 95–107.

Rai, Vineeta, Sandip Kumar Patel, Muthusivaramapandian Muthuraj, Mayuri N. Gandhi, Debasish Das, and Sanjeeva Srivastava. 2021. "Systematic Metabolome Profiling and Multi-Omics Analysis of the Nitrogen-Limited Non-Model Oleaginous Algae for Biorefining." *Biofuel Research Journal* 8 (1): 1330–41. doi:10.18331/BRJ2021.8.1.4.

Ravishankar, G. A., and R. R. Ambati 2019. "Handbook of Algal Technologies and Phytochemicals." In *Handbook of Algal Technologies and Phytochemicals*, edited by G. A. Ravishankar and R. R. Ambati. CRC Press. doi: 10.1201/9780429057892.

Reijnders, M. J., R. G. van Heck, C. M. Lam, M. A. Scaife, V. A. dos Santos, A. G. Smith, and P. J. Schaap. 2014. "Green Genes: Bioinformatics and Systems-Biology Innovations Drive Algal Biotechnology." *Trends in Biotechnology* 32 (12): 617–26. doi:10.1016/J.TIBTECH.2014.10.003.

Reynolds, Sam A., Matthew P. Davey, and David C. Aldridge. 2019. "Harnessing Synthetic Ecology for Commercial Algae Production." *Scientific Reports* 9 (1): 1–9. doi:10.1038/s41598-019-46135-6.

Richmond, A., and Q. Hu. 2013. "Handbook of Microalgal Culture: Applied Phycology and Biotechnology: Second Edition." In *Handbook of Microalgal Culture: Applied Phycology and Biotechnology: Second Edition*, edited by A. Richmond and Q. Hu (Eds.). John Wiley and Sons. Hoboken, New Jersey. doi: 10.1002/9781118567166.

Rio-Chanona, Ehecatl A. del, Fabio Fiorelli, Dongda Zhang, Nur R. Ahmed, Keju Jing, and Nilay Shah. 2017. "An Efficient Model Construction Strategy to Simulate Microalgal Lutein Photo-Production Dynamic Process." *Biotechnology and Bioengineering* 114 (11): 2518–27. doi:10.1002/BIT.26373.

Rismani-Yazdi, Hamid, Berat Z. Haznedaroglu, Carol Hsin, and Jordan Peccia. 2012. "Transcriptomic Analysis of the Oleaginous Microalga Neochloris Oleoabundans Reveals Metabolic Insights into Triacylglyceride Accumulation." *Biotechnology for Biofuels* 5 (1): 1–16. doi:10.1186/1754-6834-5-74.

Rose, J. B., P. R. Epstein, E. K. Lipp, B. H. Sherman, S. M. Bernard, and J. A. Patz. 2001. "Climate Variability and Change in the United States: Potential Impacts on Water- and Foodborne Diseases Caused by Microbiologic Agents." *Environmental Health Perspectives* 109: 211–21. doi:10.1289/EHP.01109S2211.

Rybkina, A., and A. Reissell. 2017. "Arctic Territory-Data and Modeling." *Geoinformatics Research Papers* 5 (*BS1002*): 5. doi: 10.2205/CODATA2017.

Salam, Satishchandra, and Tikendra Nath Verma. 2019. "Appending Empirical Modelling to Numerical Solution for Behaviour Characterisation of Microalgae Biodiesel." *Energy Conversion and Management* 180: 496–510. doi:10.1016/J.ENCONMAN.2018.11.014.

Sanseverino, I., D. Conduto, L. Pozzoli, S. Dobricic, and T. Lettieri. 2016. Algal Bloom and Its Economic Impact. In *Eur 27905 En*. European Union. doi: 10.2788/660478.

Sarkar, Sambit, Mriganka Sekhar Manna, Tridib Kumar Bhowmick, and Kalyan Gayen. 2020. "Priority-Based Multiple Products from Microalgae: Review on Techniques and Strategies." *Critical Reviews in Biotechnology* 40 (5): 590–607. doi:10.1080/07388551.2020.1753649.

Sarrafzadeh, Mohammad H., Hyun Joon La, Seong Hyun Seo, Hashem Asgharnejad, and Hee Mock Oh. 2015. "Evaluation of Various Techniques for Microalgal Biomass Quantification." *Journal of Biotechnology* 216: 90–97. doi:10.1016/J.JBIOTEC.2015.10.010.

Sewsynker-Sukai, Yeshona, Funmilayo Faloye, and Evariste Bosco Gueguim Kana. 2016. "Artificial Neural Networks: An Efficient Tool for Modelling and Optimization of Biofuel Production (a Mini Review)." *Biotechnology & Biotechnological Equipment* 31 (2): 221–35. doi:10.1080/13102818.2016.1269616.

Sharma, Vishal, Martin Stout, Keith Pearce, Allan L. Klein, Maryam Alsharqi, Petros Nihoyannopoulos, Jamal Nasir Khan, et al. 2020. "Report from the Annual Conference of the British Society of Echocardiography, November 2017, Edinburgh International Conference Centre, Edinburgh." *Echo Research and Practice* 6 (4): M1–2. doi:10.1530/erp-19-0056.

Sharon Mano Pappu, J., G. K. Vijayakumar, and V. Ramamurthy. 2013. "Artificial Neural Network Model for Predicting Production of Spirulina Platensis in Outdoor Culture." *Bioresource Technology* 130: 224–30. doi:10.1016/J.BIORTECH.2012.12.082.

Shetty, Prateek, Attila Farkas, Bernadett Pap, Bettina Hupp, Vince Ördög, Tibor Bíró, Torda Varga, and Gergely Maróti. 2021. "Comparative and Phylogenomic Analysis of Nuclear and Organelle Genes in Cryptic Coelastrella Vacuolata MACC-549 Green Algae." *Algal Research* 58: 102380. doi:10.1016/J.ALGAL.2021.102380.

Shin, S. E., J. M. Lim, H. G. Koh, E. K. Kim, N. K. Kang, S. Jeon, S. Kwon, et al. 2016. "CRISPR/Cas9-Induced Knockout and Knock-in Mutations in Chlamydomonas Reinhardtii." *Scientific Reports* 6. doi:10.1038/SREP27810.

Singh, Amit Kumar, and Abhay K. Pandey. 2018. "Microalgae An Ecofriendly Tool for the Treatment of Industrial Wastewaters and Biofuel Production." In *Recent Advances in Environmental Management*, edited by Ram Naresh Bharagava. CRC Press. doi:10.1201/9781351011259.

Sirikhachornkit, Anchalee, Anongpat Suttangkakul, Supachai Vuttipongchaikij, and Piyada Juntawong. 2018. "De Novo Transcriptome Analysis and Gene Expression Profiling of an Oleaginous Microalga Scenedesmus Acutus TISTR8540 during Nitrogen Deprivation-Induced Lipid Accumulation." *Scientific Reports* 8 (1): 1–12. doi:10.1038/s41598-018-22080-8.

Sommer, Veronika, Ulf Karsten, and Karin Glaser. 2020. "Halophilic Algal Communities in Biological Soil Crusts Isolated from Potash Tailings Pile Areas." *Frontiers in Ecology and Evolution* 8: 46. doi:10.3389/FEVO.2020.00046.

Sörenson, Eva, Mireia Bertos-Fortis, Hanna Farnelid, Anke Kremp, Karen Krüger, Elin Lindehoff, and Catherine Legrand. 2019. "Consistency in Microbiomes in Cultures of Alexandrium Species Isolated from Brackish and Marine Waters." *Environmental Microbiology Reports* 11 (3): 425–33. doi:10.1111/1758-2229.12736.

Sunoj, S., Ademola Hammed, C. Igathinathane, Sulaymon Eshkabilov, and Halis Simsek. 2021. "Identification, Quantification, and Growth Profiling of Eight Different Microalgae Species Using Image Analysis." *Algal Research* 60: 102487. doi:10.1016/J.ALGAL.2021.102487.

Supriyanto, Ryozo Noguchi, Tofael Ahamed, Devitra Saka Rani, Kai Sakurai, Muhammad Ansori Nasution, Dhani S. Wibawa, Mikihide Demura, and Makoto M. Watanabe. 2019. "Artificial Neural Networks Model for Estimating Growth of Polyculture Microalgae in an Open Raceway Pond." *Biosystems Engineering* 177 (January). Academic Press: 122–29. doi:10.1016/J.BIOSYSTEMSENG.2018.10.002.

Suresh, S., V. C. Srivastava, and I. M. Mishra. 2009. "Critical Analysis of Engineering Aspects of Shaken Flask Bioreactors." *Critical Reviews in Biotechnology* 29 (4): 255–78. doi:10.3109/07388550903062314.

Susanna, D., R. Dhanapal, R. Mahalingam, and V. Ramamurthy. 2019. "Increasing Productivity of Spirulina Platensis in Photobioreactors Using Artificial Neural Network Modeling." *Biotechnology and Bioengineering* 116 (11): 2960–70. doi:10.1002/BIT.27128.

Takouridis, S. J., D. E. Tribe, S. L. Gras, and G. J. O. Martin. 2015. "The Selective Breeding of the Freshwater Microalga Chlamydomonas Reinhardtii for Growth in Salinity." *Bioresource Technology* 184: 18–22. doi:10.1016/J.BIORTECH.2014.10.120.

Thiébaut, Gabrielle. 2006. "Aquatic Macrophyte Approach to Assess the Impact of Disturbances on the Diversity of the Ecosystem and on River Quality." *International Review of Hydrobiology* 91 (5): 483–97. doi:10.1002/IROH.200610868.

Thiriet-Rupert, S., G. Carrier, C. Trottier, D. Eveillard, B. Schoefs, G. Bougaran, J. P. Cadoret, B. Chénais, and B. Saint-Jean. 2018. "Identification of Transcription Factors Involved in the Phenotype of a Domesticated Oleaginous Microalgae Strain of Tisochrysis Lutea." *Algal Research* 30: 59–72. doi:10.1016/J.ALGAL.2017.12.011.

Tian, Y., and M. Huang. 2019. "An Integrated Web-Based System for the Monitoring and Forecasting of Coastal Harmful Algae Blooms: Application to Shenzhen City, China." *Journal of Marine Science and Engineering* 7 (9): 314. doi: 10.3390/JMSE7090314.

Tian, Y., and D. R. Smith. 2016. "Recovering Complete Mitochondrial Genome Sequences from RNA-Seq: A Case Study of Polytomella Non-Photosynthetic Green Algae." *Molecular Phylogenetics and Evolution* 98: 57–62. doi:10.1016/J.YMPEV.2016.01.017.

Tibocha-Bonilla, Juan D., Cristal Zuñiga, Rubén D. Godoy-Silva, and Karsten Zengler. 2018. "Advances in Metabolic Modeling of Oleaginous Microalgae Mike Himmel." *Biotechnology for Biofuels* 11 (1). doi:10.1186/S13068-018-1244-3.

Tyler, Christina R. Steadman, Blake T. Hovde, Hajnalka E. Daligault, Xiang Li Zhang, Yuliya Kunde, Babetta L. Marrone, Scott N. Twary, and Shawn R. Starkeburg. 2019. "High-Quality Draft Genome Sequence of the Green Alga Tetraselmis Striata (Chlorophyta) Generated from PacBio Sequencing." *Microbiology Resource Announcements* 8 (43). doi:10.1128/MRA.00780-19.

Udayan, A., A. K. Pandey, R. Sirohi, N. Sreekumar, B. I. Sang, S. J. Sim, S. H. Kim, and A. Pandey. 2022. "Production of Microalgae with High Lipid Content and Their Potential as Sources of Nutraceuticals." *Phytochemistry Reviews* 1–28. doi: 10.1007/S11101-021-09784-Y/FIGURES/3.

Viegas, Catarina, Margarida Gonçalves, Liliana Soares, and Benilde Mendes. 2016. "Bioremediation of Agro-Industrial Effluents Using Chlorella Microalgae." *IFIP Advances in Information and Communication Technology* 470: 523–30. doi:10.1007/978-3-319-31165-4_49.

Villa-Henriksen, Andrés, Gareth T. C. Edwards, Liisa A. Pesonen, Ole Green, and Claus Aage Grøn Sørensen. 2020. "Internet of Things in Arable Farming: Implementation,

Applications, Challenges and Potential." *Biosystems Engineering* 191: 60–84. doi:10.1016/J.BIOSYSTEMSENG.2019.12.013.

Wang, Yu, Rui Tan, Guoliang Xing, Xiaobo Tan, Jianxun Wang, and Ruogu Zhou. 2014. "Spatiotemporal Aquatic Field Reconstruction Using Cyber-Physical Robotic Sensor Systems." *ACM Transactions on Sensor Networks* 10 (4): 1–27. doi:10.1145/2505767.

Wei, Bin, Norio Sugiura, and Takaaki Maekawa. 2001. "Use of Artificial Neural Network in the Prediction of Algal Blooms." *Water Research* 35 (8): 2022–28. doi:10.1016/S0043-1354(00)00464-4.

Wirth, H., M. von Bergen, J. Murugaiyan, U. Rösler, T. Stokowy, and H. Binder. 2012. "MALDI-Typing of Infectious Algae of the Genus Prototheca Using SOM Portraits." *Journal of Microbiological Methods* 88 (1): 83–97. doi:10.1016/J.MIMET.2011.10.013.

Yao, L., T. W. Tan, Y. K. Ng, K. H. Ban, H. Shen, H. Lin, and Y. K. Lee. 2015. "RNA-Seq Transcriptomic Analysis with Bag2D Software Identifies Key Pathways Enhancing Lipid Yield in a High Lipid-Producing Mutant of the Non-Model Green Alga Dunaliella Tertiolecta." *Biotechnology for Biofuels* 8 (1). doi:10.1186/S13068-015-0382-0.

Yuan, Shuo, Xinping Zhou, Reccab M. Ochieng, and Xiangdong Zhou. 2016. "A Numerical Thermal Growth Model for Prediction of Microalgae Production in Photobioreactors." *Journal of Renewable and Sustainable Energy* 8 (2). doi:10.1063/1.4942875.

Yun, Jin-Ho, Val H. Smith, Hyun-Joon La, and Yong Keun Chang. 2016. "Towards Managing Food-Web Structure and Algal Crop Diversity in Industrial-Scale Algal Biomass Production." *Current Biotechnology* 5 (2): 118–29. doi:10.2174/221155010566616 0127002552.

Zullaikah, Siti, Adi Tjipto Utomo, Medina Yasmin, Lu Ki Ong, and Yi Hsu Ju. 2019. "9 - Ecofuel Conversion Technology of Inedible Lipid Feedstocks to Renewable Fuel." In *Advances in Eco-Fuels for a Sustainable Environment*, edited by Kalam Azad, 237–76. Woodhead Publishing. doi:10.1016/b978-0-08-102728-8.00009-7.

8 Smart Factory of Microalgae in Environmental Biotechnology

Shazia Ali[1], Kuan Shiong Khoo[2] Hooi Ren Lim[1], Hui Suan Ng[2] and Pau Loke Show[1]

[1] Department of Chemical and Environmental Engineering, Faculty of Science and Engineering, University of Nottingham Malaysia, Jalan Broga, Semenyih, Selangor Darul Ehsan, Malaysia

[2] Faculty of Applied Sciences, UCSI University, Cheras, Kuala Lumpur, Malaysia

CONTENTS

8.1 Introduction ... 264
8.2 Contribution of Microalgae in Obtaining Value-Added Products ... 266
 8.2.1 Microalgae Biomass Processing ... 266
 8.2.2 Application of Microalgae Toward the Twenty-First Century ... 269
8.3 Recent Technologies Implementation for Smart Technology System ... 271
 8.3.1 Pretreatment Method for Waste ... 271
 8.3.2 Types of Technologies Implemented Using Microalgae ... 273
8.4 Sustainability of Microalgae ... 275
 8.4.1 Environmental Assessment of Microalgae ... 276
 8.4.2 Economic and Societal Effects ... 277
8.5 Challenges and Improvements ... 278
8.6 Future Prospects ... 280
8.7 Conclusions ... 282
Acknowledgments ... 282
References ... 282

DOI: 10.1201/9781003202196-8

8.1 INTRODUCTION

As technology and industry advance, there is an increase in waste emission from various sectors, which causes environmental concerns and potential harm to the ecosystem (Ejaz et al. 2010). Waste comes in a variety of shapes and sizes, and its properties such as ignitability, corrosivity, reactivity, and toxicity can be portrayed in a variety of ways (Taboada-González et al. 2011). Some of the common factors used in waste classification are physical attributes, physical states, biodegradable potentials, reusable potentials, degree of environmental impact, and source of production (Demirbas 2011).

Solid wastes consist of food, paper, plastics, metals, glass, and wood that are sent to material recovery facilities to identify those which can be reused while disposing the rest to landfills (Abdel-Shafy and Mansour 2018). On the other hand, liquid wastes consist of wastewater, fats, oil or grease, and sludges. Leachate is also a fluid waste that percolates through landfills and is made up of liquids in the trash as well as outside water, such as rainwater, which percolates through the trash (Jayawardhana et al. 2016). These wastes are further categorized into either hazardous or nonhazardous waste. Hazardous waste can be treated using biological, thermal, chemical, and physical methods. Chemical procedures include precipitation, ion exchange, reduction, oxidation, and neutralization. One of the thermal methods that can not only clean but also destroy some organic wastes is high-temperature incineration. Thermal equipment is used to burn solid, liquid, and sludge waste (Mmereki et al. 2016). Landfarming is a way of biologically treating hazardous waste. On a suitable area of land, this procedure entails correctly combining waste with surface soil (Lukić et al. 2017). It's possible to add microbes that can break down rubbish as well as nutrition. A bacteria strain that has been genetically engineered is used in some cases.

Microalgae are photosynthesising microorganisms that can be prokaryotic or eukaryotic (Vale et al. 2020). Microalgae have a basic cell structure and grow on light, carbon dioxide, water, and nutrients such as phosphorus and nitrogen. The major chemical constituents of microalgae include lipids, proteins, and carbohydrates of varied compositions, which are stored in the microalgae cell (Khoo et al. 2020). Polysaccharides are polymers of monosaccharides joined together by glycosidic linkages. Microalgae cell is composed of a cellulosic cell wall that functions as a barrier against osmotic stress and provides tensile strength. The pigments that give the thallus its color come in a variety of forms such as chlorophyll, carotenoids, xanthophylls, phycobilin, and pyrenoids (Begum et al. 2016).

Microalgae contain polar lipids including phospholipids and glycolipids, as well as neutral storage lipids like diglycerides (DAGs), monoglycerides (MAGs), and triglycerides (TAGs), as well as hydrocarbons and free fatty acids (Pignolet et al. 2013). Polar lipids are only found in cell organelle membranes, such as the thylakoid membranes of chloroplasts. More specifically, microalgae have thylakoid membranes, which are the locations of photosynthetic processes that require light energy absorption to divide the water and create oxygen, ATP, and reductants (Osanai, Park, and Nakamura 2017). Microalgae include mitochondria, which are encased in a double membrane envelope. An aqueous matrix of solutes, soluble enzymes, and mitochondrial glucose is contained within the inner membrane of plant mitochondria. The matrix is extremely proteinaceous and coarsely granular. The organelle is semiautonomous in nature because it has its own circular DNA and ribosomes, which it uses to synthesize some of its proteins. Figure 8.1 shows the biological structure of microalgae cell.

The term Smart Factory refers to the end goal and objective of factory digitization. (Chew et al. 2021). The particular circumstances to the scarcity of fossil fuels, a clean, cost-effective, and long-term energy source are in high demand. Algal industrial revolutions are crucial in fulfilling the world's growing energy demand and achieving sustainable development goals. According to algal biorefineries integrated with the Industry 4.0 method, downstream capital is always secured for optimal returns in an efficient biorefinery that allows for the production of the greatest number of commodities and coproducts with the smallest residual quantity (Sheehan John et al. 1998). The transition from algae-based bioenergy to high-value bioproducts, as well as the model of algae-based biorefineries, has been documented in studies (Laurens, Chen-Glasser, and McMillan 2017).

This book chapter aims to investigate the use of genetic engineering techniques in algal cultivation for the production of bioenergy and by-products. Besides, various areas of focus such as the identification of the desired algal strain through automated genetic manipulation and genetic alteration were evaluated critically. This book chapter focuses on the microalgae harvesting and regeneration of high-value products which are further utilized in the various applications that contribute to the economic circle. The technologies and challenges with the utilization of microalgae are also discussed for future developments.

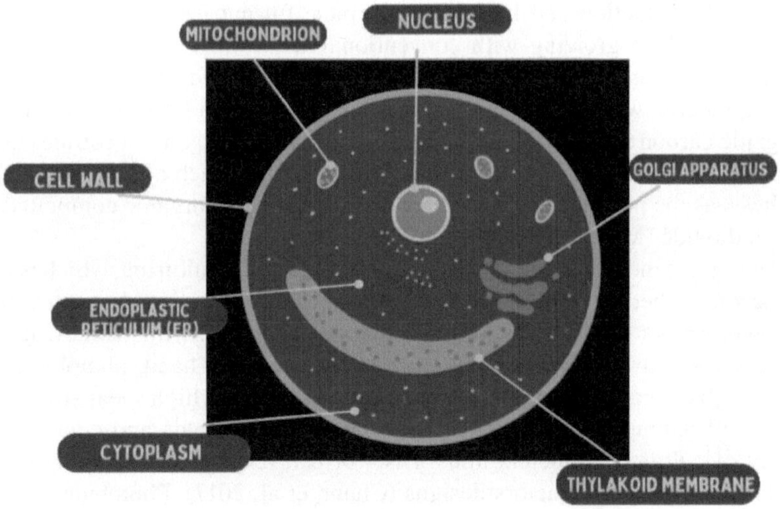

FIGURE 8.1 Composition of microalgae cell structure.

8.2 CONTRIBUTION OF MICROALGAE IN OBTAINING VALUE-ADDED PRODUCTS

Microalgae have inspired a lot of interest as a biofuel feedstock in response to the growing energy problem, global warming, and natural resource depletion. Biofuels are produced as a result of biomass harvesting and cultivation by microalgae. High-value products are also developed to improve the profitability of a microalgae biorefinery by removing a percentage of algae (Koyande et al. 2019; Low et al. 2021). Proteins, carbohydrates, lipids, pigments, antioxidants, and vitamins are examples of high-value products with uses in the cosmetics, nutritional, and pharmaceutical industries (Gujar, Kubar, and Li 2019). An innovative microalgae biorefinery structure is used to boost the process's sustainability by producing a variety of high-value goods and biofuel.

8.2.1 Microalgae Biomass Processing

Microalgal biomass is regarded as a biofuel feedstock that is both sustainable and renewable. These photosynthetic organisms naturally collect carbohydrates, primarily in the form of starch, which can be utilized as raw materials for sugar-based biofuels such as bioethanol and biobutanol. Microalgal biomass is a feasible alternative to conventional fuel sources as it has a narrower growth period than terrestrial plants or agricultural residues, which produce more biomass and greater harvesting indices and have the maximum level of carbon fixation (A. Z. Khan et al. 2018). Microalgae may also be cultivated in underdeveloped areas with effluent as a growth medium, obviating the need for a large amount of arable land (Miranda et al. 2017). For a variety of reasons, wastewater is the ideal source for microalgae biomass generation, including low-cost growth media, support for biofuel production and bulk biomass, plentiful nutrients, and the flexibility to combine algae growing with conventional wastewater treatment equipment. (Roostaei and Zhang 2017). Microalgae can be cultured in three ways: photo autotrophically, which uses light as an energy source and carbon dioxide as an inorganic carbon source; heterotrophically, which uses organic substrates as both an energy and a carbon source; and mixotrophically, which not only uses photosynthesis as the primary energy source but also requires organic compounds and carbon dioxide (Moreno-Garcia et al. 2017).

Microalgae industrialization necessitates large-scale culturing, which is generally accomplished utilizing open systems such as ponds or closed systems such as photobioreactors to avoid contamination and evaporation while achieving larger biomass concentrations (Chen et al. 2011). On the other hand, photobioreactors have a high capital cost, are difficult to scale up, and have high shear stress resulting in cell damage. However, they provide excellent yields and control over a variety of culture parameters, thus it is worth investing in developing efficient and low-cost photobioreactors designs (Chang et al. 2017). Photobioreactors are bioreactors designed specifically for the cultivation of phototrophic organisms. As a bioreactor can control carbon dioxide, temperature, mixing of nutrients and

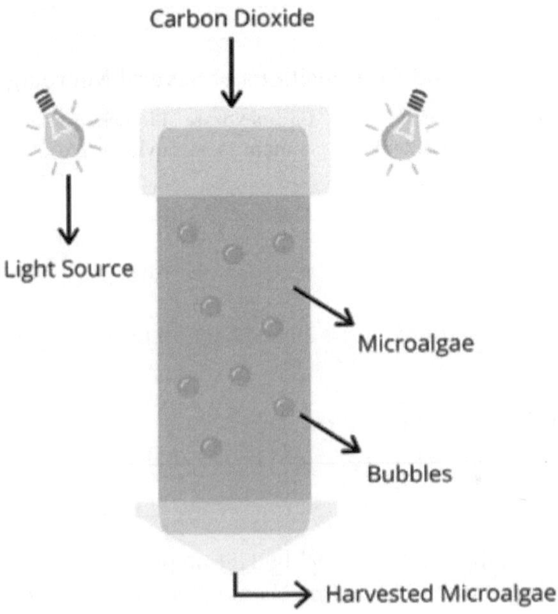

FIGURE 8.2 Schematic diagram of photobioreactor.

light intensity, the organism can attain considerably higher biomass than it could in nature. Figure 8.2 shows the schematic description of a photobioreactor system.

There are over 50,000 species of microalgae on the planet, showing a huge diversity of strains that can adapt to a wide range of environmental conditions (Richmond 2003). Chlorella has been grown to be commercially produced in circular ponds or similar raceway ponds, particularly in Asia, as well as in the Czech Republic and Bulgaria using open photobioreactors with inclined surfaces. Most recently, it has been chosen for the usage of closed tubular photobioreactor systems in Germany (Pulz 2001). *Botryococcus braunii*, an energy microalgal with high hydrocarbon content, is one of the most promising. *Dunaliella* species are photosynthetic, motile, unicellular biflagellate microalgae with no rigid cell wall which belong to the family of *Polyblepharidaceae*, phylum of *Chlorophyta*, and *Volvocales* (Hosseini Tafreshi and Shariati 2009). Table 8.1 shows various microalgae species that are used in cultivating microalgae under phototrophic conditions.

Microalgae biomass economic recovery is a difficult task, and, before microalgae-based biofuels may be commercially viable, expenses must be reduced to an acceptable level. The nature of the microalgae cells, the low concentration of biomass, and expensive build-up equipment costs, primarily when salty systems are used for microalgae cultivation, are all factors contributing to the challenges with harvesting and, as a result, the high biomass recovery costs. Harvesting is a crucial step after bulk algal production, which plays a critical

TABLE 8.1
Biomass Productivity and Compositions of Several Microalgae Species

Microalgae species	Protein content (%)	Lipid content (%)	Carbohydrate content (%)	Biomass productivity (g/L/d)	Reference
Chlorella sp.	50	32	14.5	0.45	(Enyidi 2017)
Chlorella vulgaris	15.67	41.51	20.99	0.10	(Yoo et al. 2010)
Botryococcus braunii	34.7	27.4	48	0.03	(Gouveia et al. 2017)
Dunaliella tertiolectra	2.87	61.32	21.69	0.12	(Shuping et al. 2010)
Scenedesmus obliquus	4.66	30.38	13.41	0.06	(Gupta et al. 2016)

role in determining the algal biofuels' budgeting procedure. For harvesting algal biomass, a variety of chemical and mechanical processes can be used, including centrifugation, screening, flotation, filtering, flocculation, and electrophoresis, as well as gravity sedimentation (Uduman et al. 2010). Bulk harvesting separates microalgae from bulk suspension, whereas thickening concentrates the slurry of microalgae following bulk harvesting (Lam, Khoo, and Lee 2019). The centrifuge technique is the most essential and effective way for separating algal biomass in harvesting, although due to the high operational and functional costs, it is primarily used on high-value goods (Molina Grima et al. 2003).

To form a large biomass collection, flocculation is the process that should be considered, and this procedure goes by circulating algal cells for settling. Auto-flocculation occurs when carbonates precipitate at a high pH with algal cells due to carbon dioxide intake by the algae (Vandamme, Foubert, and Muylaert 2013). Chemical flocculant that contaminates biomass is a significant problem in microalgal harvesting by flocculation as it can affect the end product (Branyikova et al. 2018). Gravity sedimentation is a straightforward method for separating algae from water and wastewater. Flotation is a sort of gravity sedimentation that is more effective and advantageous than sedimentation as it can capture pieces with thickness of less than 500 μm (de Souza Leite and Daniel 2020). Membrane filtration eliminates the need for chemicals, eliminates phase transitions, and ensures 100% solid retention (Mat Aron et al. 2021). Modifying the configuration of the filtration system itself by integrating auxiliaries such as air scouring and vibration into the filtration systems is one of the most popular ways now employed in anti-fouling harvesting (Farahah Mohd Khairuddin, Idris, and Irfan 2019). Centrifugation harvesting is characterized by high-capture efficiency under significant consumption of energy and limited flow rates, which accounts for around 90%. Centrifugation is a separating process for particles from a solution where then the particles are captured in the centrifuge tube (Dassey and Theegala 2013).

Microalgae are also utilized as a green method to extract compounds of bioactive nature such as pigments, proteins, polysaccharides, and fatty acids (Herrero et al. 2015). Carotenoids are organic pigments found in chloroplasts and chromoplasts that belong to the tetraterpenes class. They are made up of eight isoprene units (Waldron 2014). The extraction of carotenoids from biomass in solid form is, thereafter, isolating carotenoids from other contaminating compounds in a purification step. The rupture of microalgae cell walls is one of the most significant tasks in the extraction of carotenoids from biomass (Günerken et al. 2015). High-pressure cell disruption was the most efficient approach for protein extraction into an aqueous medium, but it was insufficient for green microalgae, since only about half of the proteins were recovered, indicating that more energy and hence greater expenditures would be necessary for green algae (Tang et al. 2020). Extracellular polysaccharides have been extracted and purified from a broad range of species of microalgae, including *Ankistrodesmus angustus* and *Amphora* sp., using novel technologies such as membrane filtrations (Patel et al. 2013). Extraction of lipids and, in particular, saturated fats from microalgae is comparable to that of pigments, with a first phase of the cell cycle disintegration continued by the use of organic solvents to recover the lipids component from the algae (Khoo et al. 2020).

8.2.2 Application of Microalgae Toward the Twenty-First Century

During bioenergy production, microalgae could produce a variety of value-added products with industrial, commercial, and environmental applications. Although these value-added products have the potential to counterbalance the high cost of downstream renewable fuel processing, their production has not been thoroughly investigated. This section provides a critical assessment of the current status of microalgae biotechnological applications for human benefit, as well as potential topics for future research and development.

The present depletion of global fossil-fuel supplies and rising greenhouse-gas emissions, which are causing global warming, polar ice melting, and rising sea levels, are important global concerns (Olguín 2012). As a result of the growing demand for alternative fuels that emit little greenhouse gases, biofuels have become a very appealing option for energy customers. Biofuels, or fuels made from biomass, provide new potential to diversify sources of revenue and fuel, create long-term fossil fuel substitutes, minimize greenhouse gas emissions, accelerate transportation fuel decarbonization, and increase energy supply security (Mata, Martins, and Caetano 2010). Biodiesel and bioethanol, the two most prevalent and successful biofuels, are designed to replace conventional liquid fuels such as diesel and gasoline. The processes of converting lipids to biodiesel by transesterification and primary carbohydrates to ethanol by saccharification are functions of algal components. Furthermore, there are several more roles involved in applications such as producing hydrogen gas by biosynthesis, the output of gasoline by cracking, and gasification of biomass to syngas (Amin and Prabandono 2017).

Due to their high protein content, microalgae are a valuable source of active ingredients for the cosmetic industry. Microalgae contain amino acids of around 5–20%; carbohydrates up to 20%; lipids; vitamins primarily B, C, and A; and traces of elements such as zinc, copper, and iron (Couteau and Coiffard 2018). Melanin, a complex polymer pigment that gives human skin its colour and also serves as a protective barrier for human skin cells, absorbs UV rays when skin is exposed to them for a prolonged period (Brenner and Hearing 2008). Skin pigmentation is caused by an excessive amount of melanin being produced, which must be controlled. Fucoxanthin, which is an algal pigment from the brown algae *Macrocystis*, *Alaria chorda*, and *Laminaria japonica*, can assist in inhibiting skin diseases promoters such as tyrosinase activity and melanogenesis (Foo et al. 2021). Sun protection, skincare, and hair care products contain the algae *Arthrospira* and *Chlorella* (Spolaore et al. 2006). The extract of *C. vulgaris* increases collagen production in the skin, resulting in the reduction of wrinkles and the formation of tissue regeneration (Thiyagarasaiyar et al. 2020).

Microalgae for human nutrition are currently available in a variety of formats, including pills, capsules, and liquids. Sterols are produced by microalgae and are used to treat cardiovascular disorders. *Spirulina* sp. is said to create clionasterol, which helps vascular cells avoid illness (Barrow and Shahidi 2017). Microalgae produce a variety of antioxidant chemicals, including taxanthin, β-carotene, dimethylsulfoniopropionate, and mycosporines (Shanab et al. 2012). These antioxidant substances can help protect against oxidative damage. *Chlorella* is said to be a food source as it is high in proteins denoted approximately at 51–58% dry weight, carotenoids, and other vitamins (Becker 2003). Microalgae also include β-glucan, an immunological stimulant that helps to lower free radicals and blood lipids (Barkia, Saari, and Manning 2019). Microalgae have gained prominence as a natural source of bioactive molecules due to their ability to manufacture bioactive chemicals via biological processes that are otherwise difficult to synthesize using chemical approaches. The pharmaceutical industry extracts bioactive microalgal substances including linolenic acid, palmitoleic acid, and cyanovirin, oleic acid, which have been shown to produce antibacterial, antioxidant, anticoagulant, antitumoral, anti-inflammatory, and antitumoral properties, with the potential to reduce and prevent disease (Mobin, Chowdhury, and Alam 2019).

For aquaculture animals, the source of nutrients is a principal element for the food chain. In the food chain, microalgae are the major producers as they contain nutrients that are necessary for larval and adult animal growth and survival. *Chlorella*, *Tetraselmis*, *Isochrysis*, *Pavlova*, *Phaeodactylum*, *Chaetoceros*, *Nannochloropsis*, *Skeletonema*, and *Thalassiosira* are the most commonly utilized species, with a combination of several species frequently used to offer the correct balance of protein, lipids, and vital micronutrients (Charoonnart, Purton, and Saksmerprome 2018). Microalgae are employed as fish feed in aquaculture by coloring farmed salmonids, for stabilizing, enhancing the quality of culture medium, and inducing key biological processes in red aquatic species, as well as for boosting fish immune systems (Khoo et al. 2019). Microalgae can be operated as a source of protein in chicken crops at a rate of 5–10% (Becker 2003). However, feeding microalgae to chickens for an extended period and at larger concentrations will have negative consequences. Microalgae has a negative impact on the color of the skin and shanks of a broiler, as well as the yolk of an egg (Świątkiewicz, Arczewska-Włosek, and Józefiak 2015).

Microalgae biomass can be pyrolyzed at 350–700 °C in the absence of oxygen to produce bio-oil, charcoal, and syngas (Basu 2018). The particular result of this procedure is that there is the production of biochar. Biochar which is a carbon-stable, rich charcoal can be utilized as a biofertilizer as well as a carbon sequestration provenance (Z. Khan et al. 2021). Blue–green algae can improve soil physicochemical qualities while also increasing biomass yield and reducing nitrogen fertilizer use. Microalgae can also boost residual nitrogen and carbon in the soil, as well as improve pH and electrical conductivity (Dineshkumar et al. 2018). Furthermore, environmental toxicants' monitoring, bioremediation, and bioassay all use microalgae in environmental biotechnology (Pacheco et al. 2020). Contaminants released into wastewater without adequate treatment is a major public health concern. Microalgae may consume organic substances like nitrogen and phosphorus that are commonly found in industrial wastewater (Chai et al. 2021). Furthermore, microalgae can help to mitigate the detrimental effects of industrial effluent and sewage containing nitrogenous waste from water treatment or fish farming/aquaculture (Abdel-Raouf, Al-Homaidan, and Ibraheem 2012).

8.3 RECENT TECHNOLOGIES IMPLEMENTATION FOR SMART TECHNOLOGY SYSTEM

Industry 4.0 is a new industrial stage in which several developing technologies are coming together to deliver digital solutions. However, there is a paucity of knowledge about how businesses use these technologies. As a result, this report emphasizes more on how microalgae applications are using Industry 4.0 technologies. Industry 4.0 is associated with the widespread adoption of front-end technology, with smart manufacturing playing a key role. This section focuses on how smart technology systems are used for pretreatment of wastes before recovery and implementation of microalgae. Moreover, challenges with the utilization of microalgae are pointed out with some improvements that can be done.

8.3.1 Pretreatment Method for Waste

Pretreatment of liquid and solid wastes is an important aspect of any decommissioning procedure (Chong et al. 2021). The goal of the pretreatment program is to prevent contaminants from entering the wastewater collection and transmission system, causing equipment damage, and interfering with the wastewater treatment process (Wheeler and de Rome 2002). The program is critical for protecting workers, the general public, and the environment. Mechanical, biological, and chemical methods can all be classified as pretreatment procedures (Galbe, Wallberg, and Zacchi 2011). Shredding, screening, sorting, and separation of ferrous components are all part of the former. The waste's particular density can be increased as a result of the volume reduction and increase in specific areas (Abdel-Shafy and Mansour 2018). Screening, washing compactor cleans for easier handling, grit removal, sand classifier, and, finally, grease and oil removal are all also part of the pretreatment process.

Wastewater mainly contains significant solid residues that must be filtered out efficiently. To achieve this, membrane technologies such as ultrafiltration, microfiltration, and nanofiltration will be explored further. Instantaneous retention of species and product flow through the semipermeable membrane are characterize as membrane separation techniques. High-permeate flow and selectivity; strong mechanical, chemical, and thermal stability of membrane materials; minimum fouling during operation; and good compatibility with the working environment all contribute to membrane performance (Bowen and Jenner 1995).

Membranes are classified as being porous or nonporous based on the process through which separation is achieved. Nonporous membrane separation relies on physicochemical interactions between permeating components and the membrane's material to achieve the best selectivity (Rackley 2017). Porous membranes, on the other hand, are mechanically separated by size exclusion as in a molecular sieve (Grosgogeat et al. 1991). Membranes with various pore-size distributions and physical qualities are used to filter out a variety of contaminants. As microfiltration membranes have the biggest pore size of all of the membrane technologies, they reject huge particles as well as germs and bacteria (Ismail and Goh 2015). On the other hand, ultrafiltration membranes have fewer pores than microfiltration membranes; hence, they can reject viruses and other macromolecules that are just minimally soluble, such as proteins (Pal 2003). Nanofiltration membranes are rather porous, and their performance can be beneficial in some applications, such as the manufacture of drinking water, by lowering post-treatment costs such as demineralization (Abdel-Fatah 2018).

For the treatment of polluted effluent, biological pretreatment technologies have historically been applied. The potential of various bacterial strains to produce a variety of by-products that can be advantageous or harmful to scientific study has increased the need to better understand the impact of harnessing biological processes rather than chemical or physical processes (Quezada et al. 2007). Anaerobic bacteria are responsible for the fermentation of methane gas from sewage sludge, as well as assisting in the degradation of macromolecular organic materials into smaller compounds (Achinas, Achinas, and Euverink 2020). As a result, they serve a significant role in wastewater treatment procedures. In technologies like the dissolved air floatation pretreatment tank, aerobic treatment has been employed to pretreat wastewater. As this treatment process uses a lot of energy for aeration and produces a lot of sludge, the aerobic bacteria that had access to oxygen in the pretreatment tank are unable to continue replicating in oxygen-depleted bioreactors (Kosseva 2020).

Bio-flocculants are extracellular biopolymeric compounds released by algae, yeast, and bacteria that are nontoxic and biodegradable. Bio-flocculants, in general, cause particle and cell aggregation through bridging and charge

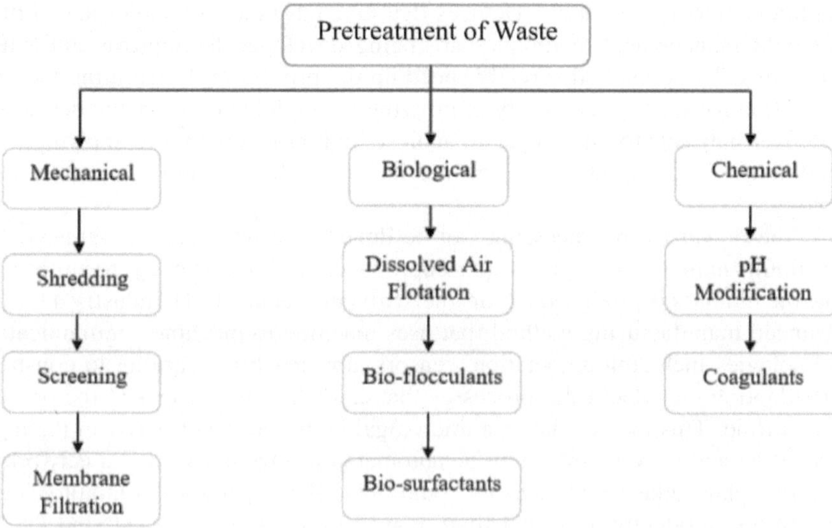

FIGURE 8.3 Pretreatment method of waste.

neutralization. Bio-flocculant composition is made up of macromolecular compounds like protein and polysaccharide-protein and is determined by the type of microorganisms that produce bio-flocculants (Maćczak, Kaczmarek, and Ziegler-Borowska 2020). Furthermore, due to the environmental and ecological benefits, biosurfactants have gotten a lot of attention. The uptake of hydrocarbons is aided by biosurfactants produced by microorganisms in the environment (Karlapudi et al. 2018).

Oil emulsions, surfactants, suspended particles, and dissolved metals are separated from a liquid water phase during the chemical pretreatment stage. pH modification and/or polymer treatment are used to achieve chemical precipitation. These processes bind the contaminants together using a coagulant, resulting in an insoluble particle that is heavier than water. The negative electrical charge on particles is neutralized during coagulation therapy, which destabilizes the forces that keep colloids apart. Coagulants used in water treatment are made up of positively charged molecules that, when combined with water, neutralize the charge (Woodard & Curran, Inc. 2006). These pretreatment technologies are necessary to avoid contaminating microalgae during harvesting and to improve the quality of biomass produced. Figure 8.3 summarizes the several pretreatment methods that are implemented before microalgae can act as a recovery agent.

8.3.2 Types of Technologies Implemented Using Microalgae

Recent technical advancements are attempting to overcome obstacles such as photosynthesis with a low solar energy conversion efficiency and ideal cultivation conditions that are unknown. Devices that take advantage of sophisticated fluid and light management techniques are being developed to improve cultivation conditions. These methods greatly speed up the process of determining the best cultivation parameters. A variety of growing methodologies are being developed to deliver nutrients to microalgae to achieve high productivities to apply optimal conditions. The methods discussed are industry 4.0, phenomics, and synthetic biology.

If the downstream processing can be done in an integrated biorefinery, the maximum number of products and coproducts can be retrieved, regardless of how the biomass is produced (Rodríguez-Sifuentes et al. 2021). Industry 4.0 is an advanced manufacturing method that uses machine-to-machine communication technologies including automation, sensors, and machine learning to construct self-adapting manufacturing processes that can adjust to changes in the process in real time. This means that in a microalgal biorefinery, not only can the algal cultivation and harvesting system be automated to save money, but a network of plug-and-play Industry 4.0 sensors could also allow operators to monitor algae growth and productivity in real time as well (Whitmore, Agarwal, and Da Xu 2015). Industry 4.0 takes it a step further by using sensor data to create a simulation, or digital twin, of the facility and the algae culture. The simulation can anticipate future cellular yield in real time and alter operations to match expected product demand while reducing waste (Tao et al. 2018).

Collecting high-dimensional phenotypic data on an organism-wide scale is referred to as phenomics (Houle, Govindaraju, and Omholt 2010). Although algal phenomics is still in its infancy, it has enormous potential in microalgal agriculture for food security, bioproduct sourcing, bioremediation, and carbon sequestration (Furbank and Tester 2011). The potential influence of phenomics techniques and technology on microalgae has been highlighted by recent breakthroughs in the field of plant phenomics. A recent phenomics-based study on *Arabidopsis thaliana* generated a mutant with improved pathogen resistance and photosynthetic growth, shattering a near-dogma among plant experts about the purported trade-off between growth and defense (Campos et al. 2016).

The lack of searchable phenomics databases is a fundamental constraint of today's microalgal phenomics. Investments in developing a comprehensive phenotypic database will be required to bring this power to the field of algae research. Standards for data sharing, knowledge retrieval, and ontology annotation have already been established in the field of plant phenomics for microalgae (Neveu et al. 2019). Standard microbiology sensors including fluorescence and absorption spectrophotometers, hyperspectral cameras, and flow cytometers can be utilized to assess phenotype data from high-throughput algal culture formats utilizing data-processing techniques currently used for model bacteria (Fernandez-Ricaud et al. 2016). Choosing growth settings that lead microalgae to display a variety of phenotypes based on their genetic predispositions will be one hurdle for establishing microalgal phenomics databases. Poorly designed algal phenomics reference habitats may result in insufficient segregation, or, worse, the conditions may be so unlike from large-scale cultivation that the measured phenotypes are

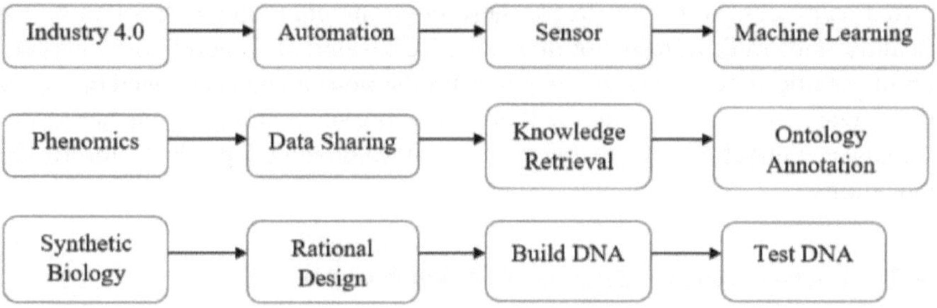

FIGURE 8.4 Smart technologies of microalgae.

misleading and useless to large-scale operations (Rawat et al. 2013). Once these obstacles have been overcome, algal phenomics will have a significant impact on algal biotechnology, allowing microalgae to be developed as new bioproducts and pharmaceutical workhorses.

Engineering ideas are applied to the rational design of biological entities in synthetic biology. A biological system is considered to be a collection of defined genetic elements that can be manipulated and reassembled to change existing functionalities or develop them from scratch in different host species in this subject. To acquire optimum metabolic configurations for biotechnological applications, genetic designs are changed through iterations of a Design–Build–Test–Learn cycle (Nielsen and Keasling 2016). Synthetic biology in microalgae will combine this powerful new technique with the advantages of a photosynthetic microbial host to create unique production strains tailored to future environmental issues.

The greater availability of sequenced genomes across many algal lineages has aided the rapid evolution of tools for the genetic engineering of microalgae. Natural transformation, electroporation, bead beating, biolistic transformation, and conjugative plasmid transfer are among the strategies for genetic transformation in microalgae that have been optimized for numerous species (Qin, Lin, and Jiang 2012). Genetic engineers are now limited in the number of designs they can assemble and test. Increased integration of computational design and automation with biology, on the other hand, is expected to quickly transform this paradigm (Fabris et al. 2020). Figure 8.4 shows the summary of smart technologies implemented using microalgae.

8.4 SUSTAINABILITY OF MICROALGAE

At the moment, sustainability is a key component in the development of manufacturing systems aimed at supplying future energy and material resources. Sustainability covers several aspects such as environmental, economic, and societal impacts that create a balance towards the technology. Engineering this natural process on an industrial scale necessitates the processing and transportation of land, energy, water, nutrients, and materials. If seawater or brackish water can be used as process water for crops, these environmental credentials can be significantly improved. In terms of nutrients, freshwater, and energy, the long-term viability of microalgae growth for energy production can be questioned (Guieysse and Plouviez 2021). This section will provide more sustainable approaches toward microalgae cultivation and recovery of high value-added products for further applications.

8.4.1 Environmental Assessment of Microalgae

Indicators such as water quality requirements and water usage must be considered when evaluating the environmental sustainability of an aquatic-biomass-based growing system. Water quality requirements for microalgae growth vary depending on alga strains. To relieve pressure on natural water supplies, low-grade wastes might be used as a supply of water. To relieve pressure on natural water supplies, low-grade wastes might be used as a supply of water (Christenson and Sims 2011). Using household and industrial wastewater as a supply of water and nutrients, as well as wastewater clean-up, could be economically and environmentally beneficial for large-scale microalgae growth.

Microalgae are becoming more well recognized to be one of the most prolific biological systems for creating biomass and absorbing carbon. In open ponds, carbon dioxide or bicarbonate capture efficiency of up to 90% has been reported (Sayre 2010). The ability to trap carbon dioxide in ponds in a nongaseous state as bicarbonate to nourish algal growth is one advantage of aquatic carbon capture and biomass production systems. Ponds located near carbon dioxide point sources provide several potential cost and energy savings benefits. Integrated powered plant–algal pond facilities would lower carbon dioxide transportation costs, create limited waste heat from the power plant for warming ponds in the winter, and perhaps grant the utility carbon credits (Wang et al. 2016). Heterotrophic microalgae, on the other hand, cannot assimilate carbon in the same way and must rely on an organic carbon supply (Bošnjaković and Sinaga 2020). By-products of bacterial decomposition of organic matter, such as acetate, or other highly biodegradable organic compounds, such as sugars from industrial sources, would often be found in wastewater streams that contain high carbon content.

Nitrogen is a crucial nutrient for microalgae (Yaakob et al. 2021). However, an overabundance of nitrogen in an aquatic environment can cause unmanageable microalgal blooms, which can lead to hazardous circumstances; thus, nitrogen recovery from wastewater via biological uptake for algal biomass formation could help to mitigate such harmful effects (Collet et al. 2011). Phosphorus is a nonrenewable material that can only be found in an inorganic state and must be mined or recovered from waste. As phosphorus is controlled by a small number of countries, supply is regulated by international policy (Garske and Ekardt 2021). Microalgae mass culture can be thought of as a regulated eutrophication process that must be effectively managed by adequate air supply and regular harvesting (Handler et al. 2012). Eutrophication, on the other hand, remains one of the most serious threats to biodiversity. Decomposition of dead algal biomass depletes oxygen in the water column, causing species that rely on oxygen for respiration to perish.

Pathogens and microalgae will coexist. Although algae growing in open ponds can inactivate pathogens, pathogens are likely to be present in harvested biomass or final process effluent if water is derived from waste streams, notably municipal or animal waste (Curtis, Mara, and Silva 1992). This will have an impact on the microalgal product's final use or at the very least the post-treatment it requires before it can be utilized in any product that could pose a health concern (Systems 2013). Those in charge of the algae farms face occupational health risks as well. Contamination is a risk in open ponds. This risk can be reduced by changing cultural conditions to make them unsuitable for native species. However, releasing non-native species may cause problems, especially if they outcompete native species (Usher et al. 2014).

The development of ponds may also result in the displacement of local animals due to habitat loss. Environmental Impact Assessment surveys can be performed to determine the magnitude of the impact that large-scale pond development might have (Geneletti 2002). While the direct effects of microalgae culture are most visible in water and land systems, large-scale microalgae production has a variety of possible atmospheric effects (Slade and Bauen 2013). Methane is another potent greenhouse gas, with a global warming potential of 84°C over 20 years, making large-scale emissions a worry in the context of climate change (Hatzikiriakos and Englezos 1993). Methane also contributes to the development of background ozone, which has ramifications for both air quality and climate. Methane is created via anaerobic breakdown by methanogenic bacteria (Lin et al. 2021). As of the continual oxygenation of the water in a well-managed microalgae system, there should be no anaerobic situations. This indicates that during the time of assessing the potential greenhouse gas emissions from microalgae farming, the production of aerobic methane is of special relevance (Keppler et al. 2006). Extensive research is needed to understand the potential of atmospheric pollution by microalgae with leading discoveries such as nitrous oxide, ammonia, and volatile organic compounds.

8.4.2 Economic and Societal Effects

In the biofuel sector, microalgae are becoming more important. They produce economically important metabolites for the biotech, food, pharmaceutical, agricultural, and cosmetic industries (Cardozo et al. 2007). The majority of techno-economic assessments concentrate solely on manufacturing costs. They frequently highlight the high production costs of microalgae biodiesel from oil transesterification and green diesel from oil hydrotreating, which range from 5.28 RM/L to 37.05 RM/L (Rosenberg et al. 2011). The dilution of microalgae biomass, which is typically approximately 0.5–1 kg/m^3 in open ponds and 5–10 kg/m^3 in photobioreactors, is one of these technological hurdles (Delrue et al. 2012). During cultivation, harvesting, and especially drying, massive amounts of water must be treated (Pugazhendhi et al. 2020). As a result, there is a lot of electricity and heat used.

One of the fundamental drawbacks of traditional techno-economic analysis is that it often examines only one technology for each phase of the process, preventing the comparison of several technologies and their optimized process routes. Through economical assessments, microalgae are the best alternative feedstock for third- generation biofuel production; however, the high cost of biomass production and biorefinery renders them currently uncompetitive with fossil fuels and established to be renewable energy resources (Behera et al. 2015). The commercialization of microalgae culture technologies is thus limited to a few profitable facilities around the world that produce very high-value products such as carotenoids, polyunsaturated fatty acids, and immune-stimulant polysaccharides, all of which have a high selling price that ensures a return on investment (Barsanti and Gualtieri 2018). Maintenance of the ponds, harvesting and de-watering, algal nutrient charges, as well as power and additional heating costs, are all part of the microalgae biotechnology operational costs which contribute to a large sum of money. Despite this, microalgae are known to be a promising and alternative method that depends highly on the location and regulations of culturing.

The term "social well-being" refers to the satisfaction of obtaining basic human requirements such as food, health, and shelter. Employment, income, and food security are indices of social well-being for bioenergy, with units of full-time equivalent jobs and daily household income in respective currencies (Boccard 2020). When it comes to algae biofuel facilities, employment and food security are particularly essential considerations. Providing cost-effective supply chains to investors attracts them to a region and creates new jobs (Henry and Bryan 2019).

Companies may be offered incentives by local governments to relocate to a specific location based on predicted hires. Government initiatives to promote the use of microalgae and to create workshops for its technology are essential as part of educating our society. Technical characteristics such as biomass productivity, lipid content, project duration, lipid extraction rate, and lipid conversion rate were found to be sensitive to employment numbers in the algal diesel sector in China, according to an investigation (Y. Yang et al. 2015). The deployment of enabling technologies could pave the way for social benefits such as increased employment and income in a region linked to algal-based biofuels. For example, in locations where freshwater is scarce, such as Australia, north-western Africa, and the Arabian Gulf, the use of saline water for growth could lead to the development of algae biofuel production (Nair and Paulose 2014). Figure 8.5 summarizes the sustainable utilization and impact of microalgae technologies for the production of useful bio-products.

8.5 CHALLENGES AND IMPROVEMENTS

As a result of the irreversible depletion of traditional fossil fuel reserves, widespread usage of fossil fuels has contributed to climate change in the global state, contamination to the environment, health difficulties, and a bottleneck in the supply of energy (Wheeler and de Rome 2002). The circumstance that complies with prohibitive capital and operating expenses results in the microalgal biofuels

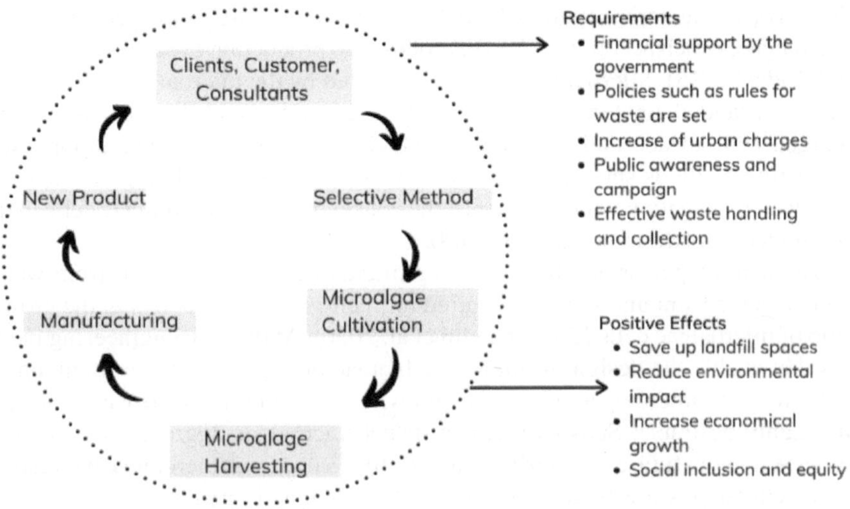

FIGURE 8.5 Sustainability Implementation of microalgae technology.

business facing a clear and serious problem which leads them to no economic production on a large scale for bulk use (Zhu 2015). Unacceptably high research and production costs will continue to be caused by a lack of understanding of microalgae, notably microalgal biofuel. For many years, exorbitant expenses, method of culture for microalgal biology, dewatering, harvesting, and demand for extraction have impeded the advancements of this technology (Ziolkowska and Simon 2014). Furthermore, the production costs may be considerably improved if the alga could be chosen, changed, or otherwise persuaded into generating more lipids or other high-value-added substances.

Open ponds were the first proposed microalgal production system, and they are still the most popular due to the usage of less energy and are more likely to be less expensive to build and operate. However, these bioreactors have poor control over culture conditions and contamination, limiting their use to microalgae that can withstand harsh environmental conditions such as high salinity or pH (Costa et al. 2019). Only several microalgae have been successfully cultivated in open ponds throughout the period, including Dunaliella, Chlorella, and Spirulina (ElFar et al. 2020). Since not all algal species flourish in highly selective settings, future advances in microalgal growth may proceed to closed systems instead of open systems. Carbon dioxide, which stimulates microalgae development, is the most expensive consumable. If flue gases are abundantly available, one viable option is to employ flue gases emitted by industrial sources, which can bring carbon dioxide's cost down to zero (Bhola et al. 2014).

Optimal harvesting techniques are those that are independent of the farmed species, utilize less energy and chemicals, and do not harm important goods during extraction (Barros et al. 2015). Various technologies such as ultrasound, electrolysis, magnetic separation, and immobilization are employed to recover microalgae, which account for 20–30% of overall production expenses (Thangavel and Sridevi 2015). Microalgal recovery is limited by the microalgae cell's diminutive size and their concentration in liquid culture, as well as the negatively charged cell surface and cell density, which hinder gravity precipitation. New harvesting methods should, in theory, rely on the constants of cell size, density, and electric potential, which influence the concentration of microalgae cells in the liquid medium (Branyikova et al. 2018).

This can be performed by improving microalgae cell separation from water, decreasing the amount of water required for their growth and boosting the growth ratio of microalgae cells (Pugazhendhi et al. 2020). At the end, engineering methods that enable the design of innovative bioreactors that use the least amount of water to grow microalgae while also allowing for the best cell-density detection are highly desirable. Sensors that can detect microalgae aggregations could be employed in the future (Maliaritsi et al. 2006). To improve microalgae harvesting, nanotechnologies are being deployed widely. Magnetic nanoparticles are presently being researched biotechnologically for microalgae cell harvesting, with promising results (Suparmaniam et al. 2019).

Flocculation and centrifugation with ferric chloride, alum, chitosan, and hydrophobic absorbents or adsorbents are commonly used in the production of high-value microalgae products (Ananthi et al. 2021). However, harvesting algal biomass using these technologies is reasonably priced and energy-intensive. For the evolution of harvesting techniques, there are several options. Compelling algae to settle through forced flocculation is one of the low-energy methods (Pugazhendhi et al. 2019). Furthermore, another biotech issue is using microalgae to extract lipids, which is challenging due to the thick cell wall, which makes oil extraction problematic.

The goal of current research is to achieve this through a combination of biological and engineering approaches. There are several more challenges that should be considered and improvements allocated to tackle such problems. This includes large-scale outdoor microalgae culture productivity which is shown to be relatively low (Fuentes et al. 2020). There is a deficiency in preventing contamination by predators and other algae species, as well as reducing temperature fluctuations and water loss owing to evaporation. Cheap advancements of an efficient reactor with optimizing the supply of light and oxygen would promote activity with the technology as well as helping employees to have accurate parameters regulations. There is also a need to improve resource use and productivity using a biorefinery strategy, as well as reduce environmental impact with the water, energy, and nutrient recycling (Griffiths et al. 2011).

8.6 FUTURE PROSPECTS

It is widely acknowledged that wastes, either solid waste that causes the presence of leachate or directly from wastewater, contain important nutritives that could be restored and utilized to produce products that are regarded as commercial, industrial, and environmental applications. Extraction technologies have advanced in recent years; nonetheless, various obstacles must be solved to attain long-term sustainability. To date, several studies have been carried out to determine the viability of biorefinery production using waste active sludge. Protein and enzyme extraction is known to be costly due to their small size, which makes it more difficult, and it demands biotechnology that has been greatly improved, such as ultrafiltration and microfiltration (Peter et al. 2021). Additional assessment of operating settings and nontoxic strain selection is required to make progress on bioproduct recovery. To boost production and economic development, rather than lab-scale research, biorefinery advances might be done on a bigger platform (M. Yang et al. 2020).

Microalgae have the potential to deliver additional added-value goods due to their diversity and adaptability, as well as their capacity to grow on the non-cultivable ground and their comparatively nontoxic nature. This technique will surely meet the demand for innovative biopharmaceuticals, such as antibodies and vaccines (Olaizola 2003). At the end, microalgal biodiversity may provide practical remedies to the environment's increasing pollution. As a result, if sufficient resources are committed in fundamental science, the future may well see considerably better, fast-growing, novel, or genetically modified microalgal species with significantly improved capacities to carry out successful photosynthesis under high light.

The goal of research should be to integrate value-product recovery technologies including anaerobic digestion and chemical crystallization, as well as a sustainable microalgae approach based on full-scale implementation experience, in a seamless manner. The particular reason for this circumstance is that economically viable paths differ at the state, national, and international levels, and the economic analysis must consider the overall position of the integrated recovery process. Reduced nutrient recovery running costs should be another research target for achieving sustainability. A range of applicable technologies are predicted to increase end-user requests for high nutrition content, low humidity, and low heavy metal and pathogen levels (Ngo et al. 2017). To address faults and management issues, data gathering and frequent checks on the technologies installed are recommended. Because different solutions and the latest automation are accessible to extend the life cycle of technologies, efficiency improvement is an ongoing activity (Stephens et al. 2010).

It is critical to combine cutting-edge biotechnology with industrial development to engineer the massive production of microalgal biomass (Chowdury, Nahar, and Deb 2020). The prospects for microalgal biotechnology may lead to a wide range of technical solutions for microalgal production. Bio-prospecting is expected to be useful in identifying desired microalgal characteristics with high-value co-products while growing on a low-cost basis. Ecological, genetic, and biochemical advancements of microalgal species connected with the amalgamation of co-located inoculation sustainability of entire operation should be researched and combined with commercial applications to optimize the economic viability of algal-based VAPs (Kothari et al. 2017).

8.7 CONCLUSIONS

This book chapter focuses on smart microalgae biotechnology and recent advancements in automation with a prime focus on sustainability in terms of operation. Wastes that were analyzed as part of the assessments are mainly domestic and industrial wastes that contain high-value products recovered by microalgae into biomass. Open ponds for microalgae cultivation are the most used methods around the world; however, few concerns have arisen such as contamination and how much it takes up space. On the other hand, the photobioreactor offers effective production of microalgae biomass and has efficient parameter control. Pretreatment of waste is necessary to ease the process and efficiency of microalgae which also increases the quality of biomass produced. Wide applications of microalgae biomass are to produce bio-fuels, medicines, cosmetic products, animal feed, and bio-fertilizers. Environmental assessments are done to ensure that microalgae exhibit higher positive environmental impact such as application in carbon sequestration and act as eutrophication control. The complete resource recovery from microalgae adds to a technique that is both efficient and long-lasting. In the near future, the market for microalgae-based commodities is likely to grow, necessitating the progress of microalgae-based technology.

ACKNOWLEDGMENTS

The main authors acknowledge the comments and ideas contributed by all the coauthors throughout this review. This work was also supported by the UCSI University Research and Innovation Grant under project code [REIG-FAS-2020/028].

REFERENCES

Abdel-Fatah, Mona A. 2018. "Nanofiltration Systems and Applications in Wastewater Treatment: Review Article." *Ain Shams Engineering Journal* 9 (4): 3077–92. https://doi.org/10.1016/j.asej.2018.08.001.

Abdel-Raouf, N., A. A. Al-Homaidan, and I. B. M. Ibraheem. 2012. "Microalgae and Wastewater Treatment." *Saudi Journal of Biological Sciences* 19 (3). King Saud University: 257–75. doi:10.1016/j.sjbs.2012.04.005.

Abdel-Shafy, Hussein I., and Mona S. M. Mansour. 2018. "Solid Waste Issue: Sources, Composition, Disposal, Recycling, and Valorization." *Egyptian Journal of Petroleum* 27 (4). Egyptian Petroleum Research Institute: 1275–90. doi:10.1016/j.ejpe.2018.07.003.

Achinas, Spyridon, Vasileios Achinas, and Gerrit Jan Willem Euverink. 2020. "Chapter 2 - Microbiology and Biochemistry of Anaerobic Digesters: An Overview." In *Bioreactors*, edited by Lakhveer Singh, Abu Yousuf, and Durga Madhab Mahapatra, 17–26. Elsevier. https://doi.org/10.1016/B978-0-12-821264-6.00002-4.

Amin, Sarmidi, and Kurniadhi Prabandono. 2017. "Biodiesel Production from Microalgae." *Algal Biofuels* 182 (January): 66–102. doi:10.1201/9781315152547.

Ananthi, V., P. Balaji, Raveendran Sindhu, Sang-Hyoun Kim, Arivalagan Pugazhendhi, and A. Arun. 2021. "A Critical Review on Different Harvesting Techniques for Algal Based Biodiesel Production." *Science of The Total Environment* 780: 146467. https://doi.org/10.1016/j.scitotenv.2021.146467.

Barkia, Ines, Nazamid Saari, and Schonna R. Manning. 2019. "Microalgae for High-Value Products towards Human Health and Nutrition." *Marine Drugs* 17 (5): 1–29. doi:10.3390/md17050304.

Barros, Ana I., Ana L. Gonçalves, Manuel Simões, and José C. M. Pires. 2015. "Harvesting Techniques Applied to Microalgae: A Review." *Renewable and Sustainable Energy Reviews* 41: 1489–1500. doi:10.1016/j.rser.2014.09.037.

Barrow, C., and F. Shahidi. 2017. *Marine Nutraceuticals and Functional Foods*. CRC Press. doi:10.1201/9781420015812.

Barsanti, Laura, and Paolo Gualtieri. 2018. "Is Exploitation of Microalgae Economically and Energetically Sustainable?" *Algal Research* 31: 107–15. https://doi.org/10.1016/j.algal.2018.02.001.

Basu, Prabir. 2018. "Chapter 5 - Pyrolysis." In *Biomass Gasification, Pyrolysis and Torrefaction (Third Edition)*, edited by Prabir Basu, 3rd ed., 155–87. Academic Press. https://doi.org/10.1016/B978-0-12-812992-0.00005-4.

Becker, Wolfgang. 2003. "Microalgae in Human and Animal Nutrition." In *Handbook of Microalgal Culture*, 312–51. John Wiley & Sons, Ltd. https://doi.org/10.1002/9780470995280.ch18.

Begum, Hasina, Fatimah M. D. Yusoff, Sanjoy Banerjee, Helena Khatoon, and Mohamed Shariff. 2016. "Availability and Utilization of Pigments from Microalgae." *Critical Reviews in Food Science and Nutrition* 56 (13): 2209–22. doi:10.1080/10408398.2013.764841.

Behera, Shuvashish, Richa Singh, Richa Arora, Nilesh Kumar Sharma, Madhulika Shukla, and Sachin Kumar. 2015. "Scope of Algae as Third Generation Biofuels." *Frontiers in Bioengineering and Biotechnology* 2: 90. doi:10.3389/fbioe.2014.00090.

Bhola, V., F. Swalaha, R. Ranjith Kumar, M. Singh, and F. Bux. 2014. "Overview of the Potential of Microalgae for CO2 Sequestration." *International Journal of Environmental Science and Technology* 11 (7): 2103–18. doi:10.1007/s13762-013-0487-6.

Boccard, Nicolas. 2020. "Social Indicators of Wellbeing." *SSRN Electronic Journal* (April). doi:10.2139/ssrn.2891720.

Bošnjaković, Mladen, and Nazaruddin Sinaga. 2020. "The Perspective of Large-Scale Production of Algae Biodiesel." *Applied Sciences* 10 (22): 1–26. doi:10.3390/app10228181.

Bowen, W. Richard, and Frank Jenner. 1995. "Theoretical Descriptions of Membrane Filtration of Colloids and Fine Particles: An Assessment and Review." *Advances in Colloid and Interface Science* 56: 141–200. https://doi.org/10.1016/0001-8686(94)00232-2.

Branyikova, Irena, Gita Prochazkova, Tomas Potocar, Zuzana Jezkova, and Tomas Branyik. 2018. "Harvesting of Microalgae by Flocculation." *Fermentation* 4 (4). doi:10.3390/fermentation4040093.

Brenner, Michaela, and Vincent J. Hearing. 2008. "The Protective Role of Melanin against UV Damage in Human Skin." *Photochemistry and Photobiology* 84 (3): 539–49. doi:10.1111/j.1751-1097.2007.00226.x.

Campos, Marcelo L., Yuki Yoshida, Ian T. Major, Dalton de Oliveira Ferreira, Sarathi M. Weraduwage, John E. Froehlich, Brendan F. Johnson, et al. 2016. "Rewiring of Jasmonate and Phytochrome B Signalling Uncouples Plant Growth-Defense Tradeoffs." *Nature Communications* 7 (1): 12570. doi:10.1038/ncomms12570.

Cardozo, Karina H. M., Thais Guaratini, Marcelo P. Barros, Vanessa R. Falcão, Angela P. Tonon, Norberto P. Lopes, Sara Campos, et al. 2007. "Metabolites from Algae with Economical Impact." *Comparative Biochemistry and Physiology. Toxicology & Pharmacology: CBP* 146 (1–2). United States: 60–78. doi:10.1016/j.cbpc.2006.05.007.

Chai, Wai Siong, Wee Gee Tan, Heli Siti Halimatul Munawaroh, Vijai Kumar Gupta, Shih-Hsin Ho, and Pau Loke Show. 2021. "Multifaceted Roles of Microalgae in the

Application of Wastewater Biotreatment: A Review." *Environmental Pollution* 269: 116236. doi:10.1016/j.envpol.2020.116236.

Chang, J. S., P. L. Show, T. C. Ling, C. Y. Chen, S. H. Ho, C. H. Tan, D. Nagarajan, and W. N. Phong. 2017. *Photobioreactors. Current Developments in Biotechnology and Bioengineering: Bioprocesses, Bioreactors and Controls*. doi:10.1016/B978-0-444-63663-8.00011-2.

Charoonnart, Patai, Saul Purton, and Vanvimon Saksmerprome. 2018. "Applications of Microalgal Biotechnology for Disease Control in Aquaculture." *Biology* 7 (2). MDPI: 24. doi:10.3390/biology7020024.

Chen, Chun Yen, Kuei Ling Yeh, Rifka Aisyah, Duu Jong Lee, and Jo Shu Chang. 2011. "Cultivation, Photobioreactor Design and Harvesting of Microalgae for Biodiesel Production: A Critical Review." *Bioresource Technology* 102 (1). Elsevier Ltd: 71–81. doi:10.1016/j.biortech.2010.06.159.

Chew, Kit Wayne, Kuan Shiong Khoo, Hui Thung Foo, Shir Reen Chia, Rashmi Walvekar, and Siew Shee Lim. 2021. "Algae Utilization and Its Role in the Development of Green Cities." *Chemosphere* 268 (April). England: 129322. doi:10.1016/j.chemosphere.2020.129322.

Chong, Jun Wei Roy, Guo Yong Yew, Kuan Shiong Khoo, Shih-Hsin Ho, and Pau Loke Show. 2021. "Recent Advances on Food Waste Pretreatment Technology via Microalgae for Source of Polyhydroxyalkanoates." *Journal of Environmental Management* 293: 112782. https://doi.org/10.1016/j.jenvman.2021.112782.

Chowdury, Kamrul Hasan, Nurun Nahar, and Ujjwal Kumar Deb. 2020. "The Growth Factors Involved in Microalgae Cultivation for Biofuel Production: A Review." *Computational Water, Energy, and Environmental Engineering* 9 (4): 185–215. doi:10.4236/cweee.2020.94012.

Christenson, Logan, and Ronald Sims. 2011. "Production and Harvesting of Microalgae for Wastewater Treatment, Biofuels, and Bioproducts." *Biotechnology Advances* 29 (6): 686–702. https://doi.org/10.1016/j.biotechadv.2011.05.015.

Collet, Pierre, Arnaud Hélias, Laurent Lardon, Monique Ras, Romy-Alice Goy, and Jean-Philippe Steyer. 2011. "Life-Cycle Assessment of Microalgae Culture Coupled to Biogas Production." *Bioresource Technology* 102 (1): 207–14. https://doi.org/10.1016/j.biortech.2010.06.154.

Costa, Jorge Alberto Vieira, Bárbara Catarina Bastos Freitas, Thaisa Duarte Santos, Bryan Gregory Mitchell, and Michele Greque Morais. 2019. "Chapter 9 - Open Pond Systems for Microalgal Culture." In *Biofuels from Algae (Second Edition)*, edited by Ashok Pandey, Jo-Shu Chang, Carlos Ricardo Soccol, Duu-Jong Lee, and Yusuf Chisti, 2nd ed., 199–223. Biomass, Biofuels, Biochemicals. Elsevier. https://doi.org/10.1016/B978-0-444-64192-2.00009-3.

Couteau, Céline, and Laurence Coiffard. 2018. "Chapter 15 - Microalgal Application in Cosmetics." In *Microalgae in Health and Disease Prevention*, edited by Ira A Levine and Joël Fleurence, 317–23. Academic Press. https://doi.org/10.1016/B978-0-12-811405-6.00015-3.

Curtis, T. P., D. D. Mara, and S. A. Silva. 1992. "The Effect of Sunlight on Faecal Coliforms in Ponds: Implications for Research and Design." *Water Science and Technology* 26 (7–8): 1729–38. doi:10.2166/wst.1992.0616.

Dassey, Adam J., and Chandra S. Theegala. 2013. "Harvesting Economics and Strategies Using Centrifugation for Cost Effective Separation of Microalgae Cells for Biodiesel Applications." *Bioresource Technology* 128: 241–45. https://doi.org/10.1016/j.biortech.2012.10.061.

Delrue, F., P.-A. Setier, C. Sahut, L. Cournac, A. Roubaud, G. Peltier, and A.-K. Froment. 2012. "An Economic, Sustainability, and Energetic Model of Biodiesel Production from Microalgae." *Bioresource Technology* 111: 191–200. https://doi.org/10.1016/j.biortech.2012.02.020.

Demirbas, Ayhan. 2011. "Waste Management, Waste Resource Facilities and Waste Conversion Processes." *Energy Conversion and Management* 52 (2): 1280–87. https://doi.org/10.1016/j.enconman.2010.09.025.

Dineshkumar, R., R. Kumaravel, J. Gopalsamy, Mohammad Nurul Azim Sikder, and P. Sampathkumar. 2018. "Microalgae as Bio-Fertilizers for Rice Growth and Seed Yield Productivity." *Waste and Biomass Valorization* 9 (5): 793–800. doi:10.1007/s12649-017-9873-5.

Ejaz, N., N. Akhtar, H. Nisar, and U. Ali Naeem. 2010. "Environmental Impacts of Improper Solid Waste Management in Developing Countries: A Case Study of Rawalpindi City." *WIT Transactions on Ecology and the Environment* 142: 379–87. doi:10.2495/SW100351.

ElFar, Omar Ashraf, Chih Kai Chang, Hui Yi Leong, Angela Paul Peter, Kit Wayne Chew, and Pau Loke Show. 2020. "Prospects of Industry 5.0 in Algae: Customization of Production and New Advance Technology for Clean Bioenergy Generation." *Energy Conversion and Management: X* 10: 100048. doi:10.1016/j.ecmx.2020.100048.

Enyidi, Uchechukwu D. 2017. "Chlorella Vulgaris as Protein Source in the Diets of African Catfish Clarias Gariepinus." *Fishes* 2 (4). doi:10.3390/fishes2040017.

Fabris, M., R. M. Abbriano, M. Pernice, D. L. Sutherland, A. S. Commault, C. C. Hall, L Labeeuw, et al. 2020. "Emerging Technologies in Algal Biotechnology: Toward the Establishment of a Sustainable, Algae-Based Bioeconomy." *Frontiers in Plant Science* 11 (March): 279. doi:10.3389/fpls.2020.00279.

Farahah Mohd Khairuddin, Nur, Ani Idris, and Muhammad Irfan. 2019. "Towards Efficient Membrane Filtration for Microalgae Harvesting: A Review." *Jurnal Kejuruteraan SI* 2 (1): 103–12.

Fernandez-Ricaud, Luciano, Olga Kourtchenko, Martin Zackrisson, Jonas Warringer, and Anders Blomberg. 2016. "PRECOG: A Tool for Automated Extraction and Visualization of Fitness Components in Microbial Growth Phenomics." *BMC Bioinformatics* 17 (June): 249. doi:10.1186/s12859-016-1134-2.

Foo, Su Chern, Kuan Shiong Khoo, Chien Wei Ooi, Pau Loke Show, Nicholas M. H. Khong, and Fatimah Md. Yusoff. 2021. "Meeting Sustainable Development Goals: Alternative Extraction Processes for Fucoxanthin in Algae." *Frontiers in Bioengineering and Biotechnology* 8: 1371. doi:10.3389/fbioe.2020.546067.

Fuentes, Juan-Luis, Zaida Montero, María Cuaresma, Mari-Carmen Ruiz-Domínguez, Benito Mogedas, Inés Garbayo Nores, Manuel del Valle, and Carlos Vílchez. 2020. "Outdoor Large-Scale Cultivation of the Acidophilic Microalga Coccomyxa Onubensis in a Vertical Close Photobioreactor for Lutein Production." *Processes* 8 (3). doi:10.3390/pr8030324.

Furbank, Robert T., and Mark Tester. 2011. "Phenomics—Technologies to Relieve the Phenotyping Bottleneck." *Trends in Plant Science* 16 (12): 635–44. https://doi.org/10.1016/j.tplants.2011.09.005.

Galbe, M., O. Wallberg, and G. Zacchi. 2011. "6.41 - Techno-Economic Aspects of Ethanol Production from Lignocellulosic Agricultural Crops and Residues." In *Comprehensive Biotechnology (Third Edition)*, edited by Murray Moo-Young, 3rd ed., 519–31. Pergamon. https://doi.org/10.1016/B978-0-444-64046-8.00380-3.

Garske, Beatrice, and Felix Ekardt. 2021. "Economic Policy Instruments for Sustainable Phosphorus Management: Taking into Account Climate and Biodiversity Targets." *Environmental Sciences Europe* 33 (1): 56. doi:10.1186/s12302-021-00499-7.

Geneletti, Davide. 2002. *Ecological Evaluation for Environmental Impact Assessment* (PhD Dissertation). University of Twente, Department of Earth System Analysis, Faculty of Geo-Information Science and Eath Observation, Amsterdam, Netherlands. http://www.itc.nl/library/Papers/phd_2002/geneletti.pdf.

Gouveia, Joao Diogo, Jesus Ruiz, Lambertus A. M. van den Broek, Thamara Hesselink, Sander Peters, Dorinde M. M. Kleinegris, Alison G. Smith, Douwe van der Veen, Maria J. Barbosa, and Rene H. Wijffels. 2017. "Botryococcus Braunii Strains Compared for Biomass Productivity, Hydrocarbon and Carbohydrate Content." *Journal of Biotechnology* 248: 77–86. https://doi.org/10.1016/j.jbiotec.2017.03.008.

Griffiths, Melinda J., Reay G. Dicks, Christine Richardson, and Susan T. L. Harrison. 2011. "Advantages and Challenges of Microalgae as a Source of Oil for Biodiesel." In *Biodiesel—Feedstocks and Processing Technologies*, edited by Margarita Stoytcheva and Gisela Montero. IntechOpen. doi:10.5772/30085.

Grosgogeat, Eric J., Joel R. Fried, Robert G. Jenkins, and S. T. Hwang. 1991. "A Method for the Determination of the Pore Size Distribution of Molecular Sieve Materials and Its Application to the Characterization of Partially Pyrolyzed Polysilastyrene/Porous Glass Composite Membranes." *Journal of Membrane Science* 57 (2): 237–55. https://doi.org/10.1016/S0376-7388(00)80681-X.

Guieysse, Benoit, and Maxence Plouviez. 2021. "Chapter 13 - Sustainability of Microalgae Cultivation." In *Cultured Microalgae for the Food Industry*, edited by Tomás Lafarga and Gabriel Acién, 343–65. Academic Press. https://doi.org/10.1016/B978-0-12-821080-2.00013-7.

Gujar, Asadullah, Saleem Kubar, and Runzhi Li. 2019. "Development, Production and Market Value of Microalgae Products." *Applied Microbiology: Open Access* 5: 162. doi:10.35248/2471-9315.19.5.162.

Günerken, E., E. D'Hondt, M. H. M. Eppink, L. Garcia-Gonzalez, K. Elst, and R. H. Wijffels. 2015. "Cell Disruption for Microalgae Biorefineries." *Biotechnology Advances* 33 (2): 243–60. https://doi.org/10.1016/j.biotechadv.2015.01.008.

Gupta, Sanjay Kumar, Faiz Ahmad Ansari, Amritanshu Shriwastav, Narendra Kumar Sahoo, Ismail Rawat, and Faizal Bux. 2016. "Dual Role of Chlorella Sorokiniana and Scenedesmus Obliquus for Comprehensive Wastewater Treatment and Biomass Production for Bio-Fuels." *Journal of Cleaner Production* 115: 255–64. https://doi.org/10.1016/j.jclepro.2015.12.040.

Handler, Robert M., Christina E. Canter, Tom N. Kalnes, F. Stephen Lupton, Oybek Kholiqov, David R. Shonnard, and Paul Blowers. 2012. "Evaluation of Environmental Impacts from Microalgae Cultivation in Open-Air Raceway Ponds: Analysis of the Prior Literature and Investigation of Wide Variance in Predicted Impacts." *Algal Research* 1: 83–92. doi:10.1016/j.algal.2012.02.003.

Hatzikiriakos, Savvas G., and Peter Englezos. 1993. "The Relationship between Global Warming and Methane Gas Hydrates in the Earth." *Chemical Engineering Science* 48 (23): 3963–69. https://doi.org/10.1016/0009-2509(93)80375-Z.

Henry, Loewendahl, and Scott Bryan. 2019. *Guidelines for Developing an Investment Promotion Strategy*. Canada–Indonesia Trade and Private Sector Assistance Project.

Herrero, Miguel, Andrea del Pilar Sánchez-Camargo, Alejandro Cifuentes, and Elena Ibáñez. 2015. "Plants, Seaweeds, Microalgae and Food by-Products as Natural Sources of Functional Ingredients Obtained Using Pressurized Liquid Extraction and Supercritical Fluid Extraction." *TrAC Trends in Analytical Chemistry* 71: 26–38. https://doi.org/10.1016/j.trac.2015.01.018.

Hosseini Tafreshi, A., and M. Shariati. 2009. "Dunaliella Biotechnology: Methods and Applications." *Journal of Applied Microbiology*. doi:10.1111/j.1365-2672.2009.04153.x.

Houle, David, Diddahally R. Govindaraju, and Stig Omholt. 2010. "Phenomics: The next Challenge." *Nature Reviews. Genetics* 11 (12): 855–66. doi:10.1038/nrg2897.

Ismail, A. F., and P. S. Goh. 2015. "Microfiltration Membrane." In *Encyclopedia of Polymeric Nanomaterials*, edited by Shiro Kobayashi and Klaus Müllen, 1250–55. Springer Berlin Heidelberg. doi:10.1007/978-3-642-29648-2_159.

Jayawardhana, Y., P. Kumarathilaka, I. Herath, and M. Vithanage. 2016. "Chapter 6 - Municipal Solid Waste Biochar for Prevention of Pollution From Landfill Leachate." In *Environmental Materials and Waste*, edited by M. N. V. Prasad and Kaimin Shih, 117–48. Academic Press. https://doi.org/10.1016/B978-0-12-803837-6.00006-8.

Karlapudi, Abraham Peele, T. C. Venkateswarulu, Jahnavi Tammineedi, Lohit Kanumuri, Bharath Kumar Ravuru, Vijaya ramu Dirisala, and Vidya Prabhakar Kodali. 2018. "Role of Biosurfactants in Bioremediation of Oil Pollution-a Review." *Petroleum* 4 (3): 241–49. https://doi.org/10.1016/j.petlm.2018.03.007.

Keppler, Frank, John T. G. Hamilton, Marc Brass, and Thomas Röckmann. 2006. "Methane Emissions from Terrestrial Plants under Aerobic Conditions." *Nature* 439 (7073): 187–91. doi:10.1038/nature04420.

Khan, Aqib Zafar, Ayesha Shahid, Hairong Cheng, Shahid Mahboob, Khalid A. Al-Ghanim, Muhammad Bilal, Fang Liang, and Muhammad Zohaib Nawaz. 2018. "Omics Technologies for Microalgae-Based Fuels and Chemicals: Challenges and Opportunities." *Protein and Peptide Letters* 25 (2). Netherlands: 99–107. doi:10.2174/0929866525666180122100722.

Khan, Zarmeena, Muhammad Habib Ur Rahman, Ghulam Haider, Rabia Amir, Rao Muhammad Ikram, Shakeel Ahmad, Hannah Kate Schofield, et al. 2021. "Chemical and Biological Enhancement Effects of Biochar on Wheat Growth and Yield under Arid Field Conditions." *Sustainability* 13 (11): 1–18. doi:10.3390/su13115890.

Khoo, Kuan Shiong, Kit Wayne Chew, Guo Yong Yew, Wai Hong Leong, Yee Ho Chai, Pau Loke Show, and Wei-Hsin Chen. 2020. "Recent Advances in Downstream Processing of Microalgae Lipid Recovery for Biofuel Production." *Bioresource Technology* 304: 122996. https://doi.org/10.1016/j.biortech.2020.122996.

Khoo, Kuan Shiong, Sze Ying Lee, Chien Wei Ooi, Xiaoting Fu, Xiaoling Miao, Tau Chuan Ling, and Pau Loke Show. 2019. "Recent Advances in Biorefinery of Astaxanthin from Haematococcus Pluvialis." *Bioresource Technology* 288: 121606. https://doi.org/10.1016/j.biortech.2019.121606.

Kosseva, Maria R. 2020. "Chapter 3 - Sources, Characteristics and Treatment of Plant-Based Food Waste." In *Food Industry Wastes (Second Edition)*, edited by Maria R. Kosseva and Colin Webb, 2nd ed., 37–66. Academic Press. https://doi.org/10.1016/B978-0-12-817121-9.00003-6.

Kothari, Richa, Arya Pandey, Shamshad Ahmad, Ashwani Kumar, Vinayak V. Pathak, and V. V. Tyagi. 2017. "Microalgal Cultivation for Value-Added Products: A Critical Enviro-Economical Assessment." *3 Biotech* 7 (4): 1–15. doi:10.1007/s13205-017-0812-8.

Koyande, Apurav Krishna, Pau Loke Show, Ruixin Guo, Bencan Tang, Chiaki Ogino, and Jo-Shu Chang. 2019. "Bio-Processing of Algal Bio-Refinery: A Review on Current Advances and Future Perspectives." *Bioengineered* 10 (1): 574–92. doi:10.1080/21655979.2019.1679697.

Lam, Man Kee, Choon Gek Khoo, and Keat Teong Lee. 2019. "Chapter 19 - Scale-up and Commercialization of Algal Cultivation and Biofuels Production." In *Biofuels from Algae (Second Edition)*, edited by Ashok Pandey, Jo-Shu Chang, Carlos Ricardo Soccol, Duu-Jong Lee, and Yusuf Chisti, 2nd ed., 475–506. Biomass, Biofuels, Biochemicals. Elsevier. https://doi.org/10.1016/B978-0-444-64192-2.00019-6.

Laurens, Lieve M. L., Melodie Chen-Glasser, and James D. McMillan. 2017. "A Perspective on Renewable Bioenergy from Photosynthetic Algae as Feedstock for Biofuels and Bioproducts." *Algal Research* 24 (March): 261–64. doi:10.1016/j.algal.2017.04.002.

Lin, Chiu-Yue, Wai S. Chai, Chyi-How Lay, Chin-Chao Chen, Chun-Yi Lee, and Pau L. Show. 2021. "Optimization of Hydrolysis-Acidogenesis Phase of Swine Manure for Biogas Production Using Two-Stage Anaerobic Fermentation." *Processes* 9 (8): 1324. doi:10.3390/pr9081324.

Low, Sze Shin, Kien Xiang Bong, Muhammad Mubashir, Chin Kui Cheng, Man Kee Lam, Jun Wei Lim, Yeek Chia Ho, Keat Teong Lee, Heli Siti Halimatul Munawaroh, and Pau Loke Show. 2021. "Microalgae Cultivation in Palm Oil Mill Effluent (POME) Treatment and Biofuel Production." *Sustainability* 13 (6): 3247. doi:10.3390/su13063247.

Lukić, Borislava, Antonio Panico, David Huguenot, Massimiliano Fabbricino, Eric D. van Hullebusch, and Giovanni Esposito. 2017. "A Review on the Efficiency of Landfarming Integrated with Composting as a Soil Remediation Treatment." *Environmental Technology Reviews* 6 (1): 94–116. doi:10.1080/21622515.2017.1310310.

Maćczak, Piotr, Halina Kaczmarek, and Marta Ziegler-Borowska. 2020. "Recent Achievements in Polymer Bio-Based Flocculants for Water Treatment." *Materials* 13 (18). doi:10.3390/ma13183951.

Maliaritsi, E., L. Zoumpoulakis, J. Simitzis, P. Vassiliou, and E. Hristoforou. 2006. "Coagulation Sensors Based on Magnetostrictive Delay Lines for Biomedical and Chemical Engineering Applications." *Journal of Magnetism and Magnetic Materials* 299 (1): 41–52. https://doi.org/10.1016/j.jmmm.2005.03.095.

Mat Aron, Nurul Syahirah, Kuan Shiong Khoo, Kit Wayne Chew, Ashokkumar Veeramuthu, Jo-Shu Chang, and Pau Loke Show. 2021. "Microalgae Cultivation in Wastewater and Potential Processing Strategies Using Solvent and Membrane Separation Technologies." *Journal of Water Process Engineering* 39: 101701. doi:10.1016/j.jwpe.2020.101701.

Mata, Teresa M., António A. Martins, and Nidia S. Caetano. 2010. "Microalgae for Biodiesel Production and Other Applications: A Review." *Renewable and Sustainable Energy Reviews* 14 (1): 217–32. doi:10.1016/j.rser.2009.07.020.

Miranda, Ana F., Narasimhan Ramkumar, Constandino Andriotis, Thorben Höltkemeier, Aneela Yasmin, Simone Rochfort, Donald Wlodkowic, et al. 2017. "Applications of Microalgal Biofilms for Wastewater Treatment and Bioenergy Production." *Biotechnology for Biofuels* 10 (1): 120. doi:10.1186/s13068-017-0798-9.

Mmereki, Daniel, Andrew Baldwin, Liu Hong, and Baizhan Li. 2016. "The Management of Hazardous Waste in Developing Countries." In *Management of Hazardous Wastes*, edited by Hosam El-Din M. Saleh and Rehab O. Abdel Rahman. IntechOpenx. doi:10.5772/63055.

Mobin, Saleh M. A., Harun Chowdhury, and Firoz Alam. 2019. "Commercially Important Bioproducts from Microalgae and Their Current Applications-A Review." *Energy Procedia* 160 (2018): 752–60. doi:10.1016/j.egypro.2019.02.183.

Molina Grima, E., E. H. Belarbi, F. G. Acién Fernández, A. Robles Medina, and Yusuf Chisti. 2003. "Recovery of Microalgal Biomass and Metabolites: Process Options and Economics." *Biotechnology Advances* 20 (7–8): 491–515. doi:10.1016/S0734-9750(02)00050-2.

Moreno-Garcia, L., K. Adjallé, S. Barnabé, and G. S. V. Raghavan. 2017. "Microalgae Biomass Production for a Biorefinery System: Recent Advances and the Way towards Sustainability." *Renewable and Sustainable Energy Reviews* 76: 493–506. doi:10.1016/j.rser.2017.03.024.

Nair, Sujith, and Hanna Paulose. 2014. "Emergence of Green Business Models: The Case of Algae Biofuel for Aviation." *Energy Policy* 65: 175–84. https://doi.org/10.1016/j.enpol.2013.10.034.

Neveu, Pascal, Anne Tireau, Nadine Hilgert, Vincent Nègre, Jonathan Mineau-Cesari, Nicolas Brichet, Romain Chapuis, et al. 2019. "Dealing with Multi-Source and Multi-Scale Information in Plant Phenomics: The Ontology-Driven Phenotyping Hybrid Information System." *The New Phytologist* 221 (1): 588–601. doi:10.1111/nph.15385.

Ngo, Huu Hao, Wenshan Guo, Rao Y. Surampalli, and Tian Cheng Zhang. 2017. *Green Technologies for Sustainable Water Management*. CRC Press.

Nielsen, Jens, and Jay D. Keasling. 2016. "Engineering Cellular Metabolism." *Cell* 164 (6): 1185–97. https://doi.org/10.1016/j.cell.2016.02.004.

Olaizola, Miguel. 2003. "Commercial Development of Microalgal Biotechnology: From the Test Tube to the Marketplace." *Biomolecular Engineering* 20 (4): 459–66. https://doi.org/10.1016/S1389-0344(03)00076-5.

Olguín, Eugenia J. 2012. "Dual Purpose Microalgae—Bacteria-Based Systems That Treat Wastewater and Produce Biodiesel and Chemical Products within a Biorefinery." *Biotechnology Advances* 30 (5): 1031–46. https://doi.org/10.1016/j.biotechadv.2012.05.001.

Osanai, Takashi, Youn-Il Park, and Yuki Nakamura. 2017. "Editorial: Biotechnology of Microalgae, Based on Molecular Biology and Biochemistry of Eukaryotic Algae and Cyanobacteria." *Frontiers in Microbiology* 8 (February). Frontiers Media S.A.: 118. doi:10.3389/fmicb.2017.00118.

Pacheco, Diana, Ana Cristina Rocha, Leonel Pereira, and Tiago Verdelhos. 2020. "Microalgae Water Bioremediation: Trends and Hot Topics." *Applied Sciences* 10 (5). doi:10.3390/app10051886.

Pal, D. 2003. "MEMBRANE TECHNIQUES | Applications of Ultrafiltration." In *Encyclopedia of Food Sciences and Nutrition (Second Edition)*, edited by Benjamin Caballero, 2nd ed., 3842–48. Academic Press. https://doi.org/10.1016/B0-12-227055-X/00764-1.

Patel, Anil Kumar, Celine Laroche, Alain Marcati, Alina Violeta Ursu, Sébastien Jubeau, Luc Marchal, Emmanuel Petit, Gholamreza Djelveh, and Philippe Michaud. 2013. "Separation and Fractionation of Exopolysaccharides from Porphyridium Cruentum." *Bioresource Technology* 145: 345–50. https://doi.org/10.1016/j.biortech.2012.12.038.

Peter, Angela Paul, Kuan Shiong Khoo, Kit Wayne Chew, Tau Chuan Ling, Shih-Hsin Ho, Jo-Shu Chang, and Pau Loke Show. 2021. "Microalgae for Biofuels, Wastewater Treatment and Environmental Monitoring." *Environmental Chemistry Letters* 19 (4): 2891–904. doi:10.1007/s10311-021-01219-6.

Pignolet, Olivier, Sébastien Jubeau, Carlos Vaca-Garcia, and Philippe Michaud. 2013. "Highly Valuable Microalgae: Biochemical and Topological Aspects." *Journal of Industrial Microbiology and Biotechnology* 40 (8): 781–96. doi:10.1007/s10295-013-1281-7.

Pugazhendhi, Arivalagan, Senthil Nagappan, Rahul R. Bhosale, Pei-Chien Tsai, Shakunthala Natarajan, Saravanan Devendran, Lamya Al-Haj, Vinoth Kumar Ponnusamy, and Gopalakrishnan Kumar. 2020. "Various Potential Techniques to Reduce the Water Footprint of Microalgal Biomass Production for Biofuel—A Review." *Science of The Total Environment* 749: 142218. https://doi.org/10.1016/j.scitotenv.2020.142218.

Pugazhendhi, Arivalagan, Sutha Shobana, Peter Bakonyi, Nándor Nemestóthy, Ao Xia, Rajesh Banu J., and Gopalakrishnan Kumar. 2019. "A Review on Chemical Mechanism of Microalgae Flocculation via Polymers." *Biotechnology Reports* 21: e00302. https://doi.org/10.1016/j.btre.2018.e00302.

Pulz, O. 2001. "Photobioreactors: Production Systems for Phototrophic Microorganisms." *Applied Microbiology and Biotechnology* 57 (3): 287–93. doi:10.1007/s002530100702.

Qin, Song, Hanzhi Lin, and Peng Jiang. 2012. "Advances in Genetic Engineering of Marine Algae." *Biotechnology Advances* 30 (6): 1602–13. https://doi.org/10.1016/j.biotechadv.2012.05.004.

Quezada, Maribel, Germán Buitrón, Iván Moreno-Andrade, Gloria Moreno, and Luz M. López-Marín. 2007. "The Use of Fatty Acid Methyl Esters as Biomarkers to Determine Aerobic, Facultatively Aerobic and Anaerobic Communities in Wastewater Treatment Systems." *FEMS Microbiology Letters* 266 (1): 75–82. doi:10.1111/j.1574-6968.2006.00509.x.

Rackley, Stephen A. 2017. "8 - Membrane Separation Systems." In *Carbon Capture and Storage (Second Edition)*, edited by Stephen A. Rackley, 2nd ed., 187–225. Boston: Butterworth-Heinemann. https://doi.org/10.1016/B978-0-12-812041-5.00008-8.

Rawat, I., R. Ranjith Kumar, T. Mutanda, and F. Bux. 2013. "Biodiesel from Microalgae: A Critical Evaluation from Laboratory to Large Scale Production." *Applied Energy* 103: 444–67. https://doi.org/10.1016/j.apenergy.2012.10.004.

Richmond, Amos. 2003. *Handbook of Microalgal Culture: Biotechnology and Applied Phycology*. doi:10.1002/9780470995280.

Rodríguez-Sifuentes, Lucio, Jolanta Elzbieta Marszalek, Gerardo Hernández-Carbajal, and Cristina Chuck-Hernández. 2021. "Importance of Downstream Processing of Natural Astaxanthin for Pharmaceutical Application." *Frontiers in Chemical Engineering* 2: 29. doi:10.3389/fceng.2020.601483.

Roostaei, Javad, and Yongli Zhang. 2017. "Spatially Explicit Life Cycle Assessment: Opportunities and Challenges of Wastewater-Based Algal Biofuels in the United States." *Algal Research* 24: 395–402. https://doi.org/10.1016/j.algal.2016.08.008.

Rosenberg, Julian N., Ashrith Mathias, Karen Korth, Michael J. Betenbaugh, and George A. Oyler. 2011. "Microalgal Biomass Production and Carbon Dioxide Sequestration from an Integrated Ethanol Biorefinery in Iowa: A Technical Appraisal and Economic Feasibility Evaluation." *Biomass and Bioenergy* 35 (9): 3865–76. doi:10.1016/j.biombioe.2011.05.014.

Sayre, Richard. 2010. "Microalgae: The Potential for Carbon Capture." *BioScience* 60 (9): 722–27. doi:10.1525/bio.2010.60.9.9.

Shanab, Sanaa M. M., Soha S. M. Mostafa, Emad A. Shalaby, and Ghada I. Mahmoud. 2012. "Aqueous Extracts of Microalgae Exhibit Antioxidant and Anticancer Activities." *Asian Pacific Journal of Tropical Biomedicine* 2 (8): 608–15. doi:10.1016/S2221-1691(12)60106-3.

Sheehan John, Dunahay Terri, Benemann John, and Roessler Paul. 1998. "A Look Back at the U.S. Department of Energy's Aquatic Species." *The European Physical Journal C* 72 (6): 14.

Shuping, Zou, Wu Yulong, Yang Mingde, Li Chun, and Tong Junmao. 2010. "Pyrolysis Characteristics and Kinetics of the Marine Microalgae Dunaliella Tertiolecta Using Thermogravimetric Analyzer." *Bioresource Technology* 101 (1): 359–65. https://doi.org/10.1016/j.biortech.2009.08.020.

Slade, Raphael, and Ausilio Bauen. 2013. "Micro-Algae Cultivation for Biofuels: Cost, Energy Balance, Environmental Impacts and Future Prospects." *Biomass and Bioenergy* 53: 29–38. doi:10.1016/j.biombioe.2012.12.019.

Souza Leite, Luan de, and Luiz Antonio Daniel. 2020. "Optimization of Microalgae Harvesting by Sedimentation Induced by High PH." *Water Science and Technology* 82 (6): 1227–36. doi:10.2166/wst.2020.106.

Spolaore, Pauline, Claire Joannis-Cassan, Elie Duran, and Arsène Isambert. 2006. "Commercial Applications of Microalgae." *Journal of Bioscience and Bioengineering* 101 (2): 87–96. doi:10.1263/JBB.101.87.

Stephens, Evan, Ian L. Ross, Jan H. Mussgnug, Liam D. Wagner, Michael A. Borowitzka, Clemens Posten, Olaf Kruse, and Ben Hankamer. 2010. "Future Prospects of Microalgal Biofuel Production Systems." *Trends in Plant Science* 15 (10): 554–64. doi:10.1016/j.tplants.2010.06.003.

Suparmaniam, Uganeeswary, Man Kee Lam, Yoshimitsu Uemura, Jun Wei Lim, Keat Teong Lee, and Siew Hoong Shuit. 2019. "Insights into the Microalgae Cultivation Technology and Harvesting Process for Biofuel Production: A Review." *Renewable and Sustainable Energy Reviews* 115: 109361. https://doi.org/10.1016/j.rser.2019.109361.

Świątkiewicz, S., A. Arczewska-Włosek, and D. Józefiak. 2015. "Application of Microalgae Biomass in Poultry Nutrition." *World's Poultry Science Journal* 71 (4): 663–672. doi:10.1017/S0043933915002457.

Systems, Environmental. 2013. "Sustainable Development of Algal Biofuels in the United States." In National Research Council. 2012. *Sustainable Development of Algal Biofuels in the United States*. Washington, DC: The National Academies Press. https://doi.org/10.17226/13437.

Taboada-González, Paul, Quetzalli Aguilar-Virgen, Sara Ojeda-Benítez, and Carolina Armijo. 2011. "Waste Characterization and Waste Management Perception in Rural Communities in Mexico: A Case Study." *Environmental Engineering and Management Journal* 10 (11): 1751–59. doi:10.30638/eemj.2011.238.

Tang, Doris Ying, Kuan Shiong Khoo, Kit Wayne Chew, Yang Tao, Shih-Hsin Ho, and Pau Loke Show. 2020. "Potential Utilization of Bioproducts from Microalgae for the Quality Enhancement of Natural Products." *Bioresource Technology* 304: 122997. https://doi.org/10.1016/j.biortech.2020.122997.

Tao, Fei, Jiangfeng Cheng, Qinglin Qi, Meng Zhang, He Zhang, and Fangyuan Sui. 2018. "Digital Twin-Driven Product Design, Manufacturing and Service with Big Data." *The International Journal of Advanced Manufacturing Technology* 94 (9): 3563–76. doi:10.1007/s00170-017-0233-1.

Thangavel, P., and G. Sridevi. 2015. "Environmental Sustainability: Role of Green Technologies." *Environmental Sustainability: Role of Green Technologies*, no. September 2016: 1–324. doi:10.1007/978-81-322-2056-5.

Thiyagarasaiyar, Krishnapriya, Bey-Hing Goh, You-Jin Jeon, and Yoon-Yen Yow. 2020. "Algae Metabolites in Cosmeceutical: An Overview of Current Applications and Challenges." *Marine Drugs* 18 (6): 323. doi:10.3390/md18060323.

Uduman, Nyomi, Y. Qi, Michael K. Danquah, G. Forde, and A. Hoadley. 2010. "Dewatering of Microalgal Cultures: A Major Bottleneck to Lgae-Based Fuels." *Journal of Renewable and Sustainable Energy* 2: 012701. doi:10.1063/1.3294480.

Usher, Philippa K., Andrew B. Ross, Miller Alonso Camargo-Valero, Alison S. Tomlin, and William F. Gale. 2014. "An Overview of the Potential Environmental Impacts of Large-Scale Microalgae Cultivation." *Biofuels*. doi:10.1080/17597269.2014.913925.

Vale, Miguel A., António Ferreira, José C. M. Pires, and Ana L. Gonçalves. 2020. "Chapter 17 - CO2 Capture Using Microalgae." In *Advances in Carbon Capture*, edited by Mohammad Reza Rahimpour, Mohammad Farsi, and Mohammad Amin Makarem, 381–405. Woodhead Publishing. https://doi.org/10.1016/B978-0-12-819657-1.00017-7.

Vandamme, D., I. Foubert, and K. Muylaert. 2013. "Flocculation as a Low-Cost Method for Harvesting Microalgae for Bulk Biomass Production." *Trends in Biotechnology* 31 (4): 233–39. doi:10.1016/j.tibtech.2012.12.005.

Waldron, K. ed. 2014. *Advances in Biorefineries: Biomass and Waste Supply Chain Exploitation*. Woodhead Publishing.

Wang, TsingHai, Chih-Lin Hsu, Chih-Hung Huang, Yi-Kong Hsieh, Chung-Sung Tan, and Chu-Fang Wang. 2016. "Environmental Impact of CO2-Expanded Fluid Extraction Technique in Microalgae Oil Acquisition." *Journal of Cleaner Production* 137: 813–20. https://doi.org/10.1016/j.jclepro.2016.07.179.

Wheeler, P. A., and L. de Rome. 2002. *Waste Pre-Treatment: A Review*. R&D Technical Report PI-344/TR, UK Environment Agency.

Whitmore, Andrew, Anurag Agarwal, and Li Da Xu. 2015. "The Internet of Things—A Survey of Topics and Trends." *Information Systems Frontiers* 17 (2): 261–74. doi:10.1007/s10796-014-9489-2.

Woodard & Curran, Inc. 2006. "7 - Methods for Treating Wastewaters from Industry." In *Industrial Waste Treatment Handbook (Second Edition)*, edited by Woodard & Curran, Inc., 2nd ed., 149–334. Butterworth-Heinemann. https://doi.org/10.1016/B978-075067963-3/50009-6.

Yaakob, Maizatul Azrina, Radin Maya Saphira Radin Mohamed, Adel Al-Gheethi, Ravishankar Aswathnarayana Gokare, and Ranga Rao Ambati. 2021. "Influence of Nitrogen and Phosphorus on Microalgal Growth, Biomass, Lipid, and Fatty Acid Production: An Overview." *Cells* 10 (2): 10–20. doi:10.3390/cells10020393.

Yang, Minliang, Nawa Raj Baral, Blake A. Simmons, Jenny C. Mortimer, Patrick M. Shih, and Corinne D. Scown. 2020. "Accumulation of High-Value Bioproducts in Planta Can Improve the Economics of Advanced Biofuels." *Proceedings of the National Academy of Sciences* 117 (15): 8639–48. doi:10.1073/pnas.2000053117.

Yang, Yanli, Bo Zhang, Jing Cheng, and Shengyan Pu. 2015. "Socio-Economic Impacts of Algae-Derived Biodiesel Industrial Development in China: An Input—Output Analysis." *Algal Research* 9: 74–81. https://doi.org/10.1016/j.algal.2015.02.010.

Yoo, Chan, So-Young Jun, Jae-Yon Lee, Chi-Yong Ahn, and Hee-Mock Oh. 2010. "Selection of Microalgae for Lipid Production under High Levels Carbon Dioxide." *Bioresource Technology* 101 (1): S71–74. https://doi.org/10.1016/j.biortech.2009.03.030.

Zhu, Liandong. 2015. "Biorefinery as a Promising Approach to Promote Microalgae Industry: An Innovative Framework." *Renewable and Sustainable Energy Reviews* 41: 1376–84. https://doi.org/10.1016/j.rser.2014.09.040.

Ziolkowska, Jadwiga R., and Leo Simon. 2014. "Recent Developments and Prospects for Algae-Based Fuels in the US." *Renewable and Sustainable Energy Reviews* 29: 847–53. https://doi.org/10.1016/j.rser.2013.09.021.

9 Biomanufacturing in Microalgae Industry

Leong Wei[1], *Wen Yi Chia*[1], *Kit Wayne Chew*[2] *and Pau Loke Show*[1]

[1] Department of Chemical and Environmental Engineering, Faculty of Science and Engineering, University of Nottingham Malaysia, Jalan Broga, Semenyih, Selangor Darul Ehsan, Malaysia

[2] School of Energy and Chemical Engineering, Xiamen University Malaysia, Jalan Sunsuria, Bandar Sunsuria, Sepang, Selangor Darul Ehsan, Malaysia

CONTENTS

9.1 Introduction 294
9.2 Manufacturing Processes of Microalgae 296
 9.2.1 Upstream Processes of Microalgal Biomanufacturing 297
  9.2.1.1 Open Production System 297
  9.2.1.2 Closed Production System 299
 9.2.2 Downstream Processes of Microalgal Biomanufacturing 300
  9.2.2.1 Harvesting 300
  9.2.2.2 Extraction 301
  9.2.2.3 Conversion 302
 9.2.3 Bioeconomy and Microalgae 304
9.3 Intelligent Manufacturing of Microalgae 305
 9.3.1 Artificial Intelligence and Biomanufacturing of Microalgae 306
  9.3.1.1 Artificial Intelligence and General Biomanufacturing of Microalgae 306
  9.3.1.2 Artificial Intelligence and Microalgal Genetics 308
  9.3.1.3 Microalgae Production Plant with Artificial Intelligence Derived Algorithm: PhotoSynthetica 309
 9.3.2 Image Analysis and Microalgae Production 310
  9.3.2.1 Image Analysis Study for Dry Cell Weight 310
  9.3.2.2 Transmission Hyperspectral Microscopic Imaging (THMI) with Machine Learning 311
  9.3.2.3 Frequency-Division-Multiplexed (FDM) Fluorescence Imaging Flow Cytometry with Machine Learning 312

DOI: 10.1201/9781003202196-9

9.3.3 Use of Nano-Additives in Biomanufacturing of Microalgae 313
9.3.3.1 Nanoparticles and Microalgae Cultivation 314
9.3.3.2 Nano-Additives and Processing of Microalgal Biomass 315
9.3.4 Genetic Engineering and Microalgae Biomanufacturing 317
9.3.4.1 Omics Approaches 318
9.3.4.2 Metabolic Engineering 318
9.4 Future Research and Development of Smart Biomanufacturing of Microalgae and Biomass 319
9.5 Conclusions 320
References 321

9.1 INTRODUCTION

Microalgae are essentially a general species of algae that are microscopic and, hence, invisible to the naked eye. However, their presence within our living atmosphere is certainly not to be overlooked as they play a huge role in industrial and agricultural sectors of our society (Randrianarison and Ashraf 2017). For example, in terms of industrial usage, the microalgae are said to be useful, and it is proven by scientists from various corners of the globe that they are essential toward the production of a new type of fuel or source of energy called the biofuel (Low et al. 2021). As for the agricultural side of things, microalgae can be biofertilizers for our food crops in the near future, even more efficient than current natural ones that we can rely on the market (Guo et al. 2020).

Due to the fact that most microalgae live in an environment where resources needed to survive are scarce and usually extreme, they have evolved to be more efficient in their metabolism to be able to grow and reproduce rapidly when the time comes when the conditions are favorable (Randrianarison and Ashraf 2017). Then, microalgae, or algae as general, would be more efficient photosynthetically than other species of flora, which means a higher rate of biomass generation. If not known, biomass could be simply defined as material sourced from animals and plants to provide energy.

Since microalgae have such spotlight as an alternative of energy production, it is highly undeniable that they have been taken notice of in the past decade and always needed proper research and development so that their role at the stage of energy generation may be secured and hence our present and future, where energy is a real part of our lives. Up to today, as research on this organism carries on and its credibility as an alternative energy source for humanity cementing, microalgae manufacturing has started, with methods of the process, be it either conducted at the exterior, in a pond under the sun (Costa and de Morais 2014), or in the interior of a building, in a photobioreactor (Tan et al. 2020). Furthermore, being relatively more survivable and capable than other species of plants also means that the microalgae can be cultivated on non-arable land or area, near a factory wastewater storage. Figure 9.1 shows different types of

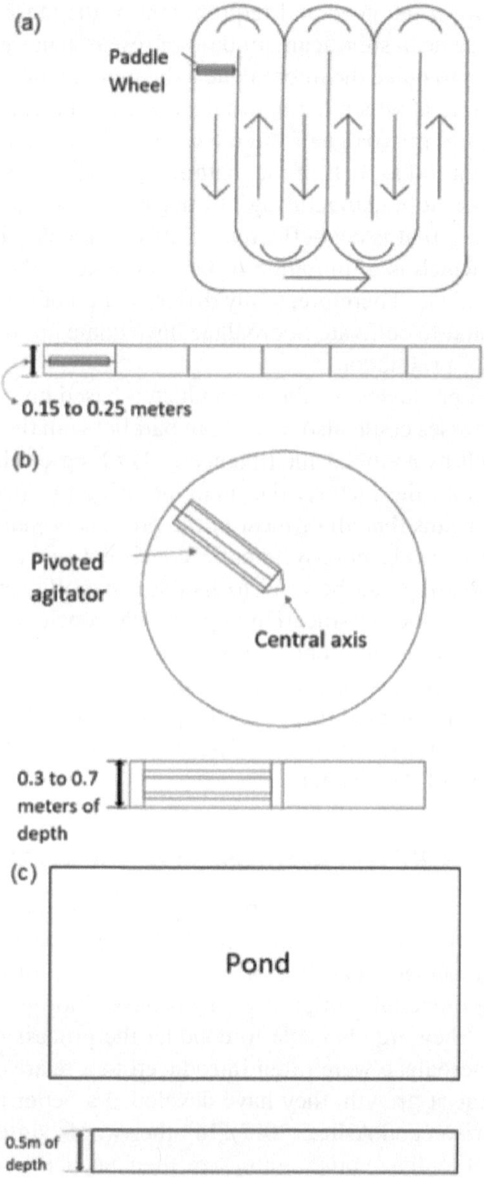

FIGURE 9.1 Open ponds for microalgae biomanufacturing: (a) raceway pond, (b) circular pond, and (c) unstirred pond.

open ponds for microalgae biomanufacturing, including raceway, circular, and unstirred pond.

However, in such environment where eutrophication occurs and population of algae would thrive, it is not enough as a method to mass culture the species.

Hence, in order to meet the need and requirement of the market through manufacturing in a large scale, a significant amount of research has gone into studying and reinventing ways to make the microalgae to be more metabolically efficient or rather enhance the rate at which the microalgae are produced, all to result in the maximal production of microalgae from a fixed pool of microalgae (Fabris et al. 2020). Besides the natural aspects of the matter, organizations and various bodies regarding the large-scale-manufacturing of microalgae have too tried to create a manufacturing system that is cost-effective relative to the continuous production of the microalgae, which is a limitation to be overcome in the name of efficient production of microalgae. Therefore, many different methods with different operating costs and capital to cultivate microalgae have come up to produce different results of microalgae production.

As the suggested processes would be conducted based on current and/or past research, these processes could also be done in parallel with the technology available nowadays, such as artificial intelligence (AI) (Teng et al. 2020). With the introduction of AI into the picture, the manufacturing of microalgae would be digitalized, which means that all or most of the process of manufacturing would be conducted in a way that is more systematic, consistent, and errorless. Moreover, various forms of imaging can be used to increase the efficiency of microalgae manufacturing through the classification of particular species of algae and prediction of concentration of the microalgae population, for example. Besides AI and advanced imaging, other fascinating technologies such as that of genetic engineering and nanoparticles could be part of the party as well. Such modern amenities surrounding Industry 4.0 would be further discussed and elaborated in the following paragraphs of the chapter.

9.2 MANUFACTURING PROCESSES OF MICROALGAE

The microalgae, for a significant number of years now, have been acting as a close alternative for fuel and agricultural catalyst, as their recognized advantages outweigh the disadvantages that they bring in the context of biomanufacturing. As these species of unicellular microorganisms have chlorophyll as other majority of green plants, they are also able to conduct the process of photosynthesis. However, as the microalgae were often introduced to a relatively harsh environment as their habitat of growth, they have developed a better property of adaptability (Randrianarison and Ashraf 2017). In other words, unlike other carefully cultivated flora, microalgae could synthesize their needed glucose (food) more effectively under conditions with reduced sunlight and carbon dioxide. Hence, if they were to be placed under normal yet controlled environment, the rate of growth and production of microalgae could easily surpass that of the other plants.

In order to further increase the efficiency of the microalgae manufacturing, the production units are usually of large surface area to ensure even distribution of fluid required for microalgae cultivation and increased exposure of the microalgae to light (Cruz et al. 2018). There, as the overall manufacturing of microalgae could be more effectively conducted in a carefully controlled environment, the

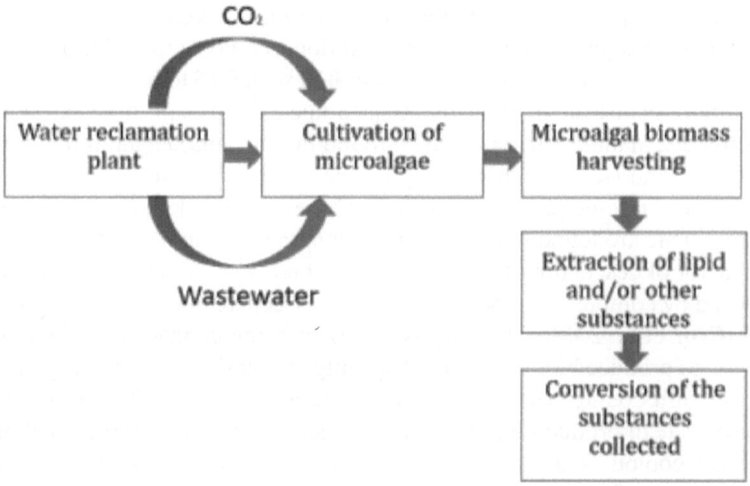

FIGURE 9.2 A simplified chart of process of microalgal biomanufacturing.

growth of microalgae would be greatly enhanced. Then, the microalgae cultivated would be harvested by batch and be processed accordingly based on the specific species, as there are species of microalgae which are particularly genetically engineered to produce certain products such as lipids and protein (Faried et al. 2017). Furthermore, to be more specific, the process of microalgal biomanufacturing (Figure 9.2) has two larger parts: the upstream process including the cultivation and the downstream process including the harvesting, extraction, and conversion.

9.2.1 Upstream Processes of Microalgal Biomanufacturing

The upstream process of microalgae production consists of mainly the ones involving cultivation of microalgae, which would then later be harvested in the form of biomass. Throughout the years, many manufacturing techniques have been experimented and used in the name of efficiency of microalgae manufacturing. However, they are generally separated and characterized into two large groups, namely the open production systems, where the manufacturing processes are being carried out outside a building, and the closed production systems, where the production happenings of the algae would occur in the interior of a factory (Chaumont 1993). Besides the operating production systems, the manufacturing processes themselves of the microalgae play an important role as a major component of the system.

9.2.1.1 Open Production System

In the case of manufacturing systems under the sun, the manufacturing processes are often conducted in reactors which are referred to as ponds, which also can be classified into types such as circular, thin- and multi-layered, and so on. Besides pools, the microalgae, under open-air conditions, could be manufactured in tanks or raceways, which are all commonly operated with 0.15 to half a meter in depth, but not more than 1 meter (Murthy 2011). Some examples of these exterior plants of microalgae manufacturing could be found in Yaeyama, Japan (round tank with axial flows), and Hainan, China (longitudinal raceways) (Raut et al. 2015).

Moreover, in some particular yet recent cases, also under an open atmosphere, the microalgae are found to be able to be manufactured in a fiberglass-made photobioreactor of descending-film type, where conditions are closely controlled and monitored by computers at the Laboratory of Green Technologies, GreenTec/EQ/UFRJ (Cruz et al. 2018). The purpose of them being conducted in the photobioreactor is to produce biomass more efficiently from microalgae than their slightly fewer modern counterparts. While being seemingly more efficient and with an increased rate of production, the photobioreactor system is designed with the aim to produce copious and sufficient amount of biomass to help produce biofuel.

In fact, up to this day and age, most microalgae, up to 95% of them, are manufactured in open-production systems, with the horizontal yet slightly shallow raceway being the best option to cultivate microalgae at the moment (Cruz et al. 2018). In this ‘raceway’ of microalgae culture, each of its unit would be around hundreds, up to few thousands of squared meters in covered area, consisting of two or more ‘tracks’ which then would be up to 10 meters wide and with an average depth of about 20 centimeters, separated by dividers, forming a somewhat meandering system, which is multi-loop configured. And in those ‘tracks’, the cell suspension would flow at a speed not faster than 0.2 meter per second, with the driving force provided by pumps (Béchet et al. 2017). Then, the cycle of operation for the system would be to extract the cell suspension from the mixture of fluid, and the excess fluid would flow back into the designated pond after that.

However, with that said, there are limitations, too, for these open manufacturing systems of microalgae. For example, with the system being relied solely on the natural atmosphere as the source of carbon dioxide, a crucial component in microalgae manufacturing, the amount of the gas in the atmosphere would be not sufficient for each microalga in the pond for most of the time. This then causes the maximum production rate in open systems to be much lower than biodynamic systems. The limitation of light availability could too be of concern, as the microalgae need sunlight to undergo photosynthesis, which they cannot in times such as night (Cruz et al. 2018). Moreover, due to these cases, which ultimately contribute to the inefficiency of the open system, the economy would not look good, as the operating costs of the system per mass of harvested biomass are high. This is then due to the systems’ need for large amount of labor, water, and energy in the form of electricity to run the system altogether. Furthermore, the exposed ponds could be easily contaminated by predators and pathogens, along with loss of fluid due to evaporation (McBride et al. 2014). Therefore, the maintenance of such systems could be expensive and difficult; hence the system’s use as a large-scale microalga manufacturing plant would be limited to certain vigorous species in environments which are carefully selected and are relatively resistant to weather changes. Those vigorous species, then, are too able to adapt to extreme conditions of the environment provided, despite the slightly better open-air environment. Some examples of those species are *Spirulina* (high alkalinity), *Chlorella* (high nutrient concentrations), and *Dunaliella* (high salinity) (Borowitzka and Vonshak 2017).

While open production system seems not to be the most suitable and effective solution to manufacture microalgae, recent innovations and new technologies such as increased concentration of carbon dioxide in the system through direct injection and stricter atmospheric control have been introduced to improve the situation (Cruz et al. 2018). Despite it all, however, the productivity of these 'traditional' systems is still considered to be lower than the new-in-the-market photobioreactor systems, which are used and characterized as part of the closed systems as well.

9.2.1.2 Closed Production System

The closed production system, however, is a manufacturing system, quite literally, conducted in a closed environment. In the case of closed manufacturing systems, processes of microalgae manufacturing are often carried out in machines called the photobioreactors (Cruz et al. 2018). These reactors are commonly built from materials which are transparent such as glass and are typically in the form of a flat panel or a curve. Moreover, there are various types of photobioreactors to be chosen from according to the needs of the customer, such as horizontal tubes and agitated tanks (Belz et al. 2013).

With the use of photobioreactors in a closed environment such as a building, the conditions of the atmosphere for microalgae manufacturing would be much more stable and could be easily controlled. For example, common atmospheric components such as temperature, pH, and light can be altered accordingly to meet the needs of the time or just to simply increase the rate at which the biomass is produced from microalgae. Besides that, the use of photobioreactor itself has ensured biosecurity for the microalgae against pathogens and other microorganisms, hence preventing any contamination from occurring (Cruz et al. 2018). As for the third advantage of using the photobioreactor as a tool for microalgae manufacturing, as computers and electrical components are mostly controlling and monitoring the environment and processes of manufacturing, labor could be required less and number of incidents caused by human error could be significantly reduced.

However, as there are always two sides to a coin, there are disadvantages of the closed manufacturing system as well, though not as significant as it does for the open production systems. For one, as the processes of production are conducted in a closed environment, there are difficulties to amplify the production of microalgae (Alishah Aratboni et al. 2019). Moreover, as glass and other similar transparent materials are not the most durable substance as themselves, in time they can deteriorate and corrode, thus causing the efficiency of the production system to be affected. Besides, due to the increased and extensive use of sensors to monitor conditions of the environment and that of other technological devices such as electric pumps in the operation of the closed microalgae production system, the maintenance and construction cost of the closed system would be considerably higher than that of open systems (Cruz et al. 2018). However, it would not be an issue when the closed manufacturing system greatly shadows the open production system in terms of overall efficiency in biomass and microalgae production.

With their efficiency confirmed and justified, the closed production systems are often given the green light to operate in an industrial scale, which the open production systems could never achieve. Moreover, the closed systems have proven to be able to produce high-value and quality products, such as nutritional supplements (*Chlorella*) and food for the aquaculture sector (Khan, Shin, and Kim 2018).

9.2.2 Downstream Processes of Microalgal Biomanufacturing

In the downstream process of microalgae biomanufacturing, the process is commonly broken down into three steps: harvest, extraction, and conversion. The downstream processes are required to be conducted at an industrial level of microalgal biomanufacturing so that the accumulated microalgal biomass could be processed accordingly to meet the commercial needs and rival against its counterparts which are commonly sourced from nonrenewable resources.

9.2.2.1 Harvesting

In the harvesting step, a proper and efficient way to do so has not yet been determined due to several reasons. First of all, as each microalgal cell has an average size of less than 10 micrometer in diameter and a density similar to that of water, the overall production of biofuel and other substances such as protein from microalgae would be difficult to maintain, alongside the economic aspects of microalgal biomanufacturing. Besides, the fact that the surfaces of microalgae are negatively charged meant that they are not easy to settle under the force of gravity (J. Kim et al. 2013). Hence, in order to create a new form of energy source alternatively and convince the general society for it to be a reasonable alternative for fuel, a harvest method which is able to increase the efficiency of the process and to improve the bioeconomy of microalgal manufacturing has to be developed.

Presently, the techniques concerning microalgal recovery and harvesting have been designed and studied based on water purification technologies. Although it seems that there are similarities between water purification and the harvest of microalgae, technical approaches need to develop that are catered for microalgal harvesting to ensure the efficiency of the process. Moreover, there are some points that are required to be considered to do so:

1) The characteristics of individual microalgal species and the types of products desired to be manufactured and accumulated.
2) The advantageous synergy that the combination of the harvesting techniques may provide to resolve the weaknesses of individual techniques.

3) The necessity to design a method that is able to separate the cells in a dilute suspension and recycle the nutrients and water used efficiently to reduce the overall manufacturing cost.
4) The efforts to reduce effects that may be caused by the following processes of biofuel conversion and algal lipid extraction.

The methods of microalgal harvesting generally consist of the following methods: centrifugation, flocculation, filtration, flotation, magnetic separation, electrolysis, ultrasound, and, last but not least, immobilization (J. Kim et al. 2013).

9.2.2.2 Extraction

The extraction of algal lipid, along with the biomass dewatering process, is a process which requires a significant amount of energy in microalgal biofuel production. This is generally caused by different kinds of factors: the cost needed to dry the algal biomass accumulated, pressure, and temperature conditions during the extraction and others. However, the major reason for such consumption of energy is the fact that microalgae themselves have a cell wall which has a high resistance against chemical substances and high mechanical strength (J. Kim et al. 2013). Hence, as high consumption of energy means high manufacturing cost, especially that which is concerning microalgal cell wall, microalgal biofuel production is still not regarded as economically feasible as its non-bio-counterparts. Therefore, an extraction process that does not require drying of microalgae should be considered to encourage alternative fuel production using microalgae as the fundamental material.

Commonly, there are two routes to which the extraction process could be conducted: the dry route, which has high yield of extraction but as mentioned earlier, requires excessive amount of energy to dry the microalgae and the wet route, in which although the overall yield is reduced and cell disruption pretreatment is required, there is better energy balance in the process itself (J. Kim et al. 2013). Hence, wet route process of algal lipid extraction would be further discussed and elaborated in the following paragraph. In the wet route, there are two steps which are typically involved: cell disruption pretreatment and lipid extraction.

In order to disrupt the cells, three general types of mechanical, biological, and chemical methods are used. In the case of mechanical methods, the use of such methods involves directly breaking the cells of microalgae through physical force and may include the use of various technologies such as ultrasonication, high pressure homogenization (HPH), and microwave. For biological methods, however, it typically involves degrading the cell wall using suitable enzymes. Although phages and autolysis are used in some cases of application of biological methods, enzymes would still be more viable as they are much commercially available and easily accessible. Last but not least, the application of chemical methods would usually mean disrupting microalgal cells by means which are chemical, such as treatments with surfactants and acids, which can degrade chemical linkages on the cell membrane and cell wall of the microalgae cell (J. Kim et al. 2013). Besides, a method called osmotic shock would be used to cause a sudden decrease or increase in the inner and outer osmotic pressure of the microalgae cell to damage the microalgal cell wall.

After the cell wall would be disrupted or damaged by the application of the methods given before, the extraction process would follow. There are currently three types of extraction methods which help to generate substances such as lipid from the microalgal cells as efficient as possible: Soxhlet extraction, direct transesterification, and milking. In the case of Soxhlet extraction, although quite similar to the simpler solid–liquid extraction technique, the extraction solvent in a Soxhlet extractor would be evaporated, undergone repeated condensation, and dropped and accumulated in the prepared sample container (Halim et al. 2012). This enables the solvent to be recyclable, therefore making the process of extraction to be more efficient. Besides that, in the case of direct transesterification, or otherwise known as in-situ transesterification, it is a method which combines both processes of extracting lipids and biofuel conversion. In the direct transesterification method, when the processed biomass and other needed substances such as enzymes would initially be mixed together and heated up to required temperature, the desired biofuel would then be directly produced after the processes of lipid extraction and transesterification were completed (J. Kim et al. 2013). Both of the aforementioned processes occur simultaneously and therefore significantly reduce the energy needed to conduct the extraction process. Furthermore, the case of the milking method, as it is slightly different from the other extraction methods which involve the disruption of cells, involves the extraction of the needed substances directly from the existing microalgal cells without killing them (Hejazi and Wijffels 2004). The simplest process of milking typically involves a reactor that consists of two parts of different phases. In the reactor, the microalgae would be cultivated in a medium which is aqueous under an organic phase, which consequently helps to extract the target substances.

9.2.2.3 Conversion

After lipid extraction is completed, in the case of algal biofuel, the extracted lipids would then be converted into other forms of fuel such as biodiesel. The conversion is necessary because the accumulated algal lipid droplets are too viscous to be used as common fuel (Fuls, Hawkins, and Hugo 1984). Besides, when an oil sample which has high viscosity is used and run in an engine, oil grudge would eventually accumulate and cause the engine to fail as a result. Hence, the overall viscosity of the unprocessed microalgal oil should be reduced to enable it to be more user-friendly and be competitive against its counterparts through the transesterification process (J. Kim et al. 2013).

In the transesterification process, catalysts in the form of enzymes and acids and excess alcohol are used to accelerate the reaction and shift the chemical equilibrium toward the right side to produce more of the desired biofuel products. However, the use of such catalysts also meant that it is difficult to recover the target products from the catalysts which may be toxic toward the environment (J. Kim et al. 2013). Besides that, as the transesterification process is being species-specific, the process also has its free fatty acid and moisture content to be taken care of in order to produce a high-quality biofuel for the general use. Therefore, strategies for the process of microalgal lipid transesterification should free of the aforementioned concerning factors to ensure that the overall production of biofuel would be more economically feasible.

In the present time, there are a few approaches to conduct the transesterification process, as demonstrated in Table 9.1 which could be: catalytic, which generally uses homogeneous catalysts like acid, base and other heterogeneous catalysts such as calcium and magnesium oxides (Umdu, Tuncer, and Seker 2009) to increase the rate at which the microalgal lipid or fatty acids is processed; non-catalytic, which typically uses methanol to simultaneously extract and transesterify algal lipids in one reaction at a critical temperature; last but not least, the in-situ transesterification (Table 9.2), which involves the use of different systems such as co-solvent system and ionic liquids. After the extracted biofuel has been

TABLE 9.1
Various Catalytic Transesterification Methods for Microalgal Biomass Conversion

Catalytic transesterification	Advantages	Remark(s)	References
Homogeneous acid	- Not dependent on free fatty acids and water - Able to induce esterification and transesterification simultaneously	- Corrosive - Low rate of reaction - High temperature of reaction - Requires effort to manage the resulting waste	(Vyas, Verma, and Subrahmanyam 2010; Fukuda, Kondo, and Noda 2001; Vicente, Martínez, and Aracil 2004; Leung, Wu, and Leung 2010; Kawashima, Matsubara, and Honda 2009)
Homogeneous base	- High rate of reaction - Moderate conditions for the reaction to occur	- Dependent on water and free fatty acids - Requires effort to manage the resulting waste	(Ramadhas, Jayaraj, and Muraleedharan 2005; Schuchardt, Sercheli, and Vargas 1998)
Homogeneous catalyst/enzyme	- Not dependent on free fatty acids - Requires less effort to manage the resulting waste	- Involves high cost - Dependent on alcohol for denaturation to occur	(Suali and Sarbatly 2012; Jegannathan et al. 2008)
Heterogeneous catalyst	- Simultaneous esterification and transesterification with acid - Requires less effort to manage the resulting waste	- Dependent on water and free fatty acids with base - Involves high cost	(Umdu, Tuncer, and Seker 2009; Lam and Lee 2012; Leung, Wu, and Leung 2010)

TABLE 9.2
Various Approaches for Transesterification Conducted In-situ

In-situ transesterification	Advantage(s)	Remark(s)	Reference(s)
Co-solvent system	- Friendly toward the environment	- Dependent on water - Dependent on the reacting species - Has a low conversion yield	(Lee, Yoon, and Oh 1998; R. Xu and Mi 2011)
Mechanical assistance	- Has a high conversion yield - Only requires moderate condition for the reaction	- Dependent on water - High consumption of energy	(Ehimen, Sun, and Carrington 2012; Patil et al. 2012)
Assistance with ionic liquid	- Not dependent on free fatty acids - Has a vapor pressure which is negligible - Recyclable - Could be easily optimized for a specific reaction	- Requires high cost of operation - Dependent on water	(Y. H. Kim et al. 2012)

converted into desired forms of substances, the substances would be separated into individual chambers to be further purified.

9.2.3 Bioeconomy and Microalgae

As it were mentioned in the previous sections or paragraphs regarding the matter of microalgae, which have shown their capability to become a viable alternative for source of energy, bio-economists have come to realize the potential that the unicellular species of microorganisms could offer and have been working to expand their influence in the general economy.

While the current economy of plantation and greeneries relies heavily on normal plants such as palm trees and rice plant, microalgae or algae as a whole should be considered to be a part of the economic pie of plants and photosynthetic organisms. Besides, as the population of the world could reach as high as 9 billion by the near future of the year 2050 (Westwood et al. 2018), the applications of alternatives such as microalgae as a source of biological supplement and material for energy generation should be encouraged by many to meet the needs of the rising population. Moreover, the constant changes of environmental aspects and factors such as rise in global temperature and loss of cultivable land meant that attention should be given towards the further development of agricultural technology. Then, advanced technologies such as genetic engineering and phenomics and high-tech engineering are implemented to enhance the growth rate and the overall yield of the plantations such as wheat and rice. At the same time, plant-based alternatives for the animal produce such as dairy products and meat are invented and developed to reduce the reliance of the population on the animal-derived produces (Fabris et al. 2020). However, despite the clearly stated advantages of these food solutions for the ever-growing population, the consequential and eventual increased reliance on the usual crops for food supply means that new solutions and resources that might involve other types of organisms are required to ensure the sustainability of the food economy.

Furthermore, as for the economy regarding energy generation and production, even as sustainable and renewable source of energy is looking bright in the present and the near possible future, there are limitations which could correlate with environmental factors and issues of financial budget. As an example, for the case of solar energy, even as the night-and-day problem could be simply resolved with an energy-storage device such as a battery, issues such as that of area coverage could pose a problem for the solar-derived energy to be applied in an industrial scale. In order to produce a significant amount of electrical energy for industrial use, for example, solar plants which consist of thousands of solar panels have got to be constructed and developed to collect and extract the maximum amount of solar energy available during the daytime. Moreover, as the global warming still exists in this age and time, the weather pattern and the flow of air in the troposphere could be changing inconsistently, which would followingly affect the energy generation using wind turbine in the open fields. Therefore, other stable alternatives such as the use of microorganisms such as microalgae to produce usable energy for the increasing population are to be researched and developed accordingly to help build a sustainable future for the generation of energy (Hussian 2018).

9.3 INTELLIGENT MANUFACTURING OF MICROALGAE

In industry 4.0, while in the context of microalgae biomanufacturing, advanced and intelligent devices are used to control the devices which are then used as monitors of the growth rate of microalgae (ElFar et al. 2020). In addition, autonomous device-to-device communication is too observed to be in action to manage the biomanufacturing plant well. Operations of microalgal biomass production, such as harvest and cultivation, could be managed in detail to ensure a much efficient operation of the overall plant (Fabris et al. 2020). Moreover, the application of Industry 4.0 could extend to the downstream operations of the manufacturing plant such as extraction and purification processes.

Furthermore, as the perspective shifts in a more advanced direction, Industry 5.0 has introduced the study and engineering of phenomics of microalgae (ElFar et al. 2020). This could be regarded to be similar to the operation of genetic engineering; however, phenomics techniques do not involve any gene-related altercations, but rather combine beneficial characteristics of microalgal species to produce a mutant. For example, based on a recent analysis on the microalgae species of *Arabidopsis thaliana*, a mutant of increased protection against pathogen and photosynthesis production was found as a result of a supposed exchange between protection and production among botanic researchers (Kumar et al. 2020). Moreover, various other beneficial traits could be generated, like a sensing of strain utilizing cloture to cause auto-flocculation and product synthesis induction, when the harvest density of the microalgal culture is reached.

In the near and foreseeable future, as technology advances and improves itself through time, the infrastructures of biomanufacturing of microalgae will be significantly upgraded and updated to suit the needs of humanity and increase the efficiency of microalgae manufacturing through digitalization, particularly. In order to increase the efficiency of the operation, technologies surrounding intelligent biomanufacturing of the concerned microorganism would be taken into account in the following discussions.

9.3.1 Artificial Intelligence and Biomanufacturing of Microalgae

In this era and century, where most of the easier tasks are practiced by computers and machines, AI has emerged through countless research of information and technology experts as a type of program that is able to learn and act by itself.

9.3.1.1 Artificial Intelligence and General Biomanufacturing of Microalgae

In the processes of identifying strain–species of microalgae, AI was generally applied in three respective areas. For the first area, AI would be coupled with microscope image processing to produce a rapid microalgae identification result. As for the second area, however, AI program would be combined with measurement methods of fluorescence to quantify concentrations of microalgae in one or multiple units of culture. Third, it would be the area where AI would be used with matrix-assisted laser desorption/ionization (MALDI)-based spectrometer to identify certain microalgae species based on the respectively generated mass spectra (Andrade et al. 2015).

As for the screening of microalgae species and strains of genes of them, strains, in particular, are selected based on how a particular species of microalgae would be cultivated and are too evaluated based on factors like growth rate and biomass production. In fact, the processes of screening of microalgae are similar to that of identification in the previous paragraph, albeit with a different purpose in mind. As an example, measured fluorescence emission spectra have functioned as an information input while the concentration of microalgae cell would act as an output in the effort to test the feasibility of using artificial neural networks (ANN) in monitoring density of microalgae in the production units (Teng et al. 2020). This is due to the sporadicity of using AI to predict and estimate production of biomass. As a result, there is a high potential where in-situ fluorescence spectrometry would be combined with an ANN to estimate concentrations of microalgae cell. However, further procedures or steps should be carried out to reduce error of prediction. Third, besides screening and identifying microalgae based on their genetic traits, the cultivation of microalgae could be greatly accelerated when AI program is used.

As a brief explanation, microalgae cultivation usually means the process where microalgae would be transferred from the nature to an environment where its conditions could be controlled or not. However, to ensure that the cultivation of the unicellular microorganisms could be conducted in an efficient and effective manner, environmental parameters such as light and temperature should be taken care of and be monitored closely as the manufacturing of microalgae and biomass proceeds. For example, light is essential for the growth of microalgae as they conduct photosynthesis to survive. Therefore, light intensity and various wavelengths of it, as according to a recent study done by John Cortez-Romero and his co-workers, are able to alter or improve the rate at which biomass is produced from a certain species of microalgae. Moreover, in the study conducted by Mr. Cortez-Romero, the results show that production of biomass is directly proportional to the light intensity. In addition, biomass production of a particular species of microalgae would be significantly increased when the units of microalgae culture were being illuminated by light which has a longer wavelength such as those in red. Besides light, as temperature in a certain unit of microalgae culture would be reduced to levels below physiological optimum, amount of unsaturated lipid within the unit would increase and so would the enzyme production from the concerned species of microalgae in the unit. Nevertheless, carotenoids like astaxanthin would be produced at a condition of higher temperatures (Hurtado et al. 2019).

Hence, while it is apparent and obvious that statistical methods like response surface methodology would be of use to study and optimize the processes of microalgae cultivation, the introduction of ANN of AI into the game would greatly improve the optimization process while modelling nonlinear systems at the same time without further physical information provided. Moreover, ANN would also be used to conduct studies on the composition of culture medium and to observe how it would affect the growth of microalgae in the culture units (Wu and Shi 2007). For a particular example, for the heterotrophic cultivation of a microalgae species of *Chlorella* sp., a hybrid neural network model is proposed and programmed for study of the culture. In the proposed model, concentration of glucose in the culture unit of the species would be the input, while the analyzed specific growth rate of *Chlorella* sp. would be the output parameter of the study (Teng et al. 2020).

As for another example that is more focused toward the effect of light on microalgal growth, an ANN model has been designed and introduced to control the intensity of light at the photobioreactor, with the input being the light intensity and the output being the growth of biomass from microalgae. Furthermore, the study about the mentioned ANN program has demonstrated that the program is capable to predict biomass growth of microalgae under conditions where the intensity of light at the culture unit would vary with enough accuracy. Moreover, other environmental factors that concern the biomass production have also been studied and interpreted by ANN programs or algorithms, such as that in the case of *Spirulina* manufacturing in an open production type system, where it has been tested and proven that datasets of 12 days are all that an AI algorithm would need to predict the model for the mentioned microalgae species (Sharon Mano Pappu, Vijayakumar, and Ramamurthy 2013).

Besides controlling and managing conditions of environment for an optimum microalgae production, the strategy at which the biomass produce is being harvested would also play a huge role in the efficiency of the overall manufacturing system. As an example of it being a subject of study, an AI-based model of nonlinear autoregressive multilayer perception is designed and proposed to predict the microalgal growth of the *Spirulina platensis* species based on its productivity in a tube-type photobioreactor by searching and studying the balance between the growth of the microalgae species and the harvest cycle of the biomass produced (Susanna et al. 2019). In this study, similar to some other mentioned studies, the input would be the various conditions of environment within the *S. platensis* culture unit and the initial concentration of biomass. As a result, the designed model for the study has predicted the growth of the species up to a duration of time of 3 days with a value of coefficient of determination of more than 0.94. Furthermore, in another study of improving microalgae-based biomass manufacturing with agricultural wastewater as source of nutrients with sequestration of integrated carbon dioxide gas into the production system, a model which has combined with ANN algorithm has helped to increase the sequestration of carbon dioxide within the production system by 57% and showed that it is feasible for AI to be used to predict processes related to sequestration of carbon dioxide and wastewater treatment (Teng et al. 2020).

Last but not least, AI algorithms could also be used to fine-tune microalgae conversion technologies. In a conventional sense, conversion of microalgae or biomass from microalgae into useful products through experimental investigation could be tedious and cumbersome at times as the process of it often requires large amount of time and researcher's effort. Hence, in the effort to make the conversion of microalgae more feasible and applicable to many species of the unicellular microorganisms, various technologies such as thermochemical and biochemical methods are introduced and studied (Teng et al. 2020; Chai et al. 2021). However, as individual species of microalgae have characteristics and conversion mechanisms of their own, it would be difficult to generalize them and organize them into groups. Therefore, to resolve this issue, AI program would be used as a common tool to predict individual performances of microalgae and help determine the optimal condition of conversion for each species of microalgae. Moreover, in the more recent times, it has been considered that the genetic information of each species of microalgae should be included or installed into the algorithm of AI so that it would be able to conduct further tasks such as unraveling basic working principle of microalgae species based on the analyzed respective performances.

9.3.1.2 Artificial Intelligence and Microalgal Genetics

Nevertheless, besides maintaining the manufacturing system, AI could also be used to help alter the genetic properties of the microalgae themselves to help create species of microalgae that are more vigorous, stronger, and productive in terms of biomass production. Nowadays, as genetic modification plays a huge role in providing and enhancing certain beneficial characteristics of various species of flora and fauna, it too has applied to species of microalgae. In the case of microalgae, modern tools such as RNAi and ZFNs are used extensively to edit genes in these unicellular microorganisms (Teng et al. 2020). Besides, the introduction and use of the modern tools also mean genetic alterations could be conducted in the most accurate, fast, and effective way possible at the time. Nevertheless, as more and more modifications are introduced to be studied for validity verification, information storage and analytic systems are simultaneously required to handle the information gathered to prevent loss of valuable data about the microalgae species, with one of the examples in the form of artificial intelligence.

In AI, the program, or algorithm, is often considered by many as self-learning, with most of its source of information being the nature itself. As for the stages or directions that the AI programs take to learn, they are commonly separated into three 'chapters', namely machine learning, metaheuristics, and expert systems. Therefore, just like the example of PhotoSynthetica which would be briefly discussed in the following section, in the field of microalgae smart biomanufacturing, AI is mostly programmed to be able to collect and interpret data by itself, without requiring any external help. However, in the genetic enhancement aspect of things, genetic expressions combined with the algorithms of AI could also be of help to increase the accuracy and conduct the microalgae identification processes more effectively (Teng et al. 2020).

Besides, the key difference between conventional data collection and analysis and machine learning in microalgae genetic interpretation and sequencing is that machine learning does not need an entire book of information about genetic sequence in microalgae species to generate large collections of different combinations of new sequence of genes in a particular species of microalgae. For example, RNA sequence of a certain cross-species microalgae could be analyzed by two or more units of AI to extract the most suitable meta-genes to be combined to create a new batch of genetically modified species of microalgae, which is able to produce a higher amount of yield in the form of biomass (Teng et al. 2020).

Furthermore, AI is also being used in an experimental level to analyze species of microalgae before they are sent to the factory to be manufactured in an industrial scale. Some of the different sections of experimenting where AI would be required are species and genetic strain identification, screening, cultivation, and conversion.

9.3.1.3 Microalgae Production Plant with Artificial Intelligence Derived Algorithm: PhotoSynthetica

As it seems to be, artificial intelligence at the current stage being introduced into the biomanufacturing processes of microalgae is found to be able to manage the production system of microalgae. Recently, a company based in London, called PhotoSynthetica, has partnered with researchers with British universities, namely Bartlett University College London (UCL) and the University of Innsbruck, to develop alternative systems to allow plants and greeneries to be planted in cities or areas where spaces are limited and environment is not suitable for growth of plants (Cousins 2019).

With that said, one of the alternatives that the company is working on and developing is a machine vision algorithm that will be a part of the general software system which is created with an aim to manage the microalgae production plant. This designed algorithm will then function to help to analyze collected data on the environment aspects such as temperature, pH, and concentration of the fluid within the plant units. Besides, information about the pattern of microalgae growth in the units too would be recorded and interpreted by the AI-derived algorithm (Cousins 2019). Then, the information generated would serve as a general study regarding the microalgae and used by the designed artificial intelligence to further interpret and response accordingly, therefore increasing the overall efficiency of maintaining the designed units.

Besides that, for PhotoSynthetica, the designed software would be used on their trial units of microalgae, which are in the form of semi-transparent curtain-like structure, which still allows light pass through and makes the microalgae within able to conduct photosynthesis. Additionally, unfiltered air which mostly consists of urban carbon dioxide and other air pollutants would pass through the units from pores underneath to the panels which contain the microalgae. Moreover, as the process of the production of biomass goes on, the artificial intelligence system would then be responsible to monitor and control the conditions of the units through sensors within the units. Then, the biomass produced through photosynthesis of microalgae would then be harvested to be used industrially.

Furthermore, the trial units installed in the cities of Dublin, the Republic of Ireland, and Helsinki, Finland have proven to be able to remove an average 1 kilogram of carbon dioxide from the urban atmosphere per day. Hence, the PhotoSynthetica company has stated that they would consider and plan to introduce the system designed in a commercial scale in the near possible future (Cousins 2019). In brief, an example of PhotoSynthetica has shown that artificial intelligence nowadays is possible to be used as an alternative for maintenance of a microalgae production plant, no matter if in an open or closed environment.

9.3.2 Image Analysis and Microalgae Production

As discussed in the previous section, to improve the production of biomass from microalgae, the growth rate of microalgae should be maintained well to remain at an optimum level, and the concentration of it should too be estimated and observed from time to time. Besides, a well laid-out and conducted study of concentration of microalgae is essential to determine the harvest period of the biomass produced effectively. Hence, in order to be able to complete the study and prediction, image analysis would be applied.

9.3.2.1 Image Analysis Study for Dry Cell Weight

Recently, in a study to estimate the concentration of microalgae whereby methods of luminance and viscosity were proposed to do so, image analysis is used to measure and calculate the concentration of four different species of microalgae, namely *Scenedesmus* sp., *Desmodesmus* sp., *Dictyosphaerium* sp., and *Klebsormidium* sp. (Winata et al. 2021). To conduct the experiments of the study, different concentrations of the dry cell weight of the four species of microalgae would be used. As for the instruments for the conduct of the experiment, a dual-camera device would be used to capture images of dry cell weight of the respective microalgae species which would then be contained in individual flasks. Moreover, to confirm the viscosity within the flasks, an apparatus called the viscometer would be used to measure and determine the concentration of microalgae. Then, an analysis would be done based on the information collected from the analysis of captured image and data from the viscometer.

As a conclusion of the experiments, the viscosity method has shown to be capable to provide a more accurate result than the method of image analysis by only a small margin. Moreover, about the image analysis method, the brightness of the captured dry cell weight image of the microalgae species could have its limitations in situations where the color would be difficult to be recognized.

Therefore, based on this study, image analysis and its processing method have proven that it can be a capable alternative to help control and monitor the conditions of the culture units of microalgae manufacturing and the overall cultivation plant efficiently. Besides that, the obtained results also showed that this method could be further improved due to its potential in the field of microalgae culture analysis.

9.3.2.2 Transmission Hyperspectral Microscopic Imaging (THMI) with Machine Learning

Besides the aforementioned combination of viscosity method and analysis of microalgae image to obtain a precise prediction of microalgae growth and cultivation, another study from Ningbo Research Institute, Zhejiang University, has introduced a method of using transmission hyperspectral microscopic imaging (THMI), with combination of machine learning (Z. Xu et al. 2020). This is to acquire a more accurate result of prediction and estimation of biomass production from microalgae.

As according to the study from Zhejiang University, the designed THMI system has a spatial resolution of 4 micrometers, while the spectral resolution of it would be 3 nanometers. Also, in this study, hyperspectral imaging (HSI) has been performed on three samples of three different species of microalgae to determine and verify their characteristics of absorption. Then, the acquired transmission spectra from the performed HSI would be analyzed and followingly classified based on extracted feature and reduced dimensionality by using support vector model (SVM) which would be derived from artificial intelligence algorithm (Z. Xu et al. 2020).

Apart from that, as a further process or study to evaluate the growth rate of each tested species of microalgae, a random forest (RF) model, also derived from artificial intelligence algorithm, is used, with the obtained transmission spectra from the HSI experiment. With that, the model has been found to be able to predict the growth stage in the microalgae cycle of growth with the provided information up to an accuracy of 98.1%.

With it being briefly mentioned in the previous paragraph, the technology of HSI, which often is described with having advantages such as high efficiency and being non-invasive during experiment with other chemical or nonchemical substances, is capable to obtain spatial and spectral information simultaneously by forming a 3-D hyperspectral cube (x,y,z) (Dwight et al. 2019). Moreover, as the obtained spectra are able to provide detailed yet complex structural information like the vibration behavior of molecular bonds, HSI too has found its use for other scientific operations or processes such as assessment of food quality in the food manufacturing industry and biomedicine operations in the biochemistry-related medical field. As high spatial resolutions are required to produce images of minute objects, HSI has too been used often with optical microscopy to conduct medically important tasks such as classification of tumor and evaluation of cancerous grades (Z. Xu et al. 2020).

Information of HSI is generally obtained from two different methods: spatial-scanning and spectral-scanning. Based on the line-scanning principle, the spatial-scanning process would collect hyperspectral information from one narrow slit, and a spatial image would be constructed based on the information obtained through more specific methods of scanning called push-broom or whisk-broom scanning. Both of these more detailed scanning methods need relative movement of both camera and the sample of substance, or, as in this chapter's case, microalgae. The implementation of push-broom usually uses a stage which is motorized to produce a wider range (Z. Xu, Jiang, and He 2020), whereas whisk-broom usually uses a galvano-mirror, which is able to capture images at a high speed and efficiency (Cai et al. 2020). As opposing to the method of spatial-scanning, spectral-scanning, however, would collect information of various wavelengths from spatial-scanning to generate images. Some of the examples of spectral-scanning include filter wheels and liquid crystal tunable filters (LCTFs).

After that, computational methods such as those which were based on AI would be performed to filter and acquire a more accurate and specific set of data from the information generated from the scanning as the information is of a wide range of wavelength and may include redundant strands of data (Z. Xu et al. 2020).

9.3.2.3 Frequency-Division-Multiplexed (FDM) Fluorescence Imaging Flow Cytometry with Machine Learning

With the study regarding THMI briefly discussed, another recent study has found that it would be possible to conduct image analysis in the form of frequency-division-multiplexed (FDM) fluorescence imaging flow cytometry, along with the application of machine learning, or rather AI-derived algorithm (Harmon et al. 2020). Just like other image analysis methods in the previous paragraphs, the FDM-type image analysis is designed with an aim to characterize or classify the microalgae species present within the biomass production units. Moreover, as microalgae play a relatively beneficial role in various industries with their function as an agent in production of biofuel and treatment of any water medium, the process of individualizing or grouping the existing microalga species based on their respective characteristics would be very valuable. This is because the results of classification of microalgal species could be of use for scientists and biotechnologists to find the most suitable species to be further researched or manufactured by studying the acquired information.

In situations which are practical, even as the surrounding atmospheric conditions are relatively stable, classification of species of microalgae and its populations could be a difficult task as there are more than two species of the microorganisms' characteristics to be determined and the microalgae's morphology are all but quite similar, making it not easy to even separate them (Lee et al. 2014). Nevertheless, it is crucial to study and determine the variations in population of a certain species of microalgae, so that the route of the energy transfer and the rate at which energy is transferred is preserved within the ecosystem of the concerned body of water or other fluids, which generally consists of microalgae and other photosynthetic organisms. Besides, study of energy transfer would mean any change in the concerned ecosystem would and could be predicted or determined in a more sensible manner. However, as mentioned in the opening of this very paragraph, the process would be difficult to complete because, for one, it takes time to separate the microalgae 'pile' into smaller groups for them to be classified accordingly and eventually still needs microscope to validate whether the microorganism group is of microalgae or not; second, difficulty to determine and characterize the species of microalgae would only increase further when both aspects of similarities in morphology and the biodiversity of the unicellular microorganism were thrown into the equation, making the process of microalgae classification to be more time-consuming. Hence, to resolve this situation, other methodologies such as imaging flow cytometry were introduced and developed to reduce the time taken to classify the species of microalgae (Hildebrand et al. 2016). As different methodologies are often combined to produce a reliable result of microalgae study, imaging flow cytometry is used along with the application of time-stretched optofluidic microscopy to generate an accurate and precise classification result.

In this particular study, for the FMD aspect of things, the primary image would be captured using photodetector which is of single pixel instead of the CCD camera, which consequently improves the cytometry's throughput (Harmon et al. 2020). Then, after the parameters of microalgae morphology were extracted from the captured three-color images from before, an SVM would be used to classify the species of microalgae, as observed from the captured images, which is a common method of machine learning. Furthermore, the process could too be extended to obtain a more detailed information about the microalgae species using a cell sorter. In this way, multiple cells of microalgae group could be evaluated simultaneously while maintaining its capabilities to analyze any single-cell microorganism.

9.3.3 Use of Nano-Additives in Biomanufacturing of Microalgae

Even as the general development of nanoparticles or nanotechnology would be as advanced as it seems to be nowadays, there has not yet been a quite proven and realistic use of the nano-additives in the field of microalgae biomanufacturing, as according to the report of the study. Nevertheless, experiments regarding the matter have been underway. For example, in one of such experiments, or rather a prototype-stage production unit, magnetic nanoparticles had been installed into the cell suspension of microalgae in the photobioreactor to combine some cells to form clumps so that light and nutrients could be distributed evenly throughout the bioreactor (Hossain, Mahlia, and Saidur 2019). In addition, the cell suspension could also be further enhanced by modifying the microalgae cell using nanoparticles in the form of nano-liquid, which would also be injected into the microalgae cell culture so that bio-separation and the harvesting of microalgae could be better executed. Furthermore, in order to achieve a better accessibility of light, nano-scaled silver was used as a coating material for the surface of the photobioreactor. Besides that, in order to obtain a higher biomass and biofuel yield, methods of ultrasonication and irradiation were conducted using spheric nanoparticles in processes of microalgae culture such as lipid extraction and hydrolysis (Hossain, Mahlia, and Saidur 2019). As the mentioned usage of nano-additives given before involve methods that came up for prototype purposes in the present day based on the current findings and research, other nano-particle-related methods could also be invented or some of the discussed methods would be further refined to find more reliable methods to enhance microalgae biomass manufacturing.

9.3.3.1 Nanoparticles and Microalgae Cultivation

Despite its prominent role in mostly enabling the processes of the biomass and biofuel production to be more efficient and effective as we know it, the other aim of using nano-additives as part of the manufacturing process was to maximize the productivity of microalgae biomass and biofuel alike in the smallest area size possible. In order to do so, as it was briefly mentioned previously, technology regarding nanoparticles is being applied for the immobilization of enzyme, as nano-scaled structures are found to be capable to increase the surface area of such immobilization, thus causing the enzymes to be more stable and acquiring high loading power. As for how it would be done, various approaches such as attaching enzymes covalently to nano-scaled fibers and electrospun nanofibers were applied in the process of performing immobilization of enzyme (Hossain, Mahlia, and Saidur 2019). Besides, the process of enzyme immobilization could be investigated further using a particular substance that consists of different types of nano-scaled carbon in the form of fullerene and graphene oxide, for example. As a result of the investigation, however, it was found that oxidized multi-walled carbon nano-scaled tubes were those that had generated the highest yield of microalgae biomass, while fullerene had the lowest of all the nano-carbons which has 'participated' in the investigation (Safarik et al. 2016).

With the aspect of enzyme immobilization placed aside, during the period when microalgae were being cultured, the efficiency of light conversion within the environment of photobioreactor was greatly increased and was boosted, in fact, with again the use of nano-scaled particles. As a side note, this matter of light conversion should be currently taken noticed as light could be insufficient for the efficient cultivation of microalgae at times, even though artificial light is provided in closed production systems when light is not available. This is because even light shines directly upon the microalgae units themselves, when self-shading occur and biofilm would be formed on the surface of the photobioreactor. Hence, to resolve this issue and achieve a better level of illumination in the photobioreactor, light-emitting diodes (LEDs) which have been equipped with nano-scaled materials are placed at various points or angles in the reactor. Another recent development on the use of nano-scaled substances in biomanufacturing of microalgae biomass would be that of nano-scaled metals being integrated with localized surface plasmon resonance, which would amplify the scattering of light at a particular wavelength (Hossain, Mahlia, and Saidur 2019). As it stands, there was a study regarding the usage of such materials, where the subject would be nano-silvers, which has shown that the mentioned material could effectively backscatter blue-colored light in a smaller-scaled photobioreactor. Further into the study, moreover, the blue light is found to be able to significantly increase the overall efficiency of photosynthesis of *Chlamydomonas reinhardtii* and *Cyanothece* 51142, which are both green and blue–green microalgae, respectively (Torkamani et al. 2010).

Furthermore, the absorption of carbon dioxide in the atmosphere and sequestration systems of the manufacturing plant can be improved with the introduction of nano-sized particles. For example, nano-sized bubbles could retain their stability in the microalgae culture for a longer duration of time. In addition, nano-bubbles could float biomass into the culture units, which consequently ensured high efficiency of mass transfer and enhanced the density of biomass produced by adequate accumulation of carbon dioxide and stripping of oxygen gas (Hossain, Mahlia, and Saidur 2019).

9.3.3.2 Nano-Additives and Processing of Microalgal Biomass

As for the conversion of microalgae biomass into biofuel or other useful products, spherical nano-catalysts that are basic and acidic in nature have been found to be able to be applied on the process of substitution of chemical compounds like sodium methoxide through chemical reactions with fatty acids that are free within the complex, especially in the case of the production of biodiesel, which is also a popular mobile biofuel in the market. Moreover, due to their chemical properties, the nano-scaled catalysts are generally recyclable and have a positive impact in the economics of the microalgae biomass and biofuel manufacturing sector. With the application of such catalysts, the reaction with the oils and fatty acids could be conducted under the conditions of low pressure and temperature (Trindade 2011), which too result in a reduction in the release of contaminants into the surrounding atmosphere due to sodium methoxide being a reactant of the reaction. Additionally, a recent industrial study regarding biodiesel has shown that nano-scaled application of calcium oxides during a scaled-up process of catalytic transesterification translates into an overall 91% efficiency of biodiesel conversion (Torkamani et al. 2010). Furthermore, another study about the cultivation of microalgae using nano-spheres of silica and compounds of calcium has demonstrated that the cellular growth of microalgae could be drastically increased with the introduction of such nano-sized particles into the photobioreactor, without affecting the process of harvesting biomass and production of biofuel (Hossain, Mahlia, and Saidur 2019). With that, the cost of biofuel and biomass manufacturing could be effectively reduced, thus securing profits in the process.

In another experimental study, it was shown that certain types of nano-scaled catalysts could be more effective in catalyzing reactions than other similar-sized counterparts. Based on the report of the experimental study, the tested SBA-15, which is a mesoporous silica nano-catalyst, while having loaded with titanium, was able to present tolerance level of free fatty acids and water ten times better than other nano-catalysts of the likes of titanium dioxide silicate and titanium silicalite-1 (Chen et al. 2014). Aside from having a tolerance level that is much higher than of the other titanium-loaded nano-scaled catalysts, the mentioned SBA-15 catalyst, too, has shown that its application in the manufacturing of biofuel (biodiesel) could significantly lower the chemical cost of the process of transesterification as it is recyclable, along with its similar 'siblings', making the overall process more environment-friendlier. Furthermore, when being combined with sulfate, titanium-loaded SBA-15 is able to extend its functionality to convert vegetable oil, a co-product of the microalgae biomanufacturing, into esterified bio-lubricant. Besides the attachment of sulfate molecules, the SBA-15 nano-catalyst could generate a significant yield of biofuel with the incorporation of dinitrogen pentoxide. Apart from that, another nano-scaled substance called nano-zeolite, which is mainly composed of elements of aluminum and silicon, has been used as an absorbent for the transesterification process. In the process of transesterification, nano-zeolite would absorb unnecessary or undesirable moisture within the environment of the reactor to help produce pure glycerin, which would be a by-product of biodiesel manufacturing. Moreover, nano-scaled zeolite particles could too remove the contents of lipid from the cell membrane of microalgae (Hossain, Mahlia and Saidur 2019).

As their covered surface area is much larger than their volume, nano-additives could act as immobilizing beds for enzymes. Hence, reactions that involve large molecules such as a certain chain of microalgal complex sugar could be conducted at an increased rate, where the chain of complex sugar would be broken down or converted into simpler sugar, which would then turn into substances such as bioethanol as a product of the chemical reactions (Hossain, Mahlia, and Saidur 2019). Few experimental studies and researches have been conducted to study such use of nano-additives. For example, one such experimental study involves the application of nano-scaled dinitrogen pentoxide on sucrose, where the nano-catalyst has possessed sites of Lewis and Bronsted acid to convert the simpler sugar to 5-hydroxymethylfurfural, which in turn produced the highest yield among other similarly conducted experiments (Kreissl et al. 2016).

Furthermore, nano-additives in the form of nano-scaled catalysts are able to synthesize biomethane, a by-product of the microalgal biomass manufacturing, into pure compounds of carbon and hydrogen. Besides, the biomethane could be further reacted in the process of anaerobic digestion to produce biogas, which then could act as a raw material for the generation of electricity that focuses on the usage of bio-based products to do so (Trindade 2011). As for which nano-catalysts could best undertake the task of biomethane synthesis, however, according to recent studies and researches, any of them which would be or has been proven to be capable to effectively increase the biomass and biofuel yield should be able to synthesize microalgal biomethane. For example, in a particular experimental study, nano-scaled titanium dioxide and cerium (IV) oxide particles were manifested to increase the yield of biogas from the photobioreactors by 10–11% (Hossain, Mahlia, and Saidur 2019). Hence, they would then proceed to be used as a part of the process of biomethane production. Besides the case of biomethane synthesis and generation, some nano-scaled catalysts were used with the purpose to increase the percentage yield of microalgal biomass and biofuel in other experimental researches. As for some examples of such nano-substances, nano-metals of nickel and iron and nano-scaled metal oxides such as nano-silicon dioxide and nano-magnesium oxide are found to be able to increase the production of biomethane by up to 70%. Moreover, nano-scaled substances in the form of nano-fly ash and -bottom ash were too proven by individual studies to be able to increase the overall yield of biomethane by 3.5 times its original yield. In addition to the discussed capabilities of nano-additives, nano-hybrid catalysts were found to be capable to take up the role of a stabilizer for the industrial sector, especially for the industry of microalgal biomass and biofuel manufacturing. Therefore, they are currently highly sought after and commercialized. Some of the examples of such nano-hybrid additives would be that of silicon dioxide–magnesium oxide nano-scaled hybrid, which itself has performed as a bio-oil stabilizer in water emulsion due to its hydrophobic property. Besides that, nano-ammonium salts have been used as emulsion stabilizers for various stages of biofuel production and refining such as the purification and extraction processes (Hossain, Mahlia and Saidur 2019).

9.3.4 Genetic Engineering and Microalgae Biomanufacturing

At recent times, a lot of items have been genetically engineered to enhance certain beneficial characteristics, so have been microalgae. In the case of microalgae, the aim of genetic engineering would be to improve the production of biomass and lipid.

However, as it was briefly discussed in the previous paragraphs, although genetic engineering and modifying of microalgae are somehow related to nanotechnology and AI, other counterparts such as synthetic biology tools in the form of transcriptional terminators and promoters, for example, have been developed to advance the progress in the process of microalgal genetic modification. Moreover, several other methods such as microinjection, electroporation, and particle bombardment have come up and used to modify the genetic properties of microalgae.

Furthermore, as one of the first examples of successful genetic transformation of microalgae, the species of *C. reinhardtii* was agitated in the presence of

DNA using polyethyleneglycerol and glass beads. In the case of other species of microalgae such *as P. tricirnutum* and *C. sorokiniana*, particle bombardment has been conducted on specimens of both mentioned species using metal particles which are coated with DNA to transform the chloroplast and nuclear genomes of the unicellular organisms. Additionally, suitable selection markers such as those which are biochemical are used to develop efficient isolation of genetic-modifying agents.

9.3.4.1 Omics Approaches

In the past decades, a significant approach of 'omics' technologies has been observed in the biomedical and bioscience sectors to help construct a more comprehensive understanding of the microalgae's descriptions and characteristics so that they may be better applied in various fields in the future. Omics technology for the algal species is a technology which uses computational software and hardware to decipher and process information about the characteristics of the algae and their separate subsystems. Then, the processed information can be used to predict cell blueprint to assemble DNA sequence and analyze gene structures and expression. Followingly, processes related to algae could be studied by applying omics technology via genomics, metabolomics, proteomics, transcriptomics, and metagenomics.

Furthermore, in the last 20 years, genomic approaches have been very popular to help develop the technology of advanced automatic sequencing. In the present, eukaryotic and prokaryotic genome sequences are readily available (Mishra et al. 2019). Besides, some algal genomic sequencings have been undertaken, but projects of whole genome sequence are only generated for a few particular species of microalgae. For example, a relatively successful example of the species studied would be *C. reinhardtii*. In the year of 2003, the genome draft sequence of the concerned species was completed, and the whole genome sequence was published 4 years later (Merchant et al. 2007). Additionally, the functional annotation of *C. reinhardtii* has enabled various metabolic pathways to be identified and further studied.

Apart from that, omics technology, when used and compared to the traditional methods such as microscopy and mass spectrometry, has shown to be able to provide a more detailed and clearer understanding of the algal consortia interactions. Moreover, a combination of omics approaches with biochemical and microbiological analyses could generate more information about the genetic properties of the algal species concerning environmental factors and protein compositions, for example.

9.3.4.2 Metabolic Engineering

As the approach of metabolic engineering commonly coincides with that of genetic engineering, the aim of it is to extract the maximum amount of biofuel from microalgae in the most efficient way possible. To produce biofuel efficiently from each production unit of microalgae manufacturing plant, a biochemical pathway which is effective should be designed with an appropriate batch of host and other parameters such as pathway targeting and the modelling of the pathway toward right product formation. Moreover, as metabolic pathway could be complicating at times, the difficulties lie in the process of determining the suitable pathway to extract biofuel. Besides the complex nature of metabolic engineering, the increased interest toward studies regarding genome sequence of algal species has shifted the attention of the bioscientists and -engineers to develop ways to better identify genetic characteristics of microalgae and analyze the data collected. Moreover, the fact that the biofuel recovery from microalgae was too low to meet the commercial requirement meant that actions such as development of low-cost technologies for biomass harvesting, drying, and oil extraction should be undertaken in order to achieve the viability of microalgal biofuel in a commercial scale (Banerjee, Dubey, and Shukla 2016). That said, the production and extraction of biofuel from microalgae could be greatly enhanced by both approaches of adopting metabolic pathways engineering and genetic engineering, with the former focused toward production of augmented algal lipid.

In the conventional context, being one of the components required for the extraction of lipid from microalgae, enzyme plays an important role, and the factors that may affect its activity such as physical stress and deprivation in nutrient are to be taken into consideration. Among the various physical stresses especially the nitrogen stress is responsible in triggering the accumulation of TAG in different species of microalgae. Besides, both phosphorus and nitrogen stresses help to cause variation in phosphorus transporter system, which too consequently triggers TAG accumulation in the microalgae (Dubey et al. 2015). Hence, stresses can be constructive to increase lipid production due to inherent advantages such as requirement of no skilled labor. However, it may reduce photosynthesis activity of microalgae which results in a lower growth rate. Therefore, external or atmospheric factors like temperature are monitored and regulated to achieve the desired balance of stress and efficient lipid production.

The approach of metabolic engineering pays more attention toward the tuning and designing of microalgal metabolic pathways to induce and cause the production of the target metabolite. Some of the strategies to do so are flux base analysis, mathematical modelling, carbon partitioning, and transcription factor engineering (Banerjee, Dubey, and Shukla 2016). Besides, as it works closely with genome-related engineering and modification, it also means that omics technology could be used to combine with the metabolic approach.

9.4 FUTURE RESEARCH AND DEVELOPMENT OF SMART BIOMANUFACTURING OF MICROALGAE AND BIOMASS

Nevertheless, further events of research have been planned and conducted by bioscientists and biotechnologists to study the many other specific properties of microalgae species using existing or more modern technologies. Then, the use of microalgae in other fields of the industrial sector could be extended, making it one of the main players in town eventually. For example, even if it was proven by experimental research that agricultural wastewater would be able to provide sufficient nutrients for the cultivation of microalgae, actual usage of such medium at realistic locations such as barren lands and organic wastewater treatment plant is still to be considered and not yet put into action. Hence, detailed analysis and experiments would be done accordingly to show that areas which have agricultural wastewater flowing through them could be a reliable and sustainable site for the establishment of a microalgal biomass manufacturing plant.

The application of the aforementioned nanoparticles in the form of microalgal culture additives is currently at a stage where it could be further refined and confined into laboratorial scale so that the extent of the activities of the nano-scaled additive particles could be limited within the environment of the microalgae culture unit. Moreover, the economy of using nano-additives as an enhancement for biomanufacturing of microalgae biomass and biofuel should be carefully considered and analyzed as these nano-scaled substances are not easy to be produced and could be difficult to be accessed for larger-scale production operations. Hence, it is clear that it would be crucial for every concerned and interested party to investigate further upon the application of nano-scaled additives at an industrial scale through the conduct of advanced researches regarding the matter and development of modern technologies to generalize the use of such substances. In addition, there are words in town that says that the application of nano-additives in the biomanufacturing of microalgae would not cause harm to the environment in general. However, the claim may be true, and a comprehensive assessment of microalgae life cycle with nano-scaled additives should be undertaken to prove that the nano-scaled additives' application in the field could bring about positive effects towards the environment. Furthermore, other concerned factors such as possibility of biohazards and safety of the general public should be taken into account as well and therefore required to be extensively analyzed before the nano-scaled additive particles would be commercialized.

Furthermore, other than the case of nanoparticles, the use of AI, which is now only applied on the biomanufacturing processes of some plants, should be programmed and monitored in a careful and sensitive manner so that the designed program would act on our favor and understand their role as maintenance and testing agent for the cultivation of microalgae. Besides, like nano-additives, concerning parties regarding microalgal biomanufacturing should consider investing more into the development of AI in manufacturing plants at an experimental scale extensively, so as to show that AI could be just as reliable and stable like, if not more than, human workers.

9.5 CONCLUSIONS

As a conclusion, there would be many more available modern technologies in the distant future that may and could be further studied and used to enhance the current stage of microalgal bio-manufacture. However, not all of them are reliable enough to be used in this particular field. Hence, the currently discovered technologies such as artificial intelligence and genetic engineering should be seen as subjects of frequent research and development so that they could be used as a reliable enhancement for microalgae production.

REFERENCES

Alishah Aratboni, Hossein, Nahid Rafiei, Raul Garcia-Granados, Abbas Alemzadeh, and José Rubén Morones-Ramírez. 2019. "Biomass and Lipid Induction Strategies in Microalgae for Biofuel Production and Other Applications." *Microbial Cell Factories* 18 (1). doi:10.1186/S12934-019-1228-4.

Andrade, L. M., M. A. Mendes, P. Kowalski, and C. A. O. Nascimento. 2015. "Comparative Study of Different Matrix/Solvent Systems for the Analysis of Crude Lyophilized Microalgal Preparations Using Matrix-Assisted Laser Desorption/Ionization Time-of-Flight Mass Spectrometry." *Rapid Communications in Mass Spectrometry* 29 (3): 295–303. doi:10.1002/RCM.7110.

Banerjee, Chiranjib, Kashyap K. Dubey, and Pratyoosh Shukla. 2016. "Metabolic Engineering of Microalgal Based Biofuel Production: Prospects and Challenges." *Frontiers in Microbiology* 7: 432. doi:10.3389/FMICB.2016.00432.

Béchet, Q., M. Plouviez, P. Chambonnière, and B. Guieysse. 2017. "21 - Environmental Impacts of Full-Scale Algae Cultivation." In *Microalgae-Based Biofuels and Bioproducts: From Feedstock Cultivation to End-Products*, edited by C. Gonzalez-Fernandez and R. Muñoz, 505–25. Woodhead Publishing. doi:10.1016/B978-0-08-101023-5.00021-2.

Belz, S., B. Ganzer, E. Messerschmid, K. A. Friedrich, and U. Schmid-Staiger. 2013. "Hybrid Life Support Systems with Integrated Fuel Cells and Photobioreactors for a Lunar Base." *Aerospace Science and Technology* 24 (1): 169–76. doi:10.1016/J.AST.2011.11.004.

Borowitzka, Michael A., and Avigad Vonshak. 2017. "Scaling up Microalgal Cultures to Commercial Scale." *European Journal of Phycology* 52 (4): 407–18. doi:10.1080/09670262.2017.1365177.

Cai, Fuhong, Min Gao, Jingwei Li, Wen Lu, and Chengde Wu. 2020. "Compact Dual-Channel (Hyperspectral and Video) Endoscopy." *Frontiers in Physics* 8 (April): 110. doi:10.3389/FPHY.2020.00110.

Chai, Wai Siong, Chee Hong Chew, Heli Siti Halimatul Munawaroh, Veeramuthu Ashokkumar, Chin Kui Cheng, Young-Kwon Park, and Pau-Loke Show. 2021. "Microalgae and Ammonia: A Review on Inter-Relationship." *Fuel* 303: 121303. doi:10.1016/j.fuel.2021.121303.

Chaumont, Daniel. 1993. "Biotechnology of Algal Biomass Production: A Review of Systems for Outdoor Mass Culture." *Journal of Applied Phycology* 5 (6): 593–604. doi:10.1007/BF02184638.

Chen, Shih Yuan, Takehisa Mochizuki, Yohko Abe, Makoto Toba, and Yuji Yoshimura. 2014. "Ti-Incorporated SBA-15 Mesoporous Silica as an Efficient and Robust Lewis Solid Acid Catalyst for the Production of High-Quality Biodiesel Fuels." *Applied Catalysis B: Environmental* 148–149: 344–56. doi:10.1016/J.APCATB.2013.11.009.

Costa, Jorge Alberto Vieira, and Michele Greque de Morais. 2014. "Chapter 1 - An Open Pond System for Microalgal Cultivation." In *Biofuels from Algae*, edited by Ashok Pandey, Duu-jong Lee, Yusuf Chisti, and Carlos R. Soccol, 1–22. Elsevier. doi:10.1016/B978-0-444-59558-4.00001-2.

Cousins, Stephen. 2019. "Carbon-Eating Bio Curtains—the Answer to City Pollution?" *The RIBA Journal*. https://www.ribaj.com/products/carbon-capture-pollution-eating-algae-filled-curtains-bio-plastics-photosynthetica-ecologicstudio.

Cruz, Yordanka Reyes, Donato A. G. Aranda, Peter R. Seidl, C., Gisel Diaz, Rene G. Carliz, Mariana M. Fortes, Deusa A. M. P. da Ponte and Paula, and Rosa C. V. De. 2018. "Cultivation Systems of Microalgae for the Production of Biofuels." In *Biofuels—State of Development*. IntechOpen. doi:10.5772/INTECHOPEN.74957.

Dubey, Kashyap Kumar, Sudhir Kumar, Deepak Dixit, Punit Kumar, Dhirendra Kumar, Arshad Jawed, and Shafiul Haque. 2015. "Implication of Industrial Waste for Biomass and Lipid Production in Chlorella Minutissima Under Autotrophic, Heterotrophic, and Mixotrophic Grown Conditions." *Applied Biochemistry and Biotechnology* 176 (6): 1581–95. doi:10.1007/S12010-015-1663-6.

Dwight, Jason G., Michal E. Pawlowski, Thuc-Uyen Nguyen, and Tomasz S. Tkaczyk. 2019. "High Performance Image Mapping Spectrometer (IMS) for Snapshot Hyperspectral Imaging Applications." *Optics Express* 27 (2): 1597–1612. doi:10.1364/OE.27.001597.

Ehimen, Ehiaze A., Zhifa Sun, and Gerry C. Carrington. 2012. "Use of Ultrasound and Co-Solvents to Improve the In-Situ Transesterification of Microalgae Biomass." *Procedia Environmental Sciences* 15: 47–55. doi:10.1016/J.PROENV.2012.05.009.

ElFar, Omar Ashraf, Chih Kai Chang, Hui Yi Leong, Angela Paul Peter, Kit Wayne Chew, and Pau Loke Show. 2020. "Prospects of Industry 5.0 in Algae: Customization of Production and New Advance Technology for Clean Bioenergy Generation." *Energy Conversion and Management: X* 10: 100048. doi:10.1016/j.ecmx.2020.100048.

Fabris, Michele, Raffaela M. Abbriano, Mathieu Pernice, Donna L. Sutherland, Audrey S. Commault, Christopher C. Hall, Leen Labeeuw, et al. 2020. "Emerging Technologies in Algal Biotechnology: Toward the Establishment of a Sustainable, Algae-Based Bioeconomy." *Frontiers in Plant Science* 11 (March): 279. doi:10.3389/fpls.2020.00279.

Faried, M., M. Samer, E. Abdelsalam, R. S. Yousef, Y. A. Attia, and A. S. Ali. 2017. "Biodiesel Production from Microalgae: Processes, Technologies and Recent Advancements." *Renewable and Sustainable Energy Reviews* 79: 893–913. doi:10.1016/J.RSER.2017.05.199.

Fukuda, Hideki, Akihiko Kondo, and Hideo Noda. 2001. "Biodiesel Fuel Production by Transesterification of Oils." *Journal of Bioscience and Bioengineering* 92 (5): 405–16. doi:10.1016/S1389-1723(01)80288-7.

Fuls, J., C. S. Hawkins, and F. J. C. Hugo. 1984. "Tractor Engine Performance on Sunflower Oil Fuel." *Journal of Agricultural Engineering Research* 30 (C): 29–35. doi:10.1016/S0021-8634(84)80003-7.

Guo, Suolian, Ping Wang, Xinlei Wang, Meng Zou, Chunxue Liu, and Jihong Hao. 2020. "Microalgae as Biofertilizer in Modern Agriculture." In *Microalgae Biotechnology for Food, Health and High Value Products*, edited by Md. Asraful Alam, Jing-Liang Xu, and Zhongming Wang, 397–411. Springer. doi:10.1007/978-981-15-0169-2_12.

Halim, Ronald, Razif Harun, Michael K. Danquah, and Paul A. Webley. 2012. "Microalgal Cell Disruption for Biofuel Development." *Applied Energy* 91 (1): 116–21. doi:10.1016/J.APENERGY.2011.08.048.

Harmon, Jeffrey, Hideharu Mikami, Hiroshi Kanno, Takuro Ito, and Keisuke Goda. 2020. "Accurate Classification of Microalgae by Intelligent Frequency-Division-Multiplexed Fluorescence Imaging Flow Cytometry." *OSA Continuum* 3 (3). doi:10.1364/osac.387523.

Hejazi, M. Amin, and Rene H. Wijffels. 2004. "Milking of Microalgae." *Trends in Biotechnology* 22 (4): 189–94. doi:10.1016/J.TIBTECH.2004.02.009.

Hildebrand, Mark, Aubrey Davis, Raffaela Abbriano, Haley R. Pugsley, Jesse C. Traller, Sarah R. Smith, Roshan P. Shrestha, et al. 2016. "Applications of Imaging Flow Cytometry for Microalgae." In *Imaging Flow Cytometry*, edited by N. S. Barteneva and I. A. Vorobjev, 1389:47–67. Humana Press Inc. doi:10.1007/978-1-4939-3302-0_4.

Hossain, Nazia, T. M. I. Mahlia, and R. Saidur. 2019. "Latest Development in Microalgae-Biofuel Production with Nano-Additives." *Biotechnology for Biofuels* 12 (125). doi:10.1186/s13068-019-1465-0.

Hurtado, Diana Ximena, Claudia Lorena Garzón-Castro, John Cortés-Romero, and Edisson Tello. 2019. "Using Different Wavelengths and Irradiance on the Microalgae Acutodesmus Obliquus Batch Culture." *Journal of Chemical Technology and Biotechnology* 94 (7): 2141–47. doi:10.1002/jctb.6019.

Hussian, Abd Ellatif Mohamed. 2018. "The Role of Microalgae in Renewable Energy Production: Challenges and Opportunities." In *Marine Ecology—Biotic and Abiotic Interactions*, edited by Muhammet Türkoğlu, Umur Önal, and Ali Ismen. IntechOpen. doi:10.5772/INTECHOPEN.73573.

Jegannathan, Kenthorai Raman, Sariah Abang, Denis Poncelet, Eng Seng Chan, and Pogaku Ravindra. 2008. "Production of Biodiesel Using Immobilized Lipase—A Critical Review." *Critical Reviews in Biotechnology* 28 (4): 253–64. doi:10.1080/07388550802428392.

Kawashima, Ayato, Koh Matsubara, and Katsuhisa Honda. 2009. "Acceleration of Catalytic Activity of Calcium Oxide for Biodiesel Production." *Bioresource Technology* 100 (2): 696–700. doi:10.1016/J.BIORTECH.2008.06.049.

Khan, Muhammad Imran, Jin Hyuk Shin, and Jong Deog Kim. 2018. "The Promising Future of Microalgae: Current Status, Challenges, and Optimization of a Sustainable and Renewable Industry for Biofuels, Feed, and Other Products." *Microbial Cell Factories* 17 (1): 36. doi:10.1186/s12934-018-0879-x.

Kim, Jungmin, Gursong Yoo, Hansol Lee, Juntaek Lim, Kyochan Kim, Chul Woong Kim, Min S. Park, and Ji Won Yang. 2013. "Methods of Downstream Processing for the Production of Biodiesel from Microalgae." *Biotechnology Advances* 31 (6): 862–76. doi:10.1016/J.BIOTECHADV.2013.04.006.

Kim, Young Hoo, Yong Keun Choi, Jungsu Park, Seongmin Lee, Yung Hun Yang, Hyung Joo Kim, Tae Joon Park, Yong Hwan Kim, and Sang Hyun Lee. 2012. "Ionic Liquid-Mediated Extraction of Lipids from Algal Biomass." *Bioresource Technology* 109: 312–15. doi:10.1016/J.BIORTECH.2011.04.064.

Kreissl, Hannah Theresa, Keizo Nakagawa, Yung Kang Peng, Yusuke Koito, Junlin Zheng, and Shik Chi Edman Tsang. 2016. "Niobium Oxides: Correlation of Acidity with Structure and Catalytic Performance in Sucrose Conversion to 5-Hydroxymethylfurfural." *Journal of Catalysis* 338: 329–39. doi:10.1016/J.JCAT.2016.03.007.

Kumar, Gulshan, Ajam Shekh, Sunaina Jakhu, Yogesh Sharma, Ritu Kapoor, and Tilak Raj Sharma. 2020. "Bioengineering of Microalgae: Recent Advances, Perspectives, and Regulatory Challenges for Industrial Application." *Frontiers in Bioengineering and Biotechnology* 8: 914. doi:10.3389/FBIOE.2020.00914.

Lam, Man Kee, and Keat Teong Lee. 2012. "Microalgae Biofuels: A Critical Review of Issues, Problems and the Way Forward." *Biotechnology Advances* 30 (3): 673–90. doi:10.1016/J.BIOTECHADV.2011.11.008.

Lee, Megan L. Eisterhold, Fabio Rindi, Swaminathan Palanisami, and Paul K. Nam. 2014. "Isolation and Screening of Microalgae from Natural Habitats in the Midwestern United States of America for Biomass and Biodiesel Sources." *Journal of Natural Science, Biology and Medicine* 5 (2): 333. doi:10.4103/0976-9668.136178.

Lee, Seog June, Byung-Dae Yoon, and Hee-Mock Oh. 1998. "Rapid Method for the Determination of Lipid from the Green Alga Botryococcus Braunii." *Biotechnology Techniques* 12 (7): 553–56. doi:10.1023/A:1008811716448.

Leung, Dennis Y. C., Xuan Wu, and M. K. H. Leung. 2010. "A Review on Biodiesel Production Using Catalyzed Transesterification." *Applied Energy* 87 (4): 1083–95. doi:10.1016/J.APENERGY.2009.10.006.

Low, Sze Shin, Kien Xiang Bong, Muhammad Mubashir, Chin Kui Cheng, Man Kee Lam, Jun Wei Lim, Yeek Chia Ho, Keat Teong Lee, Heli Siti Halimatul Munawaroh, and Pau Loke Show. 2021. "Microalgae Cultivation in Palm Oil Mill Effluent (POME)

Treatment and Biofuel Production." *Sustainability* 13 (6): 3247. doi:10.3390/su13063247.

McBride, Robert C., Salvador Lopez, Chris Meenach, Mike Burnett, Philip A. Lee, Fiona Nohilly, and Craig Behnke. 2014. "Contamination Management in Low Cost Open Algae Ponds for Biofuels Production." *Industrial Biotechnology* 10 (3): 221–27. doi:10.1089/IND.2013.0036.

Merchant, Sabeeha S., Simon E. Prochnik, Olivier Vallon, Elizabeth H. Harris, Steven J. Karpowicz, George B. Witman, Astrid Terry, et al. 2007. "The Chlamydomonas Genome Reveals the Evolution of Key Animal and Plant Functions." *Science* 318 (5848): 245–50. doi:10.1126/SCIENCE.1143609.

Mishra, Arti, Kristina Medhi, Piyush Malaviya, and Indu Shekhar Thakur. 2019. "Omics Approaches for Microalgal Applications: Prospects and Challenges." *Bioresource Technology* 291: 121890. doi:10.1016/J.BIORTECH.2019.121890.

Murthy, Ganti S. 2011. "Overview and Assessment of Algal Biofuels Production Technologies." In *Biofuels Alternative Feedstocks and Conversion Processes*, edited by A. Pandey, C. Larroche, S. C. Ricke, C. G. Dussap, and E. Gnansounou, 415–37. Academic Press. doi:10.1016/B978-0-12-385099-7.00019-X.

Patil, Prafulla D., Veera Gnaneswar Gude, Aravind Mannarswamy, Peter Cooke, Nagamany Nirmalakhandan, Peter Lammers, and Shuguang Deng. 2012. "Comparison of Direct Transesterification of Algal Biomass under Supercritical Methanol and Microwave Irradiation Conditions." *Fuel* 97: 822–31. doi:10.1016/J.FUEL.2012.02.037.

Ramadhas, A. S., S. Jayaraj, and C. Muraleedharan. 2005. "Biodiesel Production from High FFA Rubber Seed Oil." *Fuel* 84 (4): 335–40. doi:10.1016/J.FUEL.2004.09.016.

Randrianarison, Gilbert, and Muhammad Aqeel Ashraf. 2017. "Microalgae: A Potential Plant for Energy Production." *Geology, Ecology, and Landscapes* 1 (2): 104–20. doi:10.1080/24749508.2017.1332853.

Raut, Nitin, Talal Al-Balushi, Surendra Panwar, R. S. Vaidya, and G. B. Shinde. 2015. "Microalgal Biofuel." In *Biofuels—Status and Perspective*, edited by Krzysztof Biernat. IntechOpen. doi:10.5772/59821.

Safarik, Ivo, Gita Prochazkova, Kristyna Pospiskova, and Tomas Branyik. 2016. "Magnetically Modified Microalgae and Their Applications." *Critical Reviews in Biotechnology* 36 (5): 931–41. doi:10.3109/07388551.2015.1064085.

Schuchardt, Ulf, Ricardo Sercheli, and Rogério Matheus Vargas. 1998. "Transesterification of Vegetable Oils: A Review." *Journal of the Brazilian Chemical Society* 9 (3): 199–210. doi:10.1590/S0103-50531998000300002.

Sharon Mano Pappu, J., G. Karthik Vijayakumar, and V. Ramamurthy. 2013. "Artificial Neural Network Model for Predicting Production of Spirulina Platensis in Outdoor Culture." *Bioresource Technology* 130: 224–30. doi:10.1016/J.BIORTECH.2012.12.082.

Suali, Emma, and Rosalam Sarbatly. 2012. "Conversion of Microalgae to Biofuel." *Renewable and Sustainable Energy Reviews* 16 (6): 4316–42. doi:10.1016/J.RSER.2012.03.047.

Susanna, Deepti, Rahulgandhi Dhanapal, Ranjithragavan Mahalingam, and Viraraghavan Ramamurthy. 2019. "Increasing Productivity of Spirulina Platensis in Photobioreactors Using Artificial Neural Network Modeling." *Biotechnology and Bioengineering* 116 (11): 2960–70. doi:10.1002/BIT.27128.

Tan, J. S., S. Y. Lee, K. W. Chew, M. K. Lam, J. W. Lim, S. H. Ho, and P. L. Show. 2020. "A Review on Microalgae Cultivation and Harvesting, and Their Biomass Extraction Processing Using Ionic Liquids." *Bioengineered* 11 (1): 116–29. doi:10.1080/21655979.2020.1711626.

Teng, Sin Yong, Guo Yong Yew, Kateřina Sukačová, Pau Loke Show, Vítězslav Máša, and Jo Shu Chang. 2020. "Microalgae with Artificial Intelligence: A Digitalized

Perspective on Genetics, Systems and Products." *Biotechnology Advances* 44: 107631. doi:10.1016/J.BIOTECHADV.2020.107631.

Torkamani, S., S. N. Wani, Y. J. Tang, and R. Sureshkumar. 2010. "Plasmon-Enhanced Microalgal Growth in Miniphotobioreactors." *Applied Physics Letters* 97 (4). doi:10.1063/1.3467263.

Trindade, Sergio C. 2011. "Nanotech Biofuels and Fuel Additives." In *Biofuel's Engineering Process Technology*, edited by Marco Aurélio dos Santos Bernardes. IntechOpen. doi:10.5772/16955.

Umdu, Emin Selahattin, Mert Tuncer, and Erol Seker. 2009. "Transesterification of Nannochloropsis Oculata Microalga's Lipid to Biodiesel on Al2O3 Supported CaO and MgO Catalysts." *Bioresource Technology* 100 (11): 2828–31. doi:10.1016/J.BIORTECH.2008.12.027.

Vicente, Gemma, Mercedes Martínez, and José Aracil. 2004. "Integrated Biodiesel Production: A Comparison of Different Homogeneous Catalysts Systems." *Bioresource Technology* 92 (3): 297–305. doi:10.1016/J.BIORTECH.2003.08.014.

Vyas, Amish P., Jaswant L. Verma, and N. Subrahmanyam. 2010. "A Review on FAME Production Processes." *Fuel* 89 (1): 1–9. doi:10.1016/J.FUEL.2009.08.014.

Westwood, James H., Raghavan Charudattan, Stephen O. Duke, Steven A. Fennimore, Pam Marrone, David C. Slaughter, Clarence Swanton, and Richard Zollinger. 2018. "Weed Management in 2050: Perspectives on the Future of Weed Science." *Weed Science* 66 (3): 275–85. doi:10.1017/WSC.2017.78.

Winata, Haikal Nando, Muhammad Ansori Nasution, Tofael Ahamed, and Ryozo Noguchi. 2021. "Prediction of Concentration for Microalgae Using Image Analysis." *Multimedia Tools and Applications* 80 (6): 8541–61. doi:10.1007/s11042-020-10052-y.

Wu, Zhengyun, and Xianming Shi. 2007. "Optimization for High-Density Cultivation of Heterotrophic Chlorella Based on a Hybrid Neural Network Model." *Letters in Applied Microbiology* 44 (1): 13–18. doi:10.1111/J.1472-765X.2006.02038.X.

Xu, Ruoyu, and Yongli Mi. 2011. "Simplifying the Process of Microalgal Biodiesel Production Through In Situ Transesterification Technology." *Journal of the American Oil Chemists' Society* 88 (1): 91–99. doi:10.1007/S11746-010-1653-3.

Xu, Zhanpeng, Yiming Jiang, and Sailing He. 2020. "Multi-Mode Microscopic Hyperspectral Imager for the Sensing of Biological Samples." *Applied Sciences* 10 (14). doi:10.3390/APP10144876.

Xu, Zhanpeng, Yiming Jiang, Jiali Ji, Erik Forsberg, Yuanpeng Li, and Sailing He. 2020. "Classification, Identification, and Growth Stage Estimation of Microalgae Based on Transmission Hyperspectral Microscopic Imaging and Machine Learning." *Optics Express* 28 (21): 30686. doi:10.1364/OE.406036.

10 Implementation of Microalgae 4.0 in Environmental Biotechnology

Akshara Ann Varghese[1], Doris Ying Ying Tang[1], Sze Shin Low[2] and Pau Loke Show[1]

[1] Department of Chemical and Environmental Engineering, Faculty of Science and Engineering, University of Nottingham Malaysia, Jalan Broga, Semenyih, Selangor Darul Ehsan, Malaysia

[2] Research Centre of Life Science and Healthcare, China Beacons Institute, University of Nottingham Ningbo China, Ningbo, Zhejiang, China

CONTENTS

10.1 Introduction....................327
10.2 Microalgae 4.0 in Environmental Biotechnology....................328
10.2.1 Bioremediation....................328
10.2.1.1 Mechanisms of Bioremediation by Microalgae....................334
10.2.2 Wastewater Treatment....................336
10.2.3 Environmental Toxicants Monitoring....................338
10.2.4 Bioassay....................343
10.2.5 Bio-Mitigation of Carbon Dioxide Emission....................345
10.2.6 Biosurfactant....................347
10.3 Challenges and Future Research Direction....................349
10.4 Conclusions....................351
References....................352

10.1 INTRODUCTION

Microalgae are a group of unicellular microorganisms that can photosynthetically synthesis high-value biomolecules such as lipids, carbohydrates, and proteins that can be applied in various industrial areas, for example in renewable energy, food and beverages, nutraceutical and biopharmaceutical industries. Due to rapid grow rate, short lifespan, and capability of microalgae to survive in harsh environment

DOI: 10.1201/9781003202196-10

(Tang et al. 2020), microalgae have become the potential sustainable candidates for environment biotechnology in the fields of environmental toxicants monitoring, wastewater treatment, mitigation of carbon dioxide emissions, bioremediation bioassay, and so on (Gavrilescu 2010; Rizwan et al. 2018).

Interestingly, the utilization of microalgae for environmental pollution control integrated with its cultivation has been reported recently. This synergistic combination is advantageous as the microalgae can be cultured in the polluted areas to synthesize valuable biocompounds and, at the same time, remediate that polluted area by utilizing the compounds present in the polluted area for their growth. Therefore, microalgae can become a potential candidate in wastewater treatment, as biological indicators to the environmental changes and for monitoring the changes or levels of the environmental pollutants through bioassay and biosensors (Omar 2010) because microalgae are shown to be sensitive to the pollutants or contaminants present in the polluted areas. As an example, *Euglena gracilis* is susceptible to the pollutants, potentially toxic elements, and persistent organic pollutants, so has been used as a pollutant-sensitive gravitational direction (Zaghloul et al. 2020). As a case in point, the study by Jiang et al. (2013) showed that *Scenedesmus dimorphus* grew well in the environment with high-concentration carbon dioxide and nitric oxide, demonstrating satisfactory carbon dioxide fixation through the use of flue gas for the cultivation purpose. Next, microalgae can be cultivated in the wastewater and aid in clearing the heavy metals or toxin compounds from the wastewater because microalgae will consume phosphorus and nitrogen for their growth and development. As a result, a cleaner effluent with high dissolved oxygen content will be produced (Gómez et al. 2013).

The use of microalgae in environmental biotechnology faces some challenges – for example the selection of suitable microalgae strains, high operating cost, contamination of culture medium, characteristics and complexity of the polluted areas (wastewater) as well as the enhancement of cultivation conditions. Therefore, the emergence of the fourth industrial revolution with the emergence of Internet of Things improves the use of resources more efficiently and sustainably. The adaptive, automation, and connected systems that emphasize in Industry 4.0 can provide predictions and diagnosis in real time. Neural network, logic, regression, and hybrid models are some of the few artificial intelligence models that have been proposed and designed for the mitigation of environmental impacts by microalgae. They aid in providing optimized custom, flexible, and cost-effective products with minimum waste and maximum efficiency. This chapter reviews the application of microalgae in environmental biotechnology integrated with Industry 4.0.

10.2 MICROALGAE 4.0 IN ENVIRONMENTAL BIOTECHNOLOGY

10.2.1 Bioremediation

Due to the industrial development over the past few centuries, there has been increasing amounts of toxic chemicals that have been discharged improperly into the environment causing several environmental issues, for instance, Union-Carbide (Dow) Bhopal disaster, contamination of the Rhine river, Exxon Valdez oil spill, decline of conifer forest and aquatic habitats of the Northeastern US, Europe, and Canada as well as the radioactive material release in the Chernobyl incident (Sasikumar and Papinazath 2003). Therefore, bioremediation, an eco-friendly approach, has been proposed to mitigate these environmental impacts. Bioremediation is defined as the method of using microorganisms, for instance, microalgae, bacteria and fungi, to degrade the environmental pollutants in the environment, especially from contaminated soil and groundwater. In other terms, bioremediation is the biological restoration or rehabilitation of contaminated environments due to the consequences of manufacture, transportation, storage, and improper discharge of inorganic or organic chemicals (Uqab et al. 2016; How et al. 2021). In the aspect of bioremediation, the microorganisms must be able to survive in harsh and toxic environments, have the necessary enzymes to break down toxins, and should be accessible to solid or aqueous surfaces that can absorb the contaminants and be able to multiply easily (Sardrood, Goltapeh, and Varma 2013).

Algae are the ideal candidates for their ability to remove toxins like heavy metals from the surroundings because they are highly tolerant to heavy metals and are able to grow autotrophically and heterotrophically along with their surface-area-to-volume ratio (Chekroun and Baghour 2013) as illustrated in Table 10.1. The use of microalgae to remove emerging pollutants is considered to be cost-effective. Microalgae remove heavy metals via the two-stage mechanism at which the first step is rapid biosorption process through the formation of covalent bonds, ionic exchange, and ions binding followed by bioaccumulation process that occurs in cytoplasm (Leong and Chang 2020). Microalgae, for example, *Chlamydomonas reinhardtii*, *Chlorella* sp., *Parachlorella kessleri-l*, *Nannochloropsis gaditana*, *Enteromorpha*, and *Cladophora* (Singh et al. 2019), demonstrated their ability in removing the organic and inorganic pollutants through bioadsorption, biouptake, and biodegradation (Sutherland and Ralph 2019). Peng et al. (2014) investigated the breakdown of two hormones, progesterone and norgestrel by *S. obliquus* and *C. pyrenoidosa*, as these two hormones will have detrimental effects on the aquatic systems. The findings revealed that the microalgae biotransformed these two hormones into other byproducts. Xiong, Kurade, and Jeon (2017) used *Chlorella vulgaris* to degrade levofloxacin in aqueous system as this microalga was tolerant to levofloxacin. Besides, *Chlorella sorokiniana* was shown to remove nutrients, paracetamol, and salicylic acid from water (Escapa et al. 2015).

The performance of heavy metal bioremediation can be strengthened by the application of various microalgae cell types (living cells, non-living cells, biochar, pellet) and chemically modified or pre-treatment prior to the removal of heavy metals. Daneshvar et al. (2019) investigated the chromium removal with various forms of microalgae *Scenedesmus quadricauda* – for example living cells, pellets, powder, biochar, chemically modified, where microalgal biochar demonstrated the highest removal efficiency. In addition, Nath et al. (2017) studied the biosorption of chromium by *Scenedesmus dimorphus*, *Chlorella* sp., *Oscillatoria* sp., and

TABLE 10.1
The Role of Microalgae in Bioremediation

Compounds		Organism involved	Efficiency of bioremediation process	Reference
Petroleum hydrocarbons		*Spongiochloris* sp. and Hydrocarbonoclastic native microbial	• The total hydrocarbon degradation efficiency was high (99.18%).	(Abid, Saidane, and Hamdi 2017)
	Water sample from a petroleum company and pre-treated with NaOH (PPW)	*Chlorella* sp., *Scenedesmus* sp., *Picochlorum* sp., *Tetraselmis* sp., *Leptolyngbya* sp., *Monoraphidium* sp.	• Among the six microalgae species, *Chlorella* sp. grew well in the PPW and removed 73% total organic carbon, 92% total nitrogen, and other heavy metals.	(Das et al. 2019)
	Petroleum-contaminated water	Freshwater microalgae biofilm	• Microalgae biofilm able to adsorb pollutants in petroleum-contaminated water.	(Ugya et al. 2021)
	Pyrene	*Chlorella sorokiniana*	• Pyrene IC_{50} induced the synthesis of lipid and at the same time remediated the pyrene pollutant.	(Jaiswal et al. 2021)
	Benzo(a)pyrene	*Selenastrum capricornutum* and *Scenedesmus acutus*	• Complete removal was achieved: ✓ *S. capricornutum*: 99% after exposure for 15 hours ✓ *S. acutus*: 95% after 72 hours of exposure	(de Llasera et al. 2016)
Herbicide, Pesticide, Insecticide	Atrazine, molinate, simazine, isoproturon, propanil, carbofuran, dimethoate, pendimethalin, metoalcholar, pyriproxin	*Chlorella vulgaris*	• Living or immobilized algal biomass had high bioremediation capability **Short-term study:** • The highest pesticides bioremoval by living cells was atrazine and the lowest was isoproturon **Long term study:** • The highest biodegradation by living biomass was showed by simazine (96.75%), while the lowest one was showed by pendimethalin (87.85%).	(Hussein et al. 2017)
	Lindane	*Nannochloris oculata*	• The lindane was decreased by 73% and 68.2% in the 0.1 and 0.5 mg L^{-1} media concentrations, respectively.	(Pérez-Legaspi et al. 2016)

	Diazinon	*Scenedesmus obliquus*, *Chlamydomonas mexicana*, *Chlorella vulgaris*, and *Chlamydomonas pitschmannii*	• *C. vulgaris* showed the highest removal capacity (94%) of diazinon at 20 mg L^{-1}. • Microalga-mediated biodegradation of diazinon generated a less toxic product which was 2-isopropyl-6-methyl-4-pyrimidinol (IMP).	(Kurade et al. 2016)
	Mecoprop, atrazine, simazine, diazinone, alachlor, chlorfenvinphos, lindane, malathion, pentachlorobenzene, chlorpyrifos, endosulfan and clofibric acid	Microalgae consortium from an experimental high-rate algal pond treating urban wastewater	• Continuous feeding operational mode was more efficient. • Microalgae increased the removal of some pesticides in the batch reactors after 10 days of incubation such as lindane, alachlor, chlorpyrifos, endosulfan and malathion. • Increasing the hydraulic retention time from 2 to 8 days increased the removal efficiency of pesticides by around 20%.	(Matamoros and Rodríguez 2016)
	Synthetic pyrethroid insecticide zeta-cypermethrin	*Chlamydomonas reinhardtii*	• In high concentration of pesticide (600 µg/L), microalgae showed better removal efficiency as compared to duckweed.	(Yılmaz and Taş 2021)
Pharmaceutical contaminants	Endocrine-disrupting compounds (EDCs)	Microalgae consortium at which main microalgae populations were made up of *Chlorella* sp. and *Nitzschia acicularis*.	• Concentration of EDCs was very low at the end of the experiment (<5%), except for 4-octylphenol, for which it was around 20%.	(Solé and Matamoros 2016)
	River water contaminated with pharmaceutical effluent	*Chlorella* sp., *Chlorococcum* sp., and *Neochloris* sp.	• The change of organic pollution was expressed in terms of BOD and COD. • *Neochloris* sp. reduced BOD from the initial concentration to 3.67 mg L^{-1} with removal efficiency of 91%. • *Chlorella* sp. reduced BOD of the water from 41.4 mg L^{-1} to 4.8 mg L^{-1} with maximum removal efficiency of 84%. • *Chlorococcum* sp. showed removal efficiency of 83% of BOD reduction.	(Singh, Ummalyma, and Sahoo 2020)

(*Continued*)

TABLE 10.1 (Continued)

Compounds	Organism involved	Efficiency of bioremediation process	Reference
Diclofenac, ibuprofen, paracetamol, metoprolol, carbamazepine and trimethoprim	*Chlorella sorokiniana*	• The removal efficiency was 60–100% for diclofenac, ibuprofen, paracetamol, and metoprolol. • Removal of carbamazepine and trimethoprim was incomplete and did not exceed 30% and 60%, respectively.	(de Wilt et al. 2016)
26 organic microcontaminants	Microalgae consortia in high-rate algal ponds	• High removal rate (>90%): caffeine, acetaminophen, ibuprofen, methyl dihydrojasmonate, and hydrocinnamic acid • Moderate-to-high removal (from 60% to 90%): oxybenzone, ketoprofen, 5-methyl/benzotriazole, naproxen, galaxolide, tonalide, tributyl phosphate, triclosan, bisphenol A, and octylphenol • Moderate-to-low removal (from 40 to 60%): diclofenac, benzotriazole, OH-benzothiazole, triphenyl phosphate, cashmeran, diazinon, benzothiazole, celestolide, 2,4-D and atrazine (and poor or no removal) <40%, carbamazepine, methyl paraben, tris(2-chloroethyl) phosphate • The removal efficiency was affected by the hydraulic retention time during the cold season.	(Matamoros et al. 2015)
Ciprofloxacin	*Chlamydomonas Mexicana*	• Low removal efficiency (13%) was achieved in the absence of any organic substrates • Addition of sodium acetate increased the removal efficiency to 56%.	(Xiong et al. 2017)
Levofloxacin	*Chlorella vulgaris*	• After addition of NaCl, removal efficiency was increased from 9.5% to 91.5%.	(Xiong, Kurade, and Jeon 2017)

Lyngbya sp. with three cell types (live cells, non-living cells, and pre-treated cells with 0.1 N NaOH and 0.01% SDS), and the result showed that pre-treated cells displayed satisfactory removal activity. Besides, cells' immobilization also can increase the removal efficiency of heavy metal. Taking one example, the introduction of beads to *Spirulina platensis* extracts enhanced the removal of chromium under acidic condition and without any addition of polymers (Kwak et al. 2015). It is important to understand that the immobilization technique may damage the cell wall, lowering the effectiveness of heavy metal removal by microalgae (Ardila, Godoy, and Montenegro 2017). Zhang et al. (2019) utilized a green immobilization approach to remove cadmium using microalgae *Selenastrum capricornutum* and *Microcystis aeruginosa* as the bioreactors to produce cadmium-based nanoparticles incorporated with selenium. This method offers the benefits of recycling the heavy metals to produce other useful products and, at the same time, removes the heavy metals from the contaminated areas.

From these bioremediation studies by microalgae, the challenge is the selection of appropriate microalgae strains as different microalgae will produce different enzymes that can degrade particular emerging pollutants, tolerant to specific contaminants. In addition, majority of bioremediation studies are conducted in laboratory scale, facing challenges when the situation is changed to full-scale actual scenario where multiple contaminants are present (Fabris et al. 2020). Thus, a screening system that allows rapid screening and selection of appropriate microalgae strain for the removal of particular emerging pollutants in biodegradation technology is needed, for example a microalgae phenomics facility which is a database that contains all information of microalgae and their respective bioremediation studies from all around the world. Next, synthetic biology approaches also can be applied to produce genetically engineered microalgae, such as the creation of genetically modified microalgae that can overexpress the degrading enzymes or metal transporter proteins (Fabris et al. 2020). Nevertheless, the performance of genetically modified strains need to be investigated and compared with the adapted strains because microalgae have been evolved to adapt to the polluted circumstances over many generations and hence may perform better in terms of metal tolerance and bioremediation (Ibuot et al. 2017). Ibuot et al. (2017) overexpressed the CrMTP4 genes in *Chlamydomonas reinhardtii* to increase its tolerance to cadmium, raising the cadmium uptake. The performance of this modified *C. reinhardtii* was compared to that of natural microalgae strains, *Chlorella luteoviridis*, *Parachlorella hussii*, and *Parachlorella kessleri*, which had been adapted to natural wastewater. The results showed that the cadmium tolerance of CrMTP4 overexpression strains was less satisfactory as compared to these three natural microalgae strains. Despite the fact that overexpression of MTP4 gene in *C. reinhardtii* increased the metal tolerance, the performance of a single gene alteration cannot keep pace with the oxidative stress tolerance adapted by the microalgae in natural multigenic environment. In short, the use of microalgae in bioremediation, especially in the heavy metal remediation, is an auguring approach due to its tolerance and ability to uptake the heavy metals.

10.2.1.1 Mechanisms of Bioremediation by Microalgae

There are three major methods involved in the bioremediation by the microalgae, which are bioadsorption, bioaccumulation, and biodegradation.

10.2.1.1.1 Bioadsorption

Bioadsorption takes place at microalgae cells by either absorbing the toxic substances from their surrounding or by the absorption of toxic elements through the cell wall components. Bioadsorption process is considered to be a passive process because the microalgae absorb the pesticides and other toxic pollutants from the environment. Bioadsorption process is the non-metabolic reaction that takes place between the toxins and the negatively charged organic substances produced by the cell or the algae cell walls, attracting the positively charged toxins. The absorption of these toxins by the microalgae cell walls is affected by the chemical build-up of the toxins. The hydrophobic toxins are attracted towards the cell walls by electrostatic interactions, and the hydrophilic toxins are repelled. The amount of toxins that can be absorbed by the cell walls is also influenced by the demand of the area and chemistry of the cell surface (Sutherland and Ralph 2019).

Electrostatic interactions, surface complexation, ion exchange, absorption, and precipitation are a few mechanisms that can take place during bioadsorption process. The rate at which the toxins are absorbed by the cell walls depends on the physio-chemical properties of the environment, for instance the temperature, the pH levels, and the redox reactions. The functional groups that are present on the microalgae cell wall are responsible for the absorption of toxins. The cell walls consist of hydroxyl, carboxyl, sulfates, and amine functional groups. The presence of carbohydrates, polysaccharides, and fibril matrix also facilitates the toxin absorption from the environment (Nie et al. 2020). As the process is non-selective, the challenge is faced when there are multiple toxins present in the environment. The binding sites on the cell walls of microalgae may get saturated with non-target toxins as well. The availability and presence of other toxins in the environment can interfere with the adsorption rate regardless of the binding site availabilities on the microalgae cell walls (Sutherland and Ralph 2019). In summary, bioadsorption can be one of the bioremediation mechanisms to adsorb the toxic elements from the environment.

10.2.1.1.2 Bioaccumulation

Bioaccumulation is an active process that is capable to uptake the toxins or organic substances and accumulate it through the surface (Zabochnicka-Świątek and Krzywonos 2014). Bioaccumulation can be influenced by the bioconcentration factors that indicate the toxin concentration in an organism with regards to its surroundings. The bioconcentration factor is mainly affected by the availability of chemicals, the concentration mechanism, physical barrier, the amount of organic matter dissolved, metabolism, interspecies variations, ionization of compounds, and environmental conditions. The microalgae species that are exposed to the toxins induce the production of reactive oxygen species from the mitochondria, peroxidases, and chloroplasts. The generation of reactive oxygen species is highly oxidative. The chemically active oxygen atom or atomic groups that may cause the oxidation of membrane lipids and DNA will lead to dysfunctional cells and even death of the microalgae cell. These toxins can induce the expression of the genes in the cell, which are responsible to produce antioxidant enzymes and activate the protection mechanisms. Many studies have proven that bioaccumulation and biodegradation of toxins can occur simultaneously.

The green algae, *Scenedesmus obliquus*, was observed in demonstrating the rapid accumulation of triadimefon while simultaneously degrading it. This combination of accumulation and degradation proved that toxins can be quickly removed from the environment by microalgae (Nie et al. 2020). Bioaccumulation can be divided into two steps. During the first phase, metal ions bind with the functional groups on the cell walls. The second phase of this method is similar to bioabsorption. The metal ions are transported through the cell, and this is only possible if the cells are metabolically active. If the environmental conditions are suitable for the growth of algae species, the quantity of biomass will increase with time. If the increase in biomass is successful, larger quantities of heavy metal ions can bind with the cell (Zabochnicka-Świątek and Krzywonos 2014) and thus facilitating the removal or uptake of heavy metals from the environment.

10.2.1.1.3 Biodegradation

Apart from biosorption and bioaccumulation, microalgae also use biodegradation which is known as biotransformation as the main pathway to remove pollutants. Unlike bioadsorption and bioaccumulation that act as the biological filters to concentrate the toxins and eliminate them from the environment, this method transforms the complex compounds into harmless, simpler, and useful molecules through metabolic breakdowns that can be used as a nutrient source for growth. There are two ways by which biodegradation can take place either by metabolic degradation where the pollutants serve as carbon sources or electron donors and acceptors for microalgae or by co-metabolism, where the toxins are broken down using enzymes that catalyze other substrates. For the first mechanism which is metabolic degradation, the species can use mixotrophic growth strategies where both organic and inorganic dissolved carbon are used up. The biodegradation process can take place either intracellularly or extracellularly or also as a combination process, at which the initial degradation occurs extracellularly, and the product is further broken down extracellularly (Sutherland and Ralph 2019).

Take for instance, in the biodegradation of pesticides present in wastewater, the efficiency of biodegradation is higher than photodegradation. The metabolism of different enzymes is responsible for the degradation of pesticides in the environment. The breakdown of pesticides is a multi-phase process because it involves the enzyme metabolism for the activation of pesticides when there are insufficient functional groups by redox reactions. They are also considered to be the enzymes' transfer from cytosol to the pesticides that are either activated or consist of any functional group. However, the degradation mechanism on the pesticides needs further investigation and studies (Nie et al. 2020). In general, biodegradation is an essential process to eliminate the contaminants from the environment with the help of microalgae through the biotransformation of pollutants via metabolic action.

10.2.2 Wastewater Treatment

Due to increased growth in population and industrialization, the increased severity in environmental contamination has also attracted the attention from the researchers worldwide. The concentration of chemical (heavy metals, toxins, agricultural waste, industrial waste) and biological contaminants have drastically increased in wastewater. Therefore, it is necessary to reduce these contaminants to protect the environment and reduce eutrophication processes (Satpal and Khambete 2016; Chai, Tan et al. 2021; Low, Bong et al. 2021). There are numerous methods for the disposal of organic and inorganic toxins from the wastewater such as precipitation, membrane filtration, reverse-osmosis, flotation, and coagulation-flocculation. These conventional methods are expensive and time- and energy-consuming. Therefore, the use of biological methods to treat wastewater is an auguring approach as it is environmentally friendly and cost-efficient. One of the biological methods to treat wastewater is by using microalgae as the biosorbent to absorb the heavy metals in the wastewater. The advantages of the microalgae as compared to other microorganisms are simple structure, growth in harsh conditions, such as sewage, salty environment, desert, or infertile land as well as high photosynthesis ability. Various studies showed that microalgae culture has the potential to treat effluents (Abdel-Raouf, Al-Homaidan, and Ibraheem 2012; Satpal and Khambete 2016; Amenorfenyo et al. 2019; Chai, Cheun et al. 2021). Consequently, microalgae cultivation system can be used for wastewater treatment because the microalgae can use the compounds present in wastewater, for example nitrogen and phosphorus, for their growth, demonstrating their capability to eliminate heavy metals and organic compounds from wastewater (Abdel-Raouf, Al-Homaidan, and Ibraheem 2012) and at the same time producing high-value biomolecules. Mubashar et al.'s (2020) study demonstrated the ability of *Chlorella vulgaris* with the help of bacteria species *Enterobacter* sp. in removing the heavy metals (copper, chromium, lead, cadmium) in the effluents from the textile industry and, at the same time, promoted the growth of microalgae. Table 10.2 shows other examples of microalgae that have been applied in wastewater treatment. There are various factors that need to be considered for the wastewater treatment that are microalgae strains, the necessity of pre-treatment of wastewater, nutrient concentration, nitrogen and phosphorous ratios, and wastewater characteristics (Satpal and Khambete 2016; Al-Jabri et al. 2021).

Wastewater treatment can be combined with microalgal growth. In wastewater system, the microalgae cultivation growth can be divided into suspended growth and attached growth in open or closed cultivation systems with batch or semi-continuous mode. Suspended growth of microalgae refers to the cultivation and scattering of small-sized microalgae in culture medium with less than 1% of overall solid content (Zhuang, Wang, and Hu 2018). The suspended cultivation can reduce contamination, water evaporation, and there is a loss of carbon dioxide with

TABLE 10.2
Examples of Heavy Metal Ions that Are Removed by Microalgae Strains in Wastewater Treatment

Algae strain	Metal ions
Chlorella sp.	• Lead (II) ions • Nitrogen • Phosphorus
Scenedesmus abundans	• Metals like cadmium and copper
Spirulina sp.	• Heavy metals like antimony and chromium
Botryococcus braunii	• Simple inorganic compounds like nitrogen and phosphorus
Dunaliella salina	• Copper • Cadmium • Cobalt • Zinc • Hypersaline wastewater
Pediastrum sp.	• Inorganic compound indicators

Source: Modified from Abdel-Raouf, Al-Homaidan, and Ibraheem (2012).

obtaining a high volume of biomass. However, the cost for this method is high due to energy- and labor-intensive processes in the recovery of microalgae biomass (Al-Jabri et al. 2021; Zhuang, Wang, and Hu 2018). On the contrary, the microalgae cells are grown crowdedly on a static surface or biofilms or mounted onto rotating paddles within attached systems, producing 20-fold more concentrated biomass than suspended systems (Zhuang, Wang, and Hu 2018). Furthermore, monoculture or consortia of microalgae strains are also an important criterion in the wastewater treatment. As compared to the cultivation of single microalgae strain, microalgae-mixed culture is advantageous because wastewater comprises different types of contaminants, resulting in high-treatment efficiency (Al-Jabri et al. 2021). Mahapatra, Chanakya, and Ramachandra (2013) on the basis of their study revealed the use of lagoons which are also known as pond systems to treat domestic sewage using algae (euglenoides and chlorophycean) in Mysore city, South India. This algae-based sewage treatment plant can reduce the particulates and suspended solids present in the wastewater, such as ammonium, nitrogen, and phosphate.

The removal of heavy metals in the aqueous medium is affected by the susceptibility of different microalgae strains to various heavy metals and the conditions or characteristics of the wastewater. Saavedra et al. (2018) evaluated the removal rate of toxic heavy metals (arsenic, manganese, copper, zinc, and boron) by *Chlamydomonas reinhardtii*, *Chlorella vulgaris*, *Scenedesmus almeriensis*, and *Chlorophyceae* sp. with respect to the pH and contact time of the solutions. The results showed that each microalgae species demonstrated various removal efficiencies of heavy metals in monometallic solution with varying pH and contact time – for example for the 3 hours of contact time, *C. vulgaris* was the most effective in removing manganese in pH value 7.0, but *Chlorophyceae* sp. was best in removing zinc in pH value 5.5. In the different circumstances where multiple heavy metals were present, *Chlorophyceae* sp. showed the satisfactory removal efficiencies and tolerance as compared to another three microalgae species.

With the industrial revolution to Industry 4.0, the application of deep learning to microalgae cultivation can strengthen the performance of wastewater treatment. Esfandian et al. (2016) employed artificial neural network in the prediction of mercury sorption using brown algae, *Sargassum bevanom*. Manu and Thalla (2017) used support vector machine and adaptive neuro-fuzzy inference system modeling approach to estimate the removal of Kjeldahl nitrogen from a domestic wastewater treatment plant in Mangalore. Next, Ansari et al. (2021) employed artificial neural network (ANN) at which microalgae were cultured in secondary treated wastewater effluents in outdoor system in pilot scale. ANN was used to predict the interaction between wastewater characteristic and biomass growth in the aspects of temperature, pH, dissolved oxygen, electrical conductivity, nitrate, and phosphate. The study by Bhagat, Tung, and Yaseen (2020) reviewed the application of artificial intelligence models on the simulation of heavy metals, for instance copper, cadmium, lead, zinc, nickel, mercury, iron, and other heavy metals. The findings revealed that there were some obstacles faced during the development of artificial intelligence model for the removal of heavy metals such as process optimization, predictors or variables selection, and normalization of algorithms which are time-consuming and require expertise. Besides, most of the data collection focuses on the removal of heavy metals from aqueous solution in laboratory scale and lacks the information regarding the removal of heavy metals in actual or natural situation and which is the real wastewater with combinations of contaminants.

10.2.3 Environmental Toxicants Monitoring

Environmental toxicants are defined as toxic materials that are present in the environment and can affect a wide range of ecosystems, like water bodies, land, or even air. The examples of environmental toxicants are industrial waste, household waste, heavy metals, and agricultural waste (Rossignol, Genuis, and Frye 2014) that are mostly found in aquatic ecosystems. Therefore, there is a need to implement the indicators to provide the data on the severity of the pollution and assess the toxicity of the contaminants that are present in the environment. One of the indicators used to monitor the environment are bioindicators using microorganisms, such as bacteria, fungi, and microalgae. The benefits of using the biomonitors are the determination of biological impacts, demonstration of potential synergistic and antagonistic effects of various contaminants on the ecosystems, monitoring of the effects of toxicants on the human and animal health as well as cost-effectiveness (Parmar, Rawtani, and Agrawal 2016).

The characteristics of rapid reproduction rate, easy cultivation in any scale, and short life cycles that are possessed by microalgae make them the suitable and

TABLE 10.3
Algae Attributes and Associated Indicators Commonly Used in Pollution-Monitoring Programs

Algae	Indicators	Attributes
Chlorella vulgaris	Ash-free-dry-weight	Biomass
Chlamydomonas moewusii	Chlorophyll *a*	
Phaeodactylum tricornutum		
Chlamydomonas moewusii	Cell biovolume/Cell morphology (changes in the cells of algae)	
Phaeodactylum tricornutum		
Dunaliella tertiolecta	Cell physiology (study of growth in algae)	Composition/Growth
Chlamydomonas moevvusii	Damage of DNA due to toxic environments	
Chlorella vulgaris	Exposure to harmful environment indicated increase in oxidative stress level.	Metabolic state

Source: Modified from Paul et al. (2017) and Cid et al. (2012).

valuable environmental biological indicators, especially in assessing the water quality assessments (Omar 2010) because they are the most abundant organisms in aquatic ecosystems and occupy the base of the food chain. According to Tables 10.3 and 10.4, they can provide valuable information on the types, effects, bioavailability, and concentration of the pollutants that are present and, thus, help in making, managing, and conducting the decision to protect the environment from the pollutants. The microalgae also can be used to evaluate the growth of phytoplankton community and manage the growth of algal bloom (Ray, Santhakumaran, and Kookal 2021). Besides, microalgae can easily degrade or remove the toxins and heavy metals from the environments. For the microalgae to be applied as indicators to monitor the environment, the microalgae must be easy to be identified and to sampled, are available all year, and found in every environment types (Conti and Cecchetti 2003).

Conti and Cecchetti (2003) showed the algae, *U. lactuca* and *P. pavonica*, had the high potential to biomonitor the contamination of heavy metals in the coastal areas by measuring the concentrations of metal present in the algae and this, assessing the contamination levels of the coastal areas. Li et al. (2014) used *Euglena gracilis* as an early warning and indicator to evaluate the genotoxicity of organic pollutants in Meiliang Bay of Taihu Lake, China. The study by Gómez-Jacinto et al. (2015) proposed the potential of microalgae *Chlorella sorokiniana* as the bioindicator of mercury pollution in aquatic ecosystems. *Chlorella sorokiniana* is able to detoxify mercury and involves the synthesis of phytochelatins. The study by Parus and Karbowska (2020) showed that the algae, *Ulva* and *Cystoseira*, can be used to indicate the cleanliness of the environment by accumulating metals such as iron, copper, lead, and thallium present in aquatic environment. The findings from the study by O'Neill and Rowan (2022) showed that *Pseudokirchneriella subcapitata* can be used as an indicator in monitoring

TABLE 10.4
Criteria of Microalgae Used for Environmental Monitoring (Peter et al. 2021)

Criteria	Application
Biochemical oxygen demand removal	Microalgae are used in the treatment of brewery sewage discharge, and their biomass production at the same time has been analyzed. Upon further analysis, the maximum dry biomass reported was 0.917 grams per liter with a reduced BOD percentage of 91.43%.
Nutrient removal	Studies conducted have proven that the use of microalgae indicates that the efficient nutrients were removed with low maintenance and implementation of cost. Such studies have also successfully compared that phytoremediation technology that uses microalgae is better than physicochemical method that is membrane separation or ion exchange.
Heavy metal removal	Large amounts of heavy metals can easily be taken up by microalgae as they depend on them for biomass growth.
Pathogen removal	The photosynthetic process that takes place in microalgae intensifies the oxygen dissolved, which then increases the pH value causing this mechanism that provides the bacteria pathogen present in the waste discharge to decline.

and measuring the quality of aquaculture wastewater in aquatic ecosystems. These various literature studies revealed that most of the studies were conducted on the aquatic ecosystems because microalgae mostly grow in aquatic habitat. Nevertheless, Bérard et al. (2004) used the pesticide sensitivity of soil algae to evaluate the soil contamination.

To monitor the environment effectively, biosensor, an analytical device, can be utilized to detect various organic and inorganic pollutants in the environment and convert into the electrical signals through the signal transduction by transducer. The use of biosensor to monitor the environmental toxicant is fast, portable, and convenient with the automatic signal detection, and the results can be viewed online at any time. An ideal biosensor should be equipped with the characteristics such as high sensitivity, high selectivity against specific pollutant types, and the ability to detect multiple range of compounds. Biosensors can be categorized into electrochemical, optical, calorimetric, and mass-based biosensors. Microalgae can be used to fabricate the biosensors because of the fact that microalgae are easy to be cultured in any environment types, have short life cycle, and exhibit high sensitivity to diverse range of pollutant, but it is important to consider the appropriate materials used in matrix to avoid leaching as well as the specificity of the biosensors in detecting and identifying different toxicants (Peña-Vázquez et al. 2009; Low, Pan et al. 2021; Low, Chen et al. 2021). Among all the microalgae species, *Chlorella* sp. is commonly used in the production of biosensors because this microalgae genus is sensitive to a broad number of potentially toxic elements, so is able to detect more than one compound. Mostly, microalgae-based biosensors are based on whole cells because they are easy to maintain and store, show fast response, have short generation time as well as can be modified by genetic engineering. Photosynthesis disturbance, especially optical A-chlorophyll fluorescence modification, is the most common mechanism involved in the manufacturing of biosensors because photosynthesis can detect various micropollutants (heavy metals, pesticide, pharmaceutical waste, polycyclic aromatic hydrocarbons) in a sensitive way (Gosset et al. 2018). Another feature in producing microalgae-based biosensor is the cell entrapment or immobilization on a matrix without affecting its ability in biomonitoring and biosensing.

Podola, Nowack, and Melkonian (2004) synthesized a non-selective biochip-based biosensor using various microalgae species (*Klebsormidium* sp. and *Chlorella* sp.) to detect two volatile toxic compounds through the signal measurement via imaging chlorophyll fluorometer. The microalgae were immobilized on a prototype of algal sensor chip. The findings revealed that formaldehyde and methanol in vapor forms were detected within minutes significantly, proposing the identification of pollutants in gaseous environment with biochip system. Similarly, to detect the cadmium in aqueous solution in relation to the alkaline phosphatase activities, Chouteau et al. (2004) synthesized a conductometric biosensor using interdigitated conductometric electrodes as a transducer. The bioreceptor was immobilized whole-cell *Chlorella vulgaris*. Through the comparison of the performance of this biosensor with the bioassays, it demonstrated that the sensitivity and stability of this biosensor were high and were able to detect low concentration of cadmium ions. Peña-Vázquez et al. (2009) utilized three microalgae (*Dictyosphaerium chlorelloides*, *Scenedesmus intermedius*, and *Scenedesmus* sp.) in the manufacturing of fiber optic biosensor to monitor herbicide. The parameters on the preparation of the sensing layers – for example pH, algae density, and concentration of glycerol – were also evaluated. The results demonstrated that these three biosensors reacted well with the herbicides (simazine, atrazine, propazine, terbuthylazine, linuron) that inhibit the photosynthesis at photosystem II at which *Dictyosphaerium chlorelloides* biosensor showed the lowest detection limits for simazine. The integrated use of the biosensors also improved sensitivity and specificity of microalgae-based biosensor. In addition, to assess the quality of aquatic environment, Ferro et al. (2012) generated a biosensor via the measurement of chlorophyll fluorescence using three microalgae species (*Chlorella vulgaris*, *Pseudokirchneriella subcapitata*, and *Chlamydomonas reinhardtii*) as the toxicity bioindicators. It was discovered that microalgae were very sensitive and efficient in the detection of herbicides such as (3-(3,4-dichlorophenyl)-1,1-dimethylurea) and atrazine at which *C. reinhardtii* expressed the best detection limit of 0.1 μM.

A good biosensor can detect various heavy metals at the same time because the polluted area consists of broad number and diverse range of toxicants (Xu et al. 2020). Accordingly, Wong et al. (2018) produced a biosensor using microalgae *Chlorella vulgaris* to identify the presence of heavy metals and light metals in water. From the study, the biosensor can detect the presence of these metals with the concentrations of 0.001–10.000 nm/L in a short period of time, which was within 15 minutes of exposure time. Next, Rathnayake et al. (2021) produced a whole-cell optical array biosensor using microalgae (*Mesotaenium* sp.) and cyanobacteria (*Synechococcus* sp.) to assess the bioavailability of different heavy metals in aquatic systems. The optimum fluorescence values were found to be 10-minute exposure time. The microalgal/cyanobacterial cultures exhibited an antagonistic effect among the multi-metals tested. The properties of the biosensors can be combined together to increase the detection efficiency of the toxicants, pollutants, or contaminants in the environment. As an instance, Tsopela et al. (2016) developed a lab-on-chip device to analyze on-site water toxicity, especially herbicide detection, by incorporating electrochemical and fluorescence properties and thus allowing double complementary detection (electrochemical and optical). The biosensor produced can be an efficient indicator of water pollution through the detection of the changes in the photosynthetic activity induced by Diuron herbicide and reflected through a modification in oxygen production rate.

Furthermore, many approaches have been used to increase the performance of biosensors, and one of the approaches is microfluidics technology through the integration of multiple processes on a single process. This method saves the cost and time as well as is of high precision (Han et al. 2019). Gosset et al. (2018) employed xurographic fabrication to produce disposable and low-cost microfluidic biosensor for in-situ water toxicity analysis using three microalgae strains (*Chlorella vulgaris*, *Pseudokirchneriella subcapitata*, and *Chlamydomonas reinhardtii*). The feasibility and sensitivity of the biosensor were tested with Diuron pesticide, proving its good reliability, reproducibility, and performance in the detection of toxic discharges. Han et al. (2019) generated a digital microfluidic diluter-based biosensor to monitor marine pollutants, and algae motion was set as the sensor signal. Microalga species *Platymonas subcordiformis* was used in the fabrication of biosensor. The biosensor response was successfully examined in the presence of commonly concerned marine contaminants. Moreover, nanoparticles can be used to enhance the performance of the biosensor to overcome the complex manufacturing procedures and the long measurement time that may cause low signal (Roxby et al. 2020). Roxby et al. (2020) reported the formation of microalgae (*Chlorella* sp.) living biosensor using copper nanocavities (copper nanoparticles and electrode) to improve the photocurrent and amplify the photoelectrical fluorescence signal. The biosensor was able to detect various light and heavy metal ions with the detection limit of 50 nM.

Internet of Things can be applied in the environment field monitoring, especially the monitoring and detection of environmental toxins by developing a framework comprising the wireless biosensors integrated with various environmental parameters. Based on the study by Teniou, Rhouati, and Marty (2021), artificial intelligence and computational hydrodynamics are integrated by fifth-generation modeling in the data analysis, comprising Expert Systems, Fuzzy Logic, Artificial Neural Networks, Chemometrics, and others. Among these artificial intelligence systems, chemometrics is commonly used in environment studies, especially in the tracking, simulation, and forecast of the environment quality (Van Leeuwen and De Boer 2008; Gros, Petrović, and Barceló 2006).

Anagu et al. (2009) employed artificial neural network for the sorption study of nine heavy metals in the soil and for comparing its performance with regression studies. The study showed that artificial neural network can be used to estimate the content of heavy metal from soil efficiently as compared to multilinear regression with the lowest modelling efficiency of 0.79 (chromium). Kadam et al. (2020) reported the use of microalgae species as an indicator of aquatic environment in Doon valley of Western Himalayas, India. In this study, ArcGIS software tools with the geostatistical study were used to produce prediction maps that displayed the properties of the water samples and microalgae community. The identity of the microalgae species was confirmed using PAST techniques through diversity indices. The biosensors to monitor the environment divided into affinity-based sensors (use receptors, such as antibodies and aptamers) and inhibition sensors (use pollutant-based inhibition enzymes). Inhibition sensors were more appropriate and suitable to use for environmental monitoring because they are simple and cheap as compared to affinity-based sensors that can be influenced by other pollutants present (Nabok, Haron, and Ray 2004). Smart sensing devices are portable, cost-effective, smart, and rapid. Biosensors can be categorized into optical, thermal, resonant, ion-sensitive, piezoelectric, and electrochemical where electrochemical and enzymatic are some examples of biosensors that are used for measuring potentially toxic elements present in environment (Halilović et al. 2019). Sensors that have more bioreceptors are more useful as they can be used to detect various pollutants simultaneously (Teniou, Rhouati, and Marty 2021). Yang et al. (2017) developed a surface enhancement Raman spectrum (SERS) biosensor to detect mercuric ions, one of the toxic metal ions that are present in water. The information about the concentration of environmental toxins can be collected and obtained at any time and at any place to be converted into signal for data processing. If the concentration of environmental toxins exceeds the certain range or high signal intensity, the sensors will notify the system to deliver notification to the clients to improve environmental conditions (Sampathkumar et al. 2020). However, the experiments of environmental biosensors are mostly conducted at laboratory scale and lack the information of applicability in real environment. In short, microalgae-based indicators are fast, low-cost, and sensitive. Therefore, the researchers are now utilizing the microalgae as effective bioindicators for monitoring the environment toxicants in the aquatic ecosystems.

10.2.4 Bioassay

An analytical way to determine the concentration or potency of a substance with regards to a living animal or plant, living cell, or tissues is called a bioassay. Bioassays work in two ways which are in vitro and in vivo (Table 10.5). In-vitro bioassay is based on the cellular mechanism for the detection of specific chemicals, whereas in-vivo bioassay evaluates the effects of the toxicant in the whole species (organisms) and provides more detailed information about its environmental effects. The difference between bioassay and biosensor is that bioassay uses biological indicators, such as cells, tissues, or whole organisms, to assess

TABLE 10.5
The Bioassay Test Sets That Were Used to Evaluate the Quality of Marine Sediment Samples and Seawater Sample

Tests	Test species	Parameters	Duration
Bacteria luminescent toxicity test	*Vibrio fischeri*	Reduction of luminescence	30 min
Marine algae toxicity test	*Phaeodactylum tricornutum Bohlin*	Inhibition of growth and reproduction	72 hours
Acute toxicity of marine sediment to amphipods	*Corophium volutator*	Mortality (survival), burrowing behaviour	10 days

an analyte's biological activity, whereas biosensor uses biorecognition elements like enzymes, proteins, or antibodies and amplifies the signal through physical or chemical means (Valera et al. 2013).

The most frequently used microalgae in bioassays for a wide range of contaminants are genera *Chlamydomonas*, *Chlorella*, *Scenedesmus* and *Selenastrum* (Yamagishi et al. 2017) and the steps for microalgae-based bioassay are shown in Figure 10.1. Different microalgae species demonstrated different sensitivity levels to the pollutants that are present in aquatic ecosystems, so the use and selection of bioassays to assess the environmental pollutants are crucial. Bi et al.'s (2018) study evaluated the sensitivity of seven algal species to triclosan, fluoxetine, and their mixture. The findings showed that *Dunaliella parva* was the most sensitive to triclosan, whereas *Chlorella* and *Chlamydomonas* were the least sensitive algae to triclosan. For fluoxetine, *Chlorella* was the least sensitive algae species. Ma's (2005) study showed that *Selenastrum capricornutun*, *Scenedesmus quadricauda*, *Scenedesmus obliqnus*, *Chlorella vulgaris*, and *Chlorella pyrenoidosa* exhibited various sensitivity levels to organotins and pyrethroids pesticides.

Moreira-Santos, Soares, and Ribeiro (2004) employed in-situ bioassay by microalgae *Pseudokirchneriella subcapitata* to keep track of the toxicity in freshwater systems. The bioassay based on the enzyme inhibition measurements can become the popular indicators of environmental stress (Franklin et al. 2001). Franklin et al. (2001) synthesized a bioassay using flow cytometry to assess the toxicity of copper based on the inhibition of esterase activity in microalgae (*Selenastrum capricornutum*, *Chlorella* sp., *Dunaliella tertiolecta*, *Phaeodactylum tricornutum*, *Tetraselmis* sp., *Entomoneis cf. punctulata*, *Nitzschia cf. paleacea*). The production of bioassay was optimized in terms of the substrate concentration, incubation time, and media pH. The findings revealed that among all the tested microalgae, *S. capricornutum* and *E. cf. punctulata* showed the rapid results. To monitor the pollutants or screen for pesticides rapidly and autonomously in aquatic ecosystem, Moro et al. (2018) developed an optical bioassay using various microalgae species to detect photosynthetic marine pesticides such as diuron, simazine, and irgarol. The bioassay worked by examining the effects of marine pesticides which act as photosynthesis inhibitors on photosystem II fluorescence

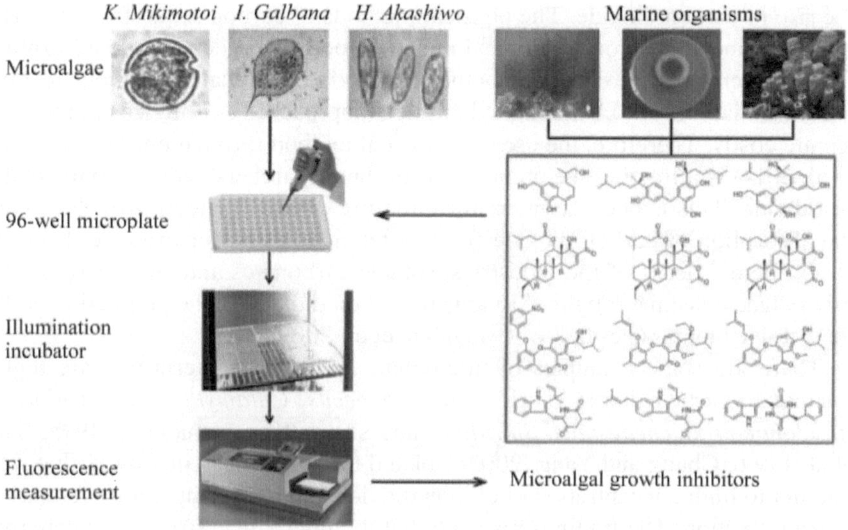

FIGURE 10.1 A diagrammatic figure indicting the step-by-step process of bioassay with permission from Zhao et al. (2018).

parameters. The bioassay could detect these pollutants in a short time, which was within 10 min, and *Chlorella mirabilis* was the most sensitive microalgae to these pollutants.

Bioassay development can be integrated with the artificial intelligence to save the time and cost. Yi et al. (2018) developed an extreme learning machine (ELM)-based prediction model to predict the concentration of chlorophyll in assessing the growth of algae because toxins from algal blooms will cause harm to the environment and do ecological harm. To predict chlorophyll *a*, two variables, weather and water quality, were involved. ELM1 model predicted the growth of algal blooms based on these two variables and included downstream chlorophyll *a* concentration whereas ELM2 involved upstream chlorophyll *a* to enhance the prediction performance. The findings revealed that two ELMs demonstrated satisfactory performance in monitoring the growth of algal bloom, and ELM2 performed better than ELM1. The same results were got by the study by Heddam, Sanikhani, and Kisi (2019) which used artificial intelligence models which were feedforward neural networks, adaptive neuro-fuzzy inference system with grid partition, adaptive neuro-fuzzy inference system with subtractive clustering, and gene expression programming to predict the concentration of phycocyanin in cyanobacterial algal bloom using water quality as the variables. All models produced satisfactory performance in the estimation of concentration of phycocyanin.

10.2.5 Bio-Mitigation of Carbon Dioxide Emission

Carbon dioxide is one of the greenhouse gases that is contributing to global warming with up to 68% emissions, and many strategies have been proposed to tackle the rise in carbon dioxide. The physical methods of carbon sequestration consist of membrane separation, geological injection, oceanic injection, and adsorption. On the other hand, the chemical methods comprise chemical adsorption and mineral carbonation. The chemical and physical approaches to mitigate carbon dioxide are costly. Therefore, the use of biological method through microalgae (Zhou et al. 2017) is attracting the attention from the researchers and environmentalist worldwide. This is because microalgae require carbon for their growth through the absorption of carbon dioxide from different sources, for example, from the atmosphere, release of the industries, soluble carbonates and etc. Theoretically, microalgae consume roughly 513 tons of carbon dioxide in the production of 280 tons of dry biomass per ha^{-1} y^{-1} (Bilanovic et al. 2009).

There are some examples of microalgae and cyanobacteria that are highly tolerant to carbon dioxide such as *Botryococcus*, *Chlamydomonas*, *Chlorella*, *Rhodobacter*, *Scenedesmus*, *Spirulina*, and so on (Sawayama et al. 1995; Sung et al. 1999). Chang and Yang (2003) isolated the microalgae in Taiwan that were tolerant to high concentration of carbon dioxide and high temperature for carbon dioxide fixation. The findings revealed that the microalgae strains that can grow well in high concentration of carbon dioxide and high temperature were *Chlorella* sp. NTU-H15 and NTU-H25, making them as the potential candidates in fixing carbon dioxide gas, especially in gaseous emission from industries, to alleviate global warming. Sydney et al. (2010) investigated the relationship between carbon dioxide fixation and the cultivation in four microalgae species which were *Dunaliella tertiolecta*, *Chlorella vulgaris*, *Spirulina platensis*, and *Botryococcus braunii*. Microalgae fixed carbon dioxide mainly for the production of biomass and the highest fixation rate was demonstrated by *B. braunii*, followed by *S. platensis*, *D. tertiolecta*, and *C. vulgaris*.

The cultivation system of microalgae can be open pond and closed photobioreactors. The microalgae cultivation system can affect the sequestration of carbon dioxide. Open raceway pond is the most common cultivation system used to sequestrate carbon dioxide because it can be scaled-up easily and simply. However, open pond system has the disadvantages of high contamination and low biomass productivity, so closed photobioreactor system with high-performance efficiency is favoured but is limited by high cost (Tang et al. 2020). Light intensity was found to have a positive relationship with the carbon dioxide fixation. Hence, the type of photobioreactor for the microalgae cultivation is crucial to maximize the sequestration of carbon dioxide. Citing an example, Fan et al. (2007) revealed that fixation rate of carbon dioxide was found to be higher in the membrane photobioreactor as compared to that in bubble column, airlift photobioreactor, and membrane contactor by studying the effect of light in terms of intensity, quality, and hollow fiber membrane modules. Dasan et al. (2020) conducted an experiment to improve the carbon dioxide fixation ability by microalgae *Chlorella vulgaris* through the sequential-flow bubble column photobioreactor. Additionally, Sun et al. (2016) installed hollow polymethyl methacrylate tubes in the flat-plate photobioreactor to amplify the light distribution, increasing the capture of carbon dioxide. The study by Abid, Saidane, and Hamdi (2017) revealed that the synergistic effects of using *Hydrocarbonoclastic* native microbial and *Spongiochloris* sp. microalgae in airlift bioreactors couples can sequestrate carbon dioxide, thus reducing the greenhouse effects and at the same time bioremediating hydrocarbon petroleum. The bio-fixation rate of carbon dioxide was 2.9205 g L^{-1} per day through the absorption of carbon dioxide during the degradation of hydrocarbon petroleum and converted by microalgae as biological biomass through increased growth. Jacob-Lopes et al.'s (2009) study showed that photoperiod will influence the carbon dioxide fixation rate, and it was discovered that continuous light supply will result in the highest carbon dioxide fixation rate. Last, bubbling the microalgae cultivation medium will enhance the utilization of carbon dioxide (Barati et al. 2021).

Genetic engineering tools, such as mutagenesis and targeted genetic modification, can be applied to modify the microalgae strains to improve the carbon dioxide fixation through the improvement of carbon dioxide fixation pathway (modifying the Calvin cycle and concentrating mechanism of carbon dioxide), changing energy harvesting complexes as well as creating additional fixation pathway (enhancing the assimilation or reducing the release of carbon dioxide) (Barati et al. 2021). Artificial intelligence can be applied in the estimation or prediction of carbon dioxide fixation rate to save the time and cost. Kasiri, Ulrich, and Prasad (2015) generated a mathematical model for carbon dioxide biofixation by incorporating parameters such as light intensity as well as the concentration of carbon dioxide and phosphate for the cultivation of *Chlorella kessleri* in oil sands processing water. The model showed that carbon dioxide, phosphate, and light intensity affected algal growth rate and thus affecting the uptake of carbon dioxide. Kushwaha, Uthayakumar, and Kumaresan (2021) created a prediction model to estimate the fixation of carbon dioxide from the microalgae. The models involved were Adaptive Neuro-Fuzzy Inference System (ANFIS) and Genetic Algorithm (GA) hybrid approach by varying the cultivation parameters – for example temperature and pH. As compared to ANFIS model, the developed hybrid model (GA-ANFIS) exhibited satisfactory prediction performance with low error for small datasets and could be used in the industrial scale in the future. In summary, the sequestering or biofixation of carbon dioxide by microalgae can be an effective way to mitigate global warming.

10.2.6 Biosurfactant

Around 3 million tons per year of surfactants are produced around the world, most of which is manufactured from petroleum, and up to 70–75% of the surfactants are being used in developed and industrialized countries. Biosurfactants are the surface-active amphophilic substances produced by microorganisms through biological processes and can be used as emulsifiers. Biosurfactants consist of a range of chemical structures which are lipopeptides, lipoproteins, phospholipids, fatty acids, and lastly glycolipids. Use of biodegradable biosurfactants and the biosurfactant-producing microorganisms is an eco-friendly method to remediate the environment by removing or degrading the pollutants (Pacwa-Płociniczak et al. 2011) because biosurfactants are non-hazardous and can be synthesized from waste substrates, minimizing the pollution. Besides, as compared to surfactants produced from petroleum, biosurfactants are biodegradable, have low toxicity and low critical micelle concentration, possess antimicrobial properties, and can tackle the depletion of fossil fuels (De Carvalho et al. 2014). Recently, there has been a marked surge of global interest in biosurfactants due to their potential in various field – for instance environment biotechnology, food, processing, and pharmaceutical (Pacwa-Płociniczak et al. 2011).

Microalgae are one of the organisms that are used to produce biosurfactants because of their ability to produce glycolipids and phospholipids. The use of microalgae to produce biosurfactants is safe, biodegradable, and non-hazardous as compared to synthetic biosurfactants, and so can be applied in the pharmaceutical and food industries. In the environmental biotechnology field, biosurfactants are used to bioremediate and for waste treatment. Biosurfactants produced from microalgae work by precipitation, adsorption and degradation to remove the pollutants from the aqueous solution (Ugya, Ajibade, and Hua 2021).

De Carvalho et al. (2014) produced biosurfactants from few microalgae strains (*Spirulina platensis Paracas*, *Spirulina* sp. LEB 18, and *Spirulina platensis* LEB 52) and evaluated the activity of biosurfactants by measuring the surface tension. The result showed that nitrogen can affect the production of biosurfactant by influencing its surface tension. The production of biosurfactants by *Spirulina* sp. LEB 18 was satisfactory as compared to other microalgae species. Furthermore, Law et al. (2018) synthesized biosurfactants from ruptured microalgae biomass that contained surface-active agents, but the emulsion stability will be affected by competitive interfacial adsorption. The findings from the study by Ugya, Ajibade, and Hua (2021) revealed that biosurfactant produced by microalgae was effective in the removal of heavy metals, solids, and ions from polluted river, River Kaduna. The biofilm produced by microalgae can be used for phycoremediation of water and is cost-effective. However, the production of biosurfactant will be affected by the organic carbon sources used in cultivation. Radmann et al.'s (2015) study showed that molasse will stimulate the production of biosurfactant.

Industry 4.0 emphasizes the digital transformation of manufacturing process; therefore, the production of biosurfactants from microalgae also can be digitalized through the optimization of the media components and prediction of biosurfactant concentration as well as surface tension that is inversely proportional to the quality of biosurfactant. Furthermore, the efficacy of the biosurfactants in heavy metal bioremediation also can be predicted using machine learning system. As an example, the media optimization of biosurfactant synthesis from *Rhodococcus erythropolis* MTCC 2794 using artificial neural network coupled with genetic algorithm was shown in study by Pal et al. (2009). Sivapathasekaran et al. (2010) optimized and maximized the synthesis of biosurfactant by using artificial neural network-genetic algorithm through the adjustment of medium components. The concentration of biosurfactant synthesized from the microalgae can also be determined using artificial intelligence models as shown in study by Santos, Ponezi, and Fileti (2017). The use of these models will allow the prediction of biosurfactant or biofilm produced from the microalgae in short time with high accuracy and precision. Virtual environment that mimics the wastewater can be created to estimate the removal efficacy of pollutants by biosurfactants produced by various microalgae species (Pattanaik et al. 2019). In short, the use of models needs to be validated on the laboratory and pilot scale for the synthesis of microalgae-based biosurfactant.

10.3 CHALLENGES AND FUTURE RESEARCH DIRECTION

The previous sections demonstrated the use of microalgae in the field of environmental biotechnology. The use of microalgae environmental biotechnology is still at the lab-scale and facing some limitations as associated (Kurade et al. 2016):

(1) *Large-scale industrial application or lack of application in real-world conditions.* Currently, the studies regarding the use of microalgae in environmental biotechnology field such as the bioremediation, removal of heavy metals, synthesis of biosensors, or biosurfactants or biofixation of carbon dioxide are conducted at laboratory scale. Next, in the case of removal of heavy metals, most of the current studies focus on particular types of pollutants or contaminants which do not mimic the field conditions. In the real world, there are some factors, such as climate, temperature, light intensity, and other microorganisms (amoeba, ciliates, rotifers, bacteria, and other microbes), influencing the environmental use of microalgae.

(2) *Selection of appropriate microalgae strains.*

Microalgae species or strains are different greatly in terms of growth rate, productivity, nutrients and light requirement, with the ability to produce useful products and adapt to harsh conditions. Therefore, it is necessary to find the suitable microalgae species for different purposes and various cultivation systems. There are multiple studies being conducted in different locations for suitable strain; however, these research works are conducted and focused on small numbers of rapid growing species. The four main features that are taken into account during the selection process are (Geada et al. 2017):

i. Growth physiology: assessed based on growth rate, cell density, tolerance to environmental characteristics (temperature, pH, salinity, oxygen, and carbon dioxide levels).
ii. Metabolite production: evaluated for the concentration and the yield of metabolite that will be useful during commercialization. The ability of microalgae species to be able to secrete metabolites in either liquid or volatile forms are features that are taken into consideration during the harvesting steps.
iii. Robustness: reasonable resilience, high culture consistency, community stability, low susceptibility towards external predators, toxins

are an extreme factor that is an important characteristic needed to be considered.

iv. Amenability to genetic manipulation: strains that are susceptible to most performance by genetic manipulation is preferable due to their high potential and flexibility.

(3) *Cost*

As a unicellular or multicellular organism, microalgae don't have complex cell structures and can grow in harsh conditions, nonviable land, and in aqueous medium. An estimation of around 50,000 species is currently being studied and evaluated. There are many studies conducted on cultivation of algae. However, each specie reacts differently with different mechanisms for adapting to its cultivation medium and the system, and therefore they need to be studied separately.

With regards to their structural properties, algae growth can differ in growth patterns within the system and medium they are cultivated in. The conditions and environment that they are grown in are dependent on the products that are needed. Microalgae are produced in closed (photobioreactor) and open (open ponds, lagoons) systems; however, depending on the products such as nutraceutical and pharmaceutical, close systems are preferred. However, the operating and maintaining costs of close systems are comparatively higher than that of open systems.

Therefore, a detailed analysis is needed to establish pilot scale systems. For the production of microalgae, both biological and non-biological factors and operating parameters are taken into consideration. Biological factors are pathogens (viruses and bacteria), and non-biological factors are light, temperature, pH level, nutrients, and salinity. Operating parameters are mixing dilution rate and harvesting frequency (Özçimen et al. 2018).

Studies and research still need to be conducted to solve problems like high installation and operation costs, unbalanced supply of nutrients, and environmental conditions. A possible solution is to develop a highly efficient and low-cost downstream process that can reduce the operational costs significantly. Also, the selection of favourable cultures can increase the productivity of the desired product and be efficient and economically feasible during microalgae cultivation systems that will also improve biomass productivity (Rizwan et al. 2018).

(4) *Ability of microalgae to detect various environmental toxicants.*

Therefore, Industry 4.0 can be a new platform of smart and autonomous manufacturing (Bai et al. 2020), but the realization of the fourth industrial revolution in the environmental biotechnology field is still in infancy stage and requires more research in the future such as the choice of suitable artificial intelligence models and creation of smart multisensor that are user-friendly. First, the use of biosensors should be categorized in the future research direction. Currently, the biosensors generated have the limitation of detecting specific types of environmental toxicants. Thus, the creation of multisensors in the future is a promising approach by combining the data from various sensors in the monitoring of potentially toxic elements in the environment. The data produced are more reliable and accurate. The advantages of the multisensors are saving the time and cost as well as increased productivity and reliability (Luo, Yih, and Su 2002).

There are some examples of artificial intelligence models for the analysis of big data, such as neural network, logic, regression, and hybrid models. These models can be applied in the environmental biotechnology in areas such as study of the removal of heavy metals, production of biosensors, and estimation of carbon dioxide fixation rate. By comparing with the classical and conventional data analysis models (for example empirical, statistical, and mathematical), artificial intelligence models are more productive, effective, and efficient. However, the selection of appropriate artificial intelligence model remains as a challenge. The application of artificial intelligence also requires expertise in the construction of models. Next, the use of artificial intelligence models needs the optimization, training, and validation procedures to obtain high accuracy of prediction value which are cost- and time-consuming. The prediction ability or accuracy of artificial intelligence models can be strengthened by the variation of important parameters of the process.

One example is the prediction of removal of the heavy metals in the aqueous solution. There are few parameters, such as the exposure time, pH, temperature, microalgae strains involved, and type of heavy metals that can be removed, which need to be optimized before the implementation of artificial intelligence models in the field condition. Some of the artificial intelligence models used in the removal of heavy metals only focus on particular heavy metals, instead of covering all fundamental heavy metals that are present in the environment, limiting the use of the model (Bhagat, Tung, and Yaseen 2020).

The hybridization of the single artificial intelligence model is another promising approach to increase the performance and reduce the operation and construction cost. Ali et al. (2015) proposed the use of hybrid artificial intelligence models to increase the productivity, estimation performance, and rapid estimation with less errors as compared to single artificial intelligence models. However, the construction of hybrid models needs a longer time, expertise, and more datasets or information. For instance Zhao et al. (2020) reviewed the economy and management of artificial intelligence towards wastewater treatment. The findings revealed that prediction accuracy on pollutant removal ranged from 0.64 to 1.00 and could reduce operational costs by up to 30%. In this study, combined AI methods were used and resulted in higher accuracy and less error. Because the experiment-based data was limited in size and range, so it is recommended that future studies will provide a bigger field or more online data to help the models become more user-friendly, operate faster, and be more reliable in real applications.

10.4 CONCLUSIONS

In summary, microalgae demonstrate satisfactory capability in mitigating the environment pollution. Compared to the conventional methods in environment biotechnology, microalgae offer advantageous properties such as high growth rate, short lifespan, and ability to survive in harsh environment. The bio-application of microalgae in environment biotechnology is eco-friendly but limited by cost and selection of appropriate microalgae strains. Therefore, Industry 4.0 with the focus on artificial intelligence, Internet of Thing, and automation comprehend with the use of microalgae. The use of genetic tools and artificial intelligence models revealed a novel sustainable aspect of microalgae-based environment biotechnology with high productivity and performance.

REFERENCES

Abdel-Raouf, N., A. A. Al-Homaidan, and I. B. M. Ibraheem. 2012. "Microalgae and wastewater treatment." *Saudi Journal of Biological Sciences* no. 19 (3):257–275. https://doi.org/10.1016/j.sjbs.2012.04.005.

Abid, Abdeldjalil, Faten Saidane, and Moktar Hamdi. 2017. "Feasibility of carbon dioxide sequestration by Spongiochloris sp microalgae during petroleum wastewater treatment in airlift bioreactor." *Bioresource Technology* no. 234:297–302. https://doi.org/10.1016/j.biortech.2017.03.041.

Al-Jabri, Hareb, Probir Das, Shoyeb Khan, Mahmoud Thaher, and Mohammed Abdul-Quadir. 2021. "Treatment of wastewaters by microalgae and the potential applications of the produced biomass—a review." *Water* no. 13 (1). doi: 10.3390/w13010027.

Ali, Jarinah Mohd, M. A. Hussain, Moses O. Tade, and Jie Zhang. 2015. "Artificial Intelligence techniques applied as estimator in chemical process systems—A literature survey." *Expert Systems with Applications* no. 42 (14):5915–5931. doi: 10.1016/j.eswa.2015.03.023.

Amenorfenyo, David K., Xianghu Huang, Yulei Zhang, Qitao Zeng, Ning Zhang, Jiajia Ren, and Qiang Huang. 2019. "Microalgae brewery wastewater treatment: Potentials, benefits and the challenges." *International Journal of Environmental Research and Public Health* no. 16 (11). doi: 10.3390/ijerph16111910.

Anagu, Ihuaku, Joachim Ingwersen, Jens Utermann, and Thilo Streck. 2009. "Estimation of heavy metal sorption in German soils using artificial neural networks." *Geoderma* no. 152 (1–2):104–112. doi: 10.1016/j.geoderma.2009.06.004.

Ansari, Faiz Ahmad, Mahmoud Nasr, Ismail Rawat, and Faizal Bux. 2021. "Artificial neural network and techno-economic estimation with algae-based tertiary wastewater treatment." *Journal of Water Process Engineering* no. 40:101761. doi: 10.1016/j.jwpe.2020.101761.

Ardila, Liliana, Rubén Godoy, and Luis Montenegro. 2017. "Sorption capacity measurement of chlorella vulgaris and scenedesmus acutus to remove chromium from tannery waste water." *IOP Conference Series: Earth and Environmental Science* no. 83 (1):012031. doi: 10.1088/1755-1315/83/1/012031.

Bai, Chunguang, Patrick Dallasega, Guido Orzes, and Joseph Sarkis. 2020. "Industry 4.0 technologies assessment: A sustainability perspective." *International Journal of Production Economics* no. 229:107776. https://doi.org/10.1016/j.ijpe.2020.107776.

Barati, Bahram, Kuo Zeng, Jan Baeyens, Shuang Wang, Min Addy, Sook-Yee Gan, and Abd El-Fatah Abomohra. 2021. "Recent progress in genetically modified microalgae for enhanced carbon dioxide sequestration." *Biomass and Bioenergy* no. 145:105927. https://doi.org/10.1016/j.biombioe.2020.105927.

Bérard, A., F. Rimet, Y. Capowiez, and C. Leboulanger. 2004. "Procedures for determining the pesticide sensitivity of indigenous soil algae: A possible bioindicator of soil contamination?" *Archives of Environmental Contamination and Toxicology* no. 46 (1):24–31. doi: 10.1007/s00244-003-2147-1.

Bhagat, Suraj Kumar, Tran Minh Tung, and Zaher Mundher Yaseen. 2020. "Development of artificial intelligence for modeling wastewater heavy metal removal: State of the art, application assessment and possible future research." *Journal of Cleaner Production* no. 250:119473. doi: 10.1016/j.jclepro.2019.119473.

Bi, Ran, Xiangfeng Zeng, Lei Mu, Liping Hou, Wenhua Liu, Ping Li, Hongxing Chen, Dan Li, Agnes Bouchez, Jiaxi Tang, and Lingtian Xie. 2018. "Sensitivities of seven algal species to triclosan, fluoxetine and their mixtures." *Scientific Reports* no. 8 (1):15361. doi: 10.1038/s41598-018-33785-1.

Bilanovic, Dragoljub, Arsema Andargatchew, Tim Kroeger, and Gedaliahu Shelef. 2009. "Freshwater and marine microalgae sequestering of CO2 at different C and N concentrations—response surface methodology analysis." *Energy Conversion and Management* no. 50 (2):262–267. doi: 10.1016/j.enconman.2008.09.024.

Chai, Wai Siong, Jie Ying Cheun, P. Senthil Kumar, Muhammad Mubashir, Zahid Majeed, Fawzi Banat, Shih-Hsin Ho, and Pau Loke Show. 2021. "A review on conventional and novel materials towards heavy metal adsorption in wastewater treatment application." *Journal of Cleaner Production* no. 296:126589. doi: 10.1016/j.jclepro.2021.126589.

Chai, Wai Siong, Wee Gee Tan, Heli Siti Halimatul Munawaroh, Vijai Kumar Gupta, Shih-Hsin Ho, and Pau Loke Show. 2021. "Multifaceted roles of microalgae in the application of wastewater biotreatment: A review." *Environmental Pollution* no. 269:116236. doi: 10.1016/j.envpol.2020.116236.

Chang, Ed-Haun, and Shang-Shyng Yang. 2003. "Some characteristics of microalgae isolated in Taiwan for biofixation of carbon dioxide." *Botanical Bulletin of Academia Sinica* no. 44:43–52.

Chekroun, K. Ben, and Mourad Baghour. 2013. "The role of algae in phytoremediation of heavy metals: A review." *Journal of Materials and Environmental Science* no. 4 (6):873–880.

Chouteau, Celine, Sergei Dzyadevych, Jean-Marc Chovelon, and Claude Durrieu. 2004. "Development of novel conductometric biosensors based on immobilised whole cell Chlorella vulgaris microalgae." *Biosensors and Bioelectronics* no. 19 (9):1089–1096. doi: 10.1016/j.bios.2003.10.012.

Cid, Ángeles, Raquel Prado, Carmen Rioboo, Paula Suarez-Bregua, and Concepción Herrero. 2012. "Use of microalgae as biological indicators of pollution: Looking for new relevant cytotoxicity endpoints." In *Microalgae: Biotechnology, Microbiology and Energy*, edited by M. N. Johnsen, 311–323. New York: Nova Science Publishers.

Conti, Marcelo Enrique, and Gaetano Cecchetti. 2003. "A biomonitoring study: Trace metals in algae and molluscs from Tyrrhenian coastal areas." *Environmental Research* no. 93 (1):99–112. doi: 10.1016/S0013-9351(03)00012-4.

Daneshvar, Ehsan, Mohammad Javad Zarrinmehr, Masoud Kousha, Atefeh Malekzadeh Hashtjin, Ganesh Dattatraya Saratale, Abhijit Maiti, Meththika Vithanage, and Amit Bhatnagar. 2019. "Hexavalent chromium removal from water by microalgal-based materials: Adsorption, desorption and recovery studies." *Bioresource Technology* no. 293:122064. doi: 10.1016/j.biortech.2019.122064.

Das, Probir, Mohammed AbdulQuadir, Mahmoud Thaher, Shoyeb Khan, Afeefa Kiran Chaudhary, Ghamza Alghasal, and Hareb Mohammed S. J. Al-Jabri. 2019. "Microalgal bioremediation of petroleum-derived low salinity and low pH produced water." *Journal of Applied Phycology* no. 31 (1):435–444. doi: 10.1007/s10811-018-1571-6.

Dasan, Yaleeni Kanna, Man Kee Lam, Suzana Yusup, Jun Wei Lim, Pau Loke Show, Inn Shi Tan, and Keat Teong Lee. 2020. "Cultivation of Chlorella vulgaris using sequential-flow bubble column photobioreactor: A stress-inducing strategy for lipid accumulation and carbon dioxide fixation." *Journal of CO2 Utilization* no. 41:101226. doi: 10.1016/j.jcou.2020.101226.

De Carvalho, Lisiane Fernandes, Mariana Souza De Oliveira, Jorge Alberto, and Vieira Costa. 2014. "Evaluation of the influence of nitrogen and phosphorus nutrients in the culture and production ofbiosurfactants by Microalga Spirulina." *Journal of Engineering Research and Applications* no. 4 (6):90–98.

de Llasera, Martha Patricia García, José de Jesús Olmos-Espejel, Gabriel Díaz-Flores, and Adriana Montaño-Montiel. 2016. "Biodegradation of benzo(a)pyrene by two freshwater microalgae selenastrum capricornutum and scenedesmus acutus: A comparative study useful for bioremediation." *Environmental Science and Pollution Research* no. 23 (4):3365–3375. doi: 10.1007/s11356-015-5576-2.

de Wilt, Arnoud, Andrii Butkovskyi, Kanjana Tuantet, Lucia Hernandez Leal, Tânia V. Fernandes, Alette Langenhoff, and Grietje Zeeman. 2016. "Micropollutant removal in an algal treatment system fed with source separated wastewater streams." *Journal of Hazardous Materials* no. 304:84–92. doi: 10.1016/j.jhazmat.2015.10.033.

Escapa, C., R. N. Coimbra, S. Paniagua, A. I. García, and M. Otero. 2015. "Nutrients and pharmaceuticals removal from wastewater by culture and harvesting of Chlorella sorokiniana." *Bioresource Technology* no. 185:276–284. doi: 10.1016/j.biortech.2015.03.004.

Esfandian, H., M. Parvini, B. Khoshandam, and A. Samadi-Maybodi. 2016. "Artificial neural network (ANN) technique for modeling the mercury adsorption from aqueous solution using Sargassum Bevanom algae." *Desalination and Water Treatment* no. 57 (37):17206–17219. doi: 10.1080/19443994.2015.1086696.

Fabris, Michele, Raffaela M. Abbriano, Mathieu Pernice, Donna L. Sutherland, Audrey S. Commault, Christopher C. Hall, Leen Labeeuw, Janice I. McCauley, Unnikrishnan Kuzhiuparambil, Parijat Ray, Tim Kahlke, and Peter J. Ralph. 2020. "Emerging technologies in algal biotechnology: Toward the establishment of a sustainable, algae-based bioeconomy." *Frontiers in Plant Science* no. 11:279.

Fan, L. H., Y. T. Zhang, L. H. Cheng, L. Zhang, D. S. Tang, and H. L. Chen. 2007. "Optimization of carbon dioxide fixation by Chlorella vulgaris cultivated in a membrane-photobioreactor." *Chemical Engineering & Technology* no. 30 (8):1094–1099. https://doi.org/10.1002/ceat.200700141.

Ferro, Yannis, Mercedes Perullini, Matias Jobbagy, Sara A. Bilmes, and Claude Durrieu. 2012. "Development of a biosensor for environmental monitoring based on microalgae immobilized in silica hydrogels." *Sensors* no. 12 (12). doi: 10.3390/s121216879.

Franklin, N. M., M. S. Adams, J. L. Stauber, and R. P. Lim. 2001. "Development of an improved rapid enzyme inhibition bioassay with marine and freshwater microalgae using flow cytometry." *Archives of Environmental Contamination and Toxicology* no. 40 (4):469–480. doi: 10.1007/s002440010199.

Gavrilescu, Maria. 2010. "Environmental biotechnology: Achievements, opportunities and challenges." *Dynamic Biochemistry, Process Biotechnology and Molecular Biology* no. 4 (1):1–36.

Geada, P., V. Vasconcelos, A. Vicente, and B. Fernandes. 2017. "Chapter 13 - microalgal biomass cultivation." In *Algal Green Chemistry*, edited by Rajesh Prasad Rastogi, Datta Madamwar and Ashok Pandey, 257–284. Amsterdam: Elsevier.

Gómez, C., R. Escudero, M. M. Morales, F. L. Figueroa, J. M. Fernández-Sevilla, and F. G. Acién. 2013. "Use of secondary-treated wastewater for the production of Muriellopsis sp." *Applied Microbiology and Biotechnology* no. 97 (5):2239–2249. doi: 10.1007/s00253-012-4634-7.

Gómez-Jacinto, Verónica, Tamara García-Barrera, José Luis Gómez-Ariza, Inés Garbayo-Nores, and Carlos Vílchez-Lobato. 2015. "Elucidation of the defence mechanism in microalgae Chlorella sorokiniana under mercury exposure. Identification of Hg—phytochelatins." *Chemico-Biological Interactions* no. 238:82–90. doi: 10.1016/j.cbi.2015.06.013.

Gosset, Antoine, Claude Durrieu, Louis Renaud, Anne-Laure Deman, Pauline Barbe, Rémy Bayard, and Jean-François Chateaux. 2018. "Xurography-based microfluidic algal biosensor and dedicated portable measurement station for online monitoring of urban polluted samples." *Biosensors and Bioelectronics* no. 117:669–677. doi: 10.1016/j.bios.2018.07.005.

Gros, Meritxell, Mira Petrović, and Damiá Barceló. 2006. "Development of a multiresidue analytical methodology based on liquid chromatography—tandem mass spectrometry (LC—MS/MS) for screening and trace level determination of pharmaceuticals in surface and wastewaters." *Talanta* no. 70 (4):678–690. doi.org/10.1016/j.talanta.2006.05.024.

Halilović, Alma, Emina Merdan, Živorad Kovačević, and Lejla Gurbeta Pokvić. 2019. "Review of biosensors for environmental field monitoring." Paper read at 2019 8th Mediterranean Conference on Embedded Computing (MECO).

Han, Shuang, Qian Zhang, Xingcai Zhang, Xianming Liu, Ling Lu, Junfeng Wei, Yuancheng Li, Yunhua Wang, and Guoxia Zheng. 2019. "A digital microfluidic diluter-based microalgal motion biosensor for marine pollution monitoring." *Biosensors and Bioelectronics* no. 143:111597. doi: 10.1016/j.bios.2019.111597.

Heddam, Salim, Hadi Sanikhani, and Ozgur Kisi. 2019. "Application of artificial intelligence to estimate phycocyanin pigment concentration using water quality data: A comparative study." *Applied Water Science* no. 9 (7):164. doi: 10.1007/s13201-019-1044-3.

How, Chee Wun, Yong Sze Ong, Sze Shin Low, Ashok Pandey, Pau Loke Show, and Jhi Biau Foo. 2021. "How far have we explored fungi to fight cancer?" *Seminars in Cancer Biology*. doi: 10.1016/j.semcancer.2021.03.009.

Hussein, M. H., A. M. Abdullah, N. I. Badr El-Din, and E. S. I. Mishaqa. 2017. "Biosorption potential of the microchlorophyte chlorella vulgaris for some pesticides." *Journal of Fertilizers & Pesticides* no. 8 (1). doi: 10.4172/2471-2728.1000177.

Ibuot, Aniefon, Andrew P. Dean, Owen A. McIntosh, and Jon K. Pittman. 2017. "Metal bioremediation by CrMTP4 over-expressing Chlamydomonas reinhardtii in comparison to natural wastewater-tolerant microalgae strains." *Algal Research* no. 24:89–96. doi: 10.1016/j.algal.2017.03.002.

Jacob-Lopes, Eduardo, Carlos Henrique Gimenes Scoparo, Lucy Mara Cacia Ferreira Lacerda, and Telma Teixeira Franco. 2009. "Effect of light cycles (night/day) on CO2 fixation and biomass production by microalgae in photobioreactors." *Chemical Engineering and Processing: Process Intensification* no. 48 (1):306–310. doi: 10.1016/j.cep.2008.04.007.

Jaiswal, Krishna Kumar, Vinod Kumar, Mikhail S. Vlaskin, and Manisha Nanda. 2021. "Impact of pyrene (polycyclic aromatic hydrocarbons) pollutant on metabolites and lipid induction in microalgae Chlorella sorokiniana (UUIND6) to produce renewable biodiesel." *Chemosphere* no. 285:131482. doi: 10.1016/j.chemosphere.2021.131482.

Jiang, Yinli, Wei Zhang, Junfeng Wang, Yu Chen, Shuhua Shen, and Tianzhong Liu. 2013. "Utilization of simulated flue gas for cultivation of Scenedesmus dimorphus." *Bioresource Technology* no. 128:359–364. doi: 10.1016/j.biortech.2012.10.119.

Kadam, Abhijeet D., Garima Kishore, Deepak Kumar Mishra, and Kusum Arunachalam. 2020. "Microalgal diversity as an indicator of the state of the environment of water bodies of Doon valley in Western Himalaya, India." *Ecological Indicators* no. 112:106077. doi: 10.1016/j.ecolind.2020.106077.

Kasiri, Sepideh, Ania Ulrich, and Vinay Prasad. 2015. "Kinetic modeling and optimization of carbon dioxide fixation using microalgae cultivated in oil-sands process water." *Chemical Engineering Science* no. 137:697–711. doi: 10.1016/j.ces.2015.07.004.

Kurade, Mayur B., Jung Rae Kim, Sanjay P. Govindwar, and Byong-Hun Jeon. 2016. "Insights into microalgae mediated biodegradation of diazinon by Chlorella vulgaris: Microalgal tolerance to xenobiotic pollutants and metabolism." *Algal Research* no. 20:126–134. doi: 10.1016/j.algal.2016.10.003.

Kushwaha, Omkar Singh, Haripriyan Uthayakumar, and Karthigaiselvan Kumaresan. 2021. "Modeling of carbon dioxide fixation rate by micro algae using hybrid artificial intelligence and fuzzy logic methods and optimization by Genetic Algorithm." *Environmental Science and Pollution Research.* doi: 10.21203/rs.3.rs-774165/v1.

Kwak, Hyo Won, Moo Kon Kim, Jeong Yun Lee, Haesung Yun, Min Hwa Kim, Young Hwan Park, and Ki Hoon Lee. 2015. "Preparation of bead-type biosorbent from water-soluble Spirulina platensis extracts for chromium (VI) removal." *Algal Research* no. 7:92–99. doi: 10.1016/j.algal.2014.12.006.

Law, Sam Q. K., Srinivas Mettu, Muthupandian Ashokkumar, Peter J. Scales, and Gregory J. O. Martin. 2018. "Emulsifying properties of ruptured microalgae cells: Barriers to lipid extraction or promising biosurfactants?" *Colloids and Surfaces B: Biointerfaces* no. 170:438–446. doi: 10.1016/j.colsurfb.2018.06.047.

Leong, Yoong Kit, and Jo-Shu Chang. 2020. "Bioremediation of heavy metals using microalgae: Recent advances and mechanisms." *Bioresource Technology* no. 303:122886. doi: j.biortech.2020.122886.

Li, Mei, Xiangyu Gao, Bing Wu, Xin Qian, John P. Giesy, and Yibin Cui. 2014. "Microalga Euglena as a bioindicator for testing genotoxic potentials of organic pollutants in Taihu Lake, China." *Ecotoxicology* no. 23 (4):633–640. doi: 10.1007/s10646-014-1214-x.

Low, Sze Shin, Kien Xiang Bong, Muhammad Mubashir, Chin Kui Cheng, Man Kee Lam, Jun Wei Lim, Yeek Chia Ho, Keat Teong Lee, Heli Siti Halimatul Munawaroh, and Pau Loke Show. 2021. "Microalgae cultivation in palm oil mill effluent (POME) treatment and biofuel production." *Sustainability* no. 13 (6):3247.

Low, Sze Shin, Zetao Chen, Yaru Li, Yanli Lu, and Qingjun Liu. 2021. "Design principle in biosensing: Critical analysis based on graphitic carbon nitride (G-C3N4) photoelectrochemical biosensor." *TrAC Trends in Analytical Chemistry* no. 145:116454. doi: 10.1016/j.trac.2021.116454.

Low, Sze Shin, Yixin Pan, Daizong Ji, Yaru Li, Yanli Lu, Yan He, Qingmei Chen, and Qingjun Liu. 2021. "Smartphone-based portable electrochemical biosensing system for detection of circulating microRNA-21 in saliva as a proof-of-concept." *Sensors and Actuators B: Chemical* no. 308:127718. doi: 10.1016/j.snb.2020.127718.

Luo, R. C., Chih-Chen Yih, and Kuo Lan Su. 2002. "Multisensor fusion and integration: Approaches, applications, and future research directions." *IEEE Sensors Journal* no. 2 (2):107–119. doi: 10.1109/JSEN.2002.1000251.

Ma, Jianyi. 2005. "Differential sensitivity of three cyanobacterial and five green algal species to organotins and pyrethroids pesticides." *Science of The Total Environment* no. 341 (1):109–117. doi: 10.1016/j.scitotenv.2004.09.028.

Mahapatra, Durga Madhab, H. N. Chanakya, and T. V. Ramachandra. 2013. "Treatment efficacy of algae-based sewage treatment plants." *Environmental Monitoring and Assessment* no. 185 (9):7145–7164. doi: 10.1007/s10661-013-3090-x.

Manu, D. S., and Arun Kumar Thalla. 2017. "Artificial intelligence models for predicting the performance of biological wastewater treatment plant in the removal of Kjeldahl Nitrogen from wastewater." *Applied Water Science* no. 7 (7):3783–3791. doi: 10.1007/s13201-017-0526-4.

Matamoros, Víctor, Raquel Gutiérrez, Ivet Ferrer, Joan García, and Josep M. Bayona. 2015. "Capability of microalgae-based wastewater treatment systems to remove emerging organic contaminants: A pilot-scale study." *Journal of Hazardous Materials* no. 288:34–42. doi: 10.1016/j.jhazmat.2015.02.002.

Matamoros, Víctor, and Yolanda Rodríguez. 2016. "Batch vs continuous-feeding operational mode for the removal of pesticides from agricultural run-off by microalgae systems: A laboratory scale study." *Journal of Hazardous Materials* no. 309:126–132. doi: 10.1016/j.jhazmat.2016.01.080.

Moreira-Santos, Matilde, Amadeu M. V. M. Soares, and Rui Ribeiro. 2004. "An in situ bioassay for freshwater environments with the microalga Pseudokirchneriella subcapitata." *Ecotoxicology and Environmental Safety* no. 59 (2):164–173. doi: 10.1016/j.ecoenv.2003.07.004.

Moro, Laura, Gianni Pezzotti, Mehmet Turemis, Josep Sanchís, Marinella Farré, Renata Denaro, Maria Grazia Giacobbe, Francesca Crisafi, and Maria Teresa Giardi. 2018. "Fast pesticide pre-screening in marine environment using a green microalgae-based optical bioassay." *Marine Pollution Bulletin* no. 129 (1):212–221. doi: 10.1016/j.marpolbul.2018.02.036.

Mubashar, Muhammad, Muhammad Naveed, Adnan Mustafa, Sobia Ashraf, Khurram Shehzad Baig, Saud Alamri, Manzer H. Siddiqui, Magdalena Zabochnicka-Świątek, Michał Szota, and Hazem M. Kalaji. 2020. "Experimental investigation of chlorella vulgaris and Enterobacter sp. MN17 for decolorization and removal of heavy metals from textile wastewater." *Water* no. 12 (11):3034. doi: 10.3390/w12113034.

Nabok, A., S. Haron, and A. K. Ray. 2004. "Registration of heavy metal ions and pesticides with ATR planar waveguide enzyme sensors." *Applied Surface Science* no. 238 (1–4):423–428. doi: 10.1016/j.apsusc.2004.05.165.

Nath, Adi, Pravin Kumar Tiwari, Awadhesh Kumar Rai, and Shanthy Sundaram. 2017. "Microalgal consortia differentially modulate progressive adsorption of hexavalent chromium." *Physiology and Molecular Biology of Plants* no. 23 (2):269–280. doi: 10.1007/s12298-017-0415-1.

Nie, Jing, Yuqing Sun, Yaoyu Zhou, Manish Kumar, Muhammad Usman, Jiangshan Li, Jihai Shao, Lei Wang, and Daniel C. W. Tsang. 2020. "Bioremediation of water containing pesticides by microalgae: Mechanisms, methods, and prospects for future research." *Science of The Total Environment* no. 707:136080. doi: 10.1016/j.scitotenv.2019.136080.

Omar, Wan Maznah Wan. 2010. "Perspectives on the use of algae as biological indicators for monitoring and protecting aquatic environments, with special reference to Malaysian freshwater ecosystems." *Tropical Life Sciences Research* no. 21 (2):51–67.

O'Neill, Emer A., and Neil J. Rowan. 2022. "Microalgae as a natural ecological bioindicator for the simple real-time monitoring of aquaculture wastewater quality including provision for assessing impact of extremes in climate variance—A comparative case study from the Republic of Ireland." *Science of The Total Environment* no. 802:149800. doi: 10.1016/j.scitotenv.2021.149800.

Özçimen, Didem, Benan İnan, Anıl Tevfik Koçer, and Meyrem Vehapi. 2018. "Bioeconomic assessment of microalgal production." In *Microalgal Biotechnology*, edited by Eduardo Jacob-Lopes, Leila Queiroz Zepka and Maria Isabel Queiroz. BoD—Books on Demand. London: IntechOpen.

Pacwa-Płociniczak, Magdalena, Grażyna A. Płaza, Zofia Piotrowska-Seget, and Swaranjit Singh Cameotra. 2011. "Environmental applications of biosurfactants: Recent advances." *International Journal of Molecular Sciences* no. 12 (1):633–654. doi: 10.3390/ijms12010633.

Pal, Moumita P., Bhalchandra K. Vaidya, Kiran M. Desai, Renuka M. Joshi, Sanjay N. Nene, and Bhaskar D. Kulkarni. 2009. "Media optimization for biosurfactant production by Rhodococcus erythropolis MTCC 2794: Artificial intelligence versus a statistical approach." *Journal of Industrial Microbiology and Biotechnology* no. 36 (5):747–756. doi: 10.1007/s10295-009-0547-6.

Parmar, Trishala K., Deepak Rawtani, and Y. K. Agrawal. 2016. "Bioindicators: The natural indicator of environmental pollution." *Frontiers in Life Science* no. 9 (2):110–118. doi: 10.1080/21553769.2016.1162753.

Parus, Anna, and Bożena Karbowska. 2020. "Marine Algae as Natural Indicator of Environmental Cleanliness." *Water, Air, & Soil Pollution* no. 231 (3):97. doi: 10.1007/s11270-020-4434-0.

Pattanaik, Archana, Lala Behari Sukla, Debabrata Pradhan, and Vinita Shukla. 2019. "Artificial intelligence and virtual environment for microalgal source for production of nutraceuticals." *Biomedical Journal of Scientific & Technical Research* no. 13 (5). doi: 10.26717/BJSTR.2019.13.002459.

Paul, Michael J., Brannon Walsh, Jacques Oliver, and Dana Thomas. 2017. *Algal Indicators in Streams: A Review of Their Application in Water Quality Management of Nutrient Pollution*. United States: Environmental Protection Agency.

Peña-Vázquez, Elena, Emilia Maneiro, Concepción Pérez-Conde, Maria Cruz Moreno-Bondi, and Eduardo Costas. 2009. "Microalgae fiber optic biosensors for herbicide monitoring using sol—gel technology." *Biosensors and Bioelectronics* no. 24 (12):3538–3543. doi: 10.1016/j.bios.2009.05.013.

Peng, Fu-Qiang, Guang-Guo Ying, Bin Yang, Shan Liu, Hua-Jie Lai, You-Sheng Liu, Zhi-Feng Chen, and Guang-Jie Zhou. 2014. "Biotransformation of progesterone and norgestrel by two freshwater microalgae (Scenedesmus obliquus and Chlorella pyrenoidosa): Transformation kinetics and products identification." *Chemosphere* no. 95:581–588. doi: 10.1016/j.chemosphere.2013.10.013.

Pérez-Legaspi, Ignacio Alejandro, Luis Alfredo Ortega-Clemente, Jesús David Moha-León, Elvira Ríos-Leal, Sergio Curiel-Ramírez Gutiérrez, and Isidoro Rubio-Franchini. 2016. "Effect of the pesticide lindane on the biomass of the microalgae Nannochloris oculata." *Journal of Environmental Science and Health, Part B* no. 51 (2):103–106. doi: 10.1080/03601234.2015.1092824.

Peter, Angela Paul, Kuan Shiong Khoo, Kit Wayne Chew, Tau Chuan Ling, Shih-Hsin Ho, Jo-Shu Chang, and Pau Loke Show. 2021. "Microalgae for biofuels, wastewater treatment and environmental monitoring." *Environmental Chemistry Letters* no. 19:2981–2904. doi: 10.1007/s10311-021-01219-6.

Podola, Björn, Eva C. M. Nowack, and Michael Melkonian. 2004. "The use of multiple-strain algal sensor chips for the detection and identification of volatile organic compounds." *Biosensors and Bioelectronics* no. 19 (10):1253–1260. doi: 10.1016/j.bios.2003.11.015.

Radmann, Elisângela Martha, Etiele Greque de Morais, Cibele Freitas de Oliveira, Kellen Zanfonato, and Jorge Alberto Vieira Costa. 2015. "Microalgae cultivation for biosurfactant production." *African Journal of Microbiology Research* no. 9 (47):2283–2289. doi: 10.5897/AJMR2015.7634.

Rathnayake, I. V. N., Thilini Munagamage, A. Pathirathne, and Mallavarapu Megharaj. 2021. "Whole cell microalgal-cyanobacterial array biosensor for monitoring Cd, Cr and Zn in aquatic systems." *Water Science and Technology* no. 84 (7):1579–1593. doi: 10.2166/wst.2021.339.

Ray, Joseph George, Prasanthkumar Santhakumaran, and Santhoshkumar Kookal. 2021. "Phytoplankton communities of eutrophic freshwater bodies (Kerala, India) in relation to the physicochemical water quality parameters." *Environment, Development and Sustainability* no. 23 (1):259–290. doi: 10.1007/s10668-019-00579-y.

Rizwan, Muhammad, Ghulam Mujtaba, Sheraz Ahmed Memon, Kisay Lee, and Naim Rashid. 2018. "Exploring the potential of microalgae for new biotechnology applications and beyond: A review." *Renewable and Sustainable Energy Reviews* no. 92:394–404. doi: 10.1016/j.rser.2018.04.034.

Rossignol, D. A., S. J. Genuis, and R. E. Frye. 2014. "Environmental toxicants and autism spectrum disorders: A systematic review." *Translational Psychiatry* no. 4 (2):e360. doi: 10.1038/tp.2014.4.

Roxby, Daniel N., Hamim Rivy, Chaoyang Gong, Xuerui Gong, Zhiyi Yuan, Guo-En Chang, and Yu-Cheng Chen. 2020. "Microalgae living sensor for metal ion detection with nanocavity-enhanced photoelectrochemistry." *Biosensors and Bioelectronics* no. 165:112420. doi: 10.1016/j.bios.2020.112420.

Saavedra, Ricardo, Raúl Muñoz, María Elisa Taboada, Marisol Vega, and Silvia Bolado. 2018. "Comparative uptake study of arsenic, boron, copper, manganese and zinc from water by different green microalgae." *Bioresource Technology* no. 263:49–57. doi: 10.1016/j.biortech.2018.04.101.

Sampathkumar, A., S. Murugan, Ahmed A. Elngar, Lalit Garg, R. Kanmani, and A. Malar. 2020. "A novel scheme for an IoT-based weather monitoring system using a wireless sensor network." In *Integration of WSN and IoT for Smart Cities*, edited by Shalli Rani, R. Maheswar, G. R. Kanagachidambaresan, and P. Jayarajan, 181–191. United States: Springer.

Santos, B., A. Ponezi, and Ana M. F. Fileti. 2017. "Development of artificial intelligence models to monitor biosurfactant concentration in real-time using waste as substrate in bioreactor through fermentation by Bacillus subtilis." *Chemical Engineering Transactions* no. 57:1009–1014. doi: 10.3303/CET1757169.

Sardrood, Babak Pakdaman, Ebrahim Mohammadi Goltapeh, and Ajit Varma. 2013. "An introduction to bioremediation." In *Fungi as bioremediators*, edited by Ebrahim Mohammadi Goltapeh, Younes Rezaee Danesh and Ajit Varma, 3–27. United States: Springer.

Sasikumar, C. Sheela, and Taniya Papinazath. 2003. "Environmental management: bioremediation of polluted environment." Paper read at Proceedings of the Third International Conference on Environment and Health, 2003, at Chennai, India.

Satpal, and A. K. Khambete. 2016. "Waste water treatment using micro-algae—a review paper." *International Journal of Engineering Technology, Management and Applied Sciences* no. 4 (2):188–192.

Sawayama, Shigeki, Seiichi Inoue, Yutaka Dote, and Shin-Ya Yokoyama. 1995. "CO2 fixation and oil production through microalga." *Energy Conversion and Management* no. 36 (6):729–731. doi: 10.1016/0196-8904(95)00108-P.

Singh, Amit Kumar, Humaira Farooqi, Malik Zainul Abdin, and Shashi Kumar. 2019. "Bioremediation of municipal wastewater and biodiesel production by cultivation of Parachlorella kessleri-I." In *The Role of Microalgae in Wastewater Treatment*, edited by Lala Behari Sukla, Enketeswara Subudhi and Debabrata Pradhan, 15–28. United States: Springer.

Singh, Anamika, Sabeela Beevi Ummalyma, and Dinabandhu Sahoo. 2020. "Bioremediation and biomass production of microalgae cultivation in river water contaminated with pharmaceutical effluent." *Bioresource Technology* no. 307:123233. doi: 10.1016/j.biortech.2020.123233.

Sivapathasekaran, C., Soumen Mukherjee, Arja Ray, Ashish Gupta, and Ramkrishna Sen. 2010. "Artificial neural network modeling and genetic algorithm based medium optimization for the improved production of marine biosurfactant." *Bioresource Technology* no. 101 (8):2884–2887. doi: 10.1016/j.biortech.2009.09.093.

Solé, Alba, and Víctor Matamoros. 2016. "Removal of endocrine disrupting compounds from wastewater by microalgae co-immobilized in alginate beads." *Chemosphere* no. 164:516–523. doi: 10.1016/j.chemosphere.2016.08.047.

Sun, Yahui, Yun Huang, Qiang Liao, Qian Fu, and Xun Zhu. 2016. "Enhancement of microalgae production by embedding hollow light guides to a flat-plate photobioreactor." *Bioresource Technology* no. 207:31–38. doi: 10.1016/j.biortech.2016.01.136.

Sung, Ki-Don, Jin-Suk Lee, Chul-Seung Shin, Soon-Chul Park, and Myung-Jae Choi. 1999. "CO2 fixation by Chlorella sp. KR-1 and its cultural characteristics." *Bioresource Technology* no. 68 (3):269–273. doi: 10.1016/S0960-8524(98)00152-7.

Sutherland, Donna L., and Peter J. Ralph. 2019. "Microalgal bioremediation of emerging contaminants—Opportunities and challenges." *Water Research* no. 164:114921. doi: 10.1016/j.watres.2019.114921.

Sydney, Eduardo Bittencourt, Wilerson Sturm, Julio Cesar de Carvalho, Vanete Thomaz-Soccol, Christian Larroche, Ashok Pandey, and Carlos Ricardo Soccol. 2010. "Potential carbon dioxide fixation by industrially important microalgae." *Bioresource Technology* no. 101 (15):5892–5896. doi: 10.1016/j.biortech.2010.02.088.

Tang, Doris Ying Ying, Kuan Shiong Khoo, Kit Wayne Chew, Yang Tao, Shih-Hsin Ho, and Pau Loke Show. 2020. "Potential utilization of bioproducts from microalgae for the quality enhancement of natural products." *Bioresource Technology* no. 304:122997. doi: 10.1016/j.biortech.2020.122997.

Teniou, Ahlem, Amina Rhouati, and Jean-Louis Marty. 2021. "Mathematical modelling of biosensing platforms applied for environmental monitoring." *Chemosensors* no. 9 (3):50. doi: 10.3390/chemosensors9030050.

Tsopela, A., A. Laborde, L. Salvagnac, V. Ventalon, E. Bedel-Pereira, I. Séguy, P. Temple-Boyer, P. Juneau, R. Izquierdo, and J. Launay. 2016. "Development of a lab-on-chip electrochemical biosensor for water quality analysis based on microalgal photosynthesis." *Biosensors and Bioelectronics* no. 79:568–573. doi: 10.1016/j.bios.2015.12.050.

Ugya, Adamu Yunusa, Fidelis Odedishemi Ajibade, and Xiuyi Hua. 2021. "The efficiency of microalgae biofilm in the phycoremediation of water from River Kaduna." *Journal of Environmental Management* no. 295:113109. doi: 10.1016/j.jenvman.2021.113109.

Ugya, Yunusa Adamu, Diya'uddeen Basheer Hasan, Salisu Muhammad Tahir, Tijjani Sabiu Imam, Hadiza Abdullahi Ari, and Xiuyi Hua. 2021. "Microalgae biofilm cultured in nutrient-rich water as a tool for the phycoremediation of petroleum-contaminated water." *International Journal of Phytoremediation* no. 23 (11):1175–1183. doi: 10.1080/15226514.2021.1882934.

Uqab, B., S. Mudasir, A. Qayoom, and R. Nazir. 2016. "Bioremediation: A management tool." *Journal of Bioremediation and Biodegradation* no. 7 (2). doi: 10.4172/2155-6199.1000331.

Valera, Enrique, Ruth Babington, Marta Broto, Salvador Petanas, Roger Galve, and Maria-Pilar Marco. 2013. "Chapter 7 - application of bioassays/biosensors for the analysis of pharmaceuticals in environmental samples." In *Comprehensive Analytical Chemistry*, edited by Mira Petrovic, Damia Barcelo and Sandra Pérez, 195–229. Amsterdam: Elsevier.

Van Leeuwen, S. P. J., and J. De Boer. 2008. "Advances in the gas chromatographic determination of persistent organic pollutants in the aquatic environment." *Journal of Chromatography A* no. 1186 (1–2):161–182. doi: 10.1016/j.chroma.2008.01.044.

Wong, L. S., S. K. Judge, B. W. N. Voon, L. J. Tee, K. Y. Tan, M. Murti, and M. K. Chai. 2018. "Bioluminescent microalgae-based biosensor for metal detection in water." *IEEE Sensors Journal* no. 18 (5):2091–2096. doi: 10.1109/JSEN.2017.2787786.

Xiong, Jiu-Qiang, Mayur B. Kurade, and Byong-Hun Jeon. 2017. "Biodegradation of levofloxacin by an acclimated freshwater microalga, Chlorella vulgaris." *Chemical Engineering Journal* no. 313:1251–1257. doi: 10.1016/j.cej.2016.11.017.

Xiong, Jiu-Qiang, Mayur B. Kurade, Jung Rae Kim, Hyun-Seog Roh, and Byong-Hun Jeon. 2017. "Ciprofloxacin toxicity and its co-metabolic removal by a freshwater microalga Chlamydomonas mexicana." *Journal of Hazardous Materials* no. 323:212–219. doi: 10.1016/j.jhazmat.2016.04.073.

Xu, Gang, Xin Li, Chen Cheng, Jie Yang, Zhaoyang Liu, Zhenghan Shi, Lihang Zhu, Yanli Lu, Sze Shin Low, and Qingjun Liu. 2020. "Fully integrated battery-free and flexible electrochemical tag for on-demand wireless in situ monitoring of heavy metals." *Sensors and Actuators B: Chemical* no. 310:127809. doi: 10.1016/j.snb.2020.127809.

Yamagishi, Takahiro, Haruyo Yamaguchi, Shigekatsu Suzuki, Yoshifumi Horie, and Norihisa Tatarazako. 2017. "Cell reproductive patterns in the green alga Pseudokirchneriella subcapitata (=Selenastrum capricornutum) and their variations under exposure to the typical toxicants potassium dichromate and 3,5-DCP." *PLoS One* no. 12 (2):e0171259. doi: 10.1371/journal.pone.0171259.

Yang, Xia, Yi He, Xueling Wang, and Ruo Yuan. 2017. "A SERS biosensor with magnetic substrate CoFe2O4@Ag for sensitive detection of Hg2+." *Applied Surface Science* no. 416:581–586. doi: 10.1016/j.apsusc.2017.04.106.

Yi, Hye-Suk, Sangyoung Park, Kwang-Guk An, and Keun-Chang Kwak. 2018. "Algal bloom prediction using extreme learning machine models at artificial weirs in the Nakdong River, Korea." *International Journal of Environmental Research and Public Health* no. 15 (10):2078. doi: 10.3390/ijerph15102078.

Yılmaz, Özlem, and Beyhan Taş. 2021. "Feasibility and assessment of the phytoremediation potential of green microalga and duckweed for zeta-cypermethrin removal." *Desalination and Water Treatment* no. 209:131–143. doi: 10.5004/dwt.2021.26484.

Zabochnicka-Świątek, Magdalena, and Małgorzata Krzywonos. 2014. "Potentials of biosorption and bioaccumulation processes for heavy metal removal." *Polish Journal of Environmental Studies* no. 23 (2):551–561.

Zaghloul, Alaa, Mohamed Saber, Samir Gadow, and Fikry Awad. 2020. "Biological indicators for pollution detection in terrestrial and aquatic ecosystems." *Bulletin of the National Research Centre* no. 44 (1):127. doi: 10.1186/s42269-020-00385-x.

Zhang, Zhengwei, Kun Yan, Ling Zhang, Qian Wang, Ruixin Guo, Zhengyu Yan, and Jianqiu Chen. 2019. "A novel cadmium-containing wastewater treatment method: Bio-immobilization by microalgae cell and their mechanism." *Journal of Hazardous Materials* no. 374:420–427. doi: 10.1016/j.jhazmat.2019.04.072.

Zhao, Lin, Tianjiao Dai, Zhi Qiao, Peizhe Sun, Jianye Hao, and Yongkui Yang. 2020. "Application of artificial intelligence to wastewater treatment: A bibliometric analysis and systematic review of technology, economy, management, and wastewater reuse." *Process Safety and Environmental Protection* no. 133:169–182. doi: 10.1016/j.psep.2019.11.014.

Zhao, Qing, An-Na Chen, Shun-Xin Hu, Qian Liu, Min Chen, Lu Liu, Chang-Lun Shao, Xue-Xi Tang, and Chang-Yun Wang. 2018. "Microalgal microscale model for microalgal growth inhibition evaluation of marine natural products." *Scientific Reports* no. 8 (1):10541. doi: 10.1038/s41598-018-28980-z.

Zhou, Wenguang, Jinghan Wang, Paul Chen, Chengcheng Ji, Qiuyun Kang, Bei Lu, Kun Li, Jin Liu, and Roger Ruan. 2017. "Bio-mitigation of carbon dioxide using microalgal systems: Advances and perspectives." *Renewable and Sustainable Energy Reviews* no. 76:1163–1175. doi: 10.1016/j.rser.2017.03.065.

Zhuang, Lin-Lan, Jing-Han Wang, and Hong-Ying Hu. 2018. "Differences between attached and suspended microalgal cells in ssPBR from the perspective of physiological properties." *Journal of Photochemistry and Photobiology B: Biology* no. 181:164–169. doi: 10.1016/j.jphotobiol.2018.03.014.

11 Industry Perspectives of Microalgae 4.0

Kreena Gada[1], Angela Paul Peter[1], Wai Siong Chai[2] and Pau Loke Show[1]

[1] Department of Chemical and Environmental Engineering, Faculty of Science and Engineering, University of Nottingham Malaysia, Jalan Broga, Semenyih, Selangor Darul Ehsan, Malaysia

[2] School of Mechanical Engineering and Automation, Harbin Institute of Technology, Shenzhen, Guangdong, China

CONTENTS

11.1 Introduction.......... 364
11.2 Algae Industry.......... 365
 11.2.1 Algae Processing in Industry 4.0.......... 365
  11.2.1.1 Why Were Industrial Revolutions Required?.......... 367
 11.2.2 A Shift from IR 3.0 to IR 4.0.......... 368
  11.2.2.1 The Difference between IR 3.0 and IR 4.0.......... 368
  11.2.2.2 Why Countries Should Shift from IR 3.0 to IR 4.0?.......... 368
 11.2.3 Energy and Economic Analysis.......... 369
 11.2.4 Algae Production.......... 371
 11.2.5 The Projections.......... 372
  11.2.5.1 Open PBR Systems.......... 372
  11.2.5.2 Closed PBR Systems.......... 373
  11.2.5.3 Future Perspectives.......... 374
11.3 Customization of Bioenergy in the Algae Industry.......... 375
 11.3.1 Bioenergy.......... 375
11.4 Algae Cultivation in IR 4.0.......... 378
 11.4.1 Tools for Alternating Cultivation Mode.......... 378
 11.4.2 Multipurpose and Specific Medium Cultivation.......... 378
 11.4.3 Cultivation Condition and Adaptation of Algae.......... 379
11.5 Algae Genetic Engineering.......... 381
 11.5.1 Genetic Modification for Desired Strain.......... 381

DOI: 10.1201/9781003202196-11

11.5.2 Automated Genetic Manipulation, Estimation, and Computing of Proper Strain Detection 383
11.5.2.1 Examples of Computer-Controlled Cultivation of Microalgae 383
11.6 Algae Industry 4.0 384
11.6.1 The New Markets 384
11.6.2 Algae Revolutionization in IR 4.0 386
11.6.3 Environmental Impacts 387
11.7 Fulfilling Sustainable Development Goals 389
11.8 Challenges and Perspectives 389
11.8.1 Large-Scale Cultivation 389
11.9 Conclusions 391
References 392

11.1 INTRODUCTION

Algae are organisms which utilize sunlight to produce nutrients, and they can be found in a variety of marine ecosystems such as oceans, lakes, ponds, rivers, and even wastewater. They are able to grow alone or in association with other species and can endure a wide temperature range, pH values, and salinities; varying light intensities; and environments in deserts or reservoirs. Algae can be divided into two major classifications depending on their sizes: macroalgae and microalgae. Macroalgae are large in size and hence multicellular which are therefore obvious to the naked eye such as seaweed. Microalgae on the other hand are singular cells that are microscopic in size (Khan et al. 2018).

Biofuels, nutraceuticals, medicines, and beautification products can all benefit from microalgae as a source of carbon compounds (Figure 11.1). They can also be used to treat wastewater and reduce the CO_2 levels in the atmosphere. Pigments, lipids, polysaccharides, vitamins, proteins, antioxidants, and bioactive compounds are only the few bioproducts produced by microalgae (Khan et al. 2018).

The use of microalgae as an inexhaustible and feasible feedstock for biofuel production has reignited curiosity in biorefinery. To strengthen their possibility as a future renewable source bioproducts, growth improvement means and genetic modification maybe applied (Khan et al. 2018).

Over the last several decades, the modern production of microalgae to generate bioproducts and biofuels has risen tremendously (Khan et al. 2018). There are clear pathways for bioproducts, produced alongside biofuels in retrofitted biorefineries, to deliver great financial value with even limited quantities of production in the coming years, as shown by the examination of the economics and sustainability of the petroleum industry (Plantlet 2019). Bioproducts are

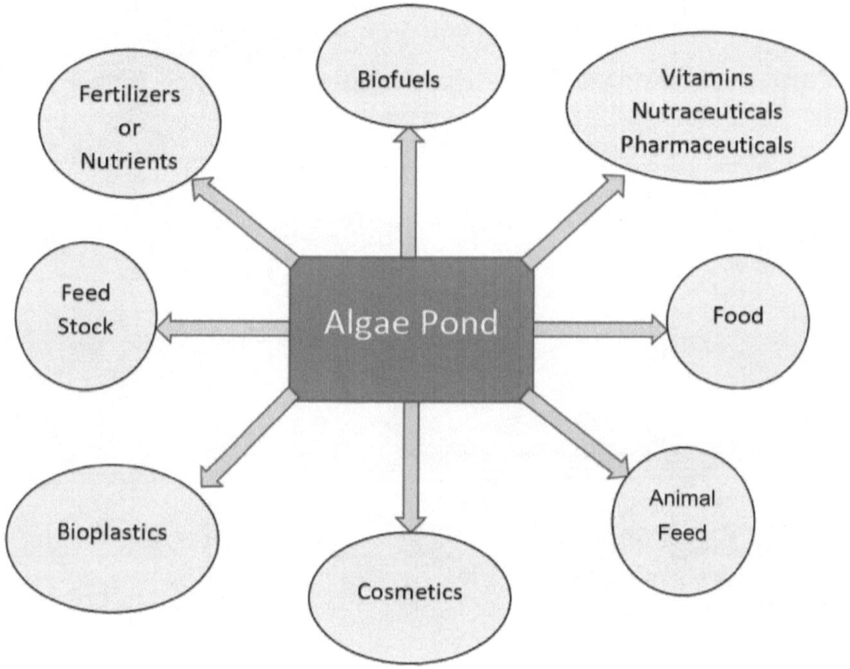

FIGURE 11.1 Beneficial products of microalgae (Khan et al. 2018).

predicted to expand in demand over the next several decades. According to a recent study conducted by the National Renewable Energy Laboratory, bioproducts would have a 25% market share in the international chemical sector by 2025. Given that bioproducts were valued at $19.7 billion in 2016 and had a market share of only 2% in 2008, it is apparent that this alternative type of energy is here to stay (Ramesh 2020).

11.2 ALGAE INDUSTRY

11.2.1 Algae Processing in Industry 4.0

Next-generation Artificial Intelligence (AI) mechanizations are facilitating the rise of a fourth industrial transformation at a time when global issues require humankind to exploit assets more effectively and continuously than ever before. The three previous industrial transformations were constituted by the emergence of steam power, electric supply chains, and computerized machinery under business frameworks that supported industrial growth (Figure 11.2). Industry 4.0 is defined by self-governing, robust, and network systems that can optimize

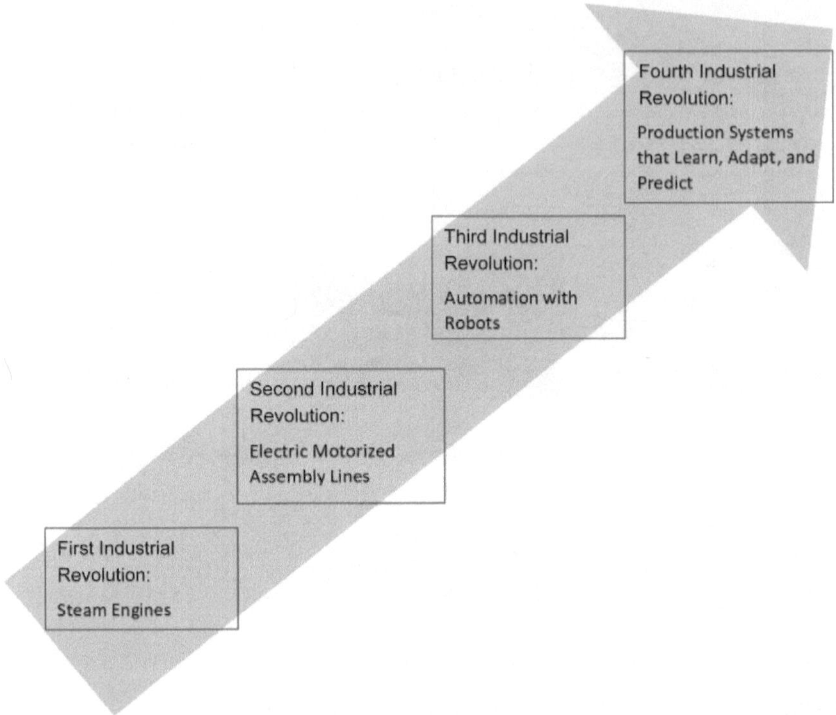

FIGURE 11.2 The various industrial revolutions (University of Sydney-Technology 2021).

production by forecasting, recognizing, and reciprocating to production needs in actual time using advanced sensors, machine learning, and shared data. Industry 4.0 will assist firms in updating their manufacturing technique by enabling personalized, really adaptable, profitable, and further sustainable manufacturing systems that reduce waste and enhance productivity (University of Sydney-Technology 2021).

Microalgae's great power for the generation of viable bioproducts can be unlocked with Industrial Revolution (IR) 4.0 technologies. IR 4.0 technology will break down culture barriers and allow for quick testing and refinement of ideal growing conditions in order to increase the output of microalgae's lucrative biological products. Microalgae have complicated lifespans and peculiar supplement specifications, demanding a progressive rather than a constant production to achieve maximum yield. IR 4.0 photobioreactors overcome these problems by monitoring microalgal growth with an array of interconnected sensors, adjusting conditions as needed, and identifying ideal growth parameters through machine learning (University of Sydney-Technology 2021).

11.2.1.1 Why Were Industrial Revolutions Required?

When agrarian cultures grew more industrialized and urbanized in the eighteenth century, the IR started. The transcontinental railroad, cotton gin, electricity, and other technologies all had a lasting impact on civilization (History.com Editors 2009). Although the first industrial revolution caused the change-over to new manufacturing mechanisms, they were quite a few problems: Poor employees were time and again kept in cramped, barbarous surroundings (LaMorte 2017). Employees were left open to a range of risks and hazards, including tight work environments with insufficient fresh air, mechanical bruises, and unhealthy exposures to heavy metals, dirt, and solvents. Subsequently, development was introduced as several new health consequences that bothered both Europe and America (Vale 2016). These were the reasons that a second IR was needed.

The second IR which was also known as the technological revolution was a time where there was rapid standardization and industrialization which occurred from the nineteenth century to the early twentieth century. With the second IR, in the industrialized countries, the economy and productivity boomed. As a result, living conditions greatly enhanced, and commodity costs plummeted. Furthermore, because rural regions were linked to big markets through transportation framework, crop failures in the farmlands no longer indicated starvation and hunger. In addition, there were less people in the fields. In addition, public health had substantially improved. This was made possible by the development of sewage structures in cities. This was led by the enactment of legislation governing refined water supply and minimum water excellence requirements. Many illnesses and deaths decreased in number as a result of these two methods (Vale 2016).

However, as previously mentioned, the second IR was a period of rapid and consistent growth. As a result, ships and other equity became outdated in a very short period of time. People lost capital, and the rate of unemployment increased dramatically (Vale 2016). Other social negative effects due to second IR included:

- The rate of urbanization increased. To be near to the industry, the people migrated into hurriedly established homes in cities.
- As labor migrated from the house to industries, families were more segregated.
- Because of the severe and hazardous working conditions at the factories, the workers' general health deteriorated.
- As the demand for products increased and decreased, the availability of jobs grew more uncertain.

With the aforementioned reasons, improvements needed to be made, and therefore a third IR had to be introduced. The third revolution gave birth to electronic devices, communications, and, notably, computers. Due to advances in technology, the third IR laid the door for space technology, research, and biotechnology (iED Team 2019). Although the third IR improved the problems caused by the second IR, the third revolution had its own shortcomings which include (Rafferty 2021):

- Overcrowding of cities and industrial towns which eventually led to the spread of infectious diseases such as cholera and smallpox.
- Pollution and other environmental problems which were mainly caused by fueling factories to produce required outputs of the manufactured goods.
- Poor working conditions as owners saw the production and profit as their main goal without considering safety of the workers or even their living conditions.

With all the cons in the third IR, it was necessary for a fourth IR. In the fourth revolution, it is where there is higher productivity and improved quality of life, along with new markets (Firican 2020). Therefore, it corrects most of the problems people were facing from the previous IR.

11.2.2 A Shift from IR 3.0 to IR 4.0

11.2.2.1 The Difference between IR 3.0 and IR 4.0

In Industry 3.0, we leverage reasoning processors and communication technology to optimize operations. Despite the fact that these activities are normally operated without human interference, they nonetheless have a human component. Industry 4.0 is defined by the provision and use of massive amounts of data on the production line (UpKeep 2019).

11.2.2.2 Why Countries Should Shift from IR 3.0 to IR 4.0?

According to Smart Industry, "Industry 4.0 should increase production flexibility, allowing a facility to respond quickly to market developments. For example, a plant control system could automatically modify output based on changing utility prices, lowering production costs (Immerman 2018)."

In comparison to Industry 3.0, the benefits of Industry 4.0 technologies provide tremendous Return on Investment (ROI) prospects. This comprises mechanization, machine-to-machine transmission, production maintenance, and managerial systems. The following are the primary advantages of Industry 4.0 (Moran 2021):

1. Improved productivity

   To express it differently, Industry 4.0 technologies support you to attain more with less. To put it simply, you can develop more and swiftly while using your resources in a more practical and economical manner. Because of upgraded machine tracking and mechanized/semi-mechanized decision-making, production lines will have less spare time (Moran 2021).

2. Collaboration and knowledge sharing will increase

    Manufacturing plants in the traditional sense work in silos. Individual facilities, like individual machines inside a facility, are silos. As a result, there is very little teamwork or knowledge sharing.

    Manufacturing plants, business operations, and departments can communicate with Industry 4.0 technologies despite of geographic section, time difference, floor, or any other circumstance. This approves knowledge gained by a sensor on a machine in one factory, for example, to be shared from one end to the other end of the company.

    The finest thing is that it can be done completely without human intervention, that is, machine-to-machine and system-to-system. To put it in other words, information from a single sensor can be used to optimize several supply chains all over the world in real time (Moran 2021).

3. Increase in operational efficiency

    As a result of Industry 4.0-related machinery, numerous sectors of the manufacturing channel will become competent. Less downtime of the machines, as well as the capability to create more items and do so faster, is an illustration of these efficiencies.

    Rapid batch switches, automated record and trace procedures, and computerized reporting are all instances of expanded efficiency. NPIs (New Product Introductions) and business decision-making become more efficient as well (Moran 2021).

4. Better products and services

    Be it quality of the product, security, or consumer experience, Industry 4.0 will provide operators with more knowledge and output, permitting to provide benefits to clients while also maintaining their enterprise (Immerman 2018).

    It is feasible to swiftly fix problems with automated track and trace capabilities, for example. In addition, the organization will have fewer product availability concerns, product quality will increase, and customers will have more options (Moran 2021).

5. Improving lives overall

As a result of modern technology, increased revenue, and economic development, people's livelihoods keep improving in general, with higher earnings, better medical treatments, and a better living standard (Immerman 2018).

For example, the company may add a second shift with minimal human costs to satisfy increased demand or compete for a new contract by fully automating the production line and applying other Industry 4.0 technology (Moran 2021).

11.2.3 Energy and Economic Analysis

Renewable liquid electricity is projected to perform a critical task in lowering greenhouse gas (GHG) emissions by substituting petroleum-derived transport fuels with a reasonable back-up. Despite the fact that biodiesel from oleaginous plants and bioethanol from sugarcane are being generated in greater quantities, demand cannot be met in a sustainable manner. As a result, alternative biomass sources are needed to meet this rising need. The use of microalgae-based oil as a biodiesel replacement is continuously being developed. When compared to specific category of microalgae, where oil measurement can outdo 60% of dry load, the oil portion in standard agricultural oil yield is quite low (about 5% of entire biomass) (Morales et al. 2019).

In terms of oil production, microalgae have considerable upper hand over farmland-based crops: high biomass yield, no controversy with feed produce, the ability to uptake factory-made carbon dioxide (CO_2), the ability to use slightly salted or ocean water, and decreased competition of area without the use of weed-killer or insecticides. In spite of these benefits, microalgae-based fuels are not generated on a large scale, owing to the high cost of manufacture. Microalgae-based energy reproduction necessitates massive, low-priced production. This suggests reactor designs that are low-cost, scalable, and produce a lot of algae. The various algal growing systems can be restricted into two groups: open and closed (Morales et al. 2019). Open photobioreactors (PBRs) have the lowest initial investment but have poorer productivity and poisoning difficulties, whereas closed photobioreactors have a higher initial investment but are easier to manage culture conditions (Genin et al. 2016)

According to Zhu et al. (2018), the total operating cost of a PBR system is $ 22.7 million/year. Figure 11.3 shows the operating cost breakdown for the PBR system.

In the projected period of 2021 to 2028, the microalgae market is expected to develop. According to Data Bridge Market Research, the market is predicted to

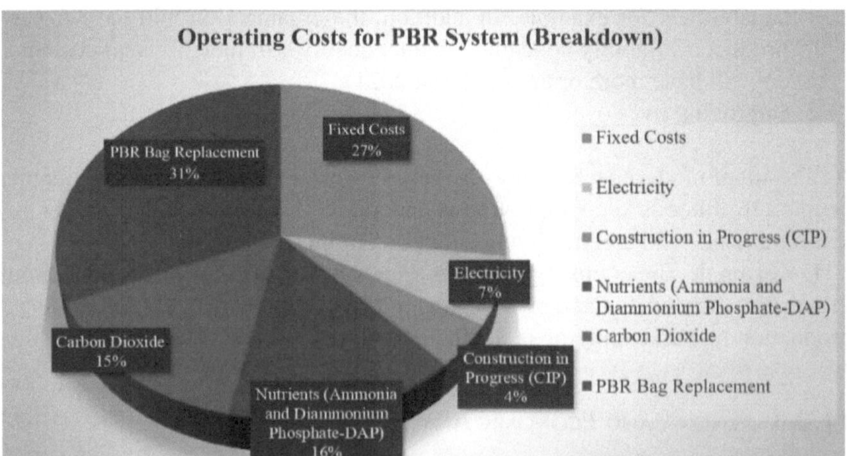

FIGURE 11.3 The breakdown allocation of total operating cost.

rise at a Compound Annual Growth Rate (CAGR) of 6.9% from 2021 to 2028, reaching USD 61,988.47 million. Consumer awareness of the health advantages of microalgae products is growing, and there is a strong need for plant-derived proteins, which is propelling the market demand in the foreseeable future (Data Bridge Market Research 2021).

The microalgae market is being propelled forward by technology and scientific advancements. The issues connected with the development of microalgae or biofuels derived from microalgae are more expensive, which may limit the growth of the microalgae industry as small businesses consider entering the microalgae market. The growing investment in algae production provides a chance for the microalgae business to expand. Microalgae can be grown in highly controlled laboratory circumstances, but problems arise when they are grown on a big scale, which is a hurdle that could obstruct the microalgae market's growth (Data Bridge Market Research 2021).

11.2.4 Algae Production

Microalgae are usually produced and used in a variety of commercial purposes, using a variety of production methods. PBRs are sometimes combined with fermenters or open ponds in some production operations. PBRs are the most popular microalgae production system (71%), whereas fermenters and open ponds account for 19% and 10% of aggregated production units, accordingly (Figure 11.4) (Araújo et al. 2021).

Particular cultivation mechanisms have different characteristics that make them better suited to the creation of particular species for profitable purposes. The production method utilized affects factors such as construction and operational expenses, land requirements, technological advancement, environmental parameter control, and maintenance.

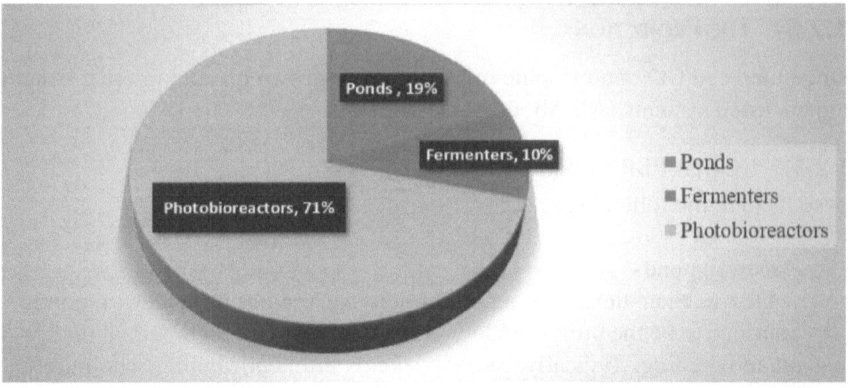

FIGURE 11.4 The percentage of production methods of microalgae (Araujo et al. 2021).

Photobioreactors, the most prevalent technique of microalgae cultivation in Europe as seen in Figure 11.4, have significant operational expenses and investment, as well as significant energy demands. However, depending on the type of PBR employed, there are significant differences within this technology. Closed PBR technology can be used to create a range of systems, ranging from green walls to tubular PBR systems made of plastic or glass with tubes deployed in parallel or perpendicular arrangements, each with its own set of leverages and flaws in terms of production stability and value. Closed PBR is a manufacturing process for cosmetics, nutraceuticals, and pharmaceuticals, which produces high-value, low-volume products. This technology improves the photosynthetic efficiency and productivity of manufacturing processes by allowing for tighter monitoring of environmental conditions and biomass quality.

Cultivation in open ponds is less expensive in terms of capital, operation, and energy, and it has the ability to make bigger biomass volumes. As a result, this form of biomass generation is more commonly used in low-value applications, albeit various technological obstacles must be overcome before its potential in algae biomass upscaling for purposes like biofuel can be realized. Open agriculture systems have the disadvantages of increased contamination risk, less control over environmental conditions, and higher water and land requirements. Furthermore, most European countries' climates are unsuitable for large-scale outdoor microalgae cultivation.

Fermenters are presently utilized by just a small percentage of European microalgae growers (Figure 11.4). These are a more recent advanced approach that enables for increased productivity and concentrations, contamination control, and biomass quality and profile alteration during the early stages of microalgae culture growth. Microalgae colonies are grown heterotrophically, with biological matter such as sugars serving as a source of carbon. Because the highly concentrated inoculum acquired could be employed to alleviate the limits of these systems, fermentation can be mixed with other approaches such as autotrophic methodology (open ponds or photobioreactors) (Araújo et al. 2021).

11.2.5 The Projections

Using enhanced CO_2, autotrophic microalgae are grown on land in open systems or in enclosed systems (All About Algae 2021).

11.2.5.1 Open PBR Systems

These include the following:

1. Raceway ponds

   Due to their flexibility and simplicity of scaling up, raceway ponds continue to be the utmost widely utilized reactors for commercial microalgae breeding. Typically, raceway ponds are kept running indefinitely by pouring fresh culture media just before the paddle wheel. Microalgal culture is harvested beneath the paddle wheel when the circulation loop is completed. Raceway pools are both economical by cost and energy due to their minor capital expenditure, minimum electricity consumption for blending, and ease of regulation of farming intensity and solvent movement. Microalgae strain versatility, land area requirement, reduced illumination efficiency, poor fuel permittivity due to restricted wet time between vapor and culture, considerable level of pollution, weak temperature regulation, and low biomass compactness, however, limit their effectiveness (Chang et al. 2017).

2. Circular ponds

   Round ponds are still one of the ancient pond shapes utilized for industrial microalgal cultivation, with the key benefit of employing abundant radiation as the microalgal blooming source of energy. These concrete ponds contain a spinning arm in the middle, which is then used to circulate the culture. They are wrapped with components like polymer sheets or neutral films as a secondary containment course of action. As the moving arm swings over larger distances, the culture can be mixed more thoroughly near the pond's boundary. As a result, round ponds are ideal for cultivating microalgae which are easy to deposit (Chang et al. 2017).

11.2.5.2 Closed PBR Systems

These include:

1. Tubular Photobioreactors

   Circular PBRs, such as stirred vessel, bubble tower, and airlift PBRs, are nowadays so often utilized enclosed mechanisms on an industrial hierarchy. They enable for more precise control of culture conditions and a reduction in contamination risk. Microalgal culture flows via solar collector tubes in tubular PBRs, where it is recycled by mechanical pumps or aeration. Furthermore, bubbling air at the bottom can provide adequate long-term blending and efficient elimination of surplus dispersed oxygen. In most circumstances, bubbling may be accomplished with air pumps in the plug movement route. Despite the potential benefits, scaling up tube-shaped PBRs for profitable mass production is difficult, and more research is vital to overpower all of the obstacles (Chang et al. 2017).

2. Flat-Plate Photobioreactors

   The most popular closed PBR is the flat-plate PBR, which has a cuboidal shape and a restricted light path. Flat-plate PBRs have a large lit enormous area-to- volume ratio as well as an open gas transmission area. The aforementioned two features make them ideal for bulk microalgae manufacturing in both indoor and in the open culture schemes. Although fouling is a significant disadvantage of these type of PBRs. The fouling happens when cells adhere to the polymer sides and the amount of light provided is reduced, increasing the chance of contamination. To expand this promising technique and minimize unnecessary coating or contamination, several units and bag replacements are required (Chang et al. 2017).

3. Internally Irradiated Photobioreactors

   The utilization of sunshine as the one and only source of light for the development of photosynthetic microphytes retains its own set of drawbacks, in the manner that diurnal and seasonal fluctuations in light intensity are caused by climate, period of the year, and geographic area. In most cases, this type of cultivation is only practicable in humid and warm climates and is restricted to the sweltering months in other regions. Incorporating an interior illumination appliance (for instance optical fibers or luminous bulbs) within the farming vessel can solve this problem. Solar light is collected, transmitted, and distributed more efficiently and uniformly inside PBRs using optical fibers, boosting standard brightness and the extent to which radiation enters inside the PBRs. The amplitude of solar and manmade light installations can also be adjusted using this type of PBR. Integrating solar and manufactured luminescence technology enables for a consistent supply of radiation to these reactors during inclement climate or during the dark. Internally lighted PBRs can also be heat uncontaminated under compulsion, reducing the danger of contaminants (Chang et al. 2017).

However, there are a few drawbacks of using optical fibers in PBR, including capital expenditures, cleaning difficulty, and light energy loss during transmission via optical fibers. As a result, determining its factual usefulness for massive use, employing PBRs with internal illumination via optical fiber remains a difficulty (Chang et al. 2017).

11.2.5.3 Future Perspectives

The effectiveness of PBRs in wastewater treatment has been demonstrated in laboratory and pilot-scale studies, and it is an up and coming development with many sensible implications. Industrial-scale PBRs, on the other hand, confront numerous hurdles. Because the developmental procedure is problematic to execute in a large-scale environment, modeling has been a key supporter of the development of PBRs. When there are more than six variables to evaluate, it is suggested that the modeling methodologies be used. For the design and construction of PBRs, however, there are some basic or widely accepted criteria. This method is still in the early stages of trial and error. The layout manuals for modernizing of flat plate and tubular PBRs have been recommended in a few additional areas. PBR upscaling has the potentiality to be a green and energy-economical technique. It is one that has piqued the scientific community's curiosity in its development and subsequent acceptance (Ngo et al. 2018).

11.3 CUSTOMIZATION OF BIOENERGY IN THE ALGAE INDUSTRY

11.3.1 Bioenergy

The world's rapidly rising population continues to raise global demand for fuel energy. Because of their unsustainable and nonrenewable character, the continuous use of fossil fuels around the world causes decline and will bring them close to exhaustion (Khan et al. 2018). Therefore, bioenergy needs to be introduced. Energy obtained from biofuels is referred to as bioenergy. Bioenergy is primarily used in residences (80%), with a smaller percentage used in industry (18%), while liquid biofuels for transportation still play a minor role (2%) as seen in Figure 11.5.

Biofuels are fuels created overtly or covertly from organic compounds, such as fauna and flora waste, which is referred to as biomass. Biofuels made from biomass have the advantages of being renewable and contributing much less to environmental damage and climate change (Khan et al. 2018). Biofuels can now be extracted from sources like wood, crops, and garbage thanks to more advanced and efficient conversion processes. Even while the term "biofuel" is frequently used in the literature in a restrictive sense to refer primarily to liquid biofuels for transportation, it can be solid, gaseous, or liquid (Greenfacts 2021).

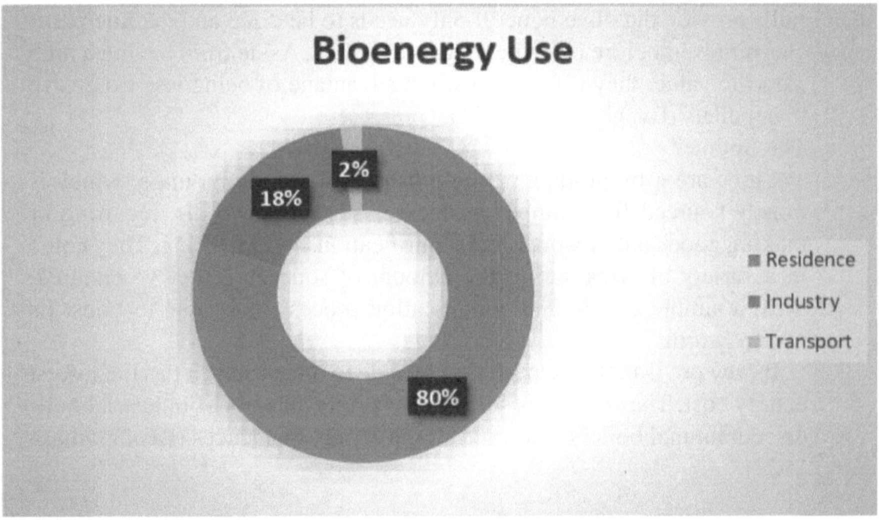

FIGURE 11.5 Percentage of bioenergy usage (Greenfacts 2021).

Biofuels can be divided into the following classes:

1. Solid Biofuels

   Solid biofuels are biological, undeposited materials of biological genesis (often referred to as biomass) that can be utilized as a source of heat or energy. Solid biofuels are defined as a product total consisting of fuelwood, charcoal, wood debris and outgrowth, bagasse, animal waste, ebony liquor, other vegetable elements and leftovers, and also the infinite proportion of industrial effluents, according to energy data (Eurostat Statistics Explained 2019).

   Given here are some of the most used solid biofuels:

- The wood pellets

  The pellet is indeed the most advanced form of biofuel. These are miniature tubes with a diameter of 6 to 8 mm and a length of 10 to 40 mm made by compressing hardwood dust. It is possible to create little cylinders without needing any additives because of lignin, a natural substance of wood that functions as an adhesive.

  One of the main advantages of using wooden pellets is that it has a high-energy content, consistency of composition, and homogeneity allowing for extremely high-energy yields. Furthermore, it also has a low ash content, hence making it easier to maintain and operate stoves and boilers (Expobiomasa 2021).
- The Olive Bone

  Its supply is determined by the annual harvest, and they are commonly employed since their size completely automates combustion. Olive oil mills provide the olive bone. It only needs to be dried and the fine particles removed before it can be used as a biofuel. Aside from having a high calorific value, they offer the distinct advantage of being less expensive than pellets (Expobiomasa 2021).
- The Splinter

  Chips are a by-product of the crushing of woody biomass, which is mostly sourced from timber businesses such as sawmills, repairing or making goods out of wood, forest purification, and trimming. They come in a variety of sizes due to the amount of squeezing they've endured. With a simple drying and classification process, it is able to boost its energy worth.

  Its key pro is that, when set side by side to other fuels, it has the lowest energy cost. The splinter is suitable for use in all sorts of industrial boilers, communal boilers, and even single-family residences (Expobiomasa 2021).

2. Liquid Biofuels

   All solvent biofuels of pure sources (e.g., derived from biomass or the environmentally friendly fraction of scrap material) that can be combined with or replaced by liquid biofuels of fossil source are classified as liquid biofuels. Liquid biofuels are a product total that includes biodiesels, biojet kerosene, biogasoline, and other liquid biofuels in energy statistics (Eurostat Statistics Explained 2019).

Ethanol and biodiesel are the most extensively utilized liquid biofuels for transportation. Ethanol is a form of alcohol that can be made from any feedstock that encompasses considerable amounts of sucrose or starch, such as sugarcane or sugar beet, or maize or wheat. While sugar can be fermented instantly to alcohol, starch must first be transformed to sugar. The fermentation process is similar to that of making wine or beer, and distillation is utilized to get pure ethanol (Greenfacts 2021).

Ethanol can be combined with gasoline or burned in practically pure form in spark-ignition engines with little modifications. A liter of ethanol incorporates around two-thirds of the energy that a liter of gasoline does. When blended with gasoline, however, it improves burning and scales down carbon monoxide and sulfur oxide emissions.

Biodiesel is prepared by blending vegetable oil or animal fat with an alcohol, which is mostly done in the European Union. Biodiesel can be combined with standard diesel fuel or used in compression ignition engines in its pure form. Its energy content is slightly lower than diesel's. Biodiesel can be made from a number of oils, such as soybean, coconut, rapeseed, jatropha or palm, and the resulting fuels can have a wider range of physical qualities than ethanol (Greenfacts 2021).

3. Gas Biofuels

Biogas is a vapor made up principally of carbon dioxide and methane that is created through thermal processes from biomass, such as waste biomass or anaerobic digestion of biomass. Biogas is an output accumulate equivalent to the sum of sewage sludge gas, landfill gas, other anaerobic digestion biogases, and biogases from thermal processes in energy statistics (Eurostat Statistics Explained 2019).

In place of coal or natural gas, reserved biogas is able to cater to a pure, inexhaustible, and lasting source of minimum power. Renewable baseload electricity can supplement more intermittent renewables. Baseload electricity is constantly produced to fulfill minimal power demands. Biogas, like gasoline, can be utilized as a source of peak energy that can be levitated up quickly. Using saved biogas reduces reliance on fossil fuels and minimizes the quantity of methane discharged into the atmosphere (Tanigawa 2017).

Anaerobic digestion, in addition to its environmental benefits, can reduce waste cleanup costs and improve local economies. Anaerobic digestion of cattle waste also minimizes stench, pathogens, and the prospect of water dirtiness. Digestate, the waste product of digestion, can be utilized or sold as manure, eliminating the demand for artificial fertilizers. The same waste product from digestion can also be sold as livestock feed or green manure to generate additional cash (Tanigawa 2017).

11.4 ALGAE CULTIVATION IN IR 4.0

11.4.1 Tools for Alternating Cultivation Mode

The two most common methods of microalgae culture are open cultivation facilities, such as open ponds, cylinders, and raceway basins, and controlled enclosed cultivating technologies, which use various forms of bioreactors. Minimal capital and running expenditures, together with a decreased energy required for culture infusion, are some of the major benefits of an open cultivation method. Open systems, on the other hand, expect a huge area to scale up and are unguarded from contamination (e.g., from birds) and bad climate. Controlling growth specifications, for instance, evaporation, culture condition in terms of temperature, and so on is still challenging in open pond systems (Narala et al. 2016).

Closed cultivation operations, also known as closed PBRs, are also productive in regard to their condition since they could be maintained under more precise circumstances and so overcome the drawbacks of open culture systems. PBRs can be built and tuned to fit the strain of your choice. This enclosed system takes up little capacity while enhancing radiation availability and significantly reducing contamination concerns. However, biofouling, heating too much, marine algae development, cleansing challenges, and a large aggregation of dissipated oxygen that results to growth limits, as well as very high capital expenses for design and operation, are some of the drawbacks of PBRs (Narala et al. 2016).

Cultivation systems fluctuate in design and principle depending on the needs. A wastewater treatment plant's open ponds might be circular or gravity driven. Similarly, during the last decade, PBRs' basic cylindrical shape has been enhanced to allow for better light availability and culture mixing, allowing for the generation of a collection of medicinal and large value nutritional products (Narala et al. 2016).

11.4.2 Multipurpose and Specific Medium Cultivation

Microalgae could be produced in a range of approaches and under a variety of environments. They require light as an energy source for photosynthesis, which converts ingested water and carbon dioxide into biomass. Photosynthetic products can make up 20–50% of overall biomass in various designs, such as storage goods or cell segments. Algae depend upon important supplements like phosphorus and nitrogen, which contribute for 10–20 percent of algae biomass. The nutrients required in large quantities sodium, magnesium, calcium, and potassium, along with microelements like molybdenum, manganese, boron, cobalt, iron, and zinc, as well as other trace elements, are all necessary for growth. The minerals essential for microalgae cultivation can be found in wastewater. Organic effluents from the agriculture and food industries can thus be used to feed microalgae (Khan et al. 2018).

11.4.3 Cultivation Condition and Adaptation of Algae

Given here are some of the main conditions required for algae cultivation.

1. Light

   In microalgae cultivation, intensity of light is among the most critical main drawbacks. Microalgae photosynthesis is directly affected by light intensity and duration, which also has an impact on the biochemical design of microalgae and biomass turnout. Light intensities fluctuate within the cultivation and decrease as culture intensity increases; this will be considered when modeling an open pond system or bioreactor (Khan et al. 2018).

   Microalgae cannot develop at very little or very big light intensity. Net growth is zero at the compensation mark, when photosynthetic carbon dioxide intake precisely matches respiratory carbon dioxide emission. Higher light intensities will boost the rate of photosynthesis to a maximum, after which it will level off until photorespiration and photoinhibition balance the rate of photosynthesis. As a result, the appropriate light intensity in each situation must be found experimentally in order to maximize carbon dioxide assimilation while minimizing photorespiration and photoinhibition. Algal photosynthesis necessitates a specified duration of light/dark intervals. The dark processes of photosynthesis that create carbon framework require light for NADPH (nicotinamide adenine dinucleotide phosphate) and ATP (adenosine triphosphate) synthesis (Khan et al. 2018).

   To avoid photo-oxidation and growth limitation, efficient light power and period are required in bioreactors for microalgae. To bypass photoinhibition, also known as the self-shading phenomenon, when algae in lower rows are shadowed from the radiation by top rows, proper light penetration and homogeneous distribution are also required. Even though fluorescent tubes can also be utilized, LED lights are an excellent alternative for this purpose (Khan et al. 2018).
2. Temperature

   Temperature is also an added key component in the cultivation of microalgae, as it has a direct impact on biochemical mechanisms in the algal unit factories, as well as photosynthesis. Respective species would have their own ideal temperature for increase in mass and size. Rise in temperature to the most favorable range aggressively boosts algal expansion, but increases or decreases in temperature apart from the ideal level slow or stop algal expansion and motion. Microalgae cultures grown at non-favorable temperatures lose a lot of biomass, especially in outdoor culture systems. Temperature is a crucial element in large-scale production, particularly in exposed-pond culture, and it requires close supervision because algae endure large temperature changes over time (Khan et al. 2018).

Cooler temperatures diminish photosynthesis by inhibiting carbon consumption, while intense heat inhibits photosynthesis by interrupting photosynthetic proteins and disturbing the cell's energy harmony. Cell size and respiration both shrink when the temperature rises. The rate of photosynthesis declines, resulting in a slower rate of growth. The main response of temperature on photosynthesis is a decrease in the movement of the dual-function stimulant ribulose-1,5-bisphosphate (Rubisco). Based on the proportional levels of oxygen and carbon dioxide in the chloroplasts, it is capable to serve as an oxygenase otherwise as a carboxylase. Rubisco-stimulant carbon dioxide fixation activity increases with increasing temperature up to a point, then decreases. As a result of its outcome on the attraction of ribulose for carbon dioxide, temperature is a restraining element for algal rate of growth and biomass output (Khan et al. 2018).

3. Mixing

   In microalgae production, mixing and oxygenating ensure homogeneous dispersion of air, carbon dioxide, and nutrients. They also allow for the seepage and even dispersion of rays throughout the culture, as well as the prevention of biomass settling and accumulating. Assuming that all other parameters are adhered to but no blending occurs, biomass yield will be drastically reduced. As a result, microalgae strains have to be constantly stirred to retain all cells suspended and exposed to light. In a photo-bioreactor, a correct mixing system not only allows for nutrient dispersion and light seepage into the culture, but it also allows for effective gaseous swap (Khan et al. 2018).

4. Mediums

   When culturing microalgae, it is very necessary to ensure the right medium is used as there are various media such as BG11 and Zarouk that can be used to grow the algae.

   For example, according to Al-Rikabey and Al-Mayah (2018), they were investigating the growth of the freshwater micro-algae; *Chlorella vulgaris* (*C. vulgaris*) using the Taguchi method in a photobioreactor. They had modified the medium slightly and observed the effects of temperature, cultivation time, pH, phosphorus concentration, nitrogen concentration, and sulfur concentration on *C. vulgaris* cultivation.

   Only certain solutions of the modified BG11 medium were able to produce significant results, showing the importance of growth medium being specific to various algae.

5. Carbon source

   Carbon sources are likely important significant factors in microalgae growth. Microalgae can be cultivated photoautotrophically, heterotrophically, or mixotrophically using a variety of carbon sources, including carbon dioxide, methanol, acetate, sugar, or other biological molecules (Yen et al. 2014).

Some microalgae could readily use organic carbon as a source of carbon in the availability or deficiency of light. Heterotrophic cultivation is the term for this type of cultivation. Other species are grown under photoautotrophic conditions, which implies that microalgae use nonbiological carbon (such as carbon dioxide or bicarbonates) as a source of carbon for photosynthesis (Yen et al. 2014).

However, because biological carbon sources would be too pricey for creating low-cost goods like biofuels, carbon dioxide or bicarbonates are still the most often employed carbon sources for microalgae expansion and biofuels manufacturing (Yen et al. 2014).

11.5 ALGAE GENETIC ENGINEERING

11.5.1 Genetic Modification for Desired Strain

Genetic change is the process of introducing a segment of DNA from one organism to another to change the characteristics of a crop, living creature, or microbe. This is performed by precisely extracting certain genes from the DNA of one creature and transplanting them to the DNA of the other. For example, this approach has been utilized to develop medicine-producing fungi and bacteria (Lotz and Smulders 2021).

The basic goal of genetic modification is to add desired qualities to living organisms, such as resistance to disease or resistance to drought. This can also be accomplished through crossing species. However, interbreeding might result in undesirable traits. The procedure of interbreeding to get rid of such undesired traits takes decades (Lotz and Smulders 2021).

Genetic engineering would hasten the process by generating strains required to provide the items we require, allowing us to avoid chopping down our last tropical rain forests or removing all marine fish from the sea (Cimons 2017).

In the recent decade, significant progress in microalgal genetics has been made. Expressed sequence tag (EST) indexes have come about to be constructed, and nuclear, mitochondrial, and chloroplast complete set of genetic information from many microalgae have been mapped. Most molecular and genetic phycological research has previously focused on the green alga *Chlamydomonas reinhardtii*. As a result, the majority of the techniques for transgenic expression and gene knockdown have been created specifically for this species. However, methods concerning "living opals" and other algae, which are of leading importance for industrial utilization, are fast being developed (Radakovits et al. 2010).

Green (Chlorophyta), red (Rhodophyta), and brown (Phaeophyta) algae, "living opals", euglenophytes, and marine protozoa consisting of two flagella have all been successfully genetically transformed. Over 30 different microalgae variants have already been successfully transformed to date (Radakovits et al. 2010).

Genetic alteration as a method for improving algal effectiveness is being more recognized as a requirement for developing novel and economically viable production systems (Organisation for Economic Cooperation and Development 2015b). There are three categories of objectives for genetic manipulation of algae:

1. Photosynthetic efficiency is being improved

   The efficiency of algal biofuel generation is directly proportional to the system's stellar photon absorption and transformation efficiency. Nonetheless, because sunshine intensity is frequently above algae's peak photosynthetic efficiency, growth is slowed, a condition known as photo inhibition (Organisation for Economic Cooperation and Development 2015b).

   According to Yang et al. (2017), an experiment was carried out to see the possibility of improving the photosynthetic capacity of the freshwater microalgae *C. vulgaris* by genetic manipulation of the Calvin cycle. They were able to conclude that by modifying the Calvin cycle in green algae, it is feasible to enhance its photosynthetic capacity.

2. Productivity of chosen products is being improved

   The growing retail demand for dyes derived from raw sources has prompted manufacturers to develop new products based on the growth of microalgae on a big scale for the production of such substances. Gene-encoding enzymes straightforwardly correlated with specific carotenoid syntheses have been studied, and also innovation of conversion approaches will allow for a dramatic rise in tetraterpenoids cellular content and thus subsidize to increased volumetric capacities of the related procedures (Organisation for Economic Cooperation and Development 2015b).

3. New products

   The establishment of genes or metabolic channels into algae to provide segments of commercial relevance that are still not currently in the wild type is a new subject in algae biotechnology (Organisation for Economic Cooperation and Development 2015b).

   Significant biosafety problems should be resolved before novel genetically engineered organisms (GEOs) are released into the environment, in order to assess potential levels of exposure and unexpected sources of harm. The idea of cultivating innovative and extremely resilient kinds of genetically engineered (GE) microalgae near natural surface waters poses concerns similar to those brought by GE crops and farm-grown GE fish, as well as non-GEOs with the plausibility to develop into being meddlesome. How often might GE algae flee from production and processing amenities, for example, spraying, wildlife carriers, stormy weather that weakens or destroys these resources, mishaps, human miscalculation, or other occurrences could all contribute to this (Snow and Smith 2012).

   Theoretically, if a disadvantageous scenario occurs, fugitive GE algae could persist and produce poisons, or they could grow so prevalent that dangerous algal blooms may occur. It would be concerning if free-living GE algae might develop into being more invasive, poisonous, or enduring of harsh abiotic environments than their natural counterparts. Most importantly, we need to know if there are any feasible situations in which GE "superalgae" or other creatures that gain specific genes from them could spread to the point of causing harm to individuals or ecological health. The perseverance of free-living GE algae is unlikely to have negative implications in many circumstances, but the scientific basis for this decision must be convincing and precise (Snow and Smith 2012).

11.5.2 Automated Genetic Manipulation, Estimation, and Computing of Proper Strain Detection

Microalgae are becoming more popular in study and industry. Microalgae cultivation as a source of energy has appeared as a viable backup to nonrenewable fuel source and biofuels. Microalgae biomass has sparked a lot of attention in recent years; however cultivating microalgae is a challenging task. The necessity to minimize infection, the availability of electricity, and the growing conditions all pose difficulties. Immense light intensities, for example, can cause light-induced reduction in photosynthetic capability, while low levels are likely to hinder photosynthesis and other cellular functions. Closed PBRs presently produce only a few hundred tons of products. The lack of sufficient PBRs to provide optimal microalgae growing conditions is one of the remarkably significant barriers to a long-term economic utilization of algal cultures (Dormido et al. 2014).

Process sensing is an important and crucial part of the PBR technology. To accurately reflect the process dynamics and to conduct relevant measures to improve process performance, industrialized PBR systems require reliable recording of many culture criterions. Many advanced computerized sensors for detecting a number of culture and operation criteria (such as survival, metabolic processes, biomass, and supplement concentrations) are accessible in addition to the basic sensors for temperature, dissipated oxygen, and neutrality. The majority of these sensors, however, are constrained by their high price, lack of everlasting stability, or both. Not to mention, the PBR system would be excessively pricey not just in terms of construction and design, but also in upkeeping, if the necessary variety of electronic hardware sensors were installed (Dormido et al. 2014).

These concerns point to the need for better simulation tools to allow researchers to test innovative approaches for increasing productivity. These tools would allow researchers to delve deeper into the layout and oversee ideal PBRs in order to facilitate manufacturing under supervised and profitably competitive settings (Dormido et al. 2014).

11.5.2.1 Examples of Computer-Controlled Cultivation of Microalgae

A bioreactor arrangement for the development of the microalgae *Synechocystis* sp. PCC6803 under regulated physiological conditions was established by Marxen et al. (2005). Inline recording of chlorophyll fluorescence characteristics was used to determine the microalgae's current physiological status. To retain the microalgae in a predetermined physiological state, a feed-back loop was used. The temporal behavior of the system was explored using variations in light conditions (as generated by modulated ultraviolet B—UVB radiation) as an input signal and chlorophyll fluorescence as an output signal for the design of this feed-back loop. The responses have a good level of repeatability. In the UVB-induced responses, kinetic analysis based on curve fitting showed two-time constants. The data of these time constants was used to create an effective feed-back loop that allows the microalgae to be grown in a specified physiological condition.

A closed integrative system (CIS) made up of lettuce, silkworms, and microalgae was built as a specific prototype of a bioregenerative life support system (BLSS) by Hu et al. (2012), with the goal of studying the gases' dynamics in the system and their closed-loop regulation and control using microalgae as a bioregenerative tool and computer simulation. In terms of biological unit components, BLSS is an artificial ecosystem that provides life support for crew members on space or terrestrial missions. Maintaining gas (oxygen—O_2 and CO_2) contents in the system to robustly stabilize at nominal levels is vital.

First, an accurate CIS kinetic model was created using system dynamics and an artificial neural network based on important ecological principles and experimental data to fully describe the dynamic properties of gas concentrations. Then, based on microalgae peculiarities such as high growth rate, metabolism flexibility, controllability, and so on, a closed-loop CIS with Linear-Quadratic Gaussian (LQG) servo controller was developed. Furthermore, the closed-loop CIS was optimized using digital modeling and preset gas dynamic reactions to control inputs. At the end, real-time simulation was used to fully evaluate and certify the effectiveness of the closed-loop CIS.

Based on actual measurements of gas concentrations, the closed-loop CIS could effectively regulate the intensity of light and aerating rate to stimulate or prevent microalgae growth and indirectly influence them to return to their nominal levels with desired dynamic response performances after deviation from originally equilibrium position.

11.6 ALGAE INDUSTRY 4.0

11.6.1 The New Markets

The usage of algae as a food root by primitive groups dates back for millennium of years. Since ancient times, countless breed of green algae have been used equally for sustenance. Microalgae cultivation began merely several decades before, when it turned out to be evident that the earth's rapidly rising residents would face a shortage of protein-rich foods. Foodstuffs and other vital bio-derived products, such as unprocessed antibiotics, are abundant in microalgae. As stated to one valuation, processing 5,000 metric tons of dehydrated algal biomass for bio-based products annually earns US\$ 1.25 × 10^9(Khan et al. 2018). Microalgae-derived polyunsaturated fatty acids (PUFAs) are also regarded high-value commercial items, with an estimated market value of \$140 USD/kg (Barkia et al. 2019).

Microalgae also create a variety of other commercially significant and important goods. They generate vitamins, which increase their value as a source of nutrition for mankind and animals. They also produce a variety of polysaccharides that are useful in medicine. Chlorophyll, beta-carotene, and other carotenoids, water-soluble proteins, and lutein are all beneficial and commercially relevant pigments produced by various species. These pigments are important in cancer, neurological disorders, and optical illnesses treatments. Protein is also abundant in microalgae. Their ability to produce critical amino acids boosts their aptitude as polypeptide-rich diets. Glycogen, roughage, hemicelluloses, and other carbohydrates are produced by microalgae from simple monomeric sugars, primarily glucose. Algal cells are an essential food source due to their greater glucose content. Microalgae also create and store huge volumes of phospholipids, which differ by categories and are influenced by a variety of circumstances. Glycerol, esterified sugars, and various forms of fatty acids are the most common lipids found in algal cells. Algal fatty acids are used in both nutrition and medicine. The majority of the chemicals generated by microalgae have medicinal properties. Microalgae are becoming more cost-effective sources of natural ingredients for food and cosmetics (Khan et al. 2018). Algae production has the potential to contribute significantly to future generations' food security, particularly in areas where available cropland is anticipated to be insufficient owing to population expansion (Ullmann and Grimm 2021).

Narcotics on the market primarily embody capsules or solvent forms of well-being-promoting items, but a recent trend in the market is the availability of different microalgae species as a supplement of various active compounds in extract form. The need for favorable algal food and well-being products is driving the growth of the microalgae industry (Khan et al. 2018).

The search for antibiotics and pharmacologically active chemicals in microalgae, particularly cyanobacteria (blue–green algae), has gotten a lot of attention in the previous decade. A huge number of antibiotic compounds have been identified and described many of which have new structures. Umpteen blue–green algae have also been discovered to give rise to drugs that eradicate viruses and anticancer chemicals. Microalgae extracts have been found to exhibit a variety of pharmacological properties. Several bioactive chemicals could be used in human or veterinary therapy, as well as farming. Others may be useful as investigation means or structural prototypes in the creation of novel medications. Microalgae are exceptionally appealing as innate sources of biologically active chemicals because they can synthesize these compounds in the cultivations, allowing for the generation of architectural complicated fragments that are nearly unimaginable to manufacture by chemical integration (Borowitzka 1995).

Microalgae pigments are up and coming natural sources of highly significant chemicals. Vitamin prototype, antioxidants, immunological boosters, and anti-inflammatory substances are among the health-promoting qualities of these dyes, which incorporate carotenoids, green photosynthetic pigment, and phycobiliproteins. Microalgae pigments, as a result, could strike gold if being used commercially as new functional component in the foodstuff and fodder industries, in addition to cosmetics and pharmaceuticals (Christaki et al. 2015).

11.6.2 Algae Revolutionization in IR 4.0

The unprecedented increase in GHG emissions is expected to result in accelerated changes in the environment, such as an overall surge in worldwide temperatures, also extreme weather surroundings, and decreased freshwater accessibility, especially in regions where fresh water is already a valuable resource (Organisation for Economic Cooperation and Development 2015a).

Global economies are being strained not just by GHG-effected climate variation scenarios, but also by the expectation of approaching or exceeding highest point of fuels and phosphorus in the near future (plausibly in the coming 15 years), which will have a detrimental impact on agriculture and industries. It is feasible that significant newly deposited oil retained can be discovered at greater extents; but the condition of these commonly named heavy lubricants is lower, as the lubricant is more thick and contains more sulfur, necessitating further refining work. Oil prices will rise as a result of these initiatives. Peak oil has an impact on agriculture, as farm machinery and insecticides are both oil-based products. Pesticide use has resulted in stable food supplies, which is directly related to population increase. In terms of peak phosphorus, projected community expansion, and limited cultivable farmland for foodstuff production, none of these factors is expected to change significantly or in unison with projected population expansion. Agriculture and aquaculture businesses will have challenges in meeting future nutritional and food supply requirements as freshwater resources become scarcer, weather conditions become more variable, and temperatures rise (Organisation for Economic Cooperation and Development 2015a).

Algae and oxygenic photosynthetic cyanobacteria are perfect fixes to the aforementioned impending difficulties since they may be grown during the whole year on non-cultivable land in a variety of effluent streams or saline to coastal waters, reducing the strain on cultivable land and lake water resources (Organisation for Economic Cooperation and Development 2015a).

When combined with AI-powered bioreactors, algae can remove CO_2 from the atmosphere 400 times more efficiently than a tree. When used effectively, it has the potential to turn a large town carbon negative without affecting its prevailing manufacturing or utilization patterns (Lamm 2019).

Carbon dioxide is naturally absorbed by trees and algae. By "absorbing" carbon into their stalks and roots and discharging oxygen back into the atmosphere, trees "consume" it as a segment of their photosynthetic process. Algae mimic the similar process, except they "absorb" the carbon in the manner of more algae. Due to their corresponding size, algae can absorb more carbon dioxide than trees as they can envelope more surface area, grow instantaneously, and can be more easily supervised by bioreactors (Lamm 2019).

Currently, the algae can be grown in various ecosystems and have the ability to produce biofuels along with acting as an efficient carbon capture agent and, hence, protecting the environment. It can therefore be seen that with the use of algae, the industry is bending in the right direction and with the aid of research in the science and engineering field, the goal can be achieved.

11.6.3 Environmental Impacts

There is indeed a pressing need to develop alternative sources of nutrients, chemical products, and electricity in the planet whereby natural materials are transpired to be uprooted and utilized at an accelerating pace. Microalgae have acquired popularity as a result of their rapid growth, adjustability to their surroundings, and capacity to accumulate essential compounds and gather supplements in a cost-effective manner. There are reasonable and unfavorable environmental implications affiliated with the production of large-scale resources using microalgae, and some of them include:

1. Contamination and Leaks

   Reactors come in a variety of designs. Large-scale farming is possible with open ponds at a reduced cost. They are, nevertheless, susceptible to contamination due to their open form. This uncertainty can be reduced by changing culture circumstances to make them unsuitable for native groups. Basins that aren't properly devised or established may constitute an immediate environmental concern due to the percolation of basin contents towards the earth. Salinization in instances where sea algae are grown on terrain or the removal of toxic substances when microalgae are employed as an effluent treatment tool are two examples. While the ponds' contents are unlikely to be hazardous, contamination of ground water is a possibility (Usher et al. 2014).

   PBRs are see-through vessels that permit light to pass through to the microalgae. Because PBRs are enclosed, they are less prone to pollution and poisoning. A leak from these containers, depending on the amount, might have a considerable repercussion, such as if they are positioned close to a natural supply of water. It would, however, be theoretically obvious to identify and, as a result, simpler to correct (Usher et al. 2014).
2. Water Footprint (WF)

   A water footprint is the entire aggregate of stream water used in the manufacturing of commodities and facilities, including surface and aquifer (blue water footprint) as well as water that has precipitated (green water footprint). WF is strongly dependent on rate of evaporation, hydraulic holding time, and photosynthetic performance, all of which are affected by weather, process design, and cell biology (Usher et al. 2014).

   Microalgae biofuels have a lower WF than other biofuels like soya or palm biodiesel, or sugarcane bioethanol, in a confined photobioreactor for biofuel production as exhibited in Table 11.1. The collection represents numbers ranging from effluent and salty water (lowest values) to

TABLE 11.1
Water Footprint of Various Transportation Fuels (Usher et al. 2014)

	Mean yearly water footprint (m^3/GJ)
Microalgae biodiesel (exposed raceway)	14–87
Microalgae biodiesel (enclosed bioreactor)	1–2
Soybean biodiesel	287
Natural gas	0.11
Sugarcane ethanol	85–139
Petroleum diesel	0.04–0.08

freshwater (highest values). This demonstrates that the effluent is critical to the long-term viability of microalgae-based biofuels, both environmentally and economically, in terms of pure water use and supplement supplies (Usher et al. 2014).

3. Consequences to aquatic biodiversity

   Microalgae mass farming is referred to as a "controlled eutrophication process," and, in itself, it must be well-controlled through proper air source and consistent collection. Eutrophication, on the other hand, continues to be one of the most serious threats to diverseness. Decay of deceased algal organic material depletes oxygen in the aqua line, causing species that rely on oxygen for respiration to die. Eutrophication has a number of consequences, including a decrease in biodiversity owing to hypoxia (low oxygen in the tissues of the species), water toxicity, and turbidity. In the anaerobic layers, methane generation can occur, resulting in pungent ejection (e.g., hydrogen sulfide) and greenhouse vapors (e.g., CH_4, CO_2, N_2O) accompanied by a high climate emergency capability. Any organisms that rely on oxygenated waters may become extinct and be replenished by powerful species (Usher et al. 2014).

   Unintentional discharge of water from agriculture areas into the wider surroundings could result in larger-scale eutrophication events, especially if farming takes place nearby a significant water basin like a lake or a coastline region. The magnitude of the repercussion is determined by the amount of the discharge and the quality of the collecting water basin. Supplement-rich aquatic waters, for example, might diminish seagrass populations, which are important for silt stabilization and contributing habitat and foodstuff for a variety of aquatic animals (Usher et al. 2014).

   Contamination is a risk in open ponds. This hazard can be reduced by changing strain circumstance to make them unsuitable for native groups. However, releasing non-native groups may cause complications, especially if they outcompete native species. Large amounts of water introduced to ordinarily dry locations may cause local climate changes in some situations. Increased evaporation rates would modify the temperature and relative humidity in these areas, as well as the biodiversity, inviting animals and feathered creatures for slurping water and providing nesting places for insects and other aquatic creatures. Therefore, it is critical that cultivation arrangements are effectively sustained and handled in either of the cases discussed earlier (Usher et al. 2014).

11.7 FULFILLING SUSTAINABLE DEVELOPMENT GOALS

The Sustainable Development Goals (SDGs), also known as the Global Goals, were recognized by the United Nations in 2015 as a worldwide positive message to eliminate poverty, preserve the environment, and ensure that almost everyone resides in unity and success by 2030 (UNDP 2021).

The 17 Sustainable Development Goals (SDGs) are interrelated, acknowledging that efforts in one category will have an impact on the success in others and that progress must find the right balance between communal, commercial, and environmental responsibility (UNDP 2021). The SDGs incorporate, where the microalgae's contribution towards these SDGs are summarized in Table 11.2.

1. Reduced hardship
2. Zero hunger
3. Proper well-being and welfare
4. Excellent schooling
5. Sexual role fairness
6. Clean water and sanitation
7. Affordable and clean energy
8. Trustworthy labor and profit-making improvement
9. Industry, modernization, and economic development
10. Reduced discrimination
11. Sustainable cities and communities
12. Efficient expenditure and manufacturing
13. Undertaking measure to improve climate
14. Survival beneath water
15. Survival above motherland
16. Truce, law, and strong institutes
17. Cooperation especially for the objectives

11.8 CHALLENGES AND PERSPECTIVES

Despite all the advantages that are there, there are some challenges with using microalgae. Given in the next section is the main challenge of using microalgae.

11.8.1 Large-Scale Cultivation

The majority of industrial microalgae production takes place in relatively simple, low-product artificial ponds, most of which are circular or 'raceway' ponds.

TABLE 11.2
Microalgae's Contribution in Sustainable Development Goals, SDGs

SDG	Description
Clean water and sanitation	Is one of the goals that can be achieved by Microalgae 4.0. Organic matter, nutrients (particularly nitrogen and phosphorus, which are the principal drivers of eutrophication in water bodies), and some hazardous pollutants and pathogens have all been removed using a microalgae–bacteria symbiosis. The grade of valuable items obtained from harvested biomass is directly connected to its composition. In order to achieve sustainable progress, we must create effective resource recovery technologies in light of global population increase, increasing consumption, and limited natural resources. Algae-based solutions are low-cost and high-efficiency choices for treating wastewater and producing usable products (Arashiro 2016).
Affordable and clean energy	Along with clean water, with Microalgae 4.0, clean and affordable energy can also be produced. According to Amir and Singh (2018), microalgae have been discovered to be a very promising and long-term feedstock for the manufacturing of biofuels. The chemical structure of algal biomass feedstock can be utilized to develop several forms of biofuels. Biodiesel, bioethanol, jet fuels, bio methanol, biobutanol, bio hydrogen, and products from thermochemical conversion such as bio-oil, syngas, and bio crude are examples of biofuels.
Zero hunger	With the rising worldwide population, achieving proper nutrition is becoming a growing global challenge. As a result, rewarding sources of supplements that can easily and quickly create vast quantities of high-nutrient-value goods are required. Algae can be a good origin of a variety of critical nutrients that are good for human health. Microalgae were being used as a man's food source and dietary supplement for centuries and centuries. A range of technologies and methodologies for mass-controlled culture of microalgae have been proposed for commercial application, in addition to harvesting from natural habitats. For instance, the United Nations World Food Conference of 1974 declared Spirulina to be the best food for the modern world, and the World Health Organization (WHO) stated that Spirulina is an exotic food for a variety of reasons, including its high iron and polypeptide content and ability to be given to minors without any risk of being hazardous. Microalgae can thus be used as an alternative to produce pure food with a low environmental significance, as they can be grown on non-cultivable ground (García et al. 2017).
Sustainable cities and communities	Microalgae growth systems might be especially useful in urban areas, where the pressing need to meet rising material and energy demands is posing a challenge to pollution control and waste generation. Because the built environment is predicted to generate 40 to 50 percent of global GHG emissions due to fossil fuel usage, reducing air pollution in cities has a significant potential for mitigating global warming. While most current attempts to improve urban environmental sustainability have centered on solutions like wind turbines, photovoltaics, and geothermal, microalgae production systems and their integration into urban places are gaining popularity. Microalgae can play a critical role in reducing GHG emissions, increasing energy efficiency, and partially substituting fossil fuel consumption in metropolitan settings. However, microalgae are introduced into urban environments not only through buildings, but also through urban canopies, sidewalks, fountains, parks, and other public and private spaces that emit high volumes of CO_2. This is a new and unique technique to improve the sustainability of cities while also expanding the use of microalgae technology (Merlo et al. 2021).

These are water-filled man-made structures that are subjected to circulation and mixing. Solely a few species of microalgae, for instance *Chlorella*, *Spirulina*, and *Dunaliella*, have proved to be successful in long-term open pond cultivation. Open ponds are generally inexpensive to build and operate and so offer numerous benefits as long as the cultivable species can be maintained (Rosch et al. 2009).

Despite the benefits of open systems, future advances in microalgae cultivation may demand closed systems as not all microalgae of relevance survive in extremely selective environments. PBRs were created to address this issue. Their hefty costs, on the other hand, have mainly prevented their commercial application until lately. PBRs, however, enable precise control of not only lighting but also temperature and other production factors. Furthermore, contamination is less likely, allowing for the production of more diverse and productive species.

Even if bioreactors with internal illumination can reach cell densities of up to 20 g/L, compared to less than 1 g/L in open ponds, making harvesting easier and increasing overall system efficiency, large-tech PBRs, on the other hand, may be difficult to scale up, resulting in a significant demand for auxiliary energy as well as high investment and operating expenses due to the energy source (Rosch et al. 2009).

Scaling up of microalgae would require a large area. For instance, scaling up the process of producing biofuel from microalgae is required for commercialization. Scaling up would necessitate a lot more area and fertilizers. Phosphorus fertilizer, which is required for microalgae, has negative effects on humans (Katiyar et al. 2017). However, these discussed problems can be resolved by using a PBR.

11.9 CONCLUSIONS

To conclude, with the help of the technology used in industry 4.0, microalgae can be exploited in a more efficient manner as their requirements can be met. Furthermore, it is seen that with a fourth industrial revolution, all the drawbacks from previous revolutions are resolved. Additionally, we saw the necessity of moving from the third to fourth revolution which includes: improvement in productivity and efficiency as well as better products and services. A prospective cost analysis was also done to see that products from microalgae have a higher compound annual growth rate as compared to other unsuitable resources. The methods by which microalgae are produced—ponds, photobioreactors, and fermenters are the most popular ones. Also, it is foreseen that in the future, it might be possible to upscale microalgae production using modelling. Furthermore, due to increasing population, it was agreed upon that bioenergy is a sustainable resource. Therefore, solid, liquid, and gas biofuels can be utilized to serve the purpose.

Additionally, the main tool to cultivate microalgae in altering conditions is the photobioreactor, and it could be open such as raceway ponds or closed as the closed photobioreactor. The light, temperature, aeration, type of medium, and carbon source are as equally important as when choosing the type of photobioreactor to cultivate the microalgae. In addition to growing condition, genetically modified microalgae would be more efficient as the performance of the algae is enhanced such as it is more efficient in photosynthesis or even, they are modified to be used in producing substances that have a large market. As a result of genetic engineering, process sensing has also become another factor which will help enhance the efficiency of the microalgae.

Furthermore, new markets such as foods, antibiotics, and pigments from microalgae are emerging. And we also see how microalgae with artificial intelligence can reduce the greenhouse gas emissions. Moreover, with using microalgae, there is a reduction in environmental impacts such as contamination and water footprints. In addition to reduce environmental impacts, cultivation of microalgae and using them in producing goods help achieve certain sustainable development goals such as providing sustainable and clean energy as well as creating sustainable cities and communities. At the end, we deduced that the only major challenge of cultivating microalgae would be scaling it up as it would require more land and fertilizer.

REFERENCES

All About Algae. 2021. "Algae basics." Accessed 13 July 2021. http://allaboutalgae.com/algae-cultivation/.

Al-Rikabey, Muna N., and Ameel Mohammed Al-Mayah. 2018. "Cultivation of chlorella vulgaris in BG-11 media using Taguchi method." *Journal of Advanced Research in Dynamical and Control Systems* 10 (7): p. 19–30.

Amir, Abuzer, and Shivani Singh. 2018. "Microalgae as promising and renewable energy source: A review." *Journal of Fundamentals of Renewable Energy and Applications* 8 (4). doi: 10.4172/2090-4541.1000266.

Arashiro, Larissa. 2016. *Microalgae as a Sustainable Alternative for Wastewater Treatment*. International Water Association. Accessed 30 June 2021. https://iwa-network.org/microalgae-sustainable-alternative-wastewater-treatment/.

Araújo, Rita, Fatima Vázquez Calderón, Javier Sánchez López, Isabel Costa Azevedo, Annette Bruhn, Silvia Fluch, Manuel Garcia Tasende, Fatemeh Ghaderiardakani, Tanel Ilmjärv, Martial Laurans, Micheal Mac Monagail, Silvio Mangini, César Peteiro, Céline Rebours, Tryggvi Stefansson, and Jörg Ullmann. 2021. "Current status of the algae production industry in Europe: An emerging sector of the blue bioeconomy." *Frontiers in Marine Science*. doi: 10.3389/fmars.2020.626389.

Barkia, Ines, Nazamid Saari, and Schonna R. Manning. 2019. "Microalgae for high-value products towards human health and nutrition." *Marine drugs* 17. doi: 10.3390/md17050304.

Borowitzka, Michael A. 1995. "Microalgae as sources of pharmaceuticals and other biologically active compounds." *Journal of Applied Phycology* 7: p 3–15. doi: 10.1007/BF00003544.

Chang, Jo-Shu, Pau-Loke Show, Tau Chuan Ling, Chun-Yen Chen, Shih-Hsin Ho, Chung Hong Tan, Dillirani Nagarajan, and Win Nee. 2017. "Photobioreactors." In *Current Developments in Biotechnology and Bioengineering*, p. 313–352. Elsevier: National Cheng Kung University, Taiwan.

Christaki, Efterpi, Eleftherios Bonos, and Panagiota Florou-Paneri. 2015. "Innovative microalgae pigments as functional ingredients in nutrition." In *Handbook of Marine Microalgae: Biotechnology Advances*, p. 233–243. Oxford, UK.

Cimore, Marlene. 2017. *Genetically modified algae could soon show up in food, fuel, and pharmaceuticals.* Accessed 16 July 2021. https://www.popsci.com/genetically-engineered-algae/#:~:text=Subscriber%20Login-,Genetically%20modified%20 algae%20could%20soon,in%20food%2C%20fuel%2C%20and%20pharmaceuticals &text=Algae%20can%20be%20used%20for,make%20it%20into%20algae%20 butter.

Data Bridge Market Research. 2021. "Global microalgae market—industry trends and forecast to 2028." Accessed 15 July 2021 https://www.databridgemarketresearch.com/reports/global-microalgae-market.

Dormido, Raquel, José Sánchez, Natividad Duro, Sebastián Dormido-Canto, María Guinaldo, and Sebastián Dormido. 2014. "An interactive tool for outdoor computer controlled cultivation of microalgae in a tubular photobioreactor system." *Sensors* 14 (3): p. 4466–4483. doi: 10.3390/s140304466.

Eurostat Statistics Explained. 2019. "Glossary: Biofuels." Last Modified 9 September 2019. Accessed 9 June 2021. https://ec.europa.eu/eurostat/statistics-explained/index.php?title=Glossary:Biofuels.

Expobiomasa. 2021. "What are the most commonly used solid biofuels?" Accessed 15 June 2021. https://expobiomasa.com/en/content/%C2%BFcu%C3%A1les-son-los-biocombustibles-s%C3%B3lidos-m%C3%A1s-utilizados.

Firican, George. 2020. "The pros and cons of the 4th industrial revolution."

García, José L., Marta de Vicente, and Beatriz Galán. 2017. "Microalgae, old sustainable food and fashion nutraceuticals." *Microbial Technology* 10 (5): p. 1017–1024. doi: 10.1111/1751-7915.12800.

Genin, Scott N., J. Stewart Aitchison, and D. Grant Allen. 2016. "Photobioreactor-based energy sources." In *Nano and Biotech Based Materials for Energy Building Efficiency*, p. 429–455. Cham: Springer.

Greenfacts—Facts on Health and the Environment. 2021. "Liquid biofuels for transport prospects, risks and opportunities." Accessed 15 June 2021 https://www.greenfacts.org/en/biofuels/l-2/1-definition.htm.

Histroy.com Editors. 2009. "Industrial revolution."

Hu, Dawei, Rui Zhou, Yi Sun, Ling Tong, Ming Li, and Houkai Zhang. 2012. "Construction of closed integrative system for gases robust stabilization employing microalgae peculiarity and computer experiment." *Ecological Engineering* 44: p. 78–87. doi: 10.1016/j.ecoleng.2012.04.001.

iED Team. 2019. "The 4 industrial revolutions." 3 July 2021. https://ied.eu/project-updates/the-4-industrial-revolutions/.

Immerman, Graham. 2018. "Industry 4.0 advantages and disadvantages." *MachineMetrics.* Accessed 5 July 2021. https://www.machinemetrics.com/blog/industry-4-0-advantages-and-disadvantages.

Katiyar, Richa, Amit Kumar, and Bhola R. Gurjar. 2017. "Microalgae based biofuel: Challenges and opportunities." In *Biofuels: Technology, Challenges and Prospects,* edited by Avinash Kumar Agarwal, Rashmi Avinash Agarwal, Tarun Gupta and Bhola Ram Gurjar, p. 157–175. Singapore: Springer.

Khan, Muhammad Imran, Jin Hyuk Shin, and Jong Deog Kim. 2018. "The promising future of microalgae: Current status, challenges, and optimization of a sustainable and renewable industry for biofuels, feed, and other products." *Microbial Cell Factories* 17: p. 1–15. doi: 10.1186/s12934-018-0879-x.

Lamm, Ben. 2019 "Algae might be a secret weapon to combatting climate change." *Quartz.* Accessed 8 July 2021. https://qz.com/1718988/algae-might-be-a-secret-weapon-to-combatting-climate-change/.

LaMorte, Wayne W. 2017. "The industrial revolution." Accessed 3 July 2021. https://sphweb.bumc.bu.edu/otlt/mph-modules/ep/ep713_history/ep713_history4.html.

Lotz, Bert, and Rene Smulders. 2021. *Genetic Modification*. Wageningen University and Research. Accessed 19 June 2021. https://www.wur.nl/en/Dossiers/file/Genetic-modification-1.htm.

Marxen, Kai, Klaus Heinrich Vanselow, Sebastian Lippemeier, Ralf Hintze, Andreas Ruser, and Ulf-Peter Hansen. 2005. "A photobioreactor system for computer controlled cultivation of microalgae." *Journal of Applied Phycology* 17: p. 535–549. doi: 10.1007/s10811-005-9004-8.

Merlo, Simone, Xavier Gabarrell Durany, Angela Pedroso Tonon, and Sergio Rossi. 2021. "Marine microalgae contribution to sustainable development." *Water* 13 (10). doi: 10.3390/w13101373.

Morales, Marjorie, Arnaud Hélias, and Olivier Bernard. 2019. "Optimal integration of microalgae production with photovoltaic panels: Environmental impacts and energy balance." *Biotechnology for Biofuels*. doi: 10.1186/s13068-019-1579-4.

Moran, Keith. 2021. "Benefits of industry 4.0." https://slcontrols.com/en/benefits-of-industry-4-0/.

Narala, Rakesh R., Sourabh Garg, Kalpesh K. Sharma, Skye R. Thomas-Hall, Miklos Deme, Yan Li, and Peer M. Schenk. 2016. "Comparison of microalgae cultivation in photobioreactor, open raceway pond, and a two-stage hybrid system." doi: 10.3389/fenrg.2016.00029.

Ngo, Huu Hao, Hoang Nhat Phong Vo, Wenshan Guo, Xuan-Thanh Bui, Phuoc Dan Nguyen, Thi Minh Hong Nguyen, and Xinbo Zhang. 2018. "Advances of photobioreactors in wastewater treatment: Engineering aspects, applications and future perspectives." In *Energy, Environment, and Sustainability*, p. 297–329. Singapore: Springer.

Organisation for Economic Cooperation and Development (OECD). 2015a. "The benefits and advantages of commercial algal biomass harvesting." In *Biosafety and the Environmental Uses of Micro-Organisms*, p. 20. Paris: OECD Publishing.

Organisation for Economic Cooperation and Development (OECD). 2015b. "The need and risks of using transgenic microalgae for the production of food, feed, chemicals and fuels "In *Biosafety and the Environmental Uses of Micro-Organisms*, p. 13. Paris: OECD Publishing.

Plantlet. 2019. "Microalgae in biofuel production." Accessed 3 June 2021. https://plantlet.org/microalgae-in-biofuel-production/.

Radakovits, Randor, Robert E. Jinkerson, Al Darzins, and Matthew C. Posewitz. 2010. "Genetic engineering of algae for enhanced biofuel production." *Eukaryotic Cell* 9: p. 486–501. doi: 10.1128/EC.00364-09.

Rafferty, John P. 2021. "The rise of the machines: Pros and cons of the industrial revolution." *Britannica*. Accessed 3 July 2021. https://www.britannica.com/story/the-rise-of-the-machines-pros-and-cons-of-the-industrial-revolution.

Ramesh, Rajat. 2020. *Bioproducts: Synthesis and Market Analysis*. Wharton Undergraduate Energy Group. Accessed 2 July 2021. http://whartonenergygroup.com/newsletter/2020/11/15/bioproducts-synthesis-and-market-analysis.

Rosch, Christine, Johannes Skarka, Andreas Patyk, and Forschungszentrum Karlsruhe. 2009. "Microalgae—opportunities and challenges of an innovative energy source.": *17th European Biomass Conference and Exhibition*, p. 1–6.

Snow, Allison A., and Val H. Smith. 2012. "Genetically engineered algae for biofuels: A key role for ecologists." *BioScience* 62 (No. 8): p. 765–768. doi: 10.1525/bio.2012.62.8.9.

Tanigawa, Sara. 2017. *Biogas: Converting Waste to Energy*. Environmental and Energy Study Institute. Accessed 15 June 2021. https://www.eesi.org/papers/view/fact-sheet-biogasconverting-waste-to-energy.

Ullmann, Jörg, and Daniel Grimm. 2021. "Algae and their potential for a future bioeconomy, landless food production, and the socio-economic impact of an algae industry." *Organic Agriculture* 11: p. 261–267. doi: 10.1007/s13165-020-00337-9.

United Nations Development Programme (UNDP). 2021. "What are the sustainable development goals?" Accessed 30 June 2021. https://www.undp.org/sustainable-develop ment-goals.

University of Technology—Sydney. 2021. "What is industry 4.0?" Accessed 6 June 2021. https://www.uts.edu.au/research-and-teaching/our-research/climate-change-cluster/ industry-engagement/industry-4.0-x-algae-biotechnology/what-industry-4.0.

UpKeep. 2019. "What is the difference between Industry 3.0 and industry 4.0?" Accessed 5 July 2021. https://www.onupkeep.com/answers/preventive-maintenance/industry-3-0-vs-industry-4-0.

Usher, Philippa K., Andrew B. Ross, Miller Alonso Camargo-Valero, Alison S. Tomlin, and William F. Gale. 2014. "An overview of the potential environmental impacts of large-scale microalgae cultivation." *Biofuels* 5 (3): p. 331–349. doi: 10.1080/17597 269.2014.913925.

Vale, Richmond. 2016. "Second industrial revolution: The technological revolution." Accessed 3 July 2021. https://richmondvale.org/en/blog/second-industrial-revolution-the-technological-revolution.

Yang, Bo, Jin Liu, Xiaonian Ma, Bingbing Guo, Bin Liu, Tao Wu, Yue Jiang, and Feng Chen. 2017. "Genetic engineering of the Calvin cycle toward enhanced photosynthetic CO2 fixation in microalgae." *Biotechnology for Biofuels* 10. doi: 10.1186/ s13068-017-0916-8.

Yen, Hong-Wei, I-Chen Hu, Chun-Yen Chen, and Jo-Shu Chang. 2014. "Design of photobioreactors for algal cultivation." In *Biofuels from Algae*, p. 23–45. Elsevier: Cambridge, United States.

Zhu, Yunhua, Susanne B. Jones, and Daniel B. Anderson. 2018. "Algae farm cost model: Considerations for photobioreactors." Accessed 15 July 2021. https://www.osti.gov/ servlets/purl/1485133.

12 Sustainability and Development of Microalgae 4.0

Henry Ng[1]*, Chung Hong Tan*[2,3]*, Saifuddin Nomanbhay*[2,3] *and Pau Loke Show*[1]

[1] Department of Chemical and Environmental Engineering, Faculty of Science and Engineering, University of Nottingham Malaysia, Jalan Broga, Semenyih, Selangor Darul Ehsan, Malaysia

[2] Institute of Sustainable Energy, Universiti Tenaga Nasional (UNITEN), Jalan IKRAM-UNITEN, Kajang, Selangor, Malaysia

[3] AAIBE Chair of Renewable Energy Universiti Tenaga Nasional (UNITEN), Kajang, Selangor, Malaysia

CONTENTS

12.1 Introduction: Microalgal Products and Background 397
12.2 Design of an Automated, Flat Panel Photobioreactor 401
12.3 Automated Photobioreactor with Remote Control 402
12.4 The Airlift Modular Automated Photobioreactor System 405
12.5 Automated Photobioreactor Constructed on the Large Scale 409
12.6 Automated, Semi-Continuous Cultivation System 412
12.7 Automated, Continuous Cultivation of Microalgae with Remote Control 415
12.8 Conclusion: Summary of the Designs Presented and Their Respective Automation Systems 418
References 419

12.1 INTRODUCTION: MICROALGAL PRODUCTS AND BACKGROUND

Over the years, microalgae have generated substantial interest owing to their potential as a sustainable source of various products. First, certain species of microalgae are able to accumulate large amounts of lipids, such as *Botryococcus braunii* which could have an oil content of up to 80%. The oil, or the whole microalgal biomass itself, can then be used to produce various products, such as biodiesel, bioethanol, biohydrogen, and biochar. Moreover, in the pharmaceutical sector, Astaxanthin, which is a carotenoid, can be extracted from various species

DOI: 10.1201/9781003202196-12

of microalgae. The substance is an effective antioxidant that helps patients with cardiovascular, inflammatory, immune, and neurological diseases (Khan, Shin, and Kim 2018; Rahman 2020).

Another pharmaceutical product that can be obtained from microalgae are omega 3 fatty acids. In fact, studies have shown that one of the omega 3 fatty acids, EPA, can be obtained in large amount with microalgae as compared to that obtained by the conventional source of marine fish oils. In total, many pharmaceutical or high-value compounds can be extracted from microalgae, including cosmetics, certain chemicals, vitamins, beta-sitosterol, microcolin-A, and beta-carotene. At the end, and instead of making a product from microalgae, the entire microalgal biomass could also be used as feed for animals in the juvenile or larval stage of life, such as oyster spat, juvenile abalone, and rotifer (Rahman 2020; Khan, Shin, and Kim 2018).

A number of processes are required if a desired compound is to be obtained from microalgae, starting with cultivation. In this first stage, the microalgae are grown into a microalgal biomass, and this process will be elaborated in the following paragraphs. From the cultivation stage, the microalgal biomass is then harvested and converted into a paste/sludge with a total suspended solid of 5–25%, but can sometimes be more. For the conversion to a paste/sludge, a number of techniques can be employed. Examples of these techniques include flotation, sedimentation, flocculation, filtration, and centrifugation. Until this day, flocculation is the easiest and most cost-effective technique due to the low energy consumption of this process. However, if high recovery rate and contamination of flocculants in the final product are of concern, centrifugation can be used instead (Ravindran et al. 2016).

After the paste/slurry has been created, the process of cell disruption ensues. In this stage, the cell walls of the cells in the microalgal biomass paste/slurry are ruptured or disrupted so that its contents 'spill' out. There are also many processes available to achieve the outcome, such as autoclaving, microwaving, bead-beating, ultrasonication, and osmotic shock. Among all of them, microwaving is currently the simplest and most effective due to the method's high mass transfer rates and lack of thermal degradation of the compounds despite high internal temperatures and pressure gradients. After this stage, the desired compound would be extracted from the now disrupted microalgal biomass. The processes that can be used for this stage, however, depend on the compound being extracted. Also, currently, the discovery and/or development of solvents that are more selective, along with more research on this process in general, is needed for the various compounds that can be extracted from the microalgae cultivated (Rösch, Roßmann, and Weickert 2019; Ravindran et al. 2016; Khan, Shin, and Kim 2018).

As for the cultivation stage, multiple requirements have to be met in order for the microalgae to grow. Not only that, these requirements may very well determine the final cell density, composition, yield of desired compounds, etc., of the microalgal biomass. The first of the many requirements is light, where light intensity and the daily duration of light exposure affect the composition and biomass yield of the microalgae. Light is important because the microalgal biomass requires it to photosynthesise. However, for any given light intensity and daily duration of exposure, the CO_2 consumption for photosynthesis will eventually match the CO_2 released due to respiration. This part of the growth is called the compensation point, where the rate of energy produced from the photosynthesis is the same as the rate of energy being consumed through respiration, which leads to a stagnated growth. This would also mean that the amount of carbohydrates accumulated before reaching the compensation point changes with light intensity and duration of exposure. There is also the problem of non-uniform distribution of light, also known as photoinhibition. All in all, experiments are needed to optimise light intensity and duration of exposure (Khan, Shin, and Kim 2018).

The second factor to consider for cultivating microalgae is the temperature of the culture. A temperature that is too low may reduce the assimilation process of carbon, which slows down photosynthesis. However, a temperature that is too high will result in the degradation of photosynthetic proteins and disruption of cell metabolism. In other words, the growth rate of the microalgae is dependent on the temperature at which the culture is maintained. Temperature also affects the size of individual cells and the rate of respiration. Besides that, most microalgal species have an optimum temperature that is from 20°C to 30°C, but thermophile microalgae may be able to handle higher temperatures. For example, microalgae that thrive in hot springs tend to tolerate temperatures of up to 80°C. The opposite is true for cryophilic microalgae, which are attracted to colder temperatures. At the end, temperature can also be used as a stressor. As an example, *Chlorella vulgaris* has been found to produce more lipids and carbohydrates when grown at 25°C instead of 30°C (Fuchs et al. 2021; Khan, Shin, and Kim 2018).

The next requirements are the nutrients, which include both micronutrients and macronutrients. Nitrogen, carbon, and phosphorus are classified under macronutrients, and the availability of all of them affects the growth rate of the microalgal biomass. However, marine species may also need silicon in large amounts. Moreover, stressing the microalgae through limiting the availability of nitrogen has been shown to increase the amount of carbohydrates and lipids accumulated, but it reduces the growth rate and productivity. The micronutrients, which are Mo, Co, K, Fe, Mn, Mg, Zn, and B, are also important as they dictate the activities of enzymes within the microalgal cells. Besides nutrients, the mixing of the culture is needed to ensure that the distribution of nutrients, CO_2, light, and air are equal throughout the culture. The agitation also prevents the settling of biomass, and not having it would reduce the productivity of the microalgal biomass significantly (Khan, Shin, and Kim 2018).

The final important factor is the pH, because microalgae will only grow well at the right pH range. Most species of microalgae thrive in the pH range of 6 to 8.76, with the exception of *C. vulgaris* which has a high tolerance of extreme pH values. However, despite the high tolerance, even *C. vulgaris* has an optimum pH to maximise growth, at a pH of 9 to 10. In general, a pH that is too high would harm the cells in the microalgae biomass due to the increased salinity that results from it. In addition to all of the possible stressors given before, such as nitrogen content and temperature, modifying the genes and strains of the microalgae can also improve their productivity and/or yield of the final product (Khan, Shin, and Kim 2018).

There are two major types of cultivation systems available for microalgae cultivation, where the first type are open pond systems. These systems usually utilise a pond that is 1 to 100-cm deep and have an area ranging from one to several acres. Popular types of open pond systems are circular ponds, shallow big ponds, and raceway ponds. The choice of open pond system is made depending on the type of microalgal species, climate, and cost of land and water. The most popular among them is the raceway pond, which features a pond that looks like a race track containing liquid that is circulated using a paddle wheel. Open pond systems are cheaper and easier to construct and operate but are disadvantaged in that large plots of land are required. Moreover, this type of cultivation system can only be used for photoautotrophic algae. Not only that, an inefficient use of sunlight in open pond systems leads to minimal production of products. There is also a risk of contamination of airborne substances and microorganisms, along with the issue of loss of dissolved CO_2 to the air and evaporation of water (Ravindran et al. 2016).

In order to prevent all of the problems that appear in open pond systems, the second type of cultivation system can be used instead. This type of cultivation system is called a closed photobioreactor or PBR for short. This can be used in an indoor and outdoor setting and can accommodate both photoautotrophic and heterotrophic microalgae. There are many types of closed photobioreactors, where the most popular types are the tubular PBR, helical PBR, airlift PBR, and flat panel, also known as flat plate, PBR. These photobioreactors come in many shapes and sizes, such as the tubular PBR that is made of glass or plaster tubes with a diameter of 10 cm or less. To add on, the tubes can be aligned horizontally or vertically, and a mechanical or airlift pump allows the microalgal biomass to flow through them. Due to their construction, tubular photobioreactors allow light to pass through, prevent microalgae from settling, and represent one of the many possibilities of how a PBR can work and look like (Ravindran et al. 2016).

However, cultivation technology does not stop there. In the age of industrial revolution 4.0, newer technology, such as automation, big-data analysis, and Internet of Things, has emerged. These advancements can be implemented into the different processes in microalgae cultivation and processing. This is especially beneficial to the cultivation stage, where automation could reduce the manpower required to monitor that crucial stage. The Internet of Things would also enable operators to monitor and control cultivation systems remotely, but this has been discussed in Chapter 4. Therefore, the following sections will present and review multiple designs of automated cultivation systems from different research papers, along with their respective performances during the test run and any suggestion for improvements. The different approaches to automation in the cultivation systems given in the subsequent sections can either be applied, or the entire cultivation system can be scaled and used.

12.2 DESIGN OF AN AUTOMATED, FLAT PANEL PHOTOBIOREACTOR

The first design to be discussed is the one presented in the research paper by Fuchs et al. (2021), which is mainly about the design of an innovative, flat panel photobioreactor by them. The flat panel photobioreactor was developed with an aim of making it easy to scale, disassemble, sterilise, and have no dead zones, which form as a result of precipitation from a microalgae biomass with high cell density during late stages of the growth phase. Figure 12.1 shows the asymmetrical, flat panel photobioreactor that was designed (Fuchs et al. 2021).

The photobioreactor, shown in Figure 12.1, was constructed using two 4-mm glass panes distanced 40 mm apart, which were attached to a metal frame through a 1-mm thick silicone seal and clamping jaws. Two 6500 K MiniMatrix lights, with a maximum output of 750 µmol m^{-2} s^{-1} and 504 LEDs, were then be mounted onto each of the two glass panes, where they were screwed onto the outer frame, to provide light for photosynthesis. Moreover, a gas mixture of air, carbon dioxide, and oxygen, mixed proportionally, was pumped into the reactor through a valve at the bottom. This can be seen as a sparger at the bottom of the photobioreactor shown in Figure 12.1. Besides the gas mixture inlet valve, there was a harvest valve which could be opened when the microalgal biomass was ready. As for

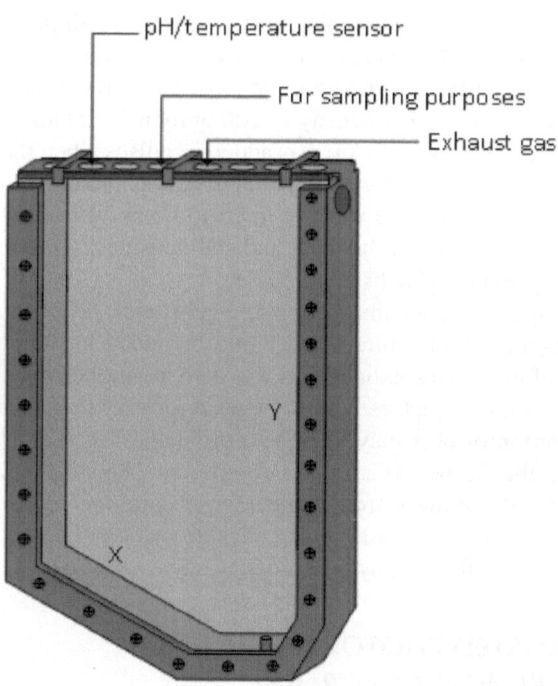

FIGURE 12.1 Design of the asymmetrical, flat panel photobioreactor (Fuchs et al. 2021).

sterilisation, all of the parts could be individually disassembled and autoclaved, or the entire photobioreactor could be autoclaved in one go. Furthermore, the fluid at point X flowed towards the sparger, which pushed the liquid towards point Y in Figure 12.1. This fluid movement prevented sedimentation, which was one of the goals of the design (Fuchs et al. 2021).

Also, a hole for exhaust gas and two ports used for sampling and pH/temperature measurement can be seen at the top of the photobioreactor, as well as the theoretical maximum of eight ports that could be carved out. What cannot be seen in the figure, however, is a cooling loop to be circulated within the walls of the metal frame. In the test run, heat from the LEDs was cooled by water that was inserted into a hole at one of the side frames and came out from another hole from the other side frame. However, another coolant could be used to cultivate cryophilic algae, including coolants that are hazardous. At the end, the control of pH was possible through the air input. The gas mixture was created by three mass flow controllers (MFCs), where one is connected to an air supply, one to an oxygen supply, and the final one to a carbon dioxide supply. The three streams then combined into one before being inserted into the photobioreactor. When a pH value that is higher or lower than the threshold set was detected, a system control unit could alter the mass flow rate of CO_2 using the respective mass flow rate controller. The oxygen supply, however, is optional as the test run was successful without it (Fuchs et al. 2021).

Before the test run, 7.2 L of yeast nitrogen base (YNB) medium, which contained 1.7 g/L YNB, 10 g/L glucose, and 5 g/L ammonium sulphate, was inserted into the photobioreactor. The photobioreactor, with the medium inside, was then autoclaved at 120°C for 20 minutes in order to sterilise it. *Chlorella sorokiniana* was then cultivated at pH 7.2, with an input of 0.133 vvm of gas, at the temperature of 26°C, and under a light intensity of 200 μmol m^{-2} s^{-1}. During certain times of the experiment, 0.028 vvm of CO_2 was added in pulses when the pH value rose above the threshold. After 7 days, a growth curve was plotted, where the period with the maximum growth rate was observed to be between Day 2 and Day 3. Unfortunately, specific data on the yield and cell densities of the final microalgal biomass was not provided (Fuchs et al. 2021).

In terms of suggestions for improvement, Fuchs et al. (2021) noted that circadian light cycles and temperature shifts could be added to the photobioreactor. The possibility of measuring exhaust gas was also mentioned (Fuchs et al. 2021). However, if the photobioreactors were to be used for very large-scale production, the even distribution of heat may become a problem. The reason is the cooling system is inside the frame, which comes into contact with the contents of the photobioreactor that are away from the center, despite the lights shining on all parts of the culture. A new cooling system design, which makes contact with the multiple areas of the culture, may be needed to solve the issue.

12.3 AUTOMATED PHOTOBIOREACTOR WITH REMOTE CONTROL

The second automated photobioreactor design is by Rahmat et al. (2020), and it contained the ability to utilise the IoT (Internet of Things) for remote operations.

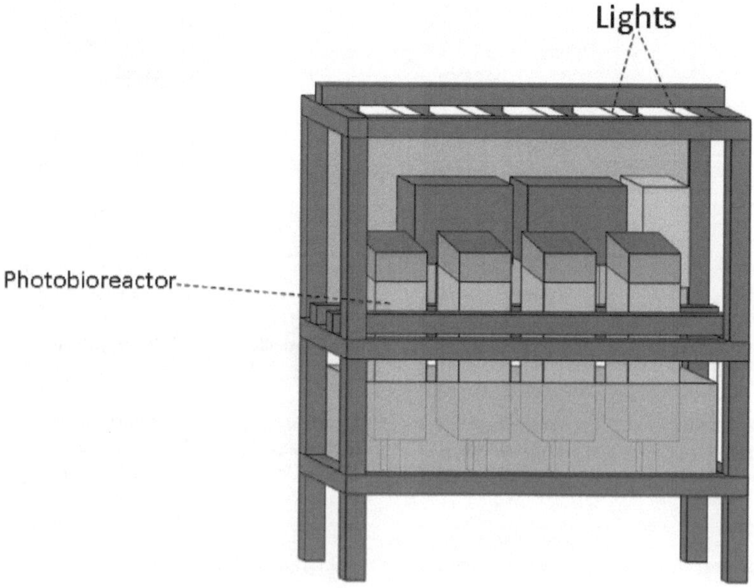

FIGURE 12.2 Sketch of the photobioreactor system (Rahmat et al. 2020).

The system that was designed used four closed, cuboid-shaped photobioreactors partially submerged in one large water tank to maintain the temperature of the photobioreactors. This set-up is similar to a water bath, as shown in Figure 12.2. For each reactor, a temperature sensor, a light sensor, and a colour sensor were used, along with an additional temperature sensor to read the temperature of the surroundings (Rahmat et al. 2020).

Each set of sensors sent data to an Arduino WeMos D1 microcontroller through a wired connection, amounting to 4 WeMos D1s in total. All of the WeMos D1s were then connected to a Raspberry Pi mini-computer, where data from the WeMos D1s were read. The Raspberry Pi mini-computer then changed the format of the data before it was sent through a router by an Internet connection. Incoming instructions by remote operators, who used a graphical user interface (GUI) through external devices connected to the Internet, were also read and stored as threshold values for temperatures and lighting. By comparing the temperature readings with these threshold values, the mini-computer could choose to turn on and off heaters that were connected to it through a relay, acting as an actuator. This was the same for the lamps, which were also connected to the mini-computer through a relay. All of the connections, wired and wireless, are shown in Figure 12.3 (Rahmat et al. 2020).

In terms of programming, Arduino sketch was used for the Arduino WeMos D1s, and Python 3.2 was used for the Raspberry Pi mini-computer, as well as Internet programming using Apache and PHP web services. In general, the programs had three functions. They were to combine the data to be sent to the database server, allow control of the reactor temperatures automatically through

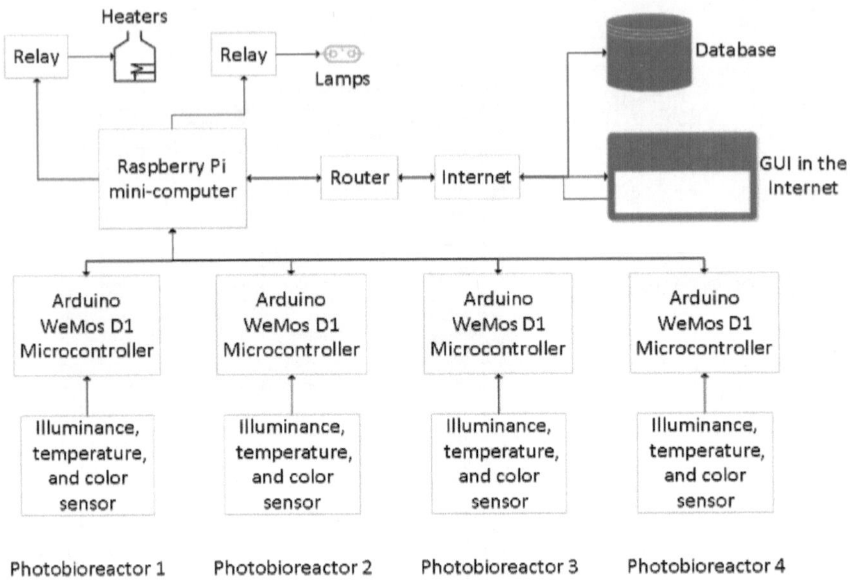

FIGURE 12.3 Connections of the system, wired and wireless (Rahmat et al. 2020).

reading and comparing sensor data with threshold values, and allowing control of the lamps by turning the lamp on or off depending on the 'on' or 'off' values set in the server by the operators. As for the functions of the Internet programming, they displayed data or graphs of the sensor readings in real time and allowed the control of lights and heaters by remote operators (Rahmat et al. 2020).

For the test run, 9 litres of water and 1 litre of microalgae were added to each of the photobioreactors, and the temperature threshold value was set to 30. The GUI was located at https://kultivasimikroalga.com/admin/index.php. Unfortunately, the test run for the design by Rahmat et al. (2020) focused on observing the performance of the sensors and equipment. Due to this, no data on microalgae yields was taken other than a note that the system may have an increased productivity of 9% compared to a photobioreactor without the IOT and automation (Rahmat et al. 2020).

As for the possible improvements that can be made, it was mentioned that past research showed the possibility of adding a decision support system that helps the operator choose the quantity and timing of nutrient feeding which would optimise and further improve the system's productivity. It was also noted that the decision support system could be developed further to include things such as a warning system that warns when an invasive species has started attacking the culture (Rahmat et al. 2020). Other than that, a water bath may not be suitable for large-scale microalgae cultivation using this photobioreactor system. Instead, small heaters could be directly integrated into each individual, cuboid-shaped photobioreactor, or a closed, heating loop that runs through the photobioreactors can be designed. Also, pH sensors and a pH control system would be very useful as most microalgal species are sensitive to changes in pH. At the end, any form of agitation would help with preventing microalgae from precipitating out from the microalgal biomass.

12.4 THE AIRLIFT MODULAR AUTOMATED PHOTOBIOREACTOR SYSTEM

Borowiak et al. (2021) designed a novel photobioreactor system which was named the airlift modular automated photobioreactor system (AMAPh-S). The modular design provided the ease of operation and clean up through a good insertion of gases into the system, swift emptying and refill of reactor, and automation of the cleaning process. The photobioreactors in the AMAPh-S worked in two different stages, which are the inoculation and production stages. Three stressors could also be applied to the microalgae in the AMAPh-S to increase the production of a favoured product. This was done in the test run using *H. pluvialis* with the aim of increasing the yield of the carotenoid, astaxanthin. The production stage consisted of 12 photobioreactors, while the inoculation stage was split further into inoculation stages 1 and 2, which both consisted of two photobioreactors each. For the test run, the photobioreactors used had a volume of 12 dm^3 in stage 1, and 90 dm^3 in stage 2 and stage 3, which added to a total of 1 m^3. However, this could be scaled depending on the situation. An outline of the system is shown in the Figure 12.4. However, the set-up of the system is explained further in the following paragraphs (Borowiak et al. 2021).

For the inoculation stage, resistance sensors, which were covered in glass, allowed a range of 0°C to 100°C for temperature measurement. These sensors were then connected to temperature transducers through an RS-485 interface and a Modbus RTU communication protocol. For the 12 photobioreactors in the production stage, DS18B20 temperature sensors, which were covered in stainless steel, were used instead. The 12 DS18B20 temperature sensors were subsequently connected to a STR-5321-D temperature converter using a 1-wire bus. The readings from the STR-5321-D were then sent to the master computer through a RS-485 interface and Modbus RTU communication protocol (Borowiak et al. 2021).

pH was measured using ERH-13x2 sensors, which were all covered in glass. However, there was a difference in the way that all of the pH sensors were connected. For the inoculation stages, the pH sensors were connected to the pH-measuring transducers through an RS-485 interface and a Modbus RTU communication protocol. However, for the pH sensors in the production module, they were connected to a Jumo ecoTrans pH 03, which is a pH meter with an analogue signal output, in the sequence of 1 to 12 using a multiplexer and its many channels. The multiplexer ensured that only one of the many signals from the inputs was sent to the pH meter at a time, switching between the channels using the RS-485 interface and a Modbus ASCII protocol. At the end, the analogue signal from the pH meter, which had a voltage of 0 V to 10 V, had to be read by an

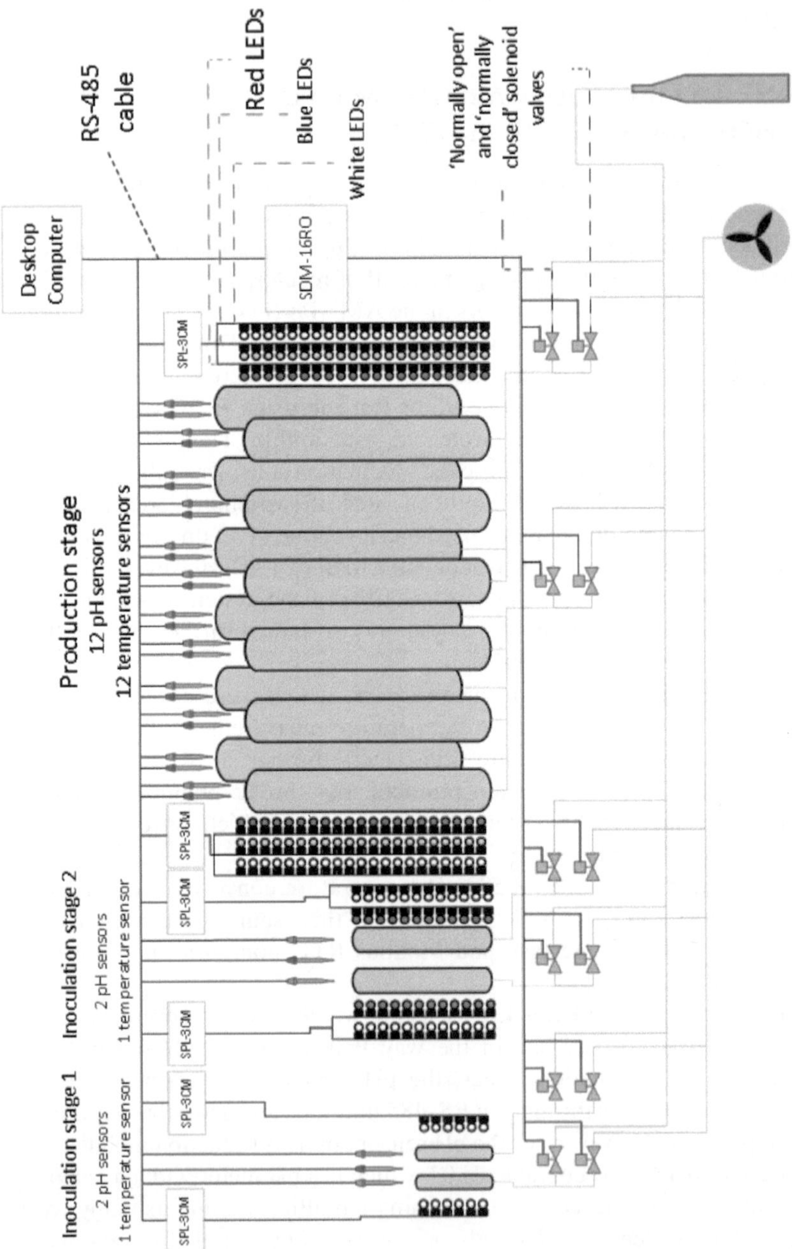

FIGURE 12.4 Outline of the AMAPh-S (Borowiak et al. 2021).

MOD-1AI measuring module that accepts a voltage and current analogue input and used an RS-485 communication interface (Borowiak et al. 2021).

The last thing that was measured by the AMAPh-S is the concentration of ammonium ions. This was done using a CPI-505 ion meter and an NH 500/2 ion selective electrode. The ammonium-ion-measuring system was calibrated using 1N-BBM medium at four different dilutions, which were 20%, 40%, 60%, and 80%. The last sample was distilled water, which represents 0% ammonium ions. The calibration curve that resulted from the calibration process was then used to determine the percentage of ammonium ions in the media used for the cultivation. However, the calibration might have been done using 1N-BBM medium because 1N-BBM medium was used for the production phase, and 2N-BBM was used in the inoculation phase. Therefore, if another medium was used, the same medium might be needed for the calibration of the ammonium measurement system (Borowiak et al. 2021).

Lighting for the photobioreactors of the AMAPh-S was done through the usage of LED panels, where all of the panels used Lumileds light-emitting diodes and required a 24 V DC input. LUXEON red LXML-PD01–0050 with a wavelength of 620–645 nm, LUXEON blue LXML-PR02–0900 with a wavelength of 440–460 nm, and LUXEON LXM3-PW51 white LEDs with a colour temperature of 4,000 K were used for the system. The power output of the lighting system was also optimised to 250 μmol/m^2/s for the test run. Moreover, only the white LEDs were used in inoculation stage 1, white and red for inoculation stage 2, and all three types of LEDs were used for the production stage. SPL-3CM diode and LED strip controllers from Dagon were connected to the groups of LEDs, also as shown in Figure 12.4, for the control of light intensity by using pulse width modulation (PWM) (Borowiak et al. 2021).

The contents of the photobioreactors were mixed by the insertion of gases through the bottom part. The AMAPh-S gets air from an air compressor and CO_2 through pressurised bottles storing it. The air from the air compressor was dried and put through sterile HEPA filters before reaching the photobioreactors. Diffusers at the bottom of each reactor ensured that the gases bubble upwards, agitating the culture. For the control of gas flow, 'normally open' (NO) and 'normally closed' (NC) solenoid valves were used, with 24 V DC coils. Sixteen SDM16RO relays connected the solenoid valves to a digital output module, which enabled control of the valves. During operation, the solenoid valves opened and closed alternately at set time intervals, where carbon dioxide flow was only enabled after closing the supply of air. This was done to ensure precise dosing of the gases into the system. An example of this is a cycle of 1 minute of CO_2 flow for every 30 minutes of air flow. Generally, this is determined by which species of the microalgae is cultivated and by the conditions needed for cultivation. At the end, the software to monitor and control the system was made in LabVIEW 2014 graphical programming environment (Borowiak et al. 2021).

In order to sterilise the AMAPh-S, the system was, first, ozonised for 4 hours. After that, 25 kg of NaOH was dissolved in 1,000 dm^3 tap water that had been passed through a set of chemical filters, namely a carbon filter, iron remover, prefilter, and a descaler, and put under UV. The resulting 2.44% NaOH solution remained in the photobioreactors for 2 hours before being emptied and rinsed away with the tap water. At the end, 1,000 dm^3 of water containing an added 20 dm^3 of HCl was inserted into the system before being taken out 2 hours later. After a final rinse with demineralised/sterilised water, the photobioreactors were ready for cultivation. For the test run, autoclaved 2 N-BBM media was used for the inoculation stages, and autoclaved, standard BBM with 1-fold nitrogen was used for the production stages. The production stages were also further separated into 'green' and 'red' phases, where 'green' phase represents the growth phase of the microalgae while the 'red' phase represents the stressing stage that produces most of the astaxanthin (Borowiak et al. 2021).

In the test run, Borowiak et al. (2021) noted that the 'green' phase lasted 21 days, and the 'red' phase lasted 9 days. The three stressors used on the *H. pluvialis* microalgae were intensity of irradiation, culture temperature, and nitrogen starvation. Starting with the first stressor, the intensity of the lighting was varied with time throughout the cultivation process. Red LEDs were the main source of light starting from the second day of cultivation. On the other hand, the intensity of blue LEDs was increased from 10% at multiple intervals until it reached 100% at Day 17. Finally, the power of white LEDs was increased gradually from 0% at Day 12 to 100% at Day 19. For the 'red' phase which lasted from Day 22 to Day 30, all of the LEDs were set to shine at full power. The second stressor, which was high temperature, is related to the light intensity of the LEDs. Day 7 to Day 23 observed a temperature range of 24°C to 26°C, while a temperature range of 27°C to 29°C was common after that. The increase in temperature was a result of the increase in power of the LEDs that caused more heat to enter the system. The final stressor came from the fact that nitrogen was almost used up by the end of Day 21 or the 'green' phase (Borowiak et al. 2021).

The dosage of CO_2 was given in a way such that the pH was maintained at or very close to 7. The end result of the test run was an average *H. pluvialis* biomass cell density of 1.68 ± 0.19 g/dm^3 and an average astaxanthin content of 2.4% ± 0.3%, where both numbers are in the format of mean ± standard deviation. Borowiak et al. (2021) also mentioned that the addition of Fe^{2+} increases the yield of astaxanthin and that Fe^{3+} ions increase the growth of the microalgal biomass. Another test run with Fe^{2+} then showed a higher content of astaxanthin at 3.2%, but Borowiak et al. (2021) stated that literature data indicates the possibility of 4% astaxanthin content in the final dry biomass (Borowiak et al. 2021). Therefore, if this design was used in the future for the production of astaxanthin from *H. pluvialis*, Fe^{2+} should be used with an optional addition of Fe^{3+}. At the end, Borowiak et al. (2021) stated that LEDs were used because they produce less heat than other light sources, which favoured astaxanthin production (Borowiak et al. 2021). However, if the designed AMAPh-S is being used to cultivate other microalgae in the future, which requires more heat, another light source may be needed, such as fluorescent light bulbs.

12.5 AUTOMATED PHOTOBIOREACTOR CONSTRUCTED ON THE LARGE SCALE

The second design is by Erbland et al. (2020), who tested a full-size photobioreactor on the cultivation of one of the microalgae species, *Tetraselmis chuii*. This photobioreactor was a fully enclosed, cone-bottom tank with an operating capacity of 1,700 litres. The reactor had a height of 2.3 metres and was 1.4 metres wide, rounded to the first decimal place. The reactor also had an acrylic window that allows visual observation of the microalgae being cultured and a flat lid that was bolted to the tank (Erbland et al. 2020).

Attached to the lids were six waterproof, manually assembled, fluorescent lamps, which were powered by a switch box, with the purpose of providing light for the microalgae to photosynthesise. Each lamp consisted of two 54 W, 6,500 K, 5000 lm T5HO fluorescent cylindrical light bulbs, which together had a length of 1.22 meters. The bulbs were contained inside clear 5.08-cm PVC pipes and standard schedule 40 and 80 PVC fittings. The lamps were then mounted to the lid, 45.7 cm from the centre of the lid, by bolting the acrylic ring top of the lamp to the lid. All six of the lamps were equally spaced in a ring shape. At the end, solid steel rods with a diameter of 4.5 cm and a length of 5 cm acted as ballasts at the bottom of the lamps (Erbland et al. 2020).

During operation, the contents of the tank were circulated by air that entered through a drain valve fitting at the base of the reactor. The input of air was provided by a 0.48-bar piston air pump, with a flow rate of 225 litres per minute using a drain valve fitting. Part of the air supply went to a ball valve to act as an airlift, providing a constant supply of the reactor's contents to a loop for monitoring. Within the loop, a PH sensor and a temperature sensor were contained into a sensor manifold made up of 2.54-cm PVC pipes. The sensors were wired to a monitoring and control unit (MCU), which was mounted to the wall of the photobioreactor. Attached to the enclosure of the MCU, and pressed against the walls of the reactor at the same time, was an optical sensor to measure the light intensity from the fluorescent lamps (Erbland et al. 2020).

The monitoring and control unit (MCU) itself is housing multiple components. These components were a microprocessor chip, a datalogger, an LCD screen, and a custom printed circuit board (PCB). Data was displayed in real time on the LCD screen and recorded into an SD card. The MCU also had a full control over a solenoid valve that was connected to a CO_2 supply line, where the CO_2 was provided by a CO_2 tank and delivered to the culture through an air-stone diffuser. In order to maintain a pH of 7, the MCU controlled the supply of CO_2 based on data from the pH sensor. At the end, Figure 12.5 shows the full diagram of the photobioreactor (Erbland et al. 2020).

Besides the design itself, the tank was sterilised through chlorination for 4 hours at a concentration for 25 ppm before operation. For the test run, the *Tetraselmis chuii* was first cultivated in a small, 20-L bubble column. After that, 1,650 L of room temperature, fresh well water, with sea salt added until the salination

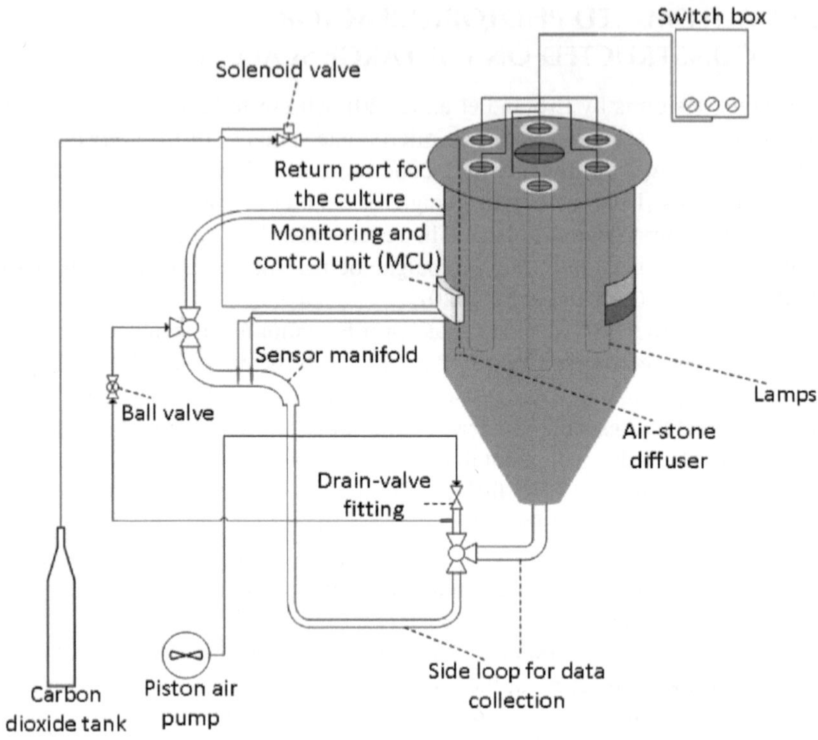

FIGURE 12.5 Diagram of the photobioreactor (Erbland et al. 2020).

level was 30 ppt, was pumped through a 5-μm filter into the photobioreactor. The contents of the well water are shown in Table 12.1. After that, chlorination at 25 ppm began, with the air supply turned on, stopping 4 hours later when sodium thiosulphate was added to the water. At the end, 0.4 mL/L of Guillard's F/2 media was added to the water before *Tetraselmis chuii* from the 20 L bubble column was inoculated into the photobioreactor. The test run of the photobioreactor showed that 300 hours of cultivation yielded 1,700 L of microalgae with a density of 2,500 cells per microlitre. In addition to that, the test run revealed that lag, growth, and stationary phases of the microalgae were 125.94 ± 36.72, 250.49 ± 62.67, and 437.21 ± 62.40 hours, respectively, for this photobioreactor. In total, the system costs $3,709.00, and Erbland et al. (2020) provided a table that compares the design to a 1,000 litre photobioreactor by Pentair called the industrial plankton, which is shown in Table 12.2 (Erbland et al. 2020).

The first improvement suggested by Erbland et al. (2020) was that the shortening of the 126-hour lag phase can be done through improving the inoculation by making the inoculum densely packed and filled with exponentially dividing cells constituting of 10% of final volume. Moreover, depending on the budget, the water can be heated or cooled to a certain, calculated temperature before it is put

TABLE 12.1
Contents of the Well Water (Erbland et al. 2020)

Parameter	Value
Chloride	27.4 mg/L
Hardness	147.5 mg/L
Iron	Less than 0.1 mg/L
Lead	Less than 0.05 mg/L
Manganese	Less than 0.05 mg/L
Nitrate	Less than 10 mg/L
Copper	Less than 0.1 mg/L
Nitrite	Less than 0.02 mg/L
Sodium	14.2 ppm
pH	8.3
Arsenic	Concentration too small to be detected

TABLE 12.2
Comparison between the Pentair's 1,000-Litre Photobioreactor and the Design Constructed by Erbland et al. (2020)

Parameter	Pentair 1,000-L photobioreactor	Photobioreactor designed (1,700 L)
PBR Cost	\$33,600	\$3,709
Cost of nutrients and CO_2	\$13.00	\$4.70
Cost of electricity (@ \$0.10/kWh)	\$28.00	\$21.10
Tetraselmis chuii cells/μL	4,500	2,500
Days to harvest	8.5	12.5

into the tank instead of waiting for it to be in equilibrium with the heat from the lamps to reduce the lag phase. Furthermore, physical factors such as salinity, pH, and temperature can affect the quantity and quality of harvested products within the microalgae. For example the levels of starch and carotenoid are related to pH, irradiance, and salinity levels of the medium for *Tetraselmis* sp. In order to take advantage of this, control of the environment can be implemented but will make the build more expensive. Besides that, LEDs could be used instead of fluorescent bulbs. Although more expensive, using LEDs instead can improve the efficiency of the system, produce less heat, and emit a specific colour or spectrum of light that is best for cultivating a particular species (Erbland et al. 2020). Other than the ones mentioned, new technology such as the Internet of Things could also be implemented into the monitoring and control unit (MCU) to enable remote operations.

12.6 AUTOMATED, SEMI-CONTINUOUS CULTIVATION SYSTEM

Sandnes et al. (2006) designed a bioreactor system that could operate semi-continuously. The bioreactor system included a tubular photobioreactor that was made up of 48 clear, PVC tubes, with an internal diameter of 30 mm and a length of 2.44 m each, stacked on top of each other with two frames at the side holding them in place. The resulting assembly, which Sandnes et al. (2006) called a biofence system, had 200 L of volume capacity in total and received culture at an inlet that could be found at the bottom of one of the frames. A centrifugal pump supplied the pressure needed to push the culture into the biofence system, and the outlet was located at the top of the same frame where the inlet is. The culture flowed through each tube, and up the biofence system, until it reached the last tube before leaving the biofence system through the outlet. The flow of the culture and the biofence system itself is shown in Figure 12.6, along with the entire bioreactor system (Sandnes et al. 2006).

The supply of culture came from and, after it reaches the outlet of the biofence system, went back to a buffer tank. Inside the buffer tank were a temperature sensor and a pH sensor, which were both connected to a controller unit. Data on temperature and pH were recorded every 15 seconds and was averaged every 5 minutes. If increased pH levels were detected, a valve that was connected to the controller could be opened in order for CO_2 gas to be supplied to the bioreactor system. The source of the CO_2 supply was not mentioned, but a pressurised CO_2 tank was likely to be used as a pressure gradient which is needed for the gas to enter the bioreactor system without any mechanical inputs. Control of temperature was also possible through a heat exchanger (HX). Besides that, the bioreactor system was also placed inside a greenhouse for the experiment to control the temperature of the environment. This part of the system is also shown in Figure 12.6 (Sandnes et al. 2006).

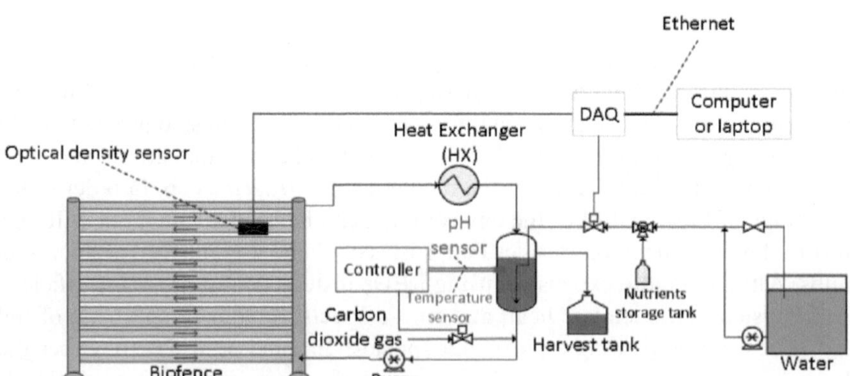

FIGURE 12.6 Diagram of the photobioreactor system designed by Sandnes et al. (2006).

In order to measure biomass densities, in g/L, Sandnes et al. (2006) also designed an optical density sensor that was to be mounted onto one of the many transparent tubes in the biofence system. The optical sensor was made up of near infrared light emitting diodes (LEDs) that were arranged in an array. To complete the optical density sensor, a photodiode detector was mounted onto the opposite side of the same PVC tube where the near infrared LEDs were located. Also, the tube of the biofence system where the sensor was mounted narrowed into an internal diameter of around 10 mm compared to the usual 30 mm. The narrowed part of the tube in the biofence that the sensor was mounted to is shown in Figure 12.6. In the test run, five near-infrared LEDs were used to ensure high sensitivity over the culture densities that were expected to be seen (Sandnes et al. 2006).

The LED's light intensity depends on its temperature, which is a function of its rate of self-heating and the temperature of the environment. A constant, high drive current was used to maintain a high and stable equilibrium temperature, and the greenhouse effect ensures that ambient temperature does not vary so much as to affect the light intensity. Calibration of the sensor was done using algal dry weights and cell number counts observed during a range of biomass densities used in the test run. After determining the voltage of the photodiode detector at a certain culture biomass density, the culture was centrifuged, resuspended in ammonium formate, and dried overnight at 100°C to determine the dry algal weight. Cell numbers were also determined, using a Coulter Multisizer particle counter. The data were then plotted on graphs, and a calibration curve was determined using a regression curve (Sandnes et al. 2006).

The voltages of the optical density sensor were recorded by a distributed data acquisition (DAQ) module and sent continuously to a computer through a local area network (LAN). The incoming data was processed, displayed, and stored using the LabView software. Due to this, data on biomass density and rate of growth of the culture could be downloaded and monitored from anywhere, provided that access to the network was possible from that location. The DAQ module was also part of a biomass density control system, which was the final component of the photobioreactor system. To maintain a set biomass density value, a water/nutrient mix was inserted into the buffer tank when the set value has been exceeded. This was done by the same DAQ module and LabView software. The voltage of the optical density sensor, which was measured on the DAQ module's analogue input channels, was compared to the set value, which corresponded to a certain biomass density, every 10 minutes. If the voltage exceeded the set value, a valve was opened using one of the DAQ module's digital output (DO) channel for a period of time, which could be in between 0 and 10 minutes (Sandnes et al. 2006).

The valve enabled the water/nutrient mix to flow into the buffer tank, where it entered near the outlet of the tank. This not only increased the total culture volume, but also enabled the upper part of the culture with a higher biomass density to be harvested. The harvesting was done using an overflow tube placed at a certain height in the buffer tank. This procedure, which led up to the opening of the water/nutrient mix valve, was repeated until the voltage from the optical density sensor was below the value set. All of the components of the biomass density control system are also shown in Figure 12.6. Sandnes et al. (2006) also derived an equation to calculate the biomass that was harvested. Using Δt as the time that the water/nutrient mix supply valve was opened for, $\Delta V = q\Delta t$, where ΔV is the volume of water/nutrient mix added and q is the constant volumetric flow rate. Assuming that the losses from evaporation are negligible, the harvested volume would be equal to the volume of water/nutrient mix added because input should be equal to output. Therefore, the harvested biomass, m, is equal to the constant biomass density in the culture, X, multiplied by the input volume, ΔV, or $m = X\Delta V$ (Sandnes et al. 2006).

For the test run, *Nannochloropsis oceanica* was used, and the velocity of the culture in the tubes was around 1 m s^{-1} which meant that volumetric flow rate was 340 L min^{-1}. Three-millimetre polymer beads were also included in the culture to ensure that sedimentation and fouling of microalgae in the plastic tubing were prevented. The nutrient mix feed was made up of agricultural fertilisers and urea, and the pH was set to between 7.3 and 7.8. Sandnes et al. (2006) used this as a cheaper alternative to Guillard's f/2 medium. The exact contents of the mixture was 1 g/L Red Superba from Norsk Hydro (4% P, 7% N, 21% K), 1 g/L of urea, 1 mL/L of the same vitamins found in Guillard's f/2 medium, and 0.5 mL/L of trace metals. After the mixture was created, salt had to be added until salt content was between 20 and 35 g/L. To do this, seawater that was filtered and treated was used (Sandnes et al. 2006).

The experiment with the system was repeated using natural sunlight and used sodium lamps that had an average light intensity throughout the period of 397 µmol m^{-2} s^{-1}. For the ones involving the artificial lights, the sodium lamps were placed in front of the biofence and covered with shading curtains. The test runs which had used sodium lamps also had daily cycles, one with 20 hours of light and 4 hours of darkness, another with 12 hours of light and darkness, and the final one with 16 hours of light and 8 hours of darkness. The experiments with the natural sunlight, 20 and 4 hours of light and darkness, respectively, and 12 hours of light and darkness, showed microalgae growth under the three different light regimes. On the other hand, the test run with the 16 hours of light and 8 hours of darkness showed the production of microalgae. At the end, the experiment with natural sunlight lasted 4 days, while the experiment using 20 hours of light and 4 hours of darkness lasted 6 days. After 6 days of using the 20 hours light and 4 hours dark regime, the culture was diluted for the test run using 12 hours of light and dark light cycle (Sandnes et al. 2006).

The results from the test run with natural sunlight showed an increase in biomass density of 1.23 g/L to 1.57 g/L after the 4 days. Moreover, there was an observation where biomass density increases during the day but falls slightly during the night. Furthermore, the same pattern of fluctuation in the biomass density was observed for the experiments using artificial lighting. For both the regimes of 20 hours of lights and 4 hours of darkness, and 12 hours of light and darkness, the biomass density increased when the lights were on and decreased slightly when the lights were off. However, on an average, the biomass density for both regimes increased over the 6 days of culturing. At the end, 29.8 litres of the culture was harvested in total over the 24-hour period of semi-continuous operation with the

16 hours of light and 8 hours of darkness cycle. The biomass density in the harvest was a constant 1.5 g/L throughout the day, which meant that the photobioreactor system produces around 44 grams of dry mass every day with this light cycle (Sandnes et al. 2006).

Sandnes et al. (2006) mentioned that it is possible to vary the set biomass density over time to increase productivity during certain situations. An example of such situations provided is when a microalgal species displays large differences in growth rate while ambient conditions change throughout the period of culture. Another situation mentioned was when a higher productivity could only be achieved through changing the biomass density of the culture in the buffer tank over the production period. Moreover, Sandnes et al. (2006) also noted that the microalgal species used in the test runs, *Nannochloropsis oceanica*, had a light absorption spectrum that covers the near-infrared light section due to its characteristics (Sandnes et al. 2006). Due to this, the optical density sensor may not be suitable for use on microalgae that do not absorb near infrared light. This may be solved by using an LED of a different colour that is within the absorption spectrum of that microalgal species.

12.7 AUTOMATED, CONTINUOUS CULTIVATION OF MICROALGAE WITH REMOTE CONTROL

$$dX/dt = \mu X - DX \tag{12.1}$$

$$dS/dt = (S_f - S)X - (\mu/Y_{S/X})X \tag{12.2}$$

$$\mu = \mu_{max}(S/(Ks+S)) \tag{12.3}$$

In a continuous photobioreactor, microalgae inside are harvested at the same time as fresh medium flows in to dilute the culture. D'Agostin et al. (2017) identified Eq. (12.1) to calculate the change in cell concentration over time, where X is the cell concentration, μ is the specific growth rate of microalgae, and D is the reactor dilution rate. Since dX/dt is equal to 0, and cell concentration is constant at steady state, the specific growth rate of the microalgae is equal to the rate of dilution. This would also mean that the specific growth rate could be controlled by changing the rate of dilution or, in other words, the flow rate of fresh medium into the bioreactor. D'Agostin et al. (2017) also identified Eqs. (12.2) and (12.3) to be relevant to microalgae culturing. For Eq. (12.2), S is the concentration of nutrients of the reactor's culture, S_f is the nutrient concentration of feed, and $Y_{S/X}$ is the substrate conversion factor. Equation (12.3) represents the Monod model, where μ_{max} is the maximum specific growth rate, μ is the specific growth rate, s is the nutrient concentration, and K_s is the saturation constant (D'Agostin et al. 2017).

With that knowledge, D'Agostin et al. (2017) designed an automated, continuous cultivation system that could be operated remotely on the laboratory scale. The system consisted of two containers and a photobioreactor tank. One of the containers stored fresh medium, while the other stored any harvested culture. To transport the fresh medium from the container to the photobioreactor, a peristaltic pump was used due to the fact that its moving parts are not in contact with the liquid being transported. The pump was also used to harvest and transport microalgae from the photobioreactor to the other container. Moreover, a 1.8 L flask was used as a photobioreactor, with two 40-W fluorescent lights providing the illumination needed for cultivation. Furthermore, air was bubbled into the system through a tube by opening a valve to provide agitation and as a supply of CO_2. D'Agostin et al. (2017) did not mention the source of the air, but it may have been an air compressor or a compressed air cylinder as a pressure difference is needed for the air to be forced into the photobioreactor. Besides that, the photobioreactor contains sensors that measured microalgae biomass density, temperature, and illuminance from the light source and pH. A third peristaltic pump also transported culture from the photobioreactor, through a loop, to an optical sensor before returning it to the reactor. The system is illustrated in Figure 12.7 (D'Agostin et al. 2017).

The optical sensor consisted of a spectrophotometer cuvette with an optical path of 10 mm, a 5-mm LED that emits light at 517 nm, and a phototransistor. The LED was placed on one side of the cuvette, while the phototransistor was on the other to measure the light going through the tube. After its construction, the optical sensor then covered in an opaque, black material. The optical sensor, along with the rest of the other sensors, was connected to an Arduino MEGA 2560™ microcontroller through the digital and analogue ports. Once data was received by the microcontroller, it was sent to a Raspberry Pi, which acted as a data server through a USB connection and serial communication. Both the microcontroller and Raspberry Pi are also shown in Figure 12.7. Using a script, data was recorded in a database and an XML file and displayed on the main page of the supervisory system. Also located in the main page, also known as the main screen, were the pump controls and other settings of the experiment. The other two pages were a data insertion page and a graph visualisation page (D'Agostin et al. 2017).

Before the test run, the optical biomass sensor had to be calibrated according to the different biomass densities. To do this, a biomass sample, with a concentration of 0.637 ± 0.04 g L^{-1}, was transferred into 10 different test tubes and diluted into 10 different concentrations. After that, a peristaltic pump extracts the diluted culture and passes it through the optical biomass sensor. The 10 dilutions were, in the form of ratios, 10 parts of the biomass sample to 0 parts of pure water, 9 parts of the biomass sample to 1 part of pure water, 8 parts of the biomass sample to 2 parts of pure water, and so on. This pattern ended with the second last dilution ratio, or the ninth dilution ratio, of 2 parts of the biomass sample to 8 parts of pure water. The last dilution ratio was 1 part of the biomass sample to 10 parts of pure water. At the end, when the calibration was over, a calibration curve was obtained (D'Agostin et al. 2017).

$$P = (X - X_0)/(t - t_0) \tag{12.4}$$

$$Pc = XD \tag{12.5}$$

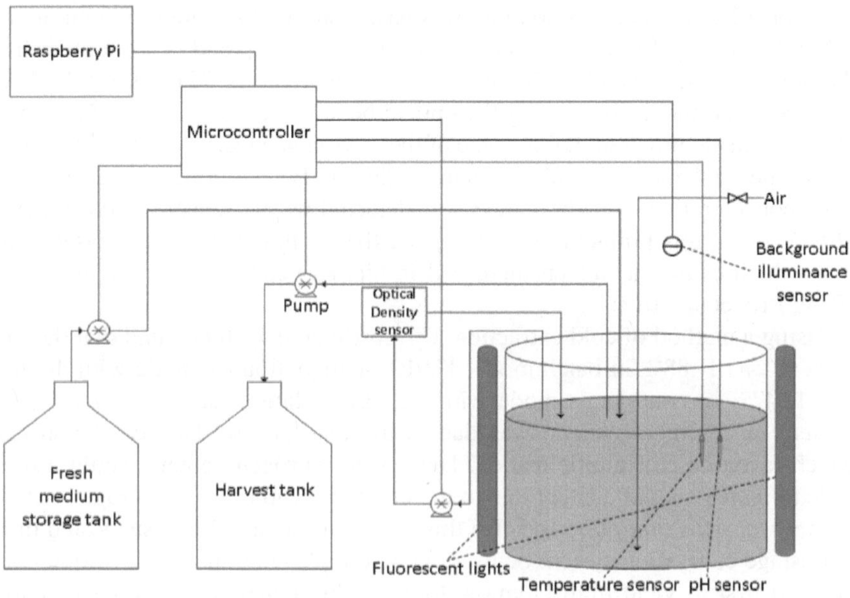

FIGURE 12.7 Sketch of the automated, continuous photobioreactor system (D'Agostin et al. 2017).

For the test run, *Acutodesmus obliquus* was cultivated using Chu's medium. The temperature of the culture was 20°C ± 2°C, with an air flow rate of 1 L min^{-1}. Moreover, three test runs were done, where the microalga was cultivated in batch, semi-continuous, and continuous modes. For the test run in batch mode, the maximum biomass concentration obtained was 0.987 g L^{-1}, with an average productivity of 49.97 mg L^{-1} day^{-1} that was calculated using Eq. (12.4). The growth curve for under this mode showed the typical characteristics of a sigmoid curve, with a lag phase, growth phase, and stationary phase. Also, Eqs. (12.2) and (12.3) were solved using the Euler method, and Microsoft Excel's Solver tool estimated μ_{max}, K_s, and $Y_{x/s}$ to be 0.48 day^{-1}, 0.2148, and 0.3146 day^{-1}, respectively. For the semi-continuous mode, the microalga was allowed to grow without any dilutions until Day 7. After Day 7, dilutions of 40% were done whenever the biomass density was at or above 0.7 g L^{-1}, and the amount of microalgal culture harvested was also the same as the amount of new culture added. The microalga was cultured for 19 days using semi-continuous mode in total, and the average productivity, which was also calculated using Eq. (12.4), was also 58.11 mg L^{-1} day^{-1} (D'Agostin et al. 2017).

At the end, in continuous mode, the microalgae were grown without dilution for the first 6 days, where biomass concentration reached 0.686 ± 0.008 g L^{-1}. After Day 6, a small dilution rate of 0.05 day^{-1} was applied in order to prevent large losses of biomass. After Day 11, the dilution rate was increased to the intended value of 0.2 day^{-1}. The dilution rate was based on the flow rate of the pump, Eq. (12.1), Eq. (12.2), Eq. (12.3), and the values of μ_{max}, K_s, and $Y_{x/s}$ obtained from the previous test runs. Overall, an average productivity of 74.6 mg L^{-1} day^{-1} was observed, and the mean error of the optical biomass sensor was 8.46%. Also, on Day 6 to Day 8, precipitation of some of the microalgae from the culture occurred due to the implementation of continuous dilution. Because of this, a drop in biomass concentration over the 2 days was detected by the sensor. D'Agostin et al. (2017) solved this problem by maximising the air flow rate for increased agitation, which caused a sudden increased in biomass concentration of the culture (D'Agostin et al. 2017).

Using a method of cold extraction, D'Agostin et al. (2017) found that the lipid content was 17.88% in batch mode, 17.91% semicontinuous mode with dilution, and 11.47% in continuous mode with 0.2 day^{-1} dilution rate. The smaller lipid content in continuous mode was due to the constant availability of nutrients, which favoured cell multiplication. Decreasing nitrogen content would also do little to increase lipid content due to the constant synthesis and storage of starch by *Acutodesmus obliquus.* To solve this, D'Agostin et al. (2017) suggested that a multistage cultivation system be used. For example continuous cultivation could be used first to yield high biomass, followed by a stage of lipid accumulation (D'Agostin et al. 2017). Moreover, when scaling the system, a different, stronger material might be used for the photobioreactor for added structural support, such as steel. If the material is opaque, it would not be possible to place the lights outside the photobioreactor. Due to this, the lights must be placed inside the photobioreactor, similar to the design shown in Section 12.5 by Erbland et al. (2020). Also, on the large scale, a cooling loop might be needed for uniform heat distribution. At the end, for the large scale, a separate CO_2 supply may be needed as the CO_2 content in air may not be sufficient to meet demands.

12.8 CONCLUSION: SUMMARY OF THE DESIGNS PRESENTED AND THEIR RESPECTIVE AUTOMATION SYSTEMS

Multiple research papers have presented different designs of an automated photobioreactor. The first one, by Fuchs et al. (2021), was an automated flat panel photobioreactor that was scalable, had no dead zones, easy to sterilise and disassemble, and asymmetrical in shape. The automation system used pH and temperature sensors to monitor data, as well as autonomously operated mass flow controllers to control the supply of gases and pH. The second design that was presented by Rahmat et al. (2020) was about automated, cuboid-shaped photobioreactors with remote control technology using the Internet of Things. The automation system of the second design used Arduino WeMos D1s as controllers; Raspberry Pi to transmit data and receive instructions; temperature, light, and colour sensors for monitoring; and heaters and lamp for the control of temperature and light.

The third design, by Borowiak et al. (2021), introduced a new photobioreactor system that was named the airlift modular automated photobioreactor system (AMAPh-S). Similar to Fuchs et al. (2021), the automation technology in the AMAPh-S consists of temperature and pH sensors for monitoring and an autonomous CO_2 dosing system for the control of pH. Next, the design by Erbland et al. (2020) was an automated photobioreactor that was built in full size. The research's approach to automation consisted of a custom-built monitor and control unit (MCU), a side loop that had a pH and temperature sensor, an optical sensor to monitor microalgal biomass density, and an autonomous CO_2 supply for the control of pH. However, all of the aforementioned designs operated in batch.

A non-batch cultivation system was designed by Sandnes et al. (2006), who designed an automated, semi-continuous cultivation system with a tubular photobioreactor. The automation system consisted of a controller, a pH and temperature sensor, a custom-made optical sensor, a distributed data acquisition (DAQ) module, a computer that was connected through a local area network (LAN), and valves which enabled the autonomous control of the supply of CO_2 and fresh water/nutrient mix. Finally, D'Agostin et al. (2017) presented an automated cultivation system, with the technology of the Internet of Things, which runs continuously. The automation system here consisted of an Arduino MEGA 2560 as a controller; a Raspberry Pi to send data and receive instructions; a pH, temperature, and optical sensor to collect data for monitoring; and peristaltic pumps and a CO_2 supply that were autonomously controlled.

REFERENCES

Borowiak, Daniel, Paweł Lenartowicz, Michał Grzebyk, Maciej Wiśniewski, Jacek Lipok, and Paweł Kafarski. 2021. "Novel, Automated, Semi-Industrial Modular Photobioreactor System for Cultivation of Demanding Microalgae That Produce Fine Chemicals—The next Story of H. Pluvialis and Astaxanthin." *Algal Research* 53: 102151. https://doi.org/10.1016/j.algal.2020.102151.

D'Agostin, D. A. L., G. M. Domene, A. S. Oliveira, M. J. C. Bonfim, and A. B. Mariano. 2017. "Automated System for Continuous Microalgae Cultivation in Photobioreactors." *Revista Da Engenharia Térmica* 16 (2): 3–9. doi:10.5380/reterm.v16i2.62204.

Erbland, Patrick, Sarah Caron, Michael Peterson, and Andrei Alyokhin. 2020. "Design and Performance of a Low-Cost, Automated, Large-Scale Photobioreactor for Microalgae Production." *Aquacultural Engineering* 90: 102103. https://doi.org/10.1016/j.aquaeng.2020.102103.

Fuchs, Tobias, Nathanael D. Arnold, Daniel Garbe, Simon Deimel, Jan Lorenzen, Mahmoud Masri, Norbert Mehlmer, Dirk Weuster-Botz, and Thomas B. Brück. 2021. "A Newly Designed Automatically Controlled, Sterilizable Flat Panel Photobioreactor for Axenic Algae Culture." *Frontiers in Bioengineering and Biotechnology* 9: 566. doi:10.3389/fbioe.2021.697354.

Khan, Muhammad Imran, Jin Hyuk Shin, and Jong Deog Kim. 2018. "The Promising Future of Microalgae: Current Status, Challenges, and Optimization of a Sustainable and Renewable Industry for Biofuels, Feed, and Other Products." *Microbial Cell Factories* 17 (1): 36. doi:10.1186/s12934-018-0879-x.

Rahman, Khondokar. 2020. "Food and High Value Products from Microalgae: Market Opportunities and Challenges." In, 3–27. doi:10.1007/978-981-15-0169-2_1.

Rahmat, Ayi, Indra Jaya, Totok Hestirianoto, Dedi Jusadi, and Kawaroe Mujizat. 2020. "Design a Photobioreactor for Microalgae Cultivation with the IOTs (Internet of Things) System." *Omni-Akuatika* 16 (July): 53. doi:10.20884/1.oa.2020.16.1.791.

Ravindran, Balasubramani, Sanjay K. Gupta, Won-Mo Cho, Jung K. Kim, Sang R. Lee, Kwang-Hwa Jeong, Dong J. Lee, and Hee-Chul Choi. 2016. "Microalgae Potential and Multiple Roles—Current Progress and Future Prospects—An Overview." *Sustainability* 8 (12). doi:10.3390/su8121215.

Rösch, Christine, Max Roßmann, and Sebastian Weickert. 2019. "Microalgae for Integrated Food and Fuel Production." *GCB Bioenergy* 11 (1): 326–34. https://doi.org/10.1111/gcbb.12579.

Sandnes, J. M., T. Ringstad, D. Wenner, P. H. Heyerdahl, T. Källqvist, and H. R. Gislerød. 2006. "Real-Time Monitoring and Automatic Density Control of Large-Scale Microalgal Cultures Using Near Infrared (NIR) Optical Density Sensors." *Journal of Biotechnology* 122 (2): 209–15. https://doi.org/10.1016/j.jbiotec.2005.08.034.

13 Remaining Challenges and Uncertainties

Sze Yin Cheng[1], *Zhi Ting Ang*[2] *and Pau Loke Show*[2]

[1] Institute of Biological Sciences, Faculty of Science, University of Malaya, Kuala Lumpur, Malaysia

[2] Department of Chemical and Environmental Engineering, Faculty of Science and Engineering, University of Nottingham Malaysia, Jalan Broga, Semenyih, Selangor Darul Ehsan, Malaysia

CONTENTS

13.1 Introduction 422
13.1.1 Overview of the Challenges 422
13.2 Challenges 422
13.2.1 Challenges in Digitalization of Microalgae Biotechnology 422
13.2.1.1 Potential Improvements 423
13.2.2 Challenges in Microalgae Cultivation 424
13.2.2.1 Potential Improvements 426
13.2.3 Challenges in Microalgae Harvesting Process 426
13.2.3.1 Potential Improvements 427
13.2.4 Challenges in Biorefineries Using Microalgae Biotechnology 430
13.2.4.1 Potential Improvements 431
13.2.5 Challenges of Waste Treatment 432
13.2.5.1 Potential Improvements 433
13.3 Uncertainty 434
13.3.1 Mass Culture of Microalgae 434
13.3.1.1 Potential Improvements 434
13.3.2 Extraction of Bioactive Compounds 435
13.3.2.1 Potential Improvements 436
13.4 Conclusion 436
References 439

DOI: 10.1201/9781003202196-13

13.1 INTRODUCTION

Microalgae are considered to be a promising material to promote resource-efficient bio-economy while meeting the regulatory drivers for environmental sustainability. Therefore, numerous studies have been conducted in the perspective of ecological and environmental services since 1960s (Alam and Zhongming 2019). The studies that have been conducted using microalgae include the areas of production of human food and animal feed, carbon dioxide (CO_2) sequestration, wastewater treatment, production of biofuels as well as generation of recombinant proteins (Posten and Chen 2016; Low et al. 2021; Chai, Tan et al. 2021; Chai, Chew et al. 2021). Despite the distinctive features that microalgae possess, the studies on microalgae mainly focus on the application of microalgae as an alternative source for clean and sustainable fuel. These studies were driven by the rise in oil prices, realization of diminishing resources, and the rising awareness on the environmental impact of conventional fuels. Countless industries and research centres have been established in more than 80 countries including the United States, China, Spain, Korea, and Australia to explore and assess the potential of microalgae for large-scale application (Alam and Zhongming 2019). Besides, it is noteworthy that the recent focus on the study of microalgae has been shifted to value-added bioproducts, bioremediation, and water reclamation (N Rashid, Park, and Selvaratnam 2018).

13.1.1 Overview of the Challenges

Although microalgae offer significant potential in different fields, there are some challenges and uncertainties that must be addressed in downstream processes of microalgal refinery. These challenges and uncertainties must be identified and tackled before making the process and the technology commercially viable. In this chapter, several case studies will be conducted to identify and address the challenges associated with the digitalization of microalgae biotechnology, biorefineries, and waste treatment. The main focuses of these studies are identifying environmental impacts of microalgae biotechnology and analyzing uncertainties that arise from the mass culture of microalgae as well as the extraction of bioactive compounds from microalgae. Further improvements and the potential technology that can be applied to mitigate these challenges and uncertainties are proposed upon identifying the respective challenges and uncertainties.

13.2 CHALLENGES

13.2.1 Challenges in Digitalization of Microalgae Biotechnology

A sustainable biological transformation is essential for the traditional manufacturing industry for both economy and society (Miehe et al. 2018). However, there are several bottlenecks in biological transformation process including demographic change, individualization, globalization, digitalization, assurance of sustainable resource availability, and minimization of environmental impacts (Miehe et al. 2018). Therefore, the visions of creating a complete circuitry of industrial metabolism or circular economy have attracted countless attentions recently to address these bottlenecks. The digitalization of traditional manufacturing industry is viewed to be a promising approach to change the face of traditional value creation. In the microalgae biotechnology context, biological transformation of industrial value creation can be defined as the systematic application of natural processes to optimize a manufacturing system regarding the associated business and societal challenges by achieving a convergence between the biosphere and techno-sphere (Miehe et al. 2018).

The convergence of biosphere and techno-sphere can be categorized into three different steps, which are known as inspiration, integration, and evolvement. The inspiration step can be defined as the transfer of natural phenomena in the form of process analysis, abstraction as well as realization of technical aspects. This approach is based on the concept of technically duplicating characteristics of nature, process of evolution, and principles of the nature. On the other hand, the integration can be defined as the integration of biotechnologies in traditional processes for value creation. One of the cases that can be studied is the growth of algae in photobioreactor using carbon dioxide and process waste heat emitted from factories. The cultivated algae can then be utilized in different value creation processes such as energy and material recovery from microalgae biomass, conversion of biomass onto biogas, production of raw material for bioplastic from microalgae oil, and the application of algal biomass for heavy metal filter (Miehe et al. 2018). The last step, evolvement, is the thorough combination of biosphere and techno-sphere in the form of cross-linking between bioprocess, manufacturing, and information technology. The bio-intelligent manufacturing system is designed based on the characteristics and nature of living beings under this step (Miehe et al. 2018).

Numerous studies focusing on the combination of biosphere and techno-sphere have been carried out to investigate the possibility of utilizing information technology on cultivation of microalgae as well as microalgae biorefinery. Borowiak et al. (2021) conducted a study on the production of astaxanthin using *Haematococcus pluvialis* microalgae in an airlift modular automated photobioreactor system (AMAPh-S). This study showed that the AMAPh-S enabled a high-level control which resulted in a significantly high microbial purity and chemical purity of metabolite. Besides, it was also proven that AMAPh-S can be scaled up to meet highly specific demands. On the other hand, Wang et al. (2021) conducted a review on application of Internet of Things (IoT) in microalgae biorefinery industries. It was concluded that the utilization of IoT is advantageous on attaining an automatic, low-cost, intelligent, and greener production process.

13.2.1.1 Potential Improvements

According to Miehe et al. (2018), there are numerous actions that must be conducted to direct the traditional processes towards bio-intelligent manufacturing including future research, initiatives of policy, societal involvement, and industrial investment. Throughout the study conducted by Yang et al. (2018), there are 10 major fields of action that must be adapted by the industries including intra-organizational challenges, social, and political challenges. The intra-organizational challenges include the development of bio-based and functional bio-intelligent materials, the configuration of bio-tech-interfaces for communication via sensors and actuators, development of bio-intelligent manufacturing technology, and the allocation of human-centred workplaces in the bio-intelligent manufacturing environment. Besides that, the encountered challenges also include the real-time data exchange between biological and technological systems, decentralization and utilization of renewable energy generation, expansion, and standardization of methodological basis of technology impact assessment, and financial support for new business models with innovative approaches.

On the other hand, the social and political challenges include the communication with the public to explain the risk and chances of these developments, creation of knowledge transfer between biologists, bio-engineers, manufacturing engineers, managers, and humanists (Yang et al. 2018). Actions stated before can be taken for the improvements of microalgae cultivation, microalgae harvesting, biorefineries as well as waste treatment using microalgae biomass. Sensors and actuators can be used to develop bio-tech-interfaces for communication during the cultivation and harvesting processes for microalgae. Apart from that, cloud servers can be utilized to address the issues related with real-time data exchange between biological and technological systems as well as exchange of knowledges between experts including biologist, engineers, and stakeholders.

13.2.2 Challenges in Microalgae Cultivation

Microalgae utilize carbon dioxide and sunlight for photosynthesis, and the algae transform the energy source into complex organic molecules using photoautotrophic metabolism. Both heterotrophic and mixotrophic algae utilize the same carbon and energy source while performing photosynthesis using light energy. Although the heterotrophic and mixotrophic algae consume the same carbon and energy source, their energy patterns are different. On the other hand, the photoheterotrophic algae transform the organic compounds as the carbon sources and utilize the light for photosynthesis. According to these observations, organic compounds and carbon dioxide are essential and necessary for the algae to carry out photosynthesis. The main cultivation practices that are used in the industries can be categorized as open pond cultivation system and closed cultivation system or photobioreactors (PBRs) (Upadhyay et al. 2019).

One of the limitations of producing algae cultivation at a large or commercial scale is that it is difficult to keep a selected species at a high density. The loss of microalgae culture may occur in a semi-sterile or less protected environment (Upadhyay et al. 2019). The growth of algae in cell culture can be influenced by abiotic factors such as temperature, light intensity, carbon dioxide, pH, toxic chemicals, dissolved oxygen, and concentration of salt and biotic factors including bacteria, fungi, virus as well as contaminated algae species. The uncertainties in maintaining the optimum light intensity as well as temperature for the microalgae cultivation are discussed in Section 13.3.1. Other than that, the potential improvements that can be made to achieve the optimum light intensity and temperature are reviewed in Section 13.3.1.1 as well.

The open pond cultivation system simulates the natural environment of algae, and it consists of natural water system, ponds, raceway ponds, and artificial large shallow tanks. The technical design for large-scale algae cultivation system consists of a structure that is commonly made up of concrete with different diameter and length with two or more circulating channels, and the typical fluid depth is 15–35 cm (Upadhyay et al. 2019). Raceway or open pond cultivation system offers several benefits. First, they are cheaper to construct and maintain as compared to other cultivation system. Evidently, the power consumption for a newly designed raceway pond was 60% lower as compared to the traditional raceway pond, and the newly designed raceway pond was able to maintain the velocity of the flow, hence providing a better mixing for photoautotrophic growth of microalgae (Brennan and Owende 2010). Second, the estimated lifespan for the open pond cultivation system is rather long, and it was estimated to be 10 years (Upadhyay et al. 2019). Open pond system utilizes direct sunlight as light source hence is able to lower construction and operating costs (Chen et al. 2011).

In addition, some open pond cultivation systems include the utilization of commercial and municipal wastewater treatment as well as flue gases such as carbon dioxide that improve sustainability of the technology. However, there are several challenges that must be addressed such as the system's requirement of continuous energy supplies for water circulation and paddlewheels, maintenance of required water depth, limited diffusion of carbon dioxide, low biomass productivity, large amount of water and nutrient, and ionic difference due to evaporation of water. As the open pond cultivation system is an excellent habitat for a large variety of microorganisms, therefore the risk of culture contamination is also one of the most challenging issues that must be addressed (Pulz and Gross 2004).

On the other hand, the closed cultivation system or photobioreactors (PBRs) are closed reactors coupled with control system. There are several types of reactors such as flat-plate reactor, bubble or tubular serpentine-type reactor, big-bag or plastic foil bioreactors, multi-fold photobioreactors, helical-type photo-bioreactor, airlift bioreactors, stirred tank, and floating photobioreactors (Richmond 2004). These reactors have been designed to accommodate the mass cultivation of specific cyanobacteria or microalgae species (Richmond 2004). The advantages of installing a closed system for mass cultivation of microalgae are such that the growth of biomass can be observed easily, and the cultivation-related parameters such as carbon dioxide, pH, temperature, and light can be controlled as compared to open pond cultivation system (Upadhyay et al. 2019). Although the closed cultivation system or photobioreactors (PBRs) possess numerous advantages, the closed cultivation system has several limitations such as bio-fouling, agitation stress, and overheating.

13.2.2.1 Potential Improvements

According to the comparison made between open pond cultivation system and closed cultivation system or photobioreactors (PBRs), the efficiency of closed cultivation system or photobioreactors (PBRs) is higher as compared to that of the open pond cultivation system in terms of quality as closed cultivation system or photobioreactors (PBRs) can be operated at closely monitored conditions, hence, overcoming the disadvantages of open pond cultivation system (Narala et al. 2016). The closed cultivation system or photobioreactors (PBRs) allow the concentration of cell to maintain within the volumetric productivities level (Brennan and Owende 2010). Besides that, the closed cultivation system or photobioreactors (PBRs) also are able to maintain the amount of carbon dioxide at optimum level for mass cultivation of diverse microalgae species while the open pond cultivation system cause a significant loss of imported carbon dioxide (Carvalho et al. 2011). On the other hand, the closed cultivation system or photobioreactor (PBRs) enables proper agitation which is essential to ensure the uniform distribution of light energy within the reactor.

Apart from that, the light and dark cycles can be regulated properly to improve the photosynthetic efficiency of microalgae growing in the closed cultivation system or photobioreactors (PBRs) (Brennan and Owende 2010). Other than light and dark cycles, the photobioreactor is also capable of controlling climatic conditions within the reactor in a continuous mode for mass cultivation of a wider range of microalgae species. On the other hand, issues related with bio-fouling, agitation stress, and overheating can be minimized by improving the design of photobioreactors (PBRs). Bio-fouling can be reduced by considering the fouling factors during the design of photobioreactors (PBRs) (Couper et al. 2009). The agitation stress issue and overheating issue can be overcome by selecting materials with higher stress-resistance and installing cooling jacket on the reactor (Sinnott and Towler 2019). As the closed cultivation system or photobioreactor (PBR) converts a large amount of energy to light source for photosynthesis, therefore, the system is very energy intensive. The energy-intensive issue can be optimized by conducting process integration to reduce the cost of utilities (Smith 2005).

13.2.3 Challenges in Microalgae Harvesting Process

The harvesting process can also be known as the recovery process where the biomass is extracted from the culture media, and it is one of the important yet challenging stages to achieve sustainable economics for the production of microalgae-based biofuels (Kumar et al. 2015). Harvesting process for commercial production of biofuels is expensive for operational costs as the harvesting process contributes around 35% of the total cost of biomass production due to dilute nature and the small sizes of the microalgae cultures (Upadhyay et al. 2019). There is no specific method that can be applied to address different forms of microalgae culture. Besides, the harvesting process of algae in water at large scale is also one of the major challenges as the microalgae and cyanobacteria are very small in size ranging from 3 to 25 μm and from 0.1 to 2 μm (Rastogi et al. 2018).

13.2.3.1 Potential Improvements

The schematic diagram of microalgae harvesting process is shown in Figure 13.1. The potential improvements that can be made for the harvesting process of microalgae include conducting pre-screening, identification as well as selection of process which are effective in harvesting the specific type of microalgae. The common physical methods that are used to harvest microalgae include filtration, floatation, gravity sedimentation, and centrifugation. On the other hand, the chemical method which can be applied for microalgae harvesting is flocculation whereas the biological method that can be used to harvest different variants/strains of algal biomass as required according to the end products is bio-flocculation. The selection of harvesting method for a specific variant of microalgae is important to ensure that the harvesting cost is refined and hence the production cost reduced. The reduction in the production cost will stimulate the economic pull toward the bioenergy derived from microalgae.

Filtration process is one the most common methods. Generally, a filtration system is made up of a pressure pump and a membrane with specific pore size. Filtration method is very useful to gather all the algal cells despite low density. The challenging issue associated with filtration process is the potential of membrane blockage, hence causing inefficient mass recovery of the cell. This issue cannot be ignored as it will enhance the economic burden on the microalgae harvesting process as it requires a large amount of energy.

Several changes can be made such as installing direct vacuum, incorporating reverse-flow vacuum coupled with a stirring blade overhead filter, and combining the membrane filtration with antifouling microfiltration to ensure that the

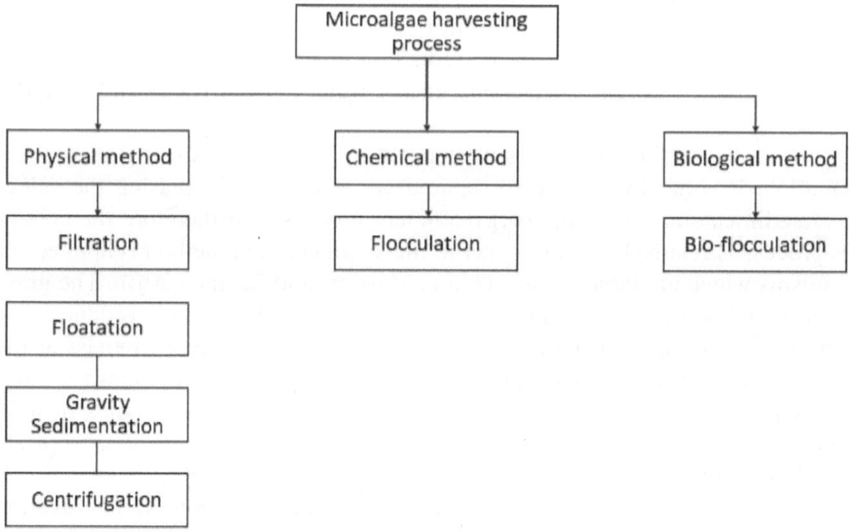

FIGURE 13.1 A schematic diagram showing microalgae harvesting process.

harvesting process using membrane filtration system can be economically feasible to a certain level (Zhang et al. 2010). Apart from that, there are also several emerging designs that are proposed such as the combination of filtration system with a photobioreactor (PBR) forming a membrane photobioreactor (PBR), separating the harvesting process upon cultivation of algae as well as controlling the fouling of membrane at a higher concentration of biomass (Rawat et al. 2011).

Other than the filtration process, floatation is also one of the most common methods that can be used for microalgae harvesting. Floatation can be defined as a process where the solid materials or particles are adhered to the air bubbles or gas molecules which can transport those particles to the liquid surface. Floatation possesses more advantages, and it is way effective than other harvesting methods due to the self-floating characteristics as well as low algal density. However, the floatation method is only applicable for algal cells which are smaller than 500 μm in size (Upadhyay et al. 2019). Besides that, the floatation process is dependent on the mechanisms of negative charge on the microalgae surface (Moosai and Dawe 2003). Part of the algal cells float on the media due to the hydrophobicity and action of growth phase, and some of the algal cells are induced by gas or air bubbles from super-saturated water.

The floatation process looks similar as a separation process due to gravity, where the air bubbles are attached to a solid particle and move upwards to the surface of liquid (Carvalho, Meireles, and Malcata 2006). The advantages of applying floatation process include the low requirements on working space, possibility of large-scale harvesting, relatively shorter process times, and insignificant step up cost for preliminary equipment (Rastogi et al. 2018). According to a study conducted by Laamanen, Ross and Scott (2016), floatation process has been extensively applied in water treatment system, and it has been proven to be an alternative option for the treatment of microalgae as well as other microorganisms. It is noteworthy that the salinity is a challenging issue for adhesion of bubble-cell, which will result in the reduction of floatation efficacy. This issue can be tackled by adding surfactants on the water surface (Qiao and Chandra 1997).

Another harvesting process using concept of gravitation pull is gravity sedimentation. It is mostly applied to separate algae cells by increasing the velocity of sedimentation. It is an energy-efficient process, and therefore the cost of this process is relatively low. In general, the sedimentation method is applied for organisms which are larger in size (Mata, Martins, and Caetano 2010). The gravity sedimentation process is applicable for small-scaled economic products due to the low microbial settling rates as well deterioration of algae biomass at the specific settling time (Barros et al. 2015). It is also widely used in the wastewater treatment plant as a cost-effective method to eliminate solid particles in bulk quantity. However, while applying this method, it is advisable to add appropriate flocculant before to increase algal settling rate (Udom et al. 2013).

Another widely used harvesting method is centrifugation. Centrifugation can be defined as the process of separating microalgae biomass from culture medium using a centrifugal force which is higher than gravitational force to boost suspension separation. Centrifugation process is a single-step process that can be used to harvest the microalgae cells for a set time period, and this process presents a significantly high efficiency which is close to 100% (Upadhyay et al. 2019). According to the measurement conducted by Wang et al. (2014), the energy consumption value for microalgae harvesting using centrifugation process was 8 kWh/m^3 with a microalgae suspension feed rate of 1 L/min. Despite remarkable performance, centrifugation process is not applicable to all types of algae cells, and the application of centrifugation for microalgae harvesting is dependent on the variants of algal cells.

It was recorded that the microalgae harvesting using centrifugation was able to generate 15–25% of algal biomass on average with an energy requirement of 45–75 kW (Milledge and Heaven 2013). There are several challenging issues that must be handled during the application of centrifugation process for microalgae harvesting. One of challenging issues was that the amount of algal biomass must exceed 92% to ensure that sufficient biofuel can be produced on a large scale (Upadhyay et al. 2019). Furthermore, the application of centrifugation process for harvesting of microalgae could damage the algae cells due to the large gravitational force. The valuable nutrients in the medium could be affected due to the damaged algae cells. Besides that, the operational cost will increase notably as the amount of energy required will increase to harvest the large amount of microalgae biomass.

Apart from the aforementioned physical methods, chemical method can be utilized as well for microalgae harvesting. Flocculation is an effective method for microalgae harvesting, and it can be applied to recover a large amount of microalgae cells from different groups of taxonomic (Molina Grima et al. 2003). The microalgae harvesting using flocculation is carried out by dispersing the aggregate, which thereby forms large particles by collision and adhesion to one another via the interaction of the algal cells' surface charge with the flocculent. Flocculation method improves the sedimentation process by both centrifugal force as well as gravitational force. There are several types of flocculation methods which are widely used such as physical flocculation, physicochemical flocculation, auto-flocculation, and bio-flocculation.

Inorganic salts and polymeric coagulants are utilized for harvesting microalgae during chemical flocculation, and the salts and coagulants neutralize the surface charge by introducing chemicals such as ferric ions and Alum [$Al_2(SO_4)_3$]. Other than that, cationic polymers and alkali are applied for flocculation as well to fluctuate or adjust the pH (Zahrim, Tizaoui, and Hilal 2011). Numerous researchers have stated that a maximum of 90% algae recovery can be achieved by using auto-flocculation despite the drawbacks such as low reliability, longer process time, and the need of calcium and magnesium ions (Milledge and Heaven 2013). The nature of the flocculants applied is dependent on the final product, and it is crucial for the selected flocculants to be inexpensive, effective, functional, and non-toxic at low concentrations.

Other than the chemical method, biological method can be applied as well for microalgae harvesting. Bio-flocculation can be defined as a process that assists sedimentation using either microorganisms or their polymer flocculants (Wan et al. 2015). In general, the bio-flocculation method is used for wastewater treatment (Van Den Hende et al. 2011). Bio-flocculation method is an emerging technique toward environment-friendly, sustainable, and low-cost harvest of algae cells (Venkata Mohan et al. 2016). The bio-flocculation method is cheaper for microalgae biomass harvesting in bulk quantity as compared to other approaches. The efficiency of bio-flocculation process is dependent on the flocculation productivity of used microbial flocculants associated with the behaviour of the specific algal species (Vandamme, Foubert, and Muylaert 2013). *Exo*-polysaccharides (EPSs) which can be defined as high molecular weight carbohydrate polymers play an important role during the microbial flocculation process (Satpute et al. 2010).

During bio-flocculation, EPSs secreted by bacteria and algae are necessary for cell–cell adhesion that eventually will form the floc. The productivity of microbial flocculation is dependent on the EPSs' synthesis by microorganisms as well as the capability of the microalgae to form floc. According to Barros et al. (Barros et al. 2015), bridging process by EPSs leads to flocculation process of microalgae. It was observed that the microorganisms may increase the yields of lipid under microbial flocculation, and the subsequent culture media can be reused to reduce the algal cultivation costs during the biofuel production (Brennan and Owende 2010). It is induced by the presence of nutrients such as nitrogen and carbon under stress conditions to harvest several variants/strains of algae. An improvement can be made by combining the bio-flocculation with either centrifugation or sedimentation to enhance the harvesting efficiency.

13.2.4 Challenges in Biorefineries Using Microalgae Biotechnology

Microalgae are a promising alternative source for production of renewable fuels as they contain a significant amount of nutrients including 11–56% of carbohydrates, 8–70% of lipids, 40–70% of proteins, and 3–5% of pigments (Muhammad et al. 2021). Microalgal lipids can be converted into biodiesel, whereas the carbohydrates can be used in the fermentation processes to produce bioethanol or biohydrogen. Apart from that, microalgae also contain some several valuable compounds such as polysaturated fatty acids and antioxidants. Therefore, biorefineries using microalgae biotechnology have attracted countless interests as these compounds widened the possibilities of producing biofuels as well as other value-added components from microalgae (Chew et al. 2017). Furthermore, the microalgae can utilize the carbon dioxide and other contaminants from flue gases, and the biorefineries using microalgae biotechnology do not require a large amount of land and good quality of water which can be used for other purposes such as agriculture (A. Fernández F.G., J.M. Fernández-Sevilla 2009).

In general, microalgae can be grown in open ponds or closed systems that consist of mixing and concentrating processes (Jacab-Lopes et al. 2020). The challenges of microalgae cultivation are reviewed and summarized in Section 13.2.2. The downstream processes of microalgae involve harvesting, drying, cell rupture, extraction of compounds, and conversion technologies (Jacab-Lopes et al. 2020). The challenges associated with microalgae harvesting are discussed in Section 13.2.3. The biomass-based diesel can be produced by different processes including transesterification and hydro-processing of lipids, pyrolysis of cellulose as well as conversion via carboxylic acid and gasification (Živković et al. 2017). Currently, biodiesel is mainly produced by transesterification of acylglycerols using methanol or ethanol extracted from natural resources over a catalyst (Živković et al. 2017). The resulting mixture containing methyl esters or ethyl esters is fit to be used as biodiesel if it meets the quality standards of biodiesel. The application of microalgae for biodiesel production possesses numerous economic and ecological advantages as compared to biodiesel produced from other crops including high photosynthetic efficiency, high oil yield, high growth rate, synergy with carbon dioxide bio-fixation, and wastewater treatment (Muhammad et al. 2021). Mandotra et al. (2016) stated that a biodiesel production process involves microalgae harvesting, processing, centrifugation, flocculation, filtration as well as transesterification.

Despite all the advantages that microalgae biorefineries possess, there are several challenges that must be addressed before the biorefinery process can be commercialized. One of the major challenges is that the microalgae biorefinery process is economically unfeasible (Ferreira and Gouveia 2020). The production cost of microalgae biomass is calculated to be $2.71/kg, and it is still higher for commercialization (Kang et al. 2015). The reason of high production cost is due to the fact that it requires an energy extensive process for microalgal harvesting which accounts for 20% to 30% of the total production costs (Naim Rashid et al. 2019). Apart from the high production costs, there are several challenging issues that must be tackled for microalgae-based biorefinery process including the difficulty in controlling the conditions of the culture, contamination of bacteria as well as unstable supply of light and temperature (Ferreira and Gouveia 2020). The selection of most suitable microalgal strains with respect to tolerance, target product, as well as the adaptation capability is very crucial for sustainable and stable cultivation of microalgae. Furthermore, the study on the mist-optimized culture conditions and design of the operation is important to improve the microalgae productivity (Ferreira and Gouveia 2020).

13.2.4.1 Potential Improvements

An innovative and effective technology with low energy consumption as well as minimum operational cost must be developed for harvesting microalgal cells to reduce the total production cost of microalgae biorefinery. According to Muhammad et al. (2021), a combination of bio-flocculation and wet extraction is more economically feasible as compared to other technologies such as filtration, flocculation, centrifugation, and flotation (Muhammad et al. 2021). Besides that, it was found that non-mechanical methods, with the utilization of green solvents including supercritical fluids, ionic liquids, and bio-based solvents, are more economically favourable as compared to the mechanical methods for industrial-scale production of clean and sustainable algae-based fuels (Muhammad et al. 2021).

Other than developing a cost-effective method for harvesting microalgal cells, the economic feasibility can be further improved by deriving multiple products from the biorefinery process (Bhalamurugan, Valerie, and Mark 2018). Once the biorefinery process is optimized and improved, several methods such as life-cycle analysis, cost assessment, and energy balance can be conducted to assess the feasibility of the process as well as the environmental impacts (Yen et al. 2013).

13.2.5 Challenges of Waste Treatment

The application of microalgal cultivation with wastewater treatment was first discussed and highlighted by Oswald and Gotaas in the 1950s (Oswald and Gotaas 1957). The algal–bacteria system can be utilized to support sustainable and cost-effective wastewater treatment due to the ability of microalgae to accumulate nutrients from the wastewaters for the growth of microalgae, hence reducing the production costs of microalgae (Ferreira et al. 2018). It was proven that the microalgal–bacterial system is a promising alternative to replace conventional secondary and/or tertiary treatment with less environmental impacts as well as lower associated costs (Ferreira and Gouveia 2020). The microalgal–bacterial system possesses several advantages including cost-free oxygenation and simultaneous removal of nutrients and production of valuable biomass without secondary pollution (Ferreira et al. 2018). The biomass produced can be further processed and converted into products for different applications. Besides, the wastewater treatment using microalgae generates water and nutrients that are readily available with lower cost (Cuellar-Bermudez et al. 2017).

The principle of microalgae-based wastewater treatment is based on photosynthesis, where the microalgae consume carbon dioxide released from the bacterial respiration and generate oxygen to the heterotrophic aerobic bacteria to convert and mineralize the organic pollutants. The application of intensive mechanical aeration can be avoided by using this system, thereby reducing the operation costs as well as minimizing the volatilization of pollutants (Muñoz and Guieysse 2006). According to a number of studies conducted (Ferreira et al. 2018; Gao et al. 2018; Kong et al. 2010; Martínez et al. 2000; Mata et al. 2013; Posadas et al. 2015; L. Wang et al. 2009; Chai, Tan, et al. 2021), a wide range of microalgae including *Botryococcus*, *Chlamydomonas*, *Chlorella*, *Phormidium*, *Scenedesmus*, and *Spirulina* have proven to be effective in wastewater treatment without using any fresh water and nutrients.

Throughout the studies conducted on microalgae cultivations using wastewater, several challenges are identified, and these challenges must be tackled before commercializing the wastewater treatment process using microalgae. According to the study carried out by Lowrey et al. (2015), photoautotrophic production in open ponds is an attractive option due to the minimal maintenance requirements and costs. However, the designs for photoautotrophic production are restricted by various parameters including low productivity, land requirements, water demand, specificity of climate, and potential contaminations (Lowrey, Brooks, and McGinn 2015). Apart from that, the type of organic carbons that are present in the wastewater is another concern for the wastewater treatment using microalgae. (Perez-Garcia et al. 2010) reported that some forms of carbon are toxic or not accessible to the microalgae cells, and the microalgae cells must possess metabolic capacity that can consume the organic acids as a source of organic carbon.

On the other hand, the contamination threat is also one of the major challenges that must be addressed as the microalgae may be severely affected in heterotrophic conditions which are rich in competing organisms. The results of this conditions include diminished viability of the culture, diminished formation of bioproducts, and total culture loss (Lowrey, Brooks, and McGinn 2015). Besides that, controlling the temperature is another crucial challenge that must be tackled during the wastewater treatment. Most of the optimum temperatures for microorganisms lie within a range of 15–35°C, and any deviation of temperature from this range will affect the kinetics of process (Bhatia et al. 2020). These wastewater-related issues may affect the efficiency of wastewater treatment process using microalgae as well as the growth of microalgae. Therefore, these issues must be solved in a comprehensive manner before developing an efficient and eco-friendly technology (Bhatia et al. 2020).

13.2.5.1 Potential Improvements

Several improvements have been proposed by researchers to overcome these challenges. One of the potential improvements that can be conducted is to carry out a more extensive research on the application of large-scale heterotrophic production systems for the microalgae cells instead of photoautotrophic production systems. The high cost of the heterotrophic product system can be eliminated by the significant increase in the productivity of biomass, lipid as well as the reduction in land requirements (Tabernero, Martín del Valle, and Galán 2012). The application of commercial-scale heterotrophic microalgae production system for biodiesel was proven to be comparable to the petroleum-based diesel according to life-cycle assessment conducted by Lee Chang et al. (2015). According to Perez-Garcia, Bashan, and Esther Puente (2011), the issues related with the toxic organic carbon, which are present in the wastewater, can be tackled by supplementing sterilized wastewater with different forms of organic carbon. However, this method results in an additional cost to the production of microalgae biomass. An evaluation on different microalgae strains can be performed to identify the microalgae strain that has the tolerance as well as the capacity to metabolize the organic compounds that are present in the wastewater to prevent the additional cost for organic carbon supplementation (Lowrey, Brooks, and McGinn 2015).

The contamination threat to the microalgae cells can be circumvented via several methods including large-scale sterilization, multistage cultivation systems, or excessive inoculation concentrations (Lowrey, Brooks, and McGinn 2015). Although these methods are effective in preventing contamination, these methods will result in additional expenses and a significant financial barrier to implement these technologies. Therefore, a low-cost sanitization system such as rapid ultraviolet radiation exposure can be implemented to sterilize wastewater in large volume (Lowrey, Brooks, and McGinn 2015). Other than that, the application of mixotrophic cultivation may mitigate the potential of culture contamination due to the photosynthetic component of the culture (Li et al. 2012). Besides that, Ben Chekroun et al. (2014) suggested that the microalgal tolerance to pollutants can be improved by genetic engineering. According to Ben Chekroun, Sánchez, and Baghour (2014), it is crucial to study and monitor parameters of ecosystems including the pH, temperature, and availability of nutrient during the wastewater treatment process to ensure that the absorption, accumulation, and biodegradation of pollutants can be accelerated.

13.3 UNCERTAINTY

13.3.1 Mass Culture of Microalgae

Microalgae growth is highly dependent on the environmental conditions as well as the available nutritional contents (dos Santos et al. 2021). The key environmental factors affecting the mass culture of microalgae include the light intensity and temperature (Borowitzka and Vonshak 2017). As the large-scale mass culture of microalgae is commonly carried out outdoors in natural light, it raises the uncertainties during the mass culture of microalgae due to the change in climate (Borowitzka and Vonshak 2017). The light intensity must be considered as the microalgae cultivation varies significantly based on the density and depth of the culture medium. The generalized ideal light intensity for microalgae cultivation varies from 2.5 klux to 5.0 klux (dos Santos et al. 2021). On the other hand, if the temperature of microalgae cultivation is lower than the optimal temperature, the rate of microalgae growth will be reduced. Contrastingly, if the temperature is higher than the optimal temperature, it will be lethal to the microalgae growth. Besides that, the temperature will affect the solubility of the nutrients as well as the solubility of the gases. The optimal temperature for green microalgae cultivation is reported to vary from 18°C to 24°C (dos Santos et al. 2021). Apart from the operating parameters for microalgae cultivation, the biological contamination of the microalgae gives rise to another uncertainty during the mass culture of microalgae. The biological contamination during the mass culture of microalgae is almost unavoidable (Zhu, Jian, and Fa 2020). Therefore, the types of biological contamination must be studied to develop strategies to control the biological contamination level.

13.3.1.1 Potential Improvements

The density of biomass affects the photosynthetic rate of algal cells at any given irradiance due to the self-shading effects which reduce the actual irradiance received by the algal cells. The average irradiance can be improved by manipulating the depth of culture in the ponds or alternating the frequency of harvesting. According to the study conducted by Borowitzka and Vonshak (2017), the pond depth strategy is effective in compensating the changes in irradiance as well as temperature, hence maximizing the annual productivity of the microalgae culture. However, the cell density cannot be reduced too much to prevent photoinhibition which will decrease the productivity. The temperature of microalgae culture is difficult to control for large raceway ponds and expensive to control for closed system, which involves the application of photobioreactor. Therefore, the selection of a suitable strain that can grow under the temperature at the site of production plant is planned. Besides, the upper limit of the lethal temperature is another important indicator that must be considered during the selection of the strain (Borowitzka and Vonshak 2017).

There are four control strategies that can be utilized to control the biological contamination of the microalgae culture, namely, chemical control, biological control, physical control, and environmental control. The chemical control strategy involves the utilization of pesticides, insecticides, and chemical reagents to protect the microalgae culture from biological contaminations and to accelerate the productivity of the algae. However, the screening of chemical or biological reagents is important to reduce the potential hazards to human health and environment. The biological control strategy uses specific pathogens and plant-derived pesticides to control the biological contaminants. The physical control strategy involves the application of physical filtration and ultraviolet radiation to control the biological contaminants whereas the environmental control strategy utilizes the adjustment of environmental conditions including temperature, light intensity, and environmental pH to control the biological contamination level (Zhu, Jian, and Fa 2020).

13.3.2 Extraction of Bioactive Compounds

Microalgae are well known worldwide as raw materials as they possess a large variety of bioactive compounds such as pigments, proteins, long fatty acids, and polysaccharides (Ventura et al. 2017). These bioactive compounds have high economical value and are applicable in different industries including animal feed, human food, energy, and cosmetics (Ventura et al. 2017). Besides that, the extraction of these high-value bioactive compounds from microalgae is attractive as the production of microalgae can be carried out the entire year with simple operating conditions (Ventura et al. 2017). However, the commercialization of these extraction processes has not reached the maximum due to the uncertainties and high costs of the extraction process. In general, there are four groups of extraction process to extract bioactive compounds from microalgae, including mechanical extraction, chemical extraction, physical extraction, and enzymatic lysis (Tan et al. 2020). The major uncertainty for the extraction of bioactive compounds is the cell disruption step (Patras, Moraru, and Socaciu 2018). One of the examples that can be studied is the ultrasonic treatment on *Chlamydomonas reinhardtii* to maximize the cell disruption (Gerde et al. 2012). The application of ultrasonic treatment on this species has the potential of causing free radicals' emission which will in turn affect the quality of the sample (Patras, Moraru, and Socaciu 2018).

13.3.2.1 Potential Improvements

Application of bead milling is one of the potential improvements that can be carried out to prevent the deterioration of the quality of the sample during cell disruption. The bead milling is a mechanical process which utilizes physical grinding between solid glass beads and cell suspension (Ilavarasi et al. 2012). Besides that, the utilization of different cell disruption methods including microwave treatment and pulsed electric field treatment has proven to be effective in disrupting the cell wall as well (Patras, Moraru, and Socaciu 2018). Biological methods such as cell lysis with chemicals or enzymes and osmotic shock are also effective in permeabilizing the cell walls and preserving unstable bioactive compounds. However, the application of biological methods will result in higher costs (Patras, Moraru, and Socaciu 2018). On the other hand, several emerging technologies such as explosive decompression using carbon dioxide, laser treatment, high frequency ultrasonication, micro-fluidization, and pulsed electric discharge can be applied for cell disruption during the extraction process (Patras, Moraru, and Socaciu 2018).

13.4 CONCLUSION

A summary on the respective improvements that can be carried out to mitigate the challenges and uncertainties is tabulated in Table 13.1.

In this chapter, challenges and uncertainties in various aspects of microalgae biotechnology are systematically summarized and discussed. Potential improvements that could be made to overcome the challenges and uncertainties are also proposed. All in all, it is undeniable that microalgae biotechnology plays an important role in biorefineries, wastewater treatment, and production of valuable chemicals. It is especially relevant as the entire world is currently dealing with global issues such as clean and sustainable water supply, exponential burst in population, and surge in exploitation of natural resources. The ability of microalgae to degrade pathogen and remove both organic and inorganic contaminants from the water resources also provide an added advantage to increase environmental sustainability of the technology. Moreover, it is also certain that the microalgae biotechnology has the potential in changing the energy supply patterns as well as leading a positive impact on the climate change issues. However, a lot more cooperation and information exchange between experts and researchers from different fields are required to ensure that the challenges and uncertainties in biorefineries, wastewater treatment, mass culture, and extraction of bioactive compounds can be tackled. Besides that, it is also noteworthy that the microalgae biotechnology is currently still in mechanical bottleneck, hence it is the microalgae-related biotechnology that requires more inventive research to ensure that these biotechnologies can be commercialized.

TABLE 13.1
Summary of Challenges and Uncertainties

Challenges/uncertainties	Issues	Potential improvements
Challenges in digitalization of microalgae biotechnology.	Intra-organizational, political, and social challenges, lack of financial support.	Future research, initiatives of policy, increased societal involvement, and industrial investment.
	Lack of communication with the public and knowledge transfer between experts.	Communication/organizing talks to explain to the public, creation of knowledge transfer network.
Challenges in microalgae cultivation.	The algae growth in cell culture can be affected by abiotic factors such as temperature, light intensity, carbon dioxide, toxic chemical, pH, dissolved oxygen, concentration of salt.	Installing closed cultivation system or photobioreactor (PBR) as it can be used to control concentration of cells, carbon dioxide level, light and dark cycles, and climatic conditions. Besides that, it also allows proper agitation which is important for uniform distribution of light energy within the reactor.
	The algae growth can be influenced by biotic factors such as bacteria, fungi, virus, and contaminated algae species.	
	Bio-fouling, agitation stress, and overheating of photobioreactor (PBR).	Improving the design of photobioreactor (PBR) by considering the fouling factor as well as selecting different construction material for the photobioreactor (PBR). Cooling jacket can be installed on the photobioreactor (PBR) to prevent overheating issue.
	High energy consumption by the closed cultivation system or photobioreactor (PBR).	Conducting process integration on the production site to reduce the amount of waste energy.
Challenges in microalgae harvesting process.	High operational cost due to the dilute nature and small size of microalgae.	Conducting pre-screening, identification, and selection of appropriate method for harvesting of algal culture. The methods that can be used include filtration, floatation, gravity sedimentation, centrifugation, chemical flocculation, and bio-flocculation. The associated challenges with each of these methods are discussed thoroughly in Section 13.2.3.1.
Challenges in biorefineries using microalgae biotechnology.	High cost.	Application of new technology involving combination of bio-flocculation and wet extraction to reduce costs.
		Utilization of non-mechanical methods with green solvents
		Conducting life-cycle analysis, cost assessment, and energy balance.

(Continued)

TABLE 13.1 (Continued)

Challenges/uncertainties	Issues	Potential improvements
Challenges of waste treatment using microalgae.	High costs.	Carrying out a more extensive research on the application of large-scale heterotrophic production systems.
	Emission of toxic organic carbon.	Supplementing sterilized wastewater with different forms of organic carbon.
	Contamination threat.	Large-scale sterilization, multistage cultivation systems, excessive inoculation concentrations, or rapid ultraviolet radiation.
		Application of mixotrophic cultivation.
	Tolerance towards pollutants.	Genetic engineering.
	Process efficiency.	Monitoring the pH, temperature, and availability of nutrient during the wastewater treatment.
Uncertainty in mass culture of microalgae.	Change in light intensity.	Adjusting and manipulating the depth of microalgae culture or the cell density.
	Change in temperature.	Selecting suitable strains that can grow under the temperature at the production site.
	Biological contamination.	Application of control strategies including chemical control, biological control, physical control, and environmental control.
Uncertainties in extraction of bioactive compounds.	Disruption of samples' quality during cell disruption.	Selecting the most appropriate disruption methods based on characteristics of the species, types of active compound to be extracted, costs, and the process feasibility.

REFERENCES

A. Fernández, F. G., J. M. Fernández-Sevilla, and E. Molina Grima. 2009. "Challenges in Microalgae Biofuels." *New Biotechnology* 25: S268. doi:10.1016/j.nbt.2009.06.600.

Alam, Md. Asraful, and Wang Zhongming. 2019. *Microalgae Biotechnology for Development of Biofuel and Wastewater Treatment. Microalgae Biotechnology for Development of Biofuel and Wastewater Treatment*, 1st ed. Springer. doi:10.1007/978-981-13-2264-8_23.

Barros, Ana I., Ana L. Gonçalves, Manuel Simões, and José C. M. Pires. 2015. "Harvesting Techniques Applied to Microalgae: A Review." *Renewable and Sustainable Energy Reviews* 41: 1489–1500. doi:10.1016/j.rser.2014.09.037.

Bhalamurugan, Gatamaneni Loganathan, Orsat Valerie, and Lefsrud Mark. 2018. "Valuable Bioproducts Obtained from Microalgal Biomass and Their Commercial Applications: A Review." *Environmental Engineering Research* 23 (3): 229–241. doi:10.4491/eer.2017.220.

Bhatia, Ravi Kant, Deepak Sakhuja, Shyam Mundhe, and Abhishek Walia. 2020. "Renewable Energy Products through Bioremediation of Wastewater." *Sustainability* 12 (18): 1–24. doi:10.3390/su12187501.

Borowiak, Daniel, Paweł Lenartowicz, Michał Grzebyk, Maciej Wiśniewski, Jacek Lipok, and Paweł Kafarski. 2021. "Novel, Automated, Semi-Industrial Modular Photobioreactor System for Cultivation of Demanding Microalgae That Produce Fine Chemicals—The next Story of H. Pluvialis and Astaxanthin." *Algal Research* 53: 102151.

Borowitzka, Michael A., and Avigad Vonshak. 2017. "Scaling up Microalgal Cultures to Commercial Scale." *European Journal of Phycology* 52 (4): 407–418. doi:10.1080/09670262.2017.1365177.

Brennan, Liam, and Philip Owende. 2010. "Biofuels from Microalgae-A Review of Technologies for Production, Processing, and Extractions of Biofuels and Co-Products." *Renewable and Sustainable Energy Reviews* 14 (2): 557–577. doi:10.1016/j.rser.2009.10.009.

Carvalho, Ana P., Luı´s A. Meireles, and F. Xavier Malcata. 2006. "Microalgal Reactors: A Review of Enclosed System Designs and Performances." *Biotechnology Progress* 22 (6): 1490–1506. doi:10.1021/bp060065r.

Carvalho, Ana P., Susana O. Silva, José M. Baptista, and F. Xavier Malcata. 2011. "Light Requirements in Microalgal Photobioreactors : An Overview of Biophotonic Aspects." *Applied Microbiology and Biotechnology* 89 (5): 1275–1288. doi:10.1007/s00253-010-3047-8.

Chai, Wai Siong, Chee Hong Chew, Heli Siti Halimatul Munawaroh, Veeramuthu Ashokkumar, Chin Kui Cheng, Young-Kwon Park, and Pau-Loke Show. 2021. "Microalgae and Ammonia: A Review on Inter-Relationship." *Fuel* 303: 121303. doi:10.1016/j.fuel.2021.121303.

Chai, Wai Siong, Wee Gee Tan, Heli Siti Halimatul Munawaroh, Vijai Kumar Gupta, Shih-Hsin Ho, and Pau Loke Show. 2021. "Multifaceted Roles of Microalgae in the Application of Wastewater Biotreatment: A Review." *Environmental Pollution* 269: 116236. doi:10.1016/j.envpol.2020.116236.

Chekroun, Kaoutar Ben, Esteban Sánchez, and Mourad Baghour. 2014. "The Role of Algae in Bioremediation of Organic Pollutants." *International Research Journal of Public and Environmental Health* 1 (2): 19–32.

Chen, Meng, Haiying Tang, Hongzhi Ma, Thomas C. Holland, K. Y. Simon Ng, and Steven O. Salley. 2011. "Effect of Nutrients on Growth and Lipid Accumulation in the Green Algae Dunaliella Tertiolecta." *Bioresource Technology* 102 (2): 1649–1655. doi:10.1016/j.biortech.2010.09.062.

Chew, Kit Wayne, Jing Ying Yap, Pau Loke Show, Ng Hui Suan, Joon Ching Juan, Tau Chuan Ling, Duu-jong Lee, and Jo-shu Chang. 2017. "Microalgae Biorefinery: High Value Products Perspectives." *Bioresource Technology*. doi:10.1016/j.biortech.2017.01.006.

Couper, James R., W. Roy Penney, James R. Fair, and Stanley M. Walas. 2009. *Chemical Process Equipment: Selection and Design*, 2nd ed. Butterworth-Heinemann. doi:10.1016/C2011-0-08248-0.

Cuellar-Bermudez, Sara P., Gibran S. Aleman-Nava, Rashmi Chandra, J. Saul Garcia-Perez, Jose R. Contreras-Angulo, Giorgos Markou, Koenraad Muylaert, Bruce E. Rittmann, and Roberto Parra-Saldivar. 2017. "Nutrients Utilization and Contaminants Removal. A Review of Two Approaches of Algae and Cyanobacteria in Wastewater." *Algal Research* 24: 438–449. doi:10.1016/j.algal.2016.08.018.

Ferreira, Alice, and Luisa Gouveia. 2020. *Microalgal Biorefineries. Handbook of Microalgae-Based Processes and Products*. Elsevier Inc. doi:10.1016/b978-0-12-818536-0.00028-2.

Ferreira, Alice, Paula Marques, Belina Ribeiro, Paula Assemany, Henrique Vieira de Mendonça, Ana Barata, Ana Cristina Oliveira, Alberto Reis, Helena M. Pinheiro, and Luisa Gouveia. 2018. "Combining Biotechnology with Circular Bioeconomy: From Poultry, Swine, Cattle, Brewery, Dairy and Urban Wastewaters to Biohydrogen." *Environmental Research* 164: 32–38. doi:10.1016/j.envres.2018.02.007.

Gao, Shumei, Changwei Hu, Shiqing Sun, Jie Xu, Yongjun Zhao, and Hui Zhang. 2018. "Performance of Piggery Wastewater Treatment and Biogas Upgrading by Three Microalgal Cultivation Technologies under Different Initial COD Concentration." *Energy* 165: 360–369. doi:10.1016/j.energy.2018.09.190.

Gerde, Jose A., Melissa Montalbo-Lomboy, Linxing Yao, David Grewell, and Tong Wang. 2012. "Evaluation of Microalgae Cell Disruption by Ultrasonic Treatment." *Bioresource Technology* 125: 175–181. doi:10.1016/j.biortech.2012.08.110.

Hende, Sofie Van Den, Han Vervaeren, Sem Desmet, and Nico Boon. 2011. "Bioflocculation of Microalgae and Bacteria Combined with Flue Gas to Improve Sewage Treatment." *New Biotechnology* 29 (1): 23–31. doi:10.1016/j.nbt.2011.04.009.

Ilavarasi, A., D. Pandiaraj, D. Mubarak Ali, M. H. Mohammed Ilyas, and N. Thajuddin. 2012. "Evaluation of Efficient Extraction Methods for Recovery of Photosynthetic Pigments from Microalgae." *Pakistan Journal of Biological Sciences* 15 (18): 883–888. doi:10.3923/pjbs.2012.883.888.

Jacab-Lopes, E., M. M. Maroneze, M. I. Queiroz, and L. Q. Zepka. 2020. *Handbook of Microalgae-Based Processes and Products: Fundamentals and Advances in Energy, Food, Feed, Fertilizer, and Bioactive Compounds*, 1st ed. Academic Press. doi:10.1016/C2018-0-04111-0.

Kang, Zion, Byung Hyuk Kim, Rishiram Ramanan, Jong Eun Choi, Ji Won Yang, Hee Mock Oh, and Hee Sik Kim. 2015. "A Cost Analysis of Microalgal Biomass and Biodiesel Production in Open Raceways Treating Municipal Wastewater and under Optimum Light Wavelength." *Journal of Microbiology and Biotechnology* 25 (1): 109–118. doi:10.4014/jmb.1409.09019.

Kong, Qing Xue, Ling Li, Blanca Martinez, Paul Chen, and Roger Ruan. 2010. "Culture of Microalgae Chlamydomonas Reinhardtii in Wastewater for Biomass Feedstock Production." *Applied Biochemistry and Biotechnology* 160 (1): 9–18. doi:10.1007/s12010-009-8670-4.

Kumar, Parveen, Pradeep Kumar Sharma, Pradip Kumar Sharma, and Deepansh Sharma. 2015. "Micro-Algal Lipids: A Potential Source of Biodiesel." *Journal of Innovations in Pharmaceuticals and Biological Sicences* 2 (2): 135–143.

Laamanen, Corey A., Gregory M. Ross, and John A. Scott. 2016. "Flotation Harvesting of Microalgae." *Renewable and Sustainable Energy Reviews* 58: 75–86. doi:10.1016/j.rser.2015.12.293.

Lee Chang, Kim Jye, Lucas Rye, Graeme A. Dunstan, Tim Grant, Anthony Koutoulis, Peter D. Nichols, and Susan I. Blackburn. 2015. "Life Cycle Assessment: Heterotrophic Cultivation of Thraustochytrids for Biodiesel Production." *Journal of Applied Phycology* 27 (2): 639–647. doi:10.1007/s10811-014-0364-9.

Li, Yecong, Wenguang Zhou, Bing Hu, Min Min, Paul Chen, and Roger R. Ruan. 2012. "Effect of Light Intensity on Algal Biomass Accumulation and Biodiesel Production for Mixotrophic Strains Chlorella Kessleri and Chlorella Protothecoide Cultivated in Highly Concentrated Municipal Wastewater." *Biotechnology and Bioengineering* 109 (9): 2222–2229. doi:10.1002/bit.24491.

Low, Sze Shin, Kien Xiang Bong, Muhammad Mubashir, Chin Kui Cheng, Man Kee Lam, Jun Wei Lim, Yeek Chia Ho, Keat Teong Lee, Heli Siti Halimatul Munawaroh, and Pau Loke Show. 2021. "Microalgae Cultivation in Palm Oil Mill Effluent (POME) Treatment and Biofuel Production." *Sustainability* 13 (6): 3247. doi:10.3390/su13063247.

Lowrey, Joshua, Marianne S. Brooks, and Patrick J. McGinn. 2015. "Heterotrophic and Mixotrophic Cultivation of Microalgae for Biodiesel Production in Agricultural Wastewaters and Associated Challenges—a Critical Review." *Journal of Applied Phycology* 27 (4): 1485–1498. doi:10.1007/s10811-014-0459-3.

Mandotra, S. K., P. Kumar, M. R. Suseela, S. Nayaka, and P. W. Ramteke. 2016. "Evaluation of Fatty Acid Profile and Biodiesel Properties of Microalga Scenedesmus Abundans under the Influence of Phosphorus, PH and Light Intensities." *Bioresource Technology* 201: 222–229. doi:10.1016/j.biortech.2015.11.042.

Martínez, M. E., S. Sánchez, J. M. Jiménez, F. El Yousfi, and L. Muñoz. 2000. "Nitrogen and Phosphorus Removal from Urban Wastewater by the Microalga Scenedesmus Obliquus." *Bioresource Technology* 73 (3): 263–272. doi:10.1016/S0960-8524(99)00121-2.

Mata, Teresa M., António A. Martins, and Nidia S. Caetano. 2010. "Microalgae for Biodiesel Production and Other Applications: A Review." *Renewable and Sustainable Energy Reviews* 14 (1): 217–232. doi:10.1016/j.rser.2009.07.020.

Mata, Teresa M., Ana C. Melo, Sónia Meireles, Adélio M. Mendes, António A. Martins, and Nídia S. Caetano. 2013. "Potential of Microalgae Scenedesmus Obliquus Grown in Brewery Wastewater for Biodiesel Production." *Chemical Engineering Transactions* 32: 901–906. doi:10.3303/CET1332151.

Miehe, R., T. Bauernhansl, O. Schwarz, A. Traube, A. Lorenzoni, L. Waltersmann, J. Full, J. Horbelt, and A. Sauer. 2018. "The Biological Transformation of the Manufacturing Industry—Envisioning Biointelligent Value Adding." *Procedia CIRP* 72: 739–743. doi:10.1016/j.procir.2018.04.085.

Milledge, John J., and Sonia Heaven. 2013. "A Review of the Harvesting of Micro-Algae for Biofuel Production." *Reviews in Environmental Science and Biotechnology* 12 (2): 165–178. doi:10.1007/s11157-012-9301-z.

Molina Grima, E., E. H. Belarbi, F. G. Acién Fernández, A. Robles Medina, and Yusuf Chisti. 2003. "Recovery of Microalgal Biomass and Metabolites: Process Options and Economics." *Biotechnology Advances* 20 (7–8): 491–515. doi:10.1016/S0734-9750(02)00050-2.

Moosai, R., and R. A. Dawe. 2003. "Gas Attachment of Oil Droplets for Gas Flotation for Oily Wastewater Cleanup." *Separation and Purification Technology* 33 (3): 303–314. doi:10.1016/S1383-5866(03)00091-1.

Muhammad, G., A. Alam, M. Mofijur, M. I. Jahirul, Y. Lv, W. Xiong, H. Chyuan, and J. Xu. 2021. "Modern Developmental Aspects in the Field of Economical Harvesting and Biodiesel Production from Microalgae Biomass." *Renewable and Sustainable Energy Reviews* 135: 110209. doi:10.1016/j.rser.2020.110209.

Muñoz, R., and B. Guieysse. 2006. "Algal-Bacterial Processes for the Treatment of Hazardous Contaminants: A Review." *Water Research* 40 (15): 2799–2815. doi:10.1016/j.watres.2006.06.011.

Narala, R. R., S. Garg, K. K. Sharma, S. R. Thomas-Hall, M. Deme, Y. Li, and P. M. Schenk. 2016. "Comparison of Microalgae Cultivation in Photobioreactor, Open Raceway Pond, and a Two-Stage Hybrid System." *Frontiers in Energy Research* 4 (29). doi:10.3389/fenrg.2016.00029.

Oswald, William J., and Harold B. Gotaas. 1957. "Photosynthesis in Sewage Treatment." *Transactions of the American Society of Civil Engineers*. doi:10.1061/taceat.0007483.

Patras, D., C. V. Moraru, and C. Socaciu. 2018. "Screening of Bioactive Compounds Synthesized by Microalgae: A Progress Overview on Extraction and Chemical Analysis." *Studia Universitatis Babes-Bolyai Chemia* 63 (1): 21–35. doi:10.24193/subbchem.2018.1.02.

Perez-Garcia, O., Y. Bashan, and M. Esther Puente. 2011. "Organic Carbon Supplementation of Sterilized Municipal Wastewater Is Essential for Heterotrophic Growth and Removing Ammonium by the Microalga Chlorella Vulgaris." *Journal of Phycology* 47 (1): 190–199. doi:10.1111/j.1529-8817.2010.00934.x.

Perez-Garcia, O., L. E. De-Bashan, J. P. Hernandez, and Y. Bashan. 2010. "Efficiency of Growth and Nutrient Uptake from Wastewater by Heterotrophic, Autotrophic, and Mixotrophic Cultivation of Chlorella Vulgaris Immobilized with Azospirillum Brasilense." *Journal of Phycology* 46 (4): 800–812. doi:10.1111/j.1529-8817.2010.00862.x.

Posadas, E., A. Muñoz, M. C. García-González, R. Muñoz, and P. A. García-Encina. 2015. "A Case Study of a Pilot High Rate Algal Pond for the Treatment of Fish Farm and Domestic Wastewaters." *Journal of Chemical Technology and Biotechnology* 90 (6): 1094–1101. doi:10.1002/jctb.4417.

Posten, C., and S. F. Chen. 2016. *Microalgae Biotechnology. Advances in Biochemical Engineering/Biotechnology*, 1st ed. Springer International Publishing. doi:10.1007/978-3-319-23808-1.

Pulz, O., and W. Gross. 2004. "Valuable Products from Biotechnology of Microalgae." *Applied Microbiology and Biotechnology* 65 (6): 635–648. doi:10.1007/s00253-004-1647-x.

Qiao, Y. M., and S. Chandra. 1997. "Experiments on Adding a Surfactant to Water Drops Boiling on a Hot Surface." *Proceedings of the Royal Society A: Mathematical, Physical and Engineering Sciences* 453 (1959): 673–689. doi:10.1098/rspa.1997.0038.

Rashid, Naim, Manoranjan Nayak, Bongsoo Lee, and Yong Keun Chang. 2019. "Efficient Microalgae Harvesting Mediated by Polysaccharides Interaction with Residual Calcium and Phosphate in the Growth Medium." *Journal of Cleaner Production* 234. Elsevier Ltd: 150–56. doi:10.1016/j.jclepro.2019.06.154.

Rashid, N., W. K. Park, and T. Selvaratnam. 2018. "Binary Culture of Microalgae as an Integrated Approach for Enhanced Biomass and Metabolites Productivity, Wastewater Treatment, and Bioflocculation." *Chemosphere* 194: 67–75. doi:10.1016/j.chemosphere.2017.11.108.

Rastogi, R. P., A. Pandey, C. Larroche, and D. Madamwar. 2018. "Algal Green Energy—R&D and Technological Perspectives for Biodiesel Production." *Renewable and Sustainable Energy Reviews* 82: 2946–2969. doi:10.1016/j.rser.2017.10.038.

Rawat, I., R. Ranjith Kumar, T. Mutanda, and F. Bux. 2011. "Dual Role of Microalgae: Phycoremediation of Domestic Wastewater and Biomass Production for Sustainable Biofuels Production." *Applied Energy* 88 (10): 3411–3424. doi:10.1016/j.apenergy.2010.11.025.

Richmond, A. 2004. "Principles for Attaining Maximal Microalgal Productivity in Photobioreactors : An Overview." *Hydrobiologia* 512: 33–37. doi:10.1023/B:HYDR.0000020365.06145.36.

Santos, M. G. B. dos, R. L. Duarte, A. M. Maciel, M. Abreu, A. Reis, and H. V. de Mendonça. 2021. "Microalgae Biomass Production for Biofuels in Brazilian Scenario: A Critical Review." *Bioenergy Research* 14 (1): 23–42. doi:10.1007/s12155-020-10180-1.

Satpute, S. K., I. M. Banat, P. K. Dhakephalkar, A. G. Banpurkar, and B. A. Chopade. 2010. "Biosurfactants, Bioemulsifiers and Exopolysaccharides from Marine Microorganisms." *Biotechnology Advances* 28 (4): 436–450. doi:10.1016/j.biotechadv.2010.02.006.

Sinnott, R., and G. Towler. 2019. *Chemical Engineering Design*, 6th ed. Elsevier. doi:10.1016/b978-0-08-102599-4.09980-x.

Smith, Robin. 2005. *Chemical Process Design and Integration*. John Wiley & Sons, Ltd.

Tabernero, A., E. M. Martín del Valle, and M. A. Galán. 2012. "Evaluating the Industrial Potential of Biodiesel from a Microalgae Heterotrophic Culture: Scale-up and Economics." *Biochemical Engineering Journal* 63. Elsevier B.V.: 104–115. doi:10.1016/j.bej.2011.11.006.

Tan, J. S., S. Y. Lee, K. W. Chew, M. K. Lam, J. W. Lim, S. H. Ho, and P. L. Show. 2020. "A Review on Microalgae Cultivation and Harvesting, and Their Biomass Extraction Processing Using Ionic Liquids." *Bioengineered* 11 (1). Taylor & Francis: 116–129. doi:10.1080/21655979.2020.1711626.

Udom, I., B. H. Zaribaf, T. Halfhide, B. Gillie, O. Dalrymple, Q. Zhang, and S. J. Ergas. 2013. "Harvesting Microalgae Grown on Wastewater." *Bioresource Technology* 139. Elsevier Ltd: 101–106. doi:10.1016/j.biortech.2013.04.002.

Upadhyay, A. K., R. Singh, J. S. Singh, and D. P. Singh. 2019. "Chapter 21 - Microalgae-Assisted Phyco-Remediation and Energy Crisis Solution: Challenges and Opportunity." In *New and Future Developments in Microbial Biotechnology and Bioengineering: Microbial Biotechnology in Agro-Environmental Sustainability*, edited by J. S. Singh and D. P. Singh, 295–307. Elsevier. doi:10.1016/B978-0-444-64191-5.00021-3.

Vandamme, D., I. Foubert, and K. Muylaert. 2013. "Flocculation as a Low-Cost Method for Harvesting Microalgae for Bulk Biomass Production." *Trends in Biotechnology* 31 (4): 233–239. doi:10.1016/j.tibtech.2012.12.005.

Venkata Mohan, S., G. N. Nikhil, P. Chiranjeevi, C. Nagendranatha Reddy, M. V. Rohit, A. N. Kumar, and O. Sarkar. 2016. "Waste Biorefinery Models towards Sustainable Circular Bioeconomy: Critical Review and Future Perspectives." *Bioresource Technology* 215: 2–12. doi:10.1016/j.biortech.2016.03.130.

Ventura, S. P. M., B. P. Nobre, F. Ertekin, M. Hayes, M. Garciá-Vaquero, F. Vieira, M. Koc, L. Gouveia, M. R. Aires-Barros, and A. M. F. Palavra. 2017. "19 - Extraction of Value-Added Compounds from Microalgae." In *Microalgae-Based Biofuels and Bioproducts: From Feedstock Cultivation to End-Products*, edited by C. Gonzalez-Fernandez and R. Muñoz, 461–483. Woodhead Publishing. doi:10.1016/B978-0-08-101023-5.00019-4.

Wan, C., M. A. Alam, X. Q. Zhao, X. Y. Zhang, S. L. Guo, S. H. Ho, J. S. Chang, and F. W. Bai. 2015. "Current Progress and Future Prospect of Microalgal Biomass Harvest Using Various Flocculation Technologies." *Bioresource Technology* 184: 251–257. doi:10.1016/j.biortech.2014.11.081.

Wang, K., K. S. Khoo, H. Y. Leong, D. Nagarajan, K. W. Chew, H. Y. Ting, A. Selvarajoo, Jo-shu Chang, and P. L. Show. 2021. "How Does the Internet of Things (IoT) Help in Microalgae Biorefinery ?" *Biotechnology Advances*, 107819. doi:10.1016/j.biotechadv.2021.107819.

Wang, L., M. Min, Y. Li, P. Chen, Y. Chen, Y. Liu, Yi Wang, and R. Ruan. 2009. "Cultivation of Green Algae Chlorella Sp. in Different Wastewaters from Municipal Wastewater Treatment Plant." *Applied Biochemistry and Biotechnology* 162 (4): 1174–1186. doi:10.1007/s12010-009-8866-7.

Wang, S. K., F. Wang, Y. R. Hu, A. R. Stiles, C. Guo, and C. Z. Liu. 2014. "Magnetic Flocculant for High Efficiency Harvesting of Microalgal Cells." *ACS Applied Materials and Interfaces* 6 (1): 109–115. doi:10.1021/am404764n.

Yang, G. Z., J. Bellingham, P. E. Dupont, P. Fischer, L. Floridi, R. Full, N. Jacobstein, et al. 2018. "The Grand Challenges of Science Robotics." *Science Robotics* 3 (14). doi:10.1126/scirobotics.aar7650.

Yen, H. W., I. C. Hu, C. Y. Chen, S. H. Ho, D. J. Lee, and J. S. Chang. 2013. "Microalgae-Based Biorefinery—From Biofuels to Natural Products." *Bioresource Technology* 135: 166–174. doi:10.1016/j.biortech.2012.10.099.

Zahrim, A. Y., C. Tizaoui, and N. Hilal. 2011. "Coagulation with Polymers for Nanofiltration Pre-Treatment of Highly Concentrated Dyes: A Review." *Desalination* 266 (1–3): 1–16. doi:10.1016/j.desal.2010.08.012.

Zhang, X., Q. Hu, M. Sommerfeld, E. Puruhito, and Y. Chen. 2010. "Harvesting Algal Biomass for Biofuels Using Ultrafiltration Membranes." *Bioresource Technology* 101 (14): 5297–5304. doi:10.1016/j.biortech.2010.02.007.

Zhu, Zhi, Jihong Jian, and Yun Fa. 2020. "Overcoming the Biological Contamination in Microalgae and Cyanobacteria Mass Cultivations for Photosynthetic Biofuel Production." *Molecules* 25 (22): 5220. doi:10.3390/molecules25225220.

Živković, S. B., M. V. Veljković, I. B. Banković-Ilić, I. M. Krstić, S. S. Konstantinović, S. B. Ilić, J. M. Avramović, O. S. Stamenković, and V. B. Veljković. 2017. "Technological, Technical, Economic, Environmental, Social, Human Health Risk, Toxicological and Policy Considerations of Biodiesel Production and Use." *Renewable and Sustainable Energy Reviews* 79: 222–247. doi:10.1016/j.rser.2017.05.048.

Index

A

absorption, 69–70, 75, 83, 109, 111, 112, 114, 117, 120, 132, 135, 156, 191, 217, 239, 264, 274, 311, 315, 334, 346–347, 382, 415, 434
accumulation, 34, 40, 69, 73, 80, 82, 115, 120–121, 132, 155, 208, 210, 219, 239, 315, 319, 335, 418, 434
acetaldehyde, 185
acetic acid, 112, 119
acetone-butanol-ethanol, 186
acidification, 14, 39, 121
activated carbon, 192
activated sludge, 34, 72, 110, 117–125, 134, 196
adaptability, 11, 19, 71, 117, 122, 127, 149, 281, 296
additive manufacturing / 3D printing, 178
adsorption, 75, 80, 110, 115, 117, 120, 192, 194
aeration, 74, 76, 84, 115–116, 118, 122–123, 272, 373, 391, 432
aerobic, 132–133, 144
agitation, 127
agriculture, 147, 160, 176, 181–183, 188–190, 194
airlift, 405
air pollution control, 196
alcoholic fermentation, 185
algorithms, 175
alkali pretreatment, 183
ammonia, 69, 74, 82, 109, 113, 117, 120, 122, 183, 184, 190, 277
amorphous region, 181
anaerobic, 22, 119, 121–122, 182, 185–186, 190, 272, 277, 388
anaerobic digestion, 35, 81, 182, 185, 186, 191, 249, 281, 317, 377
analysis, 110, 148, 152, 154, 157, 163
antibacterial, 188, 195
antibiotics, 3–4, 112, 114, 120–121, 123, 126, 193, 385
antioxidant, 20, 22, 112, 127–128, 135, 206, 215, 266, 270, 335, 364, 385, 398, 430
aquatic, 20, 22, 35, 71, 124, 126–130, 133–135, 190, 196, 270, 276, 329, 338–344, 388–389
artificial intelligence, 39–40, 159–160, 176–177, 180–181, 205, 207, 210, 212, 214–215, 217–219, 293, 296, 306, 308–311, 320, 328, 338, 342, 345, 347, 348, 350, 365, 392
artificial neural networks, 238, 247
assessment
  chemical hazard assessment, 124
  ecotoxicological assessment, 109
  risk assessment, 108, 126, 133–135, 178
  toxicity assessment, 133, 135
  water quality assessment, 109
assimilation, 112, 120, 124, 191, 235, 239, 247, 347, 379, 399
astaxanthin, 25, 206, 307, 397, 405, 408, 419, 423
asymptotic, 231
augmented reality, 178–179
automation, 229, 237, 279, 281, 401–402, 405, 409, 412, 415, 418
autonomous robots, 178

B

bacteria, 2, 4, 9, 22, 68, 73, 119–120, 123–125, 134, 184–186, 189, 238, 264, 272, 274, 277, 329, 336, 338, 340, 349, 350, 381, 425, 430–432, 437
beta-carotene, 206, 385
big data analysis, 178
  monte-carlo simulation, 232
  phylogenomics, 232
  subcellular localization, 232
bioaccumulation, 38, 114–115, 192, 329, 334–335
bioactive compounds, 6, 27, 207, 218, 364, 421, 422, 435–436, 438
bioadsorption, 192
bioassay, 37–38, 133, 271, 328, 343–345
bio-based economy, 181
biochar, 81–83, 182, 187–188
biochemical, 132–133, 153
biochemical conversion, 185, 187
biochemical oxygen demand (bod) 73, 119, 121, 189
biochemical oxygen demand test, 189
bioenergy, 267, 272
biofertilizer, 81–85
biofilms, 195
biofuels, 2, 6, 22, 24, 27–29, 35–36, 39, 41,

71, 78, 81, 87–88, 157, 180–181, 185, 197, 206, 208, 211–212, 216–217, 219, 237, 239, 246, 266, 277–279, 294, 298, 300–303, 312, 314–320, 364, 372, 375–376, 382, 387, 391, 429–430
bioalcohol, 185
biobutanol, 186
biodiesel, 27–29, 36, 77–79, 87, 111, 157, 163, 197, 208, 211, 216–217, 228, 239–240, 246, 250, 251, 269, 277, 282, 302, 315–316, 370, 377, 387–388, 390, 397, 430–431, 433
bioethanol, 27–28, 77–78, 80, 87, 182–186, 207, 216, 266, 269, 316, 370, 387, 390, 397, 430
biogas, 28, 35, 81, 181–183, 187, 206, 215–216, 232–233, 317, 377, 423
biohydrogen, 82, 87, 186
biomethane, 82, 87, 182–183, 186
bio-oil, 80–81, 182, 185, 187–188
bioinformatics, 180, 238–240, 245–246
biological ink, 180
biological oxidation, 189
biological pretreatment, 181, 184
biomass, 1, 24, 27–29, 31–36, 39–40, 68–71, 74, 76–86, 88, 116–117, 120, 123–124, 130, 132, 134, 145, 156, 161–162, 164, 175, 180–190, 193–194, 196, 206–208, 211, 214–215, 217, 228–235, 237–238, 246–249, 263, 266–269, 271, 273–274, 276–278, 280, 294, 297–303, 305–312, 314–317, 319, 330, 335, 337–340, 346–348, 350, 370, 372–373, 375–380, 383–384, 390, 397–402, 405, 408, 413–419, 423–434
bio-mitigation, 345
biomolecules, 176–177, 182, 189
biopolymeric, 272
bioprocessing scale-up, 86–88
bioreactor, 190–191
biorefinery, 88
bioregenerative life support system, 384
bioremediation, 8–10, 24, 35–36, 38, 73, 75, 83, 86, 88, 192, 271, 274, 328–330, 332–334, 348–349, 422
biosorption, 9, 38, 113–114, 194, 329, 335
biosurfactants, 273, 287, 347–349
biotechnology, 2–8, 10, 25, 68–69, 71, 87–88, 133, 176–177, 179–181, 188, 206, 210, 219, 228–229, 232, 240, 271, 275, 278, 281–282, 328, 348–352, 367, 382, 422–423, 430, 436–437

C

Calvin cycle, 239
carbohydrate, 68, 80
carbon, 110, 112, 117, 123, 134, 185, 187–189, 191–192, 194, 196–197, 206, 208, 210
carbon dioxide, 3, 18–19, 23, 28, 32, 38–39, 41, 67–75, 77, 80–83, 88, 110, 116, 124, 134, 154–156, 158–159, 161, 183, 206, 208, 212, 219, 238, 247, 264, 266, 268, 276, 279, 296, 298–299, 308, 310, 315, 327–328, 336, 345–347, 349, 351, 364, 370, 377–381, 386, 407, 422–426, 430–432, 436–437
carbonate, 112
carbon dioxide sequestration, 69–70
carbon fixation, 266
carboxyl, 115, 117, 334
carboxylesterase, 113
carcinogenic, 75, 191, 195
carotenoids, 20, 25, 131, 157, 206, 235, 264, 269–270, 278, 307, 382, 385, 397, 405, 411
catalytic activity, 194
cefradine, 126
ceftazidime, 121, 194
cellular dynamics, 228, 231
cellulose crystallinity, 185
central composite design, 247
chemical oxygen demand (cod) 73, 109, 113
chemical oxygen demand test, 189
chemical pretreatment, 182–183
Chlorella vulgaris, 186–188, 190, 194, 197
chlorophyll(s), 82, 127, 149, 153, 160
chloroplast, 17, 19–20, 126, 264, 269, 318, 334, 380–381
circular pond, 295
Clostridium sp., 186
cloud, 146, 149–153, 158–159, 178, 230, 233, 424
cloud computing, 39, 151–153, 162, 178, 233
coagulation, 273
communicative, 230
compound annual growth rate, 364, 371
conventional pretreatment, 181
convolutional neural network, 238
copper, 75, 111, 127, 131, 192, 270, 336–339, 342, 344, 411
corn starch, 185
cultivation, 398, 412, 415
continuous, 415
semi-continuous, 415
cultivation system
open pond, 400, 424, 425–426, 432
culture, 111, 118, 129–130, 134, 155–156, 158–164
cyanobacteria, 2, 20–22, 26, 37, 70, 82–86, 88, 115, 191, 218, 239, 342, 345–346, 385–386, 425–426
cyber, 185, 194
cyber-physical production systems, 229

cyber-physical system (cps) 176
cybersecurity, 161–162, 178
cytometry, 131, 235, 237, 312–313, 344

D

damage, 10, 12, 14, 16, 22, 71, 114, 116, 124, 126–128, 132, 157, 178, 266, 270–271, 302, 333, 339, 375, 429
data, 5, 39, 109, 124, 126–128, 145–152, 154, 158–162, 164, 176–181, 207, 210–215, 217–218, 229–232, 238–241, 244–248, 250–251, 274, 281, 309–312, 319, 338, 342–343, 351, 366, 368, 376, 384, 402–404, 408–409, 412–413, 416, 418–419, 424
data analytics, 229
Data Bridge Market Research, 370–371
data monitoring, 233
data processing, 148, 151, 152, 230, 274, 343
deciphering, 240
decision making, 230
decision support systems, 158, 211, 230, 404
decomposition, 276
degradation, 37–38, 73, 75, 81, 83–84, 114, 116, 119, 121–122, 135, 176, 184, 194–195, 272, 330, 335, 347–348, 398–399
biodegradation, 8, 38, 112, 114, 121, 128, 188, 194–195, 329–331, 333–335, 434
photodegradation, 114
denitrification, 73, 113, 120, 122
dephosphorization, 118
detection, 126, 129–131, 134, 180, 210, 213–214, 236, 280, 340–343
detoxifications, 88
digitalization, 422–423, 437
  downstream processing, 229, 246, 274
  midstream processing, 229
  upstream processing, 229
discharge, 73, 109, 118, 120, 122–123, 126–127, 135, 193, 329, 340, 388, 436
distillation, 185
drugs, 3, 5, 26, 114, 117, 120, 127, 195, 385
Dunaliella, 186, 187, 188

E

ecological, 14, 20, 24, 35, 110, 126–128, 132–135, 233, 246, 273, 281, 345, 383–384, 422, 431
ecosystem, 3, 6–11, 14–15, 20–23, 30, 35, 39, 69, 71, 75, 109, 135, 196, 229, 232, 237, 264, 313, 338–340, 343–344, 364, 384, 386, 434
effluent, 35–36, 68, 73–74, 77, 113–114, 118, 123, 125, 135, 189–193, 195, 217, 266, 271–272, 277, 328, 331, 336, 338, 376, 378, 386–388
electron, 116, 132, 197, 233–234, 335
emission, 11–12, 14–15, 22, 30–31, 33, 68–69, 77, 81, 110, 117, 123, 196, 217, 234, 248, 264, 269, 277, 306, 328, 345–346, 369, 377, 379, 386, 390, 392, 436, 438
endpoint, 128, 130–131
environment, 3–4, 7–12, 14–15, 23–24, 30, 33–34, 37–38, 41, 68, 70–71, 75–76, 80, 82, 84, 109–110, 114, 117, 119–120, 122, 124, 126–130, 132–133, 135, 147, 153, 159, 176–178, 184, 190–196, 208, 215, 218–219, 228–230, 235, 238, 248, 251, 271–273, 276, 278, 294–296, 299, 302, 304, 307–308, 310, 314, 316, 320, 327–329, 333–336, 338–343, 345, 347–352, 374, 382, 386–387, 389–390, 407, 411–413, 424–425, 430, 435
environmental biotechnology, 176, 179, 180
environmental impact, 181, 264, 277, 280, 282, 422
environmental sustainability, 14, 31, 181, 219, 276, 390, 422, 436
enzymatic hydrolysis, 183, 185
enzyme, 5, 7, 9, 19–20, 26, 37–38, 78, 84, 113, 116, 121, 128, 130–132, 184–185, 187, 194–195, 207, 239, 241, 265, 281, 301–303, 307, 314, 316, 319, 329, 333, 335, 343–344, 382, 399, 436
eutrophication, 35, 71–73, 82, 118, 188, 189, 190, 191, 276, 282, 295, 336, 388, 390
expenditure, 374, 378
exposure, 22, 37, 75–76, 109, 124, 126–127, 130–135, 153–154, 158, 160, 192, 194–195, 296, 330, 339, 342, 351, 382, 398–399, 434
expressed sequence tag, 381
extracelluar, 125, 134, 277

F

factors, 115, 119–120, 129–130, 132–135, 147, 154–155, 164
  lights, 115–118, 124, 130, 132–133, 153–155, 159
  salinity, 111, 113, 117, 120–121
  temperature, 110–111, 115–119, 130, 132, 147, 153–154, 156, 159–161
feedstock, 77–81
fermentation, 180, 185–186
first industrial revolution, 12, 176, 367
flat panel, 401
flocculation, 28, 76, 118, 121, 124, 156, 192, 268, 280, 301, 398, 427, 429–431, 437
flue gas, 70
fluorescence, 37, 127–128, 131, 159, 164, 214, 237–238, 274, 306, 312, 341–342, 344, 383–384

flux analysis, 239
food security, 279
freshwater, 20–21, 35, 68–69, 78, 120, 124, 128–129, 155, 192, 276, 278, 330, 344, 380, 382, 386, 388
fungi, 8–9, 68, 123, 184–185, 329, 338, 381, 425, 437

G

galaxolide, 196
gateway, 158
gene, 121, 126, 134–135, 210, 214, 219, 240–241, 244, 246, 318, 333, 345, 381
genetically engineered, 264, 297, 317, 333, 382
genetic modification, 82, 206, 308, 317, 347, 381
genome, 5, 40, 206, 208, 210–211, 219, 227, 229, 231, 240–244, 246, 275, 318–319
genome editing
  promoters, 241
  terminators, 241, 243
  transcription, 231, 241, 243
genome-scale metabolic models, 231
genome sequencing, 240
genotoxicity, 157–158, 162
geospatial analysis, 233
germination, 84–85
glucose, 183, 185–186
glycolytic, 185
greenhouse gases, 68, 266, 272, 369, 386, 390
growth kinetics, 236

H

hardware, 7, 147–151, 160, 318, 383
health hazards, 191
heavy metal, 71, 75, 82, 84, 188, 190, 192–193
heavy metal removal, 192–193
hemicellulose, 181, 183, 185
heterotrophic, 112, 188, 208, 266, 276, 307, 329, 372, 381, 400, 424, 432–433, 438
high-rate algal pond, 76
hormones, 120, 128
human intervention, 160
hydrogen, 114, 119, 185, 186, 269, 316
hydrolysis, 110, 119, 121, 181, 183–186
hydrothermal liquefaction, 78, 81
hydrothermal pretreatment, 181, 184, 185
hydroxylation, 194
hydroxyl groups, 185

I

ibuprofen, 120, 145–146
illumination, 156
image analysis, 39, 310–312
image cytometry, 237
immobilization, 118, 280, 301, 314, 333, 341
impact, 109, 129–130, 147–148, 152–153, 162
indicators, 132, 160
industrialization, 68, 108, 109, 117–118, 124, 176–177, 179, 266, 336, 367
industrial revolution, 176–177, 179–181, 366, 367, 391
industry, 108, 146, 154–155, 157, 160–162
industry, 4.0, 163, 175–176, 181, 206, 216, 365, 400
  cognition, 230
  configuration, 230
  connection, 228
  conversion, 230
  cyber, 230
inhibition, 116, 126, 128, 130–131, 186, 228, 343–344, 382
inorganic, 75, 110, 112–113, 117, 119–121, 123, 126, 188–191, 208, 266, 276, 329, 335–337, 340, 429, 436
inorganic nutrient, 190–191
intensity, 29, 115–117, 122, 130, 132, 154–155, 159, 195, 207–208, 231, 234, 267, 307, 343, 346–247, 349, 373–374, 379, 382, 384, 398–399, 402, 407–409, 413–414, 424–425, 434–435, 437–438
Internet of Things (IOT) 39, 145–156, 158–164, 176, 178–179, 205, 207, 211–212, 230–231, 328, 342, 352, 400, 402, 404, 411, 418–419, 423
intracellular, 115, 131–132, 135, 194, 335
ionic liquid, 181–185

K

ketoprofen, 114

L

legitimate alternative, 191
life cycle assessments, 433
light intensity, 116–117, 130, 132, 134, 155, 159, 207–208, 234, 267, 307, 346–347, 349, 374, 379, 398–399, 402, 407, 409, 413–414, 424–425, 434–435, 437–438
lignin peroxidase, 184
lignin recovery, 181
lignocellulosic, 181, 183, 184, 187
limitation, 162, 164
linear-quadratic gaussian, 384
lipid, 17–18, 22, 25–27, 75, 113, 128, 131, 157, 188, 196–197, 207, 214, 216, 218–219, 236, 239–240, 246, 264, 266, 269–270, 279–280, 297, 302–303, 327, 335, 347, 364, 385, 397, 399, 430–431

M

machine learning, 237
macromolecular, 277
manganese peroxidase, 184
mass spectrometry, 238, 240
matrix-assisted laser desorption/ionization, 238
mechanism, 9, 34, 38, 70, 73, 84, 87, 112–114, 120–121, 126, 152, 194–197, 207–208, 213, 228, 308, 329, 334–335, 340–341, 343, 347, 350, 367, 371, 373, 379, 428
membrane, 17, 21, 39, 110, 115, 117–118, 122, 126, 156–157, 264–265, 268–269, 272, 301, 316, 335–336, 340, 346, 427–428
memory, 148, 149
metabolic degradation, 194
metabolic engineering, 240–241, 246
metabolism, 112–113, 116–118, 123, 127, 130, 155, 188–189, 191, 194, 208, 210–211, 231, 237, 239–240, 245–246, 294, 334–335, 384, 399, 423–424
metaheuristic, 228
methanogens, 186
microalgae, 187–197
microalgae biomass, 314–315, 320
microalgal, 189–190, 193
microbial activity, 121, 184–185
microbiology, 231
Microcystis, 21, 37, 71, 83, 112–115, 125, 128–129, 131, 133, 333
microfluidic
  droplets, 236–237
  mechanical traps, 236–237
  microchambers, 236–237
microorganism, 4, 7–9, 26, 34, 69, 76, 84, 88, 114, 119–121, 123–124, 129, 185–186, 189, 194, 207–208, 212, 236, 264, 273, 296, 304–305, 307–309, 313, 327, 329, 336, 347, 349, 400, 425, 430, 433
micropollutants, 193–196
microprocessors, 164, 409
milling, 181–183
mixotrophic, 188, 207–208, 335, 424, 434, 438
molecular, 26–27, 40, 87, 112, 132, 189, 196, 210, 238, 241, 272, 312, 381, 430
monosaccharides, 264
morphology, 228, 234, 237–238
multi-omics
  genomics, 231, 239–241, 251
  metabolomics, 239, 241
  proteomics, 239–240
  transcriptomics, 239

N

nanofiltration, 272
nanoparticles, 34, 126, 131, 280, 296, 313–314, 320, 333, 342
nanotechnology, 313, 317
naproxen's photodegradation, 195
new product introduction, 364, 369
next generation sequencing, 240
nickel, 36, 111, 115, 126, 317, 338
nicotinamide adenine dinucleotide phosphate, 379
nitrate, 82, 111, 115, 188, 191, 197, 206, 338
nitrification, 73, 113, 120
nitrogen, 18, 20, 22, 34–35, 72–74, 79–85, 87, 110, 112–113, 116–120, 122, 124, 189, 192–193, 196–197, 206, 238, 264, 271, 276, 319, 328, 336–338, 348, 378, 380, 390, 399, 402, 408, 418, 430
noxious compounds, 192
nutrition, 265, 273

O

oil emulsions, 273
oleaginous microalgae, 28
organic, 9, 26, 36, 38, 70, 72, 74–76, 80, 83, 110–113, 116–123, 126, 131, 134, 182–183, 187–195, 197, 208, 264, 266, 269, 271–272, 276–277, 302, 320, 328–332, 334–335, 336, 339–340, 348, 375, 378, 381, 388, 424, 429, 432–433, 436, 438
organic pollutant removal, 189
organosolv, 183
oxidation110, 119–120, 122, 189, 194, 197, 264, 335, 379
oxidative, 126–128, 130–131, 270, 333, 335, 339
oxygen, 19–20, 22–23, 69, 73–74, 81, 111, 116–117, 119, 121–122, 126–127, 134, 159, 187, 189, 264, 271–272, 277, 280, 315, 328, 334–335, 338, 340, 342, 349, 373, 378, 380, 383–384, 386, 388, 401–402, 424, 432, 437

P

paradigm, 230
permeability, 125, 149
peroxidase-antioxidative, 121
personal care products, 193
pesticide, 37–38, 84, 110, 112, 114, 126, 128, 192–194, 330–331, 334–335, 340–342, 344, 386, 435
pH, 399
pharmaceutical, 5, 9, 22, 24, 26–27, 34, 41, 75,

77, 110, 114–115, 120, 127, 157, 163, 180, 188, 193, 195, 210, 213, 215, 266, 270, 275, 277, 281, 327, 331, 341, 348, 350, 372, 397–398
phenol oxidase, 184
phenomics, 279–280
Phormidium bohneri, 191
phormidium sp., 191
phosphate, 74–75, 82–83, 113, 115, 122, 191–192, 195–196, 239, 332, 337–338, 347
phosphorus, 34–35, 72–74, 80, 82–84, 110–113, 116, 118, 120–122, 189, 193, 196, 206, 238, 264, 271, 276, 319, 328, 336–337, 378, 380, 386, 390–391, 399
photoautotrophic, 20, 188, 208, 381, 400, 424–425, 432–433
photobioreactor, 28, 31–32, 39–40, 86–87, 159, 211–212, 217, 237, 247–248, 266–267, 277, 282, 294, 298–299, 307–308, 314–317, 346, 350, 366, 370, 372–374, 380, 387, 391, 400–405, 407–413, 415–419, 424–426, 428, 435, 437
photodegradation, 194–196
photodegradation mechanism, 195
photo-physiological, 231
photosynthesis, 19–20, 22–23, 39, 69–70, 73, 76, 81–82, 111–112, 116, 123–124, 130–131, 133, 155, 196, 206, 208, 215, 231, 264, 266, 273, 281, 296, 298, 306–307, 310, 315, 319, 336, 341, 344, 378–381, 383, 392, 399, 401, 409, 424, 426, 432
photosynthesis rate, 231
photosynthetic, 22, 40, 69, 83, 88, 111, 117, 124, 126, 128, 130–131, 135, 155, 186, 191, 196–197, 206, 211, 217–219, 264, 266–267, 274–275, 294, 304, 309–310, 313, 327, 340, 342, 344, 372, 374, 378–380, 382–383, 385–387, 399, 426, 431, 434
physical pretreatment, 181, 183
phytoplankton, 67
phytoremediation, 8, 340
pivoted agitator, 247
plasmodesmata, 26
pollutant, 7–9, 37–38, 41, 68, 71–74, 76, 108–114, 116–117, 119, 121–122, 124, 126–135, 188–189, 192–195, 197, 310, 328–330, 333–336, 339–345, 347–349, 351, 390, 432, 434, 438
pollution, 108, 110, 130, 143, 150
polyculture, 248
polymeric, 117, 195, 272, 429
polyphyletic, 17
polysaccharides, 113, 114, 118, 123, 157, 264, 269, 278, 334, 364, 385, 435
polyunsaturated fatty acids, 384
porosity, 181
precipitation, 265, 278
predictive modelling, 250
pretreatment methods, 181
protein, 3, 17–18, 22, 25–27, 80, 82, 85, 111, 120–121, 128, 130–131, 156–157, 188, 206–208, 210, 213, 215–216, 238, 240, 243, 245–246, 251, 264–266, 268–270, 272–273, 281, 297, 300, 318, 327, 333, 344, 364, 371, 380, 384–385, 399, 422, 430, 435
proteomics, 39
pyruvate, 185

Q

quality, 3–4, 11, 28, 37, 39, 74, 79, 109, 115–117, 120–121, 133–134, 176, 178–180, 188, 190, 208, 210–211, 213, 215, 217, 219, 229, 270, 273, 276–277, 282, 300, 303, 312, 339–342, 345–346, 348, 368–369, 372, 388, 411, 426, 430–431, 436, 438

R

raceway pond, 295
radiation, 119, 206–207, 237, 239, 373–374, 378–379, 434–435, 438, 484
real-time, 148, 154, 162, 175, 182, 230, 279
red-green-blue colour analysis, 238
regulations, 10, 68, 109, 128, 185, 188, 278, 280
removal, 35–36, 38, 41, 71–75, 111–117, 120–121, 122, 124, 127, 183, 189–196, 208, 215, 272, 329–333, 335, 337–338, 340, 348–349, 351, 387, 432
renewability, 191
renewable baseload electricity, 377
renewable energy, 217, 278
reproduction, 2, 22, 121, 124, 130, 132, 134, 193, 234, 338, 344, 370
resistance, 110, 120–121, 123, 128, 134, 193, 235, 274, 301, 381, 405, 426
respiration losses, 231
return of investment, 368
Rhodobacter sp., 186
ribulose-1,5-biphosphate, 380
robustness, 187

S

saccharomyces cerevisiae, 185
Scenedesmus sp., 190, 195
screening, 239–240

second-generation pollution, 190
security, 16, 23, 69, 135, 148–152, 155, 161–162, 180, 206, 219, 269, 274, 278, 369, 385
sedimentation, 132
self-adaptation, 229
sensitivity, 109, 124, 128–129, 133, 213, 232, 340–342, 344, 413
sensors, 148–150, 153, 155, 159, 163–164, 175, 177, 179, 195
sewage, 34, 36–37, 73–74, 108, 110, 112–114, 116–123, 134, 189, 232, 271–272, 336–337, 340, 367, 377
sigmoidal formula, 231
simulation, 178, 180
sludge, 264, 272, 281
smart factory, 263, 265
smart microalgae, 205–209, 212, 215–219, 282
smartphone, 33, 147, 150–151, 177
soil conditioner, 83
solar, 176
solid char, 185
spectra, 238
stability, 129, 131, 147
starch, 80–82
stress, 16, 33, 79, 84, 126–128, 130–131, 155, 207, 237, 264, 266, 319, 333, 339, 344, 425–426, 430, 437
sulfamethoxazole, 112, 115, 125, 195
sustainable development goals, 364, 389, 390, 392
synthetic biology, 240, 241
system dynamics, 230
system integration, 178
systems biology, 240, 246

T

Taguchi method, 380
taxonomic, 238
temperature, 2, 10, 18, 35–36, 81–82, 110–111, 115–119, 147, 153–154, 156–157, 159–161, 182–187, 191, 197, 230–231, 238–239, 280, 299, 301–304, 307, 310, 315, 319, 334, 338, 346–347, 349–351, 378–380, 399–400, 402–405, 407–413, 416–419, 424–425, 431, 433–435, 437–438
teratogenic, 191
tetracycline, 114, 115
thermochemical, 187
thermochemical conversion, 185
tolerance, 111, 113–114, 127–128, 197
toxicant, 271, 328, 338, 340–343, 350–351
toxicity, 109–110, 113, 120–121, 124, 126–135, 190, 195
transesterification, 78
transformation, 119–120, 135, 138–140, 176, 178, 180, 182, 195
transmission, 126, 128
trichlorfon, 112
triclosan, 76, 115, 125, 128, 136, 142, 195, 344

U

ultraviolet B, 384
unicellular microalgae, 70
unstirred pond, 295
urbanization, 68–69, 72, 124, 367

V

volatilization, 113, 120, 194, 196, 432

W

waste treatment, 422, 424, 432, 438
wastewater, 188–195, 206, 215, 232, 264, 266, 268, 271–272, 276, 280
  heavy metal wastewater, 110, 115
  municipal wastewater, 110
  organic wastewater, 119
  pharmaceutical wastewater, 114–115, 120, 127
wastewater treatment, 108–109, 111–114, 117–120, 123–124, 134, 188–194
  agricultural, 73, 76, 77, 188, 308, 319–320
  industrial, 73, 77
  municipal, 71, 73–74, 77
  palm oil mill effluent, 74, 77
  plant, 193
  swine slurry, 74, 77
water footprint, 387–388
web-connected, 229
wireless networks, 145, 148

Y

Yaksawa, 180